Navier–Stokes Equations and Turbulence

This book aims to bridge the gap between practicing mathematicians and the practitioners of turbulence theory. It presents the mathematical theory of turbulence to engineers and physicists as well as the physical theory of turbulence to mathematicians. The book is the result of many years of research by the authors, who analyze turbulence using Sobolev spaces and functional analysis. In this way the authors have recovered parts of the conventional theory of turbulence, deriving rigorously from the Navier–Stokes equations what had been arrived at earlier by phenomenological arguments.

The mathematical technicalities are kept to a minimum within the book, enabling the discussion to be understood by a broad audience. Each chapter is accompanied by appendices that give full details of the mathematical proofs and subtleties. This unique presentation should ensure a volume of interest to mathematicians, engineers, and physicists.

Ciprian Foias is an Emeritus Professor in the Department of Mathematics at Indiana University at Bloomington and Professor of Mathematics at Texas A&M University at College Station. He has held numerous visiting professorships, including those at Virje University (Netherlands), Israel Institute of Technology, University of California at San Diego, Université Paris-Sud, and the Collège de France. In 1995, he was awarded the Norbert Wiener prize by the American Mathematical Society.

Oscar P. Manley works as a consultant and independent researcher on the foundations of turbulent flows. He has acted as head of the U.S. Department of Energy's Engineering Research Program and as the Program Manager for the Department of Energy's research on magnetic fusion theory. Dr. Manley has held visiting professorships at the Université Paris-Sud and Indiana University.

Ricardo Rosa is a Professor of Mathematics at the Universidade Federal do Rio de Janeiro. He has also held positions as Visiting Researcher at the Institute for Scientific Computing and Applied Mathematics at Indiana University and as Visiting Professor at the Université Paris-Sud in Orsay, France.

Roger Temam is a Professor of Mathematics at the Université Paris-Sud and Senior Scientist at the Institute for Scientific Computing and Applied Mathematics at Indiana University. He has been awarded an Honorary Professorship at Fudan University (Shanghai), the French Academy of Science's Grand Prix Alexandre Joannidès, and the Seymour Cray Prize in Numerical Simulation. Professor Temam has authored or co-authored nine books and published more than 260 articles in international refereed journals. His current research interests in fluid mechanics are in the areas of control of turbulence, boundary layer theory, and geophysical fluid dynamics.

ENCYCLOPEDIA OF MATHEMATICS AND ITS APPLICATIONS

Volume 83
Navier–Stokes Equations and Turbulence

6 H. Minc *Permanents*
19 G. G. Lorentz, K. Jetter, and S. D. Riemenschneider *Birkhoff Interpolation*
22 J. R. Bastida *Field Extensions and Galois Theory*
23 J. R. Cannon *The One-Dimensional Heat Equation*
24 S. Wagon *The Banach–Tarski Paradox*
25 A. Salomaa *Computation and Automata*
27 N. H. Bingham, C. M. Goldie, and J. L. Teugels *Regular Variation*
28 P. P. Petrushev and V. A. Popov *Rational Approximation of Real Functions*
29 N. White (Ed.) *Combinatorial Geometries*
30 M. Pohst and H. Zassenhaus *Algorithmic Algebraic Number Theory*
31 J. Aczel and J. Dhombres *Functional Equations in Several Variables*
32 M. Kuczma, B. Choczewski, and R. Ger *Iterative Functional Equations*
33 R. V. Ambartzumian *Factorization Calculus and Geometric Probability*
34 G. Gripenberg, S.-O. Londen, and O. Staffans *Volterra Integral and Functional Equations*
35 G. Gasper and M. Rahman *Basic Hypergeometric Series*
36 E. Torgersen *Comparison of Statistical Experiments*
38 N. Korneichuk *Exact Constants in Approximation Theory*
39 R. Brualdi and H. Ryser *Combinatorial Matrix Theory*
40 N. White (Ed.) *Matroid Applications*
41 S. Sakai *Operator Algebras in Dynamical Systems*
42 W. Hodges *Basic Model Theory*
43 H. Stahl and V. Totik *General Orthogonal Polynomials*
45 G. Da Prato and J. Zabczyk *Stochastic Equations in Infinite Dimensions*
46 A. Björner et al. *Oriented Matroids*
47 G. Edgar and L. Sucheston *Stopping Times and Directed Processes*
48 C. Sims *Computation with Finitely Presented Groups*
49 T. Palmer *Banach Algebras and the General Theory of *-Algebras*
50 F. Borceux *Handbook of Categorical Algebra 1*
51 F. Borceux *Handbook of Categorical Algebra 2*
52 F. Borceux *Handbook of Categorical Algebra 3*
53 V. F. Kolchin *Random Graphs*
54 A. Katok and B. Hasselblatt *Introduction to the Modern Theory of Dynamical Systems*
55 V. N. Sachkov *Combinatorial Methods in Discrete Mathematics*
56 V. N. Sachkov *Probabilistic Methods in Discrete Mathematics*
57 P. M. Cohn *Skew Fields*
58 R. Gardner *Geometric Topography*
59 G. A. Baker, Jr. and P. Graves-Morris *Padé Approximants*
60 J. Krajicek *Bounded Arithmetic Propositional Logic and Complexity Theory*
61 H. Groemer *Geometric Applications of Fourier Series and Spherical Harmonics*
62 H. O. Fattorini *Infinite Dimensional Optimization and Control Theory*
63 A. C. Thompson *Minkowski Geometry*
64 R. B. Bapat and T. E. S. Raghavan *Nonnegative Matrices with Applications*
65 K. Engel *Sperner Theory*
66 D. Cvetkovic, P. Rowlinson, and S. Simic *Eigenspaces of Graphs*
67 F. Bergeron, G. Labelle, and P. Leroux *Combinatorial Species and Tree-Like Structures*
68 R. Goodman and N. R. Wallach *Representations and Invariants of the Classical Groups*
69 T. Beth, D. Jungnickel, and H. Lenz *Design Theory,* vol. 1
70 A. Pietsch and J. Wenzel *Orthonormal Systems for Banach Space Geometry*
71 G. E. Andrews, R. Askey, and R. Roy *Special Functions*
72 R. Ticciati *Quantum Field Theory for Mathematicians*
73 M. Stern *Semimodular Lattices*
74 I. Lasiecka and R. Triggiani *Control Theory for Partial Differential Equations I*
75 I. Lasiecka and R. Triggiani *Control Theory for Partial Differential Equations II*
76 A. A. Ivanov *Geometry of Sporadic Groups 1*
77 A. Schinzel *Polynomials with Special Regard to Reducibility*
78 H. Lenz, T. Beth, and D. Jungnickel *Design Theory,* vol. 2

ENCYCLOPEDIA OF MATHEMATICS AND ITS APPLICATIONS

Navier–Stokes Equations and Turbulence

C. FOIAS O. MANLEY

R. ROSA R. TEMAM

CAMBRIDGE
UNIVERSITY PRESS

CAMBRIDGE UNIVERSITY PRESS
Cambridge, New York, Melbourne, Madrid, Cape Town, Singapore, São Paulo

Cambridge University Press
The Edinburgh Building, Cambridge CB2 8RU, UK

Published in the United States of America by Cambridge University Press, New York

www.cambridge.org
Information on this title: www.cambridge.org/9780521360326

First published 2001
This digitally printed version 2008

A catalogue record for this publication is available from the British Library

Library of Congress Cataloguing in Publication data
Navier–Stokes equations and turbulence / C. Foias ... [et al.].
p. cm. – (Encyclopedia of mathematics and its applications; v. 83)
Includes bibliographical references and index.
ISBN 0-521-36032-3
1. Turbulence. 2. Navier–Stokes equations. I. Foias, Ciprian.
II. Series.

QA913.N38 2001
532′.0527 – dc21 00-053008

ISBN 978-0-521-36032-6 hardback
ISBN 978-0-521-06460-6 paperback

Contents

Preface

This monograph is an attempt to address the theory of turbulence from the points of view of several disciplines. The authors are fully aware of the limited achievements here as compared with the task of understanding turbulence. Even though necessarily limited, the results in this book benefit from many years of work by the authors and from interdisciplinary exchanges among them and between them and others. We believe that it can be a useful guide on the long road toward understanding turbulence.

One of the objectives of this book is to let physicists and engineers know about the existing mathematical tools from which they might benefit. We would also like to help mathematicians learn what physical turbulence is about so that they can focus their research on problems of interest to physics and engineering as well as mathematics. We have tried to make the mathematical part accessible to the physicist and engineer, and the physical part accessible to the mathematician, without sacrificing rigor in either case. Although the rich intuition of physicists and engineers has served well to advance our still incomplete understanding of the mechanics of fluids, the rigorous mathematics introduced herein will serve to surmount the limitations of pure intuition. The work is predicated on the demonstrable fact that some of the abstract entities emerging from functional analysis of the Navier–Stokes equations represent real, physical observables: energy, enstrophy, and their decay with respect to time.

Beside this didactic objective, one of our scientific goals – in this book and in its underlying research – was to see what we can learn about the physical properties of turbulence using Sobolev spaces and the functional analysis methods that are based on them. As we subsequently show, these spaces – which seem to be abstract mathematical inventions – are in fact representations of observable physical quantities. In this way we have recovered several parts of the conventional theory of turbulence, deriving rigorously from the Navier–Stokes equations (NSE) what had been arrived at earlier by phenomenological arguments (Kolmogorov [1941a,b]), but in addition we derive new results. We have shown that the conventional estimate of the number of degrees of freedom in homogeneous, isotropic turbulence (viz., (Reynolds number)$^{9/4}$) is at best an upper bound on the number of degrees of freedom needed for numerical simulations of real flows. We have also provided a rigorous, mathematical way to avoid the common underlying assumption of the ergodicity of turbulent flows. In fact, we show that (in a suitable sense) time averages of various turbulent flow properties equal the related ensemble averages with respect to adequate statistical solutions; we have also found a means for removing the high-wavenumber components of the flow in such a way as to yield an effective viscosity, while providing a rough upper bound on the error committed relative to the true solution of the flow equation.

Another task, the second scientific objective of this book, was to make the connection between three of the classical approaches to turbulence: the Navier–Stokes equations; the dynamical systems approach (following the work and ideas of Lorenz [1963], Smale [1967], and Ruelle and Takens [1971]); and the conventional statistical theory of turbulence (following the works and ideas of Kolmogorov [1941a,b, 1962], Batchelor [1959], Kraichnan [1967], and others – e.g., Landau and Lifshitz [1971] and Monin and Yaglom [1975]). Before the research underlying the material presented here, these classical approaches evolved largely independently. In particular, the conventional theory of turbulence is based mostly on dimensional phenomenological arguments that traditionally make little reference to the NSE (see Tennekes and Lumley [1972]). However, we believe it is useful and instructive to show that many known results can be directly derived from the Navier–Stokes equations. We develop those connections to the widest possible extent.

The level of mathematical preparation necessary for understanding this material is an elementary knowledge of partial differential equations and their solutions in terms of eigenfunction expansions. Terms and concepts beyond that level are presented in detail as needed. Also included is a brief tutorial on Sobolev spaces and inequalities. To aid readers unfamiliar with some useful classical inequalities, they are presented (without proof) in Chapter I along with the tutorial.

Mathematically oriented readers are assumed to be familiar with elementary physics and continuum mechanics, including such principles as conservation of momentum and energy and the relationship between stress and strain. For their benefit, Chapter I contains also a short tutorial on the Kolmogorov (conventional) theory of turbulence.

One of the unresolved difficulties encountered in this monograph is due to limitations in the present stage of the mathematical theory of the NSE; the theory is fairly complete in the 2-dimensional case but still incomplete in dimension 3. Thus, while we realize that natural turbulence is usually 3-dimensional, here we sometimes emphasize 2-dimensional flows, which are fully within the grasp of modern methods of functional analysis.

The word *turbulence* has different meanings to different people, which indicates that turbulence is a complex and multifaceted phenomenon. For mathematicians, outstanding problems revolve around the Navier–Stokes equations (such as well-posedness and low-viscosity behavior, especially in the presence of walls or singular vortices). For physicists, major questions include ergodicity and statistical behavior as related to statistical mechanics of turbulence. Engineers would like responses to questions simple to articulate but amazingly difficult to answer: What are the heat transfer properties of a turbulent flow? What are the forces applied by a fluid to its boundary (be it a pipe or an airfoil)? To others pursuing the dynamical system approach, of interest is the large time behavior of the flow. Another ambitious question for engineers is the control of turbulence (to either reduce or enhance it), which is already within reach. Finally, a major goal in turbulence research – of interest to all and toward which progress is constantly made – is trustworthy and reliable computation of turbulent flows (see e.g. Orszag [1970] and Ferziger, Mehta, and Reynolds [1977]).

We do not address here any computational aspects, although this problem is very much present in our thoughts; neither do we address control problems, nor most of the practical engineering problems (see Schlichting [1960]). After the introductory and tutorial Chapter I, the core of the book consists of four chapters, Chapters II–V. Each of them, in addressing a particular topic, could actually be developed into a whole independent volume. Chapter II summarizes some classical and some more recent aspects of the mathematical theory of the Navier–Stokes equations – namely, their formulation and well-posedness. We start by presenting the physical background of the mathematical theory, introducing kinetic energy and enstrophy, conservation of kinetic energy, and the Helmholtz–Leray decomposition of vector fields. We present function spaces, the spaces of finite kinetic energy and finite enstrophy vector functions, as well as some additional related abstract spaces. After recalling the weak formulation of the NSE, a starting point of their mathematical theory going back to the work of Jean Leray in the early 1930s, we recall the main theorems of existence, uniqueness, and regularity of solutions. Then we describe analyticity properties of the solutions; first, analyticity in time, which is sometimes related to intermittency (a question briefly addressed in Sections 6.2 and 6.3 of Chapter V); and second, analyticity in space and time (Gevrey class regularity), which is related in the space-periodic case to the decline of Fourier coefficients of the solution. Finally, we briefly discuss the no-slip case with moving boundaries and establish properties of the rate of dissipation of flows.

Chapter III revolves around the idea (hinted at long ago by Landau and Lifshitz) that, in the permanent regime, turbulent flows as solutions of NSE are finite-dimensional. This concept, which in fact follows easily from the Kolmogorov approach to turbulence, was novel in its time; by now it has been substantiated in many different ways and extended as well to other equations modeling other physical phenomena. In Chapter III we discuss finite dimensionality of turbulent flows in the context of determining modes and nodes, showing that such flows are fully determined by either a finite (sufficiently large) number of modes or a finite number of observation points (nodes). We discuss also the large time behavior in the context of attractors and show finite dimensionality of attractors; all these dimensions are physically relevant and related to the Landau–Lifshitz estimates. We briefly discuss approximate inertial manifolds, the initial point for multilevel numerical algorithms under development; in some sense, these algorithms produce in time what multigrid or wavelet methods produce in space. Chapter IV comes closest to the issue of ergodicity. We introduce, in space dimensions 2 and 3, stationary statistical solutions and relate them to the limits of time averages. We consider also the corresponding invariant measure and relate it to the attractor that carries it. We then apply these tools to the study of the cascade processes in turbulent flows.

Finally, in Chapter V, we study the concept of statistical solutions, the evolution of the probability distribution of the flow, and homogeneous flows. We start by introducing the time-dependent statistical solutions on bounded domains. Then we introduce the (space-invariant) homogeneous statistical solutions for space-periodic flows and flows in the whole spaces. The Reynolds averaged equations are introduced, and we then discuss self-similar homogeneous statistical solutions (SSHSS); we introduce a 2-parameter family of such solutions from which, on the one hand, we resolve a paradox on SSHSS pointed out by Hopf [1952] and, on the other hand, we recover and complete some elements of the conventional theory of turbulence. For instance, we show how the Kolmogorov spectrum follows naturally from NSE and how the intermittency of turbulent flows is related to the fractal nature (see Novikov and Stewart [1964] and Mandelbrot [1982]) of energy dissipation in 3-dimensional flows.

As with all interdisciplinary work, it is not easy to write a book that is readable by (and of equal interest to) people with differing perspectives. In order to overcome this difficulty, we have divided each of the main chapters into two parts: the main one, in which we hope the language is understandable by all, contains as few mathematical technicalities as possible yet still states the results in a rigorous way. Then, as needed, a long appendix gives the details of the proofs.

The reader should note that some of the cited original articles underlying this monograph may treat the same problem in two distinct publications: a more physically oriented treatment appearing in a physics or mechanics journal as well as a corresponding "heavy" mathematical treatment presented in a mathematics journal. That is clearly due to the idiosyncrasies of the two kinds of publications and the need for different presentation styles when addressing the different audiences.

A few remarks will conclude this Preface. First, the authors are fully aware that this book is difficult to read because, owing to the nature of the subject, it assumes the reader's familiarity with several distinct areas of knowledge. The three senior authors hope that the younger generation, more accustomed to interdisciplinary work than their predecessors, will find this work more readily accessible than will their elders. In that regard, the three senior authors are delighted that their younger colleague (RR) had agreed to involve himself so deeply in *all* the aspects of this book, and they hope that this bodes well for its future – especially insofar as its interest and accessibility to the younger generation are concerned.

Second, on the anecdotal side, we recall briefly the genesis of this interdisciplinary collaboration. For a number of years CF and RT had worked independently on the analysis of the Navier–Stokes equations; CF had learned the subject from Jacques-Louis Lions and Giovanni Prodi; he collaborated with Prodi and started to develop a rigorous theory of statistical solutions of those equations. RT learned the subject from Jacques-Louis Lions and Jean Leray, and he also worked on the stochastic solutions of the NSE. Then CF and RT met in the summer of 1970 at a meeting – organized by Giovanni Prodi – in Varenna (Italy), and CF visited RT at Orsay (France) in the fall of 1974. Their collaboration started, addressing such different aspects of the NSE as analysis, statistical solutions, and the long time behavior (dynamical systems point of view). At some point RT suggested that their collaboration would become more interesting if they could join forces with a physicist.

By chance in the spring of 1980, Jacqueline Mossino, a former student of RT, met Yvain Trève (OM's co-worker) at a conference on plasma physics in Tucson (Arizona), and contact was established by letter at a time predating e-mail; eventually they met face-to-face for the first time at a meeting in Dekalb (Illinois) in 1981. At that time, OM was working with Trève on finite–mode number approximations of thermal convection satisfying the first and second laws of thermodynamics and discovered that the qualitative nature of the numerical results depended critically on the number of modes retained (Trève and Manley [1982]). As they started to interact, CF, OM, and RT realized immediately the extent of the common ground between the two communities and perspectives that they represent. That realization was the original stimulus for much of the research reported in this volume. More specifically, the direction of that research was set by the recognition that a simple, physically based argument (conservation of energy and momentum in thermal convection; Trève and Manley [1981]) yielded a result – a bound on the sufficient number of degrees of freedom for this fluid flow – that is essentially equivalent to an elaborate mathematical exercise in Sobolev spaces (Foias, Manley, Temam, and Trève [1983b]). This collaboration has extended through the rest of the 1980s, the 1990s, and beyond.

The youngest author was a graduate student at Indiana University from 1992 to 1996, and he had many opportunities to be exposed to this research through courses, informal discussions, and seminar lectures. He enthusiastically agreed to participate in this book, which has been in process for a number of years, and eventually started to collaborate on more recent works. As indicated earlier, the three senior authors are delighted that RR has joined them in this task, and they see it as a good omen for a successful transmittal of these results to the next generation.

Beside the prolonged and extended efforts of the four authors, this book has benefitted extensively from the input and influence of many others by occasional collaborations, discussions, and other forms of interaction. It is not possible to name them all, but we want to thank them for their constructive influence on us. Also, we would like to extend our deepest thanks to those who have co-authored relevant publications with one or more of the authors of this book, works that are partially or fully reported in this monograph: Hari Bercovici, Peter Constantin (with whom three of us had an extended collaboration), Arnaud Debussche, Jean-Michel Ghidaglia, David Gottlieb, Martine Marion, Jean-Claude Saut, George Sell, Denis Serre, Edriss Titi, and Yvain Trève.

Finally, the authors are very grateful to David Tranah and Alan Harvey of Cambridge University Press for their interest, their encouragement, and their great patience in waiting for the delivery of the manuscript.

Ciprian Foias   Oscar Manley
Ricardo Rosa   Roger Temam

Acknowledgments

The research presented in this book has been supported along the years by grants from: the National Science Foundation NSF-DMS, 8802596, 9024769, 9400615, 9705229, 9706903, CDA-9601632; the U.S. Department of Energy, Office of Scientific Computing, DE-AC02-82ER12049, DE-FG02-86ER25020, DE-FG02-92ER25120; the Air Force Office of Scientific Research AFOSR-88-103; and the Office of Naval Research, N00014-91-J-1140, N00014-96-1-0-425.

This research was also partially supported by: the Research Fund of Indiana University; the Université de Paris-Sud and the Centre National de la Recherche Scientifique (CNRS) through the Laboratoire d'Analyse Numérique d'Orsay; CNPq (Brasília, Brazil), FUJB, and FAPERJ (Rio de Janeiro, Brazil); the Institute for Mathematics and Its Applications, University of Minnesota; and the Center for Nonlinear Studies, Los Alamos National Laboratory, New Mexico.

OM gratefully acknowledges the understanding and encouragement of his efforts shown by his former supervisor, Dr. J. Coleman, at the U.S. Department of Energy. Furthermore, he wishes to express his deep appreciation to the Laboratoire d'Analyse Numérique d'Orsay at Université de Paris-Sud, the Department of Mathematics at Indiana University, and Ecole Central de Lyon, where he has been a visitor on many inspiring and productive occasions.

I

Introduction and Overview of Turbulence

Introduction

In this chapter we first briefly recall, in Section 1, the derivation of the Navier–Stokes equations (NSE) starting from the basic conservation principles in mechanics: conservation of mass and momentum. Section 2 contains some general remarks on turbulence, and it alludes to some developments not presented in the book. For the benefit of the mathematically oriented reader (and perhaps others), Section 3 provides a fairly detailed account of the Kolmogorov theory of turbulence, which underlies many parts of Chapters III–V. For the physics-oriented reader, Section 4 gives an intuitive introduction to the mathematical perspective and the necessary tools. A more rigorous presentation appears in the first half of Chapter II and thereafter as needed. For each of the aspects that we develop, the present chapter should prove more useful for the nonspecialist than for the specialist.

1 Viscous Fluids. The Navier–Stokes Equations

Fluids obey the general laws of continuum mechanics: conservation of mass, energy, and linear momentum. They can be written as mathematical equations once a representation for the state of a fluid is chosen. In the context of mathematics, there are two classical representations. One is the so-called Lagrangian representation, where the state of a fluid "particle" at a given time is described with reference to its initial position. The other representation (adopted throughout this book) is the so-called Eulerian representation, where at each time t and position $\mathbf{x}$ in space the state – in particular, the velocity $\mathbf{u}(\mathbf{x}, t)$ – of the fluid "particle" at that position and time is given.

In the Eulerian representation of the flow, we also represent the density $\rho(\mathbf{x}, t)$ as a function of the position $\mathbf{x}$ and time t. The conservation of mass is expressed by the continuity equation

$$\frac{\partial \rho}{\partial t} + \operatorname{div}(\rho \mathbf{u}) = 0. \tag{1.1}$$

The conservation of momentum is expressed in terms of the acceleration $\boldsymbol{\gamma}$ and the Cauchy stress tensor $\boldsymbol{\sigma}$:

$$\rho \gamma_i = \sum_{j=1}^{3} \frac{\partial \sigma_{ij}}{\partial x_j} + f_i, \quad i = 1, 2, 3. \tag{1.2}$$

Here $\boldsymbol{\gamma} = (\gamma_1, \gamma_2, \gamma_3)$ and $\boldsymbol{\sigma} = (\sigma_{ij})_{i,j=1,2,3}$, componentwise in the 3-dimensional case. Moreover, $\mathbf{f} = (f_i, f_2, f_3)$ represents volume forces applied to the fluid.

The acceleration vector $\boldsymbol{\gamma} = \boldsymbol{\gamma}(\mathbf{x}, t)$ of the fluid at position $\mathbf{x}$ and time t can be expressed, using purely kinematic arguments, by the so-called material derivative

$$\boldsymbol{\gamma} = \frac{D\mathbf{u}}{Dt} = \frac{\partial \mathbf{u}}{\partial t} + (\mathbf{u} \cdot \nabla)\mathbf{u}, \tag{1.3}$$

or, componentwise,

$$\gamma_i = \frac{\partial u_i}{\partial t} + \sum_{j=1}^{3} u_j \frac{\partial u_i}{\partial x_j}, \quad i = 1, 2, 3.$$

Inserting this expression into the left-hand side (LHS) of equation (1.2) yields the term $\rho(\mathbf{u} \cdot \nabla)\mathbf{u}$, which is the only nonlinear term in the Navier–Stokes equations; this term is also called the *inertial term*. The Navier–Stokes equations are among the very few equations of mathematical physics for which the nonlinearity arises not from the physical attributes of the system but rather from the mathematical (kinematical) aspects of the problem.

Further transformations of the conservation of momentum equation necessitate additional physical arguments and assumptions. Rheology theory relates the stress tensor to the velocity field for different materials through the so-called stress–strain law and other constitutive equations. Assuming the fluid is Newtonian, which is the case of interest to us, amounts to assuming that the stress–strain law is linear. More precisely, for Newtonian fluids the stress tensor is expressed in terms of the velocity field by the formula

$$\sigma_{ij} = \mu \left\{ \frac{\partial u_i}{\partial x_j} + \frac{\partial u_j}{\partial x_i} \right\} + (\lambda \operatorname{div} \mathbf{u} - p)\delta_{ij}, \tag{1.4}$$

where $p = p(\mathbf{x}, t)$ is the pressure. Here, δ_{ij} is the Kronecker symbol and μ, λ are constants. The constant μ is called the shear viscosity coefficient, and $3\lambda + 2\mu$ is the dilation viscosity coefficient. For thermodynamical reasons, $\mu > 0$ and $3\lambda + 2\mu \geq 0$. Inserting the stress–strain law (1.4) into the momentum equation (1.2), we obtain

$$\rho \left\{ \frac{\partial \mathbf{u}}{\partial t} + (\mathbf{u} \cdot \nabla)\mathbf{u} \right\} = \mu \Delta \mathbf{u} + (\mu + \lambda) \nabla \operatorname{div} \mathbf{u} - \nabla p + \mathbf{f}. \tag{1.5}$$

Equations (1.1) and (1.5) govern the motion of compressible Newtonian fluids such as the air at high speeds (Mach number larger than 0.5). If we also assume that the fluid is incompressible and homogeneous, then the density is constant in space and time: $\rho(\mathbf{x}, t) \equiv \rho_0$. In this case, the continuity equation is reduced to the *divergence-free* condition:

$$\operatorname{div} \mathbf{u} = 0. \tag{1.6}$$

Because the density is constant, we may divide the momentum equation (1.5) by ρ and consider the so-called kinematic viscosity $\nu = \mu/\rho_0$; we may then replace the pressure p and the volume force $\mathbf{f}$ by the kinetic pressure p/ρ_0 and the mass density of body forces $\mathbf{f}/\rho_0$, respectively. In doing so, and taking into consideration the divergence-free condition (1.6), we obtain the *Navier–Stokes equations for a viscous, incompressible, homogeneous flow*:

$$\frac{\partial \mathbf{u}}{\partial t} - \nu\Delta\mathbf{u} + (\mathbf{u}\cdot\nabla)\mathbf{u} + \nabla p = \mathbf{f}, \tag{1.7a}$$

$$\nabla\cdot\mathbf{u} = 0, \tag{1.7b}$$

where, for notational simplicity, we represent the divergence of $\mathbf{u}$ by $\nabla\cdot\mathbf{u}$. For all pratical purposes, the density has actually been normalized to unity; even so, we may sometimes replace (1.7a) by (1.5), remembering then that $\nabla\cdot\mathbf{u} = 0$ and ρ is constant.

For more details on the physical aspects of fluid mechanics, we refer the reader to the classical books of Batchelor [1988] and Landau and Lifshitz [1971].

It is readily accepted that the Navier–Stokes equations govern the motion of common fluids such as air or water, so we are faced with the persistent challenging question of recovering from (1.7) such complex motions as that of smoke dispersion in the air and the turbulent flow of a river around a bridge pillar.

The flow of fluids at the microscopic level is governed by phenomena in the realm of statistical mechanics of fluids. The appropriate statistics is given by the solution of the Boltzmann equation. That equation represents the evolution of the governing distribution function, which is dependent on the position and velocity of the particles colliding with one another as a result of thermal excitation at any finite temperature. The collisions are described by an integral collision operator. In general, the collision operator represents simultaneous collisions among many particles, necessitating the use of a many-particle distribution. As such, it is very complicated and essentially impossible to evaluate precisely. Only in the case of dilute gases can one limit oneself to considering the evolution of a single-particle distribution and to binary collisions, since many body collisions are highly unlikely. In this idealized situation, the collision operator can be approximated by first-order and second-order spatial derivatives. The former is the familiar pressure gradient and the latter is the Laplacian operating on the velocity, multiplied by a constant known as the viscosity. With that approximation in hand, we can take the appropriate moments of the one-particle Boltzmann equation and so derive first the conservation of mass equation and second the conservation of momentum equation that we recognize as the NSE (when the incompressibility condition is a valid assumption).

Although such a derivation has been carried out for dilute gases, a corresponding exercise for liquids remains an open problem. This is because binary collisions play a relatively minor role in liquids, which are much denser than gases and hence feature collisions between clusters of particles. However, for practical reasons and lacking a better option, we use the Navier–Stokes equations with a simple constant viscosity as a reasonable model for liquid flows.

The origin of viscosity imposes a limit on the domain of validity of the Navier–Stokes equations. Thus phenomena on a length scale comparable to or smaller than the collision mean free path in air at atmospheric pressure (say, 10^{-3} cm) cannot be described by a continuum model such as the NSE. Subsequently we will learn about some natural lengths that characterize the length scale region in which flow energy dissipation is dominated by viscous phenomena. It will be important then to be sure that we are still in the regime characterized by a continuum model of the flow. A similar cautionary remark applies to the amplitude of fluctuations in turbulent flows: once we are in a regime in which those fluctuations are comparable with thermally (finite temperature) induced fluctuations, the model based on Navier–Stokes equations ceases to be relevant.

Nondimensional Form of the Navier–Stokes Equations

It is sometimes convenient, both for physical discussions and mathematical transparency, to consider a nondimensional form of the conservation of momentum equation. For that purpose we introduce a reference length L_* and a reference time T_* for the flow, and we set

$$\mathbf{x} = L_*\mathbf{x}', \quad t = T_*t', \quad p = P_*p', \quad \mathbf{u} = U_*\mathbf{u}', \quad \mathbf{f} = \frac{L_*}{T_*^2}\mathbf{f}',$$

where $P_* = U_*^2$ and $U_* = L_*/T_*$ are a reference pressure and a reference velocity, respectively. By substitution into (1.7) we obtain for $\mathbf{u}'$, p', $\mathbf{f}'$ the same equation but with ν replaced by Re^{-1}, where Re is a nondimensional number called the *Reynolds number*:

$$\mathrm{Re} = \frac{L_*U_*}{\nu}. \tag{1.8}$$

The value of the Reynolds number depends on the choice of the reference length and velocity. Usually, if Ω (the domain occupied by the fluid) is bounded then L_* can be taken as the diameter of Ω or as some other large-scale length related to Ω, such as the width of a channel. The choice of U_* (and hence of T_*) depends on the type of forcing of the flow; it can be related to the forces applied at the boundary of Ω or to a pressure gradient, for example. Various choices of L_* and U_* can be appropriate for a given flow, leading to various definitions of the Reynolds number, but turbulent flows result for all appropriate choices when Re is large. How large depends to some extent on the shape of the domain occupied by the fluid. Once the shape of the domain Ω is fixed, however, rescalings in length (L_*) and velocity (U_*) and changes in viscosity (ν) affect the equations only through the single parameter Re.

Hence, different experiments may lead to the same nondimensional equations. For example, multiplying the velocity by 2 and dividing the diameter of the domain by 2 leaves the Reynolds number unchanged, so we can pass from one experiment to another; this is the Reynolds similarity hypothesis constantly used in mechanical engineering. At a given Reynolds number, flows remote from the boundaries of the domain Ω, irrespective of the latter's shape, are similar owing to some universality properties of turbulent flows. Moreover, with flows around blunt bodies (say, a sphere), as the body's radius increases and the flow velocity and/or viscosity is adjusted so as to maintain the Reynolds number constant, the flow throughout the modified flow domain remains similar. That is what has made possible the design of aircraft by means of relatively small models tested in moderately sized wind tunnels.

In Chapter III, instead of the Reynolds number we will use another nondimensional number: the Grashof number (see Section 13 in Chapter II).

A heuristic argument illustrating the significance of the Reynolds number emerges by comparing the inertial and dissipation terms of the Navier–Stokes equations. The inertial term $(\mathbf{u} \cdot \nabla)\mathbf{u}$ has dimension

$$\frac{U_*^2}{L_*},$$

while the dissipation term has dimension

$$\nu \frac{U_*}{L_*^2}.$$

The inertial term dominates when

$$\mathrm{Re} = \frac{L_* U_*}{\nu} \gg 1.$$

However, a much more subtle analysis that is valid at each length scale is made for the Kolmogorov theory of turbulence.

By setting $\mathrm{Re} = +\infty$ (i.e., $\nu = 0$), we obtain the case of inviscid flows. In this case, the divergence-free condition is retained but the momentum equation changes, resulting in the Euler equations for inviscid perfect fluids:

$$\frac{\partial \mathbf{u}}{\partial t} + (\mathbf{u} \cdot \nabla)\mathbf{u} + \nabla p = \mathbf{f}, \tag{1.9a}$$

$$\nabla \cdot \mathbf{u} = 0. \tag{1.9b}$$

Note that some of the difficulties encountered in studying turbulent behavior, a largely inviscid regime, arise because the transition from Euler's equations to the Navier–Stokes equations necessitates a change from a first-order system to a second-order one in space (∇ to Δ), which involves a singular perturbation.

2 Turbulence: Where the Interests of Engineers and Mathematicians Overlap

Principal substantive questions related to turbulence have been raised since the beginning of the twentieth century, and a large number of empirical and heuristical results were derived – motivated principally by engineering applications. This includes the work of Lamb [1957], mostly on addressing idealized inviscid flows; Prandtl [1904], on eddy viscosity and boundary layers; Taylor [1935, 1937], on viscous flows; and von Karman [1911, 1912], on the nature of the boundary layer.

At the same time, in mathematics there appears the pioneering work of Jean Leray [1933, 1934a,b] on the Navier–Stokes equations. Leray speculated that turbulence is due to the formation of point or "line vortices" on which some component of the velocity becomes infinite.[1] To enable dealing with such a situation, he suggested the concept of weak, nonclassical solutions to the Navier–Stokes equations (1.7), and this has become the starting point of the mathematical theory of the Navier–Stokes equations to this day. We will consider this approach in Chapter II and beyond. It is noteworthy that, more generally, Leray's ideas serve also as the starting point for several important elements of the modern theory of partial differential equations. Even today, despite much effort, Jean Leray's conjecture concerning the appearance of singularities in 3-dimensional turbulent flows has been neither proved nor disproved. Let us mention, however, the result of Caffarelli, Kohn, and Nirenberg [1982] (see also Scheffer [1977]), which considerably extends an earlier result of Leray: Given the possibility that the singular points are a fractal set (assuming that such a set exists), the 1-dimensional Hausdorff measure of that set in space and time is 0. Hence the occurence of smooth line vortices is not possible, explaining our quotation marks around "line vortices." Nevertheless, for all physical purposes this powerful mathematical result leaves room for a tremendously complex set of singularities, and so we remain far from closing the issues raised by Leray's conjecture.

Before continuing with these historical notes, we remark in passing that engineers are not directly affected by such purely mathematical issues; rather, they want to calculate or measure certain physical quantities (forces, velocities, pressures, etc.). Here, however, beside the possible occurrence of singularities, another critical aspect of turbulence comes to mind: in a turbulent flow, many interesting quantities vary rapidly in time and cannot be readily measured. In practice, all that can be measured in laboratory experiments are averages (usually time averages). These averages are well-defined, reproducible quantities. This leads to the concept of ensemble averages underlying the conventional theory of turbulence, and to the concept of statistical solutions of the Navier–Stokes equations (1.7). It leads also to the idea of *ergodicity,* which is taken for granted by engineers. Loosely speaking, for all initial experimental conditions and for all sorts of reasonable ensemble averages, the experiments always yield the same measured results to within the accuracy of the measurements. We address here those questions of direct interest to engineers: the need for statistical solutions, the equivalence between ensemble averages and time averages (a question addressed in Chapter IV), and the so far unchallenged issue of the axiomatic nature of ergodicity.

We return to our brief overview of some highlights in the history of the studies of turbulent flows. It is impossible to explore here all the aspects of that history. Hence, with apologies to all whose important contributions are not mentioned here, we limit ourselves to those aspects of the history most relevant to the subject of this monograph.

[1] In fact, if such discontinuities occur then another question of physical nature needs to be raised concerning the validity of the Navier–Stokes equations themselves; indeed, at very short distances of order 10^{-3} cm (the collision mean free path of the particles), the fluid equations are no longer pertinent.

Turbulent flows have mystified people for ages, as evidenced for example by Leonardo da Vinci's sketches of the turbulent wakes downstream of some bridge columns. Beginning with careful experimental studies of flows under various experimental conditions (Reynolds [1883, 1895]) and with the subsequent formulation of the Navier–Stokes equations, turbulence became a subject of thorough scientific inquiry. For many years, two difficulties held the attention of various investigators. The first was a technical mathematical obstacle: the presence of the inertial term (a quadratic nonlinearity) precludes a straightforward use of the many available tools of perturbation methods. The structure of the equations demands that, at any given step in an approximation scheme, information from the next step is necessary. This had led to many attempts at formulating the so-called closure schemes, where at some step in the approximation sequence an assumption about the nature of the subsequent term is made, thereby terminating that sequence. Such an assumption, usually justified in terms of intuitive physical arguments, was then used to break the impasse in the approximation sequence. In principle, closure schemes by and large call for unprovable assumptions beyond those composing the basis for the Navier–Stokes equations. Some of the better-known closure schemes may be found in such texts as Tennekes and Lumley [1972], Leslie [1973], and Lesieur [1997], although further attempts (and controversies) in this area continue. As we shall find in the present work, the invention of the so-called inertial manifolds in the context of the rigorous theory of NSE (as well as of other nonlinear partial differential equations) opens the door to mathematically more soundly based schemes for computational approaches, offering an alternative to the conventional closure schemes.

The second obstacle to progress in the theory of turbulence was largely conceptual. Namely, how was it possible for a system described by perfectly deterministic equations to exhibit behavior that was undeniably statistical in nature? This aspect of turbulent flows, both from the experimental side and from the nascent theoretical side, is dealt with at length in the monumental work of Monin and Yaglom [1975]. Hopf [1952], followed by Foias and Prodi [1976] (see also Foias [1972, 1973, 1974]), studied an extension of Liouville's theorem that in principle yields the probability distribution function underlying the Navier–Stokes equation. Many of these efforts rested on the experimental and theoretical work of Taylor [1935, 1937] and von Karman and Howarth [1938], who clarified, on intuitive grounds, the nature of homogeneous isotropic turbulence. The simplifications resulting from the symmetries inherent in this idealized form of turbulence yielded the well-known von Karman–Howarth ordinary differential equation for the self-similar evolution of the two-point velocity correlation tensor. This idealization has also yielded Kolmogorov's theory for the spectrum of homogeneous isotropic turbulence in three dimensions (Kolmogorov [1941a,b]) (and later Batchelor's [1959] and Kraichnan's [1967] corresponding results for turbulence in two dimensions), a subject of the next section. All of these results were obtained without full understanding of the origin of the statistical nature of turbulence. A significant breakthrough occurred in the 1960s and 1970s with the discovery of stochastic instabilities in seemingly innocuous low-order ordinary differential equations (Lorenz [1963]) and in some nonlinear difference equations (Feigenbaum [1980]). Subsequent research (Foias and Prodi [1976], Vishik and Fursikov [1977a,b, 1978], Foias and Temam [1979]) on dynamical systems governed by nonlinear partial differential equations revealed that such dynamical systems may reside, in finite-dimensional function spaces, on compact attractors that may be characterized by chaotic behavior.

It is now appropriate to reiterate a point hinted at earlier, namely, the essential need for careful mathematical analysis when dealing with nonlinear entities such as the Navier–Stokes equations. While much of our physical intuition serves us well in the domain of linear phenomena modeled adequately by linear differential and partial differential equations, it can fail us – with potentially disastrous consequences – in nonlinear domains. A fairly instructive example, outside the realm of this book but worth mentioning here, concerns modeling sonic flow transition as a boundary value problem rather than (and more correctly) as an initial value problem (Greenberg and Trève [1960]). Although this may appear to be unnecessary pedantry, it clearly makes a lot of difference in the context of, say, nuclear reactor safety (Bilicki et al. [1987]). Unlike the case in linear systems, in nonlinear systems small causes can lead to very large effects indeed, as well as to qualitative differences. Because nonlinear equations can have multiple, qualitatively different solutions (different basins of attraction), a small change in initial conditions can sometimes lead to radically different time-asymptotic behavior. An even more dramatic, counterintuitive example is the previously mentioned possibility of chaotic behavior in what at first sight seem to be innocent deterministic systems (Lorenz [1963], Feigenbaum [1980], Smale [1967]). Here is a class of problems in which necessarily limited computer "experiments" can lead to misleading conclusions about the behavior of a system as a function of the governing parameters. Only a thorough analysis of the system can reveal its true nature. Occasionally, such an analysis will reveal, even without detailed numerical computations, an unphysical aspect of the system (e.g., infinite energy density, decreasing entropy, or other pathologies), which is a clear alert to the flawed nature of the system model.

In this work we concentrate on those aspects of turbulent fluid flows that can be represented in terms of so-called Sobolev spaces – that is, a class of functions satisfying the given boundary conditions – and the given physical constraints, such as divergence-free (incompressible) flow. The various norms (i.e., various integrals of some seemingly abstract quantities) in these function spaces are in fact readily recognized as tangible physical quantities that are more or less readily accessible to direct experimental observation. The relationships among these norms, and the rules for their manipulations, reveal some aspects of the turbulent flows that justify many ad hoc interpretations and inspire insights derived from direct observations of turbulence while also revealing some hitherto unrealized ones. As such, these mathematical endeavors can serve to enlarge our intuitive horizons beyond the limits of linear theories and models.

3 Elements of the Theories of Turbulence of Kolmogorov and Kraichnan

Turbulent flows seem to display self-similar statistical properties at length scales smaller than the scales at which energy is delivered to the flow. Kolmogorov [1941a,b] argued that, at these scales, in three dimensions, the fluids display universal statistical features. Turbulent flow is conventionally visualized as a cascade of large eddies (large-scale components of the flow) breaking up successively into ever smaller sized eddies (fine-scale components of the flow; Onsager [1945]). Such a cascade, or flow of kinetic energy from large to small scales, is taken to occur in a regime at lengths sufficiently large for the effects of viscosity to be inconsequential. The apparent energy dissipation – that is, the removal of energy from one length scale to a smaller one – is solely due to the presence of the nonlinear (inertial) term in the Navier–Stokes equations. The energy dissipation rate $\epsilon = \nu\kappa_0^3|\nabla\mathbf{u}(\mathbf{x},t)|^2$ is assumed to be constant in space and time. A further essential assumption is that the cascade proceeds so that, at every length scale (or at every corresponding wavenumber), there is an equilibrium between energy flowing in from above to a given scale and that flowing out to a lower scale. Such a picture and the associated assumptions imply that, in this range of length scales (or this range of wavenumbers), the energy density at a given wavenumber can depend only on the energy dissipation rate ϵ and the wavenumber k itself. Then dimensional analysis alone yields $\mathcal{S}(\kappa) = \text{const.} \times \epsilon^{2/3}/\kappa^{5/3}$ for the energy density. Such a cascade process cannot continue to arbitrarily small length scales because, as the norm of the Laplacian operator increases with the decreasing length scale, eventually the effects of molecular dissipation begin to dominate the nonlinear inertial term. That length, denoted by ℓ_d, is the endpoint of the inertial range and the beginning of the dissipation range.

Let us determine ℓ_d. At each scale ℓ (or wavenumber $\kappa = \ell^{-1}$), we can define by dimensional analysis, through ϵ and ℓ, a natural time scale τ and speed u. Indeed, $\epsilon = \ell^2/\tau^3$ gives $\tau = (\ell^2/\epsilon)^{1/3}$ and $u = \ell/\tau = (\ell\epsilon)^{1/3}$. Now, the dissipation length ℓ_d is where the viscous term $\nu\Delta\mathbf{u}$ starts to dominate, on average, the inertial term. Hence,

$$\nu\Delta\mathbf{u} \sim \frac{\nu u}{\ell^2} \sim \frac{\nu}{\ell\tau} > (\mathbf{u}\cdot\nabla)\mathbf{u} \sim \frac{u^2}{\ell} \sim \frac{\ell}{\tau^2}.$$

Therefore,

$$\ell^2 < \nu\tau = \nu\left(\frac{\ell^2}{\epsilon}\right)^{1/3} \iff \ell^{4/3} < \left(\frac{\nu^3}{\epsilon}\right)^{1/3}$$

and

$$\ell_d = \left(\frac{\nu^3}{\epsilon}\right)^{1/4}. \tag{3.1}$$

Kolmogorov thus inferred that, in 3-dimensional turbulent flows, the eddies of length size sensibly smaller than ℓ_d are of no dynamical consequence. As we said, the length ℓ_d as defined by (3.1) is known as the Kolmogorov *dissipation length*. The corresponding wavenumber,

$$\kappa_d = \frac{1}{\ell_d} = \left(\frac{\epsilon}{\nu^3}\right)^{1/4}, \tag{3.2}$$

is the Kolmogorov *dissipation wavenumber*.

The inertial range, within which inertial effects dominate, is the range $\ell_1 < \ell < \ell_d$, where $\ell_1 = L_1$ is the wavelength at which energy is injected in the flow. To each length ℓ in this range we can associate a Reynolds number $\mathrm{Re}_\ell = u\ell/\nu$; hence,

$$\mathrm{Re}_\ell^{3/4} = \ell\left(\frac{\epsilon}{\nu^3}\right)^{1/4}.$$

The largest of these Reynolds numbers obtained for $\ell =$ the Kolmogorov macro-scale length $L_* \simeq L_1$ is the Reynolds number Re of the flow. Hence, with (3.1),

$$\mathrm{Re} = \left(\frac{L_*}{\ell_d}\right)^{4/3}, \quad \text{or} \quad L_* = \mathrm{Re}^{3/4}\,\ell_d. \tag{3.3}$$

This relationship leads naturally to the heuristic estimate of the number of degrees of freedom in 3-dimensional flows, which is $\mathrm{Re}^{9/4}$. As we shall see, this heuristic estimate is actually an upper bound on the sufficient (but not necessary) number of degrees of freedom in 3-dimensional turbulent flows.

We now present a somewhat more elaborate derivation (but one that is still divorced from the Navier–Stokes equations) of the so-called Kolmogorov spectrum.

Let ϵ denote the average of the energy per unit mass. Then, according to the Kolmogorov theory, the length ℓ_d at which the turbulent eddies are rapidly annihilated by the viscosity should be a universal function of ϵ and the kinematic viscosity ν, namely:

$$\ell_d = f(\nu, \epsilon). \tag{3.4}$$

In particular, f should be independent of the choice of units for space and time. Thus, if we pass from $\mathbf{x}, t$ to $\mathbf{x}' = \xi\mathbf{x}$ and $t' = \tau t$ then we should still have

$$\ell_d' = f(\nu', \epsilon'). \tag{3.5}$$

Here ν' and ϵ' are not independent of ν and ϵ, and dimensional analysis yields

$$\ell_d' = \xi\ell_d, \quad \nu' = \frac{\xi^2}{\tau}\nu, \quad \epsilon' = \frac{\xi^2}{\tau^3}\epsilon; \tag{3.6}$$

that is,

$$\xi f(\nu, \epsilon) = f(\xi^2\tau^{-1}\nu, \xi^2\tau^{-3}\epsilon). \tag{3.7}$$

With the choices

$$\frac{\xi^2}{\tau} = \frac{1}{\nu} \text{ and } \frac{\xi^2}{\tau^3} = \frac{1}{\epsilon}, \quad (\text{i.e., } \tau = (\epsilon\nu)^{1/2} \text{ and } \xi = \epsilon^{1/4}/\nu^{3/4}),$$

the relation (3.7) becomes

$$f(\nu, \epsilon) = \frac{1}{\xi} f(1,1) \sim \left(\frac{\nu^3}{\epsilon}\right)^{1/4}.$$

Following Kolmogorov, one can also argue that the average energy per unit mass, $e_{\kappa,2\kappa}$, of the eddies of lengths between $\ell/2$ and ℓ (i.e., between the wavenumbers κ and 2κ, where $\kappa = 1/\ell$) should enjoy a similar universal property – namely,

$$e_{\kappa,2\kappa} = g(\epsilon, \kappa), \tag{3.8}$$

provided that $\kappa \ll \kappa_d$ (so that the effect of the viscosity can be neglected) and that κ is much larger than the wavenumber at which energy is pumped into the flow. Again the universality of g implies that

$$\xi^2\tau^{-2} g(\epsilon, \kappa) = g(\xi^2\tau^{-3}\epsilon, \xi^{-1}\kappa);$$

whence, upon taking $\xi = \kappa$ and $\tau = \epsilon^{1/3}\kappa^{2/3}$, one obtains

$$e_{\kappa,2\kappa} = c\frac{\epsilon^{2/3}}{\kappa^{2/3}}, \tag{3.9}$$

where $c = g(1,1)$, a universal constant.

Consider now the Navier–Stokes equations with periodic boundary conditions. That is, we consider the solutions of equations (1.7) that are periodic in space with period L in each direction. Using Fourier series expansions (see Sections 2 and 5 in Chapter II for details), we can write

$$\mathbf{u}(\mathbf{x}) = \sum_{\mathbf{k}\in\mathbb{Z}^3\setminus\{0\}} \hat{\mathbf{u}}_{\mathbf{k}} e^{\kappa_1 i\mathbf{k}\cdot\mathbf{x}}, \quad \hat{\mathbf{u}}_{\mathbf{k}} \cdot \mathbf{k} = 0, \ \hat{\mathbf{u}}_{-\mathbf{k}} = \bar{\hat{\mathbf{u}}}_{\mathbf{k}}. \tag{3.10}$$

For $\mathbf{k}$ in (3.10), $\kappa = \kappa_1|\mathbf{k}|$ is the corresponding wavenumber, where $\kappa_1 = 2\pi/L$. The lowest wavenumber is κ_1. The component of $\mathbf{u}$ with wavenumbers between κ' and κ'' is

$$\mathbf{u}_{\kappa',\kappa''} = \sum_{\kappa'\le\kappa<\kappa''} \hat{\mathbf{u}}_{\mathbf{k}} e^{\kappa_1 i\mathbf{k}\cdot\mathbf{x}}. \tag{3.11}$$

The energy per unit mass and the enstrophy (square of vorticity) per unit mass are, respectively,

$$|\mathbf{u}_{\kappa',\kappa''}|^2 = \sum_{\kappa'\le\kappa<\kappa''} |\hat{\mathbf{u}}_{\mathbf{k}}|^2 \tag{3.12}$$

and

$$\|\mathbf{u}_{\kappa',\kappa''}\|^2 = \kappa_1^2 \sum_{\kappa'\le\kappa<\kappa''} |\mathbf{k}|^2|\hat{\mathbf{u}}_{\kappa}|^2. \tag{3.13}$$

Note that $\mathbf{u}_{\kappa_1,\infty} = \mathbf{u}$.

Physicists and engineers assume that, for $L \gg 1$, the time averages of

$$|\mathbf{u}_{\kappa',\kappa''}(t)|^2 \quad \text{and} \quad \kappa_1^2\|\mathbf{u}_{\kappa',\kappa''}(t)\|^2 \tag{3.14}$$

exist (see Chapters IV and V). Note that the first average should be $e_{\kappa',\kappa''}$, the average of energy per unit mass of the eddies of linear size between $1/\kappa''$ and $1/\kappa'$. We denote these averages by

$$\langle|\mathbf{u}_{\kappa',\kappa''}(\cdot)|^2\rangle \quad \text{and} \quad \kappa_1^2\langle\|\mathbf{u}_{\kappa',\kappa''}(t)\|^2\rangle. \tag{3.15}$$

Moreover, it is also assumed that these values can be viewed (at least when κ_1 is small) as integrals in the wavenumbers; that is,

$$\int_{\kappa'}^{\kappa''} \mathcal{S}(\kappa)\,d\kappa \quad \text{and} \quad \int_{\kappa'}^{\kappa''} \mathcal{S}_1(\kappa)\,d\kappa. \tag{3.16}$$

Comparing (3.12) and (3.13) with (3.15), we see that if (3.16) makes sense then one is led to the relation

$$\mathcal{S}_1(\kappa) \equiv \kappa^2\mathcal{S}(\kappa). \tag{3.17}$$

The function $\mathcal{S}(\kappa)$ (≥ 0) is called the *energy spectrum* of the turbulent flow produced by the driving force $\mathbf{f}$ in (1.7). Also, the driving force is assumed to have no high-wavenumber components: $\mathbf{f} = \mathbf{f}_{\kappa_1,\bar{\kappa}}$, where $\bar{\kappa}$ is comparable in size with κ_1 (the lowest wavenumber).

So, according to Kolmogorov's theory, we have (see (3.9))

$$\int_{\kappa}^{2\kappa} \mathcal{S}(\chi)\,d\chi \sim c\frac{\epsilon^{2/3}}{\kappa^{2/3}}, \tag{3.18}$$

at least as long as

$$\bar{\kappa} \ll \kappa \ll \kappa_d. \tag{3.19}$$

Taking the derivative in (3.18) yields

$$\mathcal{S}(\kappa) - \mathcal{S}(2\kappa) \sim \frac{2c}{3}\frac{\epsilon^{2/3}}{\kappa^{5/3}},$$

whence

$$\mathcal{S}(\kappa) \sim \mathcal{S}(2^{m+1}\kappa) + \frac{2c}{3}\left(1 + \frac{1}{2^{5/3}} + \frac{1}{2^{10/3}} + \cdots + \frac{1}{2^{5m/3}}\right)\frac{\epsilon^{2/3}}{\kappa^{5/3}} \tag{3.20}$$

as long as $2^m\kappa \ll \kappa_d$. For turbulent flows, $\kappa_d \gg \kappa_1 \approx \bar{\kappa}$ and so we may take $m \gg 1$ in (3.20). Then

$$\mathcal{S}(\kappa) \sim C_K'\frac{\epsilon^{2/3}}{\kappa^{5/3}}, \tag{3.21}$$

where $C_K' = (2/3)c(1 - 2^{-5/3})^{-1}$. The form (3.21) for the energy spectrum is called the *Kolmogorov energy spectrum* of the turbulent flow. The constant C_K' is known as the *Kolmogorov constant* in energy space (there is a similar relation in which a constant C_K appears and takes the name of Kolmogorov constant in physical space; see (5.26) in Chapter V). The empirical value of C_K' is of the order of unity. The range of κ in (3.19) for which (3.21) holds is the Kolmogorov *inertial range.*

It must be noted that the estimate (3.21) is really a time average, as the amplitude of $\mathcal{S}(\kappa)$ fluctuates wildly in time. Furthermore, it is only an approximation. In reality, for a turbulent flow in a bounded domain, intermittency effects in the energy dissipation rate ϵ result in small but measurable corrections to the simple expression (3.21) (see e.g. Kolmogorov [1962] and Novikov and Stewart [1964]). These corrections depend on the size (say, L_*) of Ω, thus destroying – to some extent – the universality of turbulence.

As seen in the preceding, the arguments leading up to (3.21) are clearly divorced from the NSE itself and are applicable solely to turbulent flows in three dimensions. For flows in two dimensions (which, as stated earlier, are amenable to deep analysis), we must turn to a phenomenological theory proposed by Batchelor [1959] and Kraichnan [1967]. That theory is in the spirit of Kolmogorov's approach but does not parallel it because the physical situation is quite different; hence we offer a separate exposition.

Two-dimensional flows are not commonly encountered in nature. Examples that do come to mind are thin liquid films and (to within some approximation) the atmospheric layer on the surface of the Earth – although clearly the significant phenomena (e.g., weather and climate changes) occur on scales within which the finite thickness of that layer must be taken explicitly into account. However, some firm mathematical results derived in the study of 2-dimensional flows appear to carry over to some 3-dimensional flows, so it is instructive to follow what can be learned about 2-dimensional flows. Moreover, further advances in functional analysis of the Navier–Stokes equations in three dimensions may yield the necessary tools for solving some critical open problems. For now we turn to a summary of Kraichnan's work on the phenomenological theory of 2-dimensional turbulence.

We limit ourselves here to fluid flows in the plane, although much of the theory could carry over to more general 2-dimensional manifolds. The principal physical difference between 2-dimensional and 3-dimensional flows is that, in the 2-dimensional case, the vorticity (i.e., the curl of the velocity) has only one component – in the direction normal to the plane of the flow. This imposes a severe constraint on the kinematics and the dynamics of the turbulence. For instance, in addition to the conservation of energy, the flow must conserve enstrophy, that is, the integral of the square of the vorticity over the flow domain must be constant. The nonlinear interactions may be viewed in wavenumber (Fourier) space as three-wave interactions. They cannot simultaneously satisfy the two conservation principles. Hence, the energy cascade cannot coexist with the enstrophy cascade; they must occur in distinct portions of the wavenumber domain. As a consequence, in the turbulent regime (large Reynolds number) for the 2-dimensional case, there are two contiguous ranges: for wavenumbers lower than that at which the forcing of the flow is introduced there is an inverse energy cascade, with small eddies coalescing into larger ones and with the cascade terminating at a wavenumber determined by the size of the flow domain. The spectrum of the energy, $\mathcal{S}(\kappa)$, in that domain is the same as the Kolmogorov spectrum. Toward the higher wavenumbers, the principal cascading entity is the enstrophy – that is, the successive breakup of the vortices into ever smaller ones, with the attendant enstrophy dissipation rate η resulting from nonlinear interactions and not being affected by the molecular viscosity above the *Kraichnan cutoff length*. According to

Kraichnan, the two portions of the inertial range (i.e., the inverse cascade and the enstrophy cascade) cannot overlap.

Using arguments based on dimensional analysis along the lines followed by Kolmogorov in the 3-dimensional regime, we can now determine the energy spectrum in the enstrophy cascade region as well as the Kraichnan cutoff wavenumber. Note first that the enstrophy dissipation rate has the dimension of $(\text{time})^{-3}$. Assuming that, in the enstrophy cascade range, the energy spectrum depends only on η and the wavenumber κ, we then find that $S(\kappa) = \text{const.} \times \eta^{2/3}/\kappa^3$. Similarly, it follows from dimensional considerations that the Kraichnan cutoff wavenumber is given by

$$\kappa_\eta = \left(\frac{\eta}{\nu^3}\right)^{1/6}. \tag{3.22}$$

An extended treatment of turbulent flows in two dimensions is presented in Section 5 of Chapter IV.

4 Function Spaces, Functional Inequalities, and Dimensional Analysis

The mathematical theory of the Navier–Stokes equations is based on the use of function spaces, which are at the heart of the modern theory of partial differential equations. A formal presentation of the needed tools appears in Chapter II and thereafter as needed. However, in this section we give an informal introduction and emphasize that these spaces are not merely inventions of mathematicians; rather, they are strongly related to the physics of the problem.

The Fundamental Function Spaces

Consider the domain Ω occupied by the fluid; Ω is a domain of $\mathbb{R}^3$, which could be the whole space $\mathbb{R}^3$ in certain idealized cases. The first natural function space is $L^2(\Omega)$, the space of square integrable functions on Ω; we also have $L^2(\Omega)^3$, the space of square integrable vector fields on Ω. These spaces are endowed with a scalar product, which we denote by

$$(u, v) = \int_\Omega u(\mathbf{x}) v(\mathbf{x})\, d\mathbf{x}$$

in the scalar case and, similarly,

$$(\mathbf{u}, \mathbf{v}) = \int_\Omega \mathbf{u}(\mathbf{x}) \cdot \mathbf{v}(\mathbf{x})\, d\mathbf{x}$$

in the vector-valued case. To these scalar products correspond the following norms[2] (mean square norms):

$$|u| = \left(\int_\Omega |u(\mathbf{x})|^2\, d\mathbf{x}\right)^{1/2}, \qquad |\mathbf{u}| = \left(\int_\Omega |\mathbf{u}(\mathbf{x})|^2\, d\mathbf{x}\right)^{1/2}.$$

[2] We will use the same notation $|\cdot|$ for the Euclidean norm in $\mathbb{R}^2$ and $\mathbb{R}^3$ (and even $\mathbb{C}^2$ and $\mathbb{C}^3$), and also for various L^2 norms. This abuse of notation, for purposes of simplification, does not lead to any confusion once the context is taken into account.

The inner product and the corresponding norm are also related by the so-called Cauchy–Schwarz inequality:

$$|(u,v)| = \left| \int_\Omega u(\mathbf{x})v(\mathbf{x})\,d\mathbf{x} \right| \le \left(\int_\Omega |u(\mathbf{x})|^2\,dx \right)^{1/2} \left(\int_\Omega |v(\mathbf{x})|^2\,dx \right)^{1/2} = |u||v| \tag{4.1}$$

for all u, v in $L^2(\Omega)$. Similarly, for any $\mathbf{u}, \mathbf{v} \in L^2(\Omega)^3$, we have

$$|(\mathbf{u}, \mathbf{v})| \le |\mathbf{u}||\mathbf{v}|. \tag{4.2}$$

Now, given the velocity vector field $\mathbf{u}$ of the fluid in Ω,

$$\mathbf{u} : \mathbf{x} \in \Omega \mapsto \mathbf{u}(\mathbf{x}) \in \mathbb{R}^3,$$

we see that the square of the L^2-norm, $|\mathbf{u}|^2$, is merely twice the kinetic energy of the flow (assuming that the density of the fluid has been normalized to unity):

$$e(\mathbf{u}) = \frac{1}{2} \int_\Omega |\mathbf{u}(\mathbf{x})|^2\,d\mathbf{x} = \frac{1}{2}|\mathbf{u}|^2.$$

Without entering into the details of measure theory, we recall that $L^2(\Omega)$ and $L^2(\Omega)^3$ are Hilbert spaces for these scalar products and norms. Also, for $L^2(\Omega)$ (and the same is true for $L^2(\Omega)^3$), the following characterizaton holds:[3]

> $u \in L^2(\Omega)$ *if and only if* there exists a sequence of smooth functions u_n, compactly supported in Ω, such that $|u_n|$ (or $e(u_n)$) remains bounded and u_n is converging to u (in the distribution sense) as $n \to \infty$. (4.3)

Most of the spaces that we will consider are derived from the space $L^2(\Omega)$ and from another space that we will now introduce, the so-called Sobolev space $H^1(\Omega)$.

In Chapter II we will address the concept of enstrophy of a fluid velocity $\mathbf{u} = (u_1, u_2, u_3)$, namely,

$$E(\mathbf{u}) = \sum_{i,j=1}^{3} \int_\Omega \left| \frac{\partial u_i}{\partial x_j} \right|^2 d\mathbf{x}.$$

By comparison with (4.3), a natural question is the following:

> What can we say of a sequence of smooth velocity vector fields $\mathbf{u}_n$ such that $E(\mathbf{u}_n)$ remains bounded? (4.4)

If Ω is bounded,[4] then one can prove that u_n contains one (or more) subsequence(s) that converge in the distribution sense to some limit $\mathbf{u}$. This vector function $\mathbf{u} = (u_1, u_2, u_3)$ is in $L^2(\Omega)^3$, as are its (distributional) first derivatives

$$\frac{\partial u_i}{\partial x_j}, \quad i, j = 1, 2, 3;$$

[3] The functions u_n in this characterization of $L^2(\Omega)$ are defined, say, by some kind of approximation procedure.

[4] If Ω is not bounded, then we should also require that $e(\mathbf{u}_n)$ remain bounded as well.

in fact,

$$E(\mathbf{u}) \leq \sup_{n\in\mathbb{N}} E(\mathbf{u}_n).$$

We say that such a function **u** belongs to the Sobolev space $H^1(\Omega)^3$ (with a similar definition for $H^1(\Omega)$ in the scalar case). In some sense:

> The space $L^2(\Omega)^3$ consists of all the vector fields **u** with finite kinetic energy, and the space $H^1(\Omega)^3$ consists of all the vector fields **u** with finite enstrophy. (4.5)

From the mathematical point of view, the Sobolev spaces $H^1(\Omega)$ and $H^1(\Omega)^3$ are Hilbert spaces for the following inner products and norms (see Chapter II):

$$((u, v))_1 = \frac{1}{L^2}\int_\Omega u(\mathbf{x})v(\mathbf{x})\,d\mathbf{x} + \sum_{j=1}^{3}\int_\Omega \frac{\partial u}{\partial x_j}\frac{\partial v}{\partial x_j}\,d\mathbf{x},$$

$$((\mathbf{u}, \mathbf{v}))_1 = \frac{1}{L^2}\int_\Omega \mathbf{u}(\mathbf{x})\cdot\mathbf{v}(\mathbf{x})\,d\mathbf{x} + \sum_{i,j=1}^{3}\int_\Omega \frac{\partial u_i}{\partial x_j}\frac{\partial v_i}{\partial x_j}\,d\mathbf{x},$$

$$\|u\|_1 = [((u, u))_1]^{1/2}, \qquad \|\mathbf{u}\|_1 = [((\mathbf{u}, \mathbf{u}))_1]^{1/2},$$

where L is a typical length (e.g., the diameter of Ω). In nondimensional variables, $L = 1$. As with the space $L^2(\Omega)$ and any other Hilbert space, we have the Cauchy–Schwarz inequality, which in this context reads

$$|((u, v))_1| \leq \|u\|_1\|v\|_1, \qquad |((\mathbf{u}, \mathbf{v}))_1| \leq \|\mathbf{u}\|_1\|\mathbf{v}\|_1 \tag{4.6}$$

for all u, v in $H^1(\Omega)$ and all $\mathbf{u}, \mathbf{v}$ in $H^1(\Omega)^3$.

Most function spaces that we consider are derived from these two physically obvious spaces, the space $L^2(\Omega)^3$ of finite kinetic energy and the space $H^1(\Omega)^3$ of finite enstrophy. For instance, two central spaces V and H appear throughout the book; assuming for simplicity that Ω is bounded, a mathematically rigorous and physically intuitive definition of the spaces V and H is as follows:

> V is made up of all the limit points (in the distributional sense) of all the possible sequences of smooth vector fields $\mathbf{u}_n$ which are divergence-free, which satisfy the boundary conditions of the problem, and whose enstrophy remains bounded; that is, $E(\mathbf{u}_n) \leq$ const. $< \infty$. (4.7)

The space H is defined in a similar way, replacing the boundedness of the enstrophy by the boundedness of the kinetic energy: $e(\mathbf{u}_n) \leq$ const. $< \infty$. More details are given in Chapter II.

Functional Inequalities

The functions belonging to the space $H^1(\Omega)$ (and to other related spaces) satisfy certain inequalities, which are called Sobolev inequalities in the mathematical literature.

We make extensive use of these inequalities in the course of this book. Some of them are proven by interpolation, others by appropriate direct methods (e.g., Gagliardo–Nirenberg's, Agmon's, Ladyzhenskaya's, and Poincaré's inequalities). Our objective here is twofold:

(i) to prove the Ladyzhenskaya and the Poincaré inequalities; and
(ii) to emphasize the *physical invariance by dilatation* or *by change of scale* of all such inequalities.

The Ladyzhenskaya inequality may be described as follows. For a smooth, compactly supported scalar function u in $\mathbb{R}^2$, we have

$$\int_{\mathbb{R}^2} |u(\mathbf{x})|^4\,d\mathbf{x} \le \left(\int_{\mathbb{R}^2} |u(\mathbf{x})|^2\,d\mathbf{x}\right)\left(\int_{\mathbb{R}^2} |\nabla u(\mathbf{x})|^2\,d\mathbf{x}\right). \tag{4.8}$$

In $\mathbb{R}^3$, we have

$$\int_{\mathbb{R}^3} |u(\mathbf{x})|^4\,d\mathbf{x} \le \left(\int_{\mathbb{R}^3} |u(\mathbf{x})|^2\,d\mathbf{x}\right)^{1/2}\left(\int_{\mathbb{R}^3} |\nabla u(\mathbf{x})|^2\,d\mathbf{x}\right)^{3/2}. \tag{4.9}$$

In (4.8) and (4.9) – and in (4.10), (4.11), and (4.12) – the smoothness of the functions u and g is assumed for the sake of simplicity, since these inequalities are valid for more general (less smooth) functions.

Remark 4.1 Note the difference between space dimensions 2 and 3. It is this very discrepancy between the two cases that induces many of the difficulties in the mathematical theory of the Navier–Stokes equations in space dimension 3.

That no dimensional constant appears in the RHS of (4.8) or (4.9) is due to the fact that these inequalities are invariant by dilatation, or homothety ($\mathbf{x} \mapsto \lambda\mathbf{x}$), or (in physical terms) that both sides of these inequalities have the same dimension:

$$U^4L^2 \sim U^2L^2\left(\frac{U^2L^2}{L^2}\right) \quad \text{for (4.8)},$$

$$U^4L^3 \sim (U^2L^3)^{1/2}\left(\frac{U^2L^3}{L^2}\right)^{3/2} \quad \text{for (4.9)}.$$

Remark 4.2 This invariance of the inequality by dilatation (or homogeneity) is common to many functional inequalities. However, the lack of a multiplicative constant in the RHS of (4.8) and (4.9) is not usual. In most cases, we know (we can prove) the existence of a multiplicative constant in the right-hand side of the inequalities, which in some cases may be obtained explicitly. In general, however, such constants can be taken as of order unity.

We present now the proof of (4.8). The proof of (4.9) follows a similar idea and uses (4.8). We start with the following inequality (of Agmon's type) in space dimension 1:

$$|g(x)|^2 \le \left(\int_{\mathbb{R}} |g(\xi)|^2\,d\xi\right)^{1/2}\left(\int_{\mathbb{R}} |g'(\xi)|^2\,d\xi\right)^{1/2} \quad \text{for all } x \in \mathbb{R}, \tag{4.10}$$

for any smooth function g with compact support in $\mathbb{R}$. Since g has compact support, there exists $L > 0$ sufficiently large such that $g(x)$ vanishes for x outside $(-L, L)$. Then we can write

$$g(x)^2 = 2\int_{-L}^{x} g(\xi)g'(\xi)\,d\xi \leq 2\int_{-L}^{x} |g(\xi)||g'(\xi)|\,d\xi$$

and

$$g(x)^2 = -2\int_{x}^{L} g(\xi)g'(\xi)\,d\xi \leq 2\int_{x}^{L} |g(\xi)||g'(\xi)|\,d\xi.$$

Adding both identities, we obtain

$$g(x)^2 \leq \int_{-L}^{L} |g(\xi)||g'(\xi)|\,dx = \int_{\mathbb{R}} |g(\xi)||g'(\xi)|\,dx.$$

Then, using the Cauchy–Schwarz inequality, we find (4.10).

We now prove the 2-dimensional Ladyzhenskaya inequality (4.8). We use the Agmon inequality (4.10) twice, once in each space direction. We have

$$\begin{aligned} u(x_1, x_2)^4 &= u(x_1, x_2)^2 u(x_1, x_2)^2 \\ &\leq \Bigg[\left(\int_{\mathbb{R}} |u(\xi_1, x_2)|^2\,d\xi_1\right)\left(\int_{\mathbb{R}} |u_{\xi_1}(\xi_1, x_2)|^2\,d\xi_1\right) \\ &\qquad \left(\int_{\mathbb{R}} |u(x_1, \xi_2)|^2\,d\xi_2\right)\left(\int_{\mathbb{R}} |u_{\xi_2}(x_1, \xi_2)|^2\,d\xi_2\right)\Bigg]^{1/2}. \end{aligned}$$

Thus,

$$\begin{aligned} \int_{\mathbb{R}}\int_{\mathbb{R}} u(x_1, x_2)^4\,dx_1\,dx_2 &\leq \int_{\mathbb{R}}\left(\int_{\mathbb{R}} |u(\xi_1, x_2)|^2\,d\xi_1 \int_{\mathbb{R}} |u_{\xi_1}(\xi_1, x_2)|^2\,d\xi_1\right)^{1/2} dx_2 \\ &\quad \int_{\mathbb{R}}\left(\int_{\mathbb{R}} |u(x_1, \xi_2)|^2\,d\xi_2 \int_{\mathbb{R}} |u_{\xi_2}(x_1, \xi_2)|^2\,d\xi_2\right)^{1/2} dx_1. \end{aligned}$$

Using the Cauchy–Schwarz inequality, we obtain

$$\begin{aligned} &\int_{\mathbb{R}}\int_{\mathbb{R}} u(x_1, x_2)^4\,dx_1\,dx_2 \\ &\quad\leq \left(\int_{\mathbb{R}}\int_{\mathbb{R}} |u(\xi_1, x_2)|^2\,d\xi_1\,dx_2\right)^{1/2}\left(\int_{\mathbb{R}}\int_{\mathbb{R}} |u_{\xi_1}(\xi_1, x_2)|^2\,d\xi_1\,dx_2\right)^{1/2} \\ &\qquad \left(\int_{\mathbb{R}}\int_{\mathbb{R}} |u(x_1, \xi_2)|^2\,d\xi_2\,dx_1\right)^{1/2}\left(\int_{\mathbb{R}}\int_{\mathbb{R}} |u_{\xi_2}(x_1, \xi_2)|^2\,d\xi_2\,dx_1\right) \\ &\quad\leq \left(\int_{\mathbb{R}}\int_{\mathbb{R}} |u(x_1, x_2)|^2\,dx_1\,dx_2\right)\left(\int_{\mathbb{R}}\int_{\mathbb{R}} |\nabla u(x_1, x_2)|^2\,dx_1\,dx_2\right), \end{aligned}$$

which proves (4.8). As mentioned earlier, the proof of (4.9) follows a similar idea: one first writes $u(x_1, x_2, x_3)^4$ as a product of two squares; then one applies the Agmon inequality (4.10) with respect to one variable and the 2-dimensional Ladyzhenskaya inequality (4.8) in the remaining two variables.

With little effort we can now prove another fundamental (often used) inequality: the Poincaré inequality.[5] We assume that u is smooth and that it vanishes outside a bounded set; in fact, it suffices for u to vanish outside a slab, say $-L/2 < x_1 < L/2$. In space dimension 2 (the proof is the same in any space dimension), we write (4.10) in direction x_1 with x_2 fixed:

$$u(x_1, x_2)^2 \leq \left(\int_{\mathbb{R}} |u(\xi_1, x_2)|^2 \, d\xi_1 \right)^{1/2} \left(\int_{\mathbb{R}} \left| \frac{\partial u}{\partial \xi_1}(\xi_1, x_2) \right|^2 d\xi_1 \right)^{1/2}.$$

Observing that the RHS of this inequality does not depend on x_1, we integrate with respect to x_1, from $-L/2$ to $L/2$, and find

$$\int_{\mathbb{R}} \int_{\mathbb{R}} |u(x_1, x_2)|^2 \, dx_1 \, dx_2$$

$$\leq L \int_{\mathbb{R}} \left(\int_{\mathbb{R}} |u(\xi_1, x_2)|^2 \, d\xi_1 \right)^{1/2} \left(\int_{\mathbb{R}} \left| \frac{\partial u}{\partial \xi_1}(\xi_1, x_2) \right|^2 d\xi_1 \right)^{1/2} dx_2$$

$\leq$ (after using the Cauchy–Schwarz inequality and renaming the dummy variable)

$$\leq L \left(\int_{\mathbb{R}} \int_{\mathbb{R}} |u(x_1, x_2)|^2 \, dx_1 \, dx_2 \right)^{1/2} \left(\int_{\mathbb{R}} \int_{\mathbb{R}} \left| \frac{\partial u}{\partial x_1}(x_1, x_2) \right|^2 dx_1 \, dx_2 \right)^{1/2}.$$

This implies the Poincaré inequality

$$\left(\int_{\mathbb{R}} \int_{\mathbb{R}} |u(x_1, x_2)|^2 \, dx_1 \, dx_2 \right)^{1/2} \leq L \left(\int_{\mathbb{R}} \int_{\mathbb{R}} \left| \frac{\partial u}{\partial x_1}(x_1, x_2) \right|^2 dx_1 \, dx_2 \right)^{1/2}, \tag{4.11}$$

as well as the following form, which is used more often:

$$\left(\int_{\mathbb{R}} \int_{\mathbb{R}} |u(x_1, x_2)|^2 \, dx_1 \, dx_2 \right)^{1/2} \leq L \left(\int_{\mathbb{R}} \int_{\mathbb{R}} |\nabla u(x_1, x_2)|^2 \, dx_1 \, dx_2 \right)^{1/2}. \tag{4.12}$$

The Poincaré inequality in higher dimensions can be proved similarly. In three dimensions, the form that we will often use reads

$$\left(\int_{\mathbb{R}^3} |u(\mathbf{x})|^2 \, d\mathbf{x} \right)^{1/2} \leq L \left(\int_{\mathbb{R}^3} |\nabla u(\mathbf{x})|^2 \, d\mathbf{x} \right)^{1/2}. \tag{4.13}$$

Remark 4.3 (cf. Remark 4.2) Note that a coefficient L (with dimension of length) appears in the RHS of (4.11) and (4.12). This length is given in the assumption on u (it vanishes outside $-L/2 < x_1 < L/2$), and both sides of (4.11) and of (4.12) have the same dimension, namely UL.

Another frequently used version of the Poincaré inequality (which we state without proof) relates to functions u defined on a bounded domain Ω whose average on Ω vanishes. We will use this inequality in space dimension $d = 2$ or 3, but for any space dimension d it reads as follows: There exists a constant $c = c(\Omega)$ such that

[5] More precisely, we prove one of the forms of this inequality; another form (valid for space-periodic functions) is given later in this section.

$$\int_{\Omega} |u(x_1, \ldots, x_d)|^2 \, dx_1 \ldots dx_d \le c(\Omega) \int_{\Omega} |\nabla u(x_1, \ldots, x_d)|^2 \, dx_1 \ldots dx_d \tag{4.14}$$

for all functions u such that

$$\int_{\Omega} u(x_1, \ldots, x_d) \, dx_1 \ldots dx_d = 0. \tag{4.15}$$

This inequality will be frequently used for space-periodic functions with zero average on the period Ω. Note that the constant $c(\Omega)$, which is not easy to determine, has dimension L^2 (square of a length; see Remark 4.3). In the periodic case, we write this constant more explicitly as $c(\Omega) = \tilde{c}(\Omega)L^2$, where L is the smallest period. In this case, the constant $\tilde{c}(\Omega)$ depends only on the "shape" of Ω, in the sense that it is invariant under dilatation; see Remark 4.2.

More Inequalities

The methods of functional analysis employed throughout this volume rely heavily on the use of some relatively simple and well-known inequalities, as well as on more sophisticated ones. For the convenience of the reader, we list here (without proof) all inequalities in the first category. We shall then also list those in the second category – namely, the Sobolev inequalities and some of their variants, which extend (4.8) and (4.9) in various ways.

Schwarz's inequality:

$$ab \le \frac{1}{2}\left(\varepsilon a^2 + \frac{b^2}{\varepsilon}\right) \tag{4.16}$$

for all real numbers a, b and all $\varepsilon > 0$.

Young's inequality:

$$ab \le \frac{1}{p} a^p + \frac{1}{p'} b^{p'} \tag{4.17}$$

for all $a, b > 0$ and all $1 < p < \infty$, with $p' = p/(p-1)$ (i.e., $1/p + 1/p' = 1$). Also,

$$ab \le \frac{\varepsilon}{p} a^p + \frac{1}{\varepsilon^{1/(p-1)} p'} b^{p'} \tag{4.18}$$

for all a, b, p, p' as before and all $\varepsilon > 0$.

Hölder's inequality:

$$\int_{\Omega} u(\mathbf{x}) v(\mathbf{x}) \, d\mathbf{x} \le \left(\int_{\Omega} |u(\mathbf{x})|^p \, d\mathbf{x} \right)^{1/p} \left(\int_{\Omega} |v(\mathbf{x})|^{p'} \, d\mathbf{x} \right)^{1/p'} \tag{4.19}$$

for all measurable functions u and v for which the right-hand side is finite. Here, $1 < p < \infty$, $p' = p/(p-1)$, and Ω is an arbitrary open set in $\mathbb{R}^d$, $d \in \mathbb{N}$. Also,

$$\int_{\Omega} u(\mathbf{x}) v(\mathbf{x}) \, d\mathbf{x} \le \sup_{\mathbf{x} \in \Omega} |u(\mathbf{x})| \int_{\Omega} |v(\mathbf{x})| \, d\mathbf{x}. \tag{4.20}$$

A weaker form of (4.20) involving the essential supremum of u appears in the sequel. In addition to the inequalities just listed, we use extensively the Poincaré inequalities

(4.11), (4.12), and (4.14), as well as related forms of this inequality discussed in the main text.

Lebesgue Spaces

We have recalled already the definition of the space $L^2(\Omega)$ of square integrable functions on a domain Ω in $\mathbb{R}^d$, $d \in \mathbb{N}$. This is one of the so-called Lebesgue spaces. More generally, for any $1 \le p < \infty$, we have the Lebesgue space $L^p(\Omega)$, which is the space of measurable functions whose absolute value to the pth power is integrable:

$$\int_\Omega |u(\mathbf{x})|^p \, d\mathbf{x} < \infty.$$

This space is endowed with the norm

$$\|u\|_{L^p(\Omega)} = \left(\int_\Omega |u(\mathbf{x})|^p \, d\mathbf{x} \right)^{1/p}. \tag{4.21}$$

The space $L^2(\Omega)$ is a particular case of the preceding spaces, and it is the only one whose norm is associated with an inner product. Because of the frequent use of the L^2-norm, we denote it simply by $|\cdot|$ and denote the associated inner product by $(\cdot, \cdot)$.

The limiting case $p = \infty$ can also be considered. The space $L^\infty(\Omega)$ is the space of measurable functions that are uniformly bounded almost everywhere on Ω. More precisely, a measurable function $u = u(\mathbf{x})$ on Ω belongs to $L^\infty(\Omega)$ if and only if there is a number $M \ge 0$ such that $|u(\mathbf{x})| \le M$ for almost every $\mathbf{x}$ in Ω (i.e., except on a set of measure zero). The smallest such M is the norm of u, which is denoted $\|u\|_{L^\infty(\Omega)}$.

With those spaces in mind, the Hölder inequality (4.19) reads as

$$(u, v) \le \|u\|_{L^p(\Omega)} \|v\|_{L^{p'}(\Omega)} \tag{4.22}$$

for all $u \in L^p(\Omega)$ and all $v \in L^{p'}(\Omega)$, where $p' = p/(p-1)$ (i.e., $1/p + 1/p' = 1$), $1 \le p \le \infty$.

Higher-Order Sobolev Spaces

A number of technical function spaces will be introduced as needed, especially in Chapter II. However, at this point we would like to mention the higher-order Sobolev spaces $H^m(\Omega)$, which are central in the mathematical theory of partial differential equations.

For any domain Ω in $\mathbb{R}^d$, which may be bounded or unbounded (possibly $\Omega = \mathbb{R}^d$) and whose boundary $\partial\Omega$ may or may not be smooth:[6] for integer $m \ge 1$, we denote by $H^m(\Omega)$ the space of square integrable real functions u on Ω whose distributional derivatives of orders up to m are also square integrable. In mathematical notation,

[6] Let us mention that there are some difficulties with nonsmooth domains, a question that is not addressed in this monograph.

$$H^m(\Omega) = \{u \in L^2(\Omega);\ D^\alpha u \in L^2(\Omega),\ [\alpha] \le m\},$$

where $\alpha = (\alpha_1, \dots, \alpha_d) \in \mathbb{N}^d$ is a multi-integer, $[\alpha] = \alpha_1 + \cdots + \alpha_d$, and D^α is a short notation for

$$\frac{\partial^{\alpha_1 + \cdots + \alpha_d}}{\partial_{x_1}^{\alpha_1} \cdots \partial_{x_d}^{\alpha_d}}.$$

We endow this space with the following inner product and norm:

$$\|u\|_{H^m(\Omega)} = \{((u, u))_{H^m(\Omega)}\}^{1/2},$$

$$((u, u))_{H^m(\Omega)} = \sum_{k=0}^{m} \frac{1}{L^{2(m-k)}} \sum_{[\alpha]=k} \int_\Omega D^\alpha u(\mathbf{x}) D^\alpha v(\mathbf{x})\, d\mathbf{x}. \tag{4.23}$$

Here, L is a typical length scale associated with Ω; we introduced it in (4.23) to make the RHS dimensionally homogeneous. In the mathematical context we usually use nondimensional variables, and then $L = 1$. The space $H^m(\Omega)$ is the Sobolev space of order m; it is a Hilbert space for the inner product $((\cdot, \cdot))_{H^m(\Omega)}$ and the associated norm.

Of course, for $m = 1$, we recover the space $H^1(\Omega)$ previously mentioned. Note that, like $H^1(\Omega)$ (see (4.3) and after), we can say that

> $u \in H^m(\Omega)$ *if and only if* there exists a sequence of smooth functions u_j, compactly supported in Ω, such that $\|u_j\|_{H^m(\Omega)}$ remains bounded and u_j is converging to u (in the distribution sense) as $j \to \infty$. (4.24)

Sobolev spaces based on the Lebesgue spaces $L^p(\Omega)$ for any $1 \le p \le \infty$ can also be defined. They are very important in the mathematical theory of partial differential equations, but they will not be necessary for our purposes in this monograph.

Sobolev Embeddings and Inequalities

An important property of Sobolev spaces are the Sobolev embeddings, based on inequalities similar to (4.8) and (4.9), which are generally called the Sobolev inequalities. For instance, (4.8) implies that, in space dimension 2,

$$H^1(\mathbb{R}^2) \subset L^4(\mathbb{R}^2);$$

such inequalities are valid more generally for any (smooth but not necessarily bounded) domain $\Omega \subset \mathbb{R}^2$, and imply similarly for such a domain that

$$H^1(\Omega) \subset L^4(\Omega).$$

In the same manner, in space dimension 3, (4.9) implies that

$$H^1(\mathbb{R}^3) \subset L^4(\mathbb{R}^3)$$

and, for any (smooth but not necessarily bounded) domain $\Omega \subset \mathbb{R}^3$,

$$H^1(\Omega) \subset L^4(\Omega).$$

In general, for $1 \leq m < d/2$, where d is the space dimension,

$$H^m(\Omega) \subset L^q(\Omega), \tag{4.25}$$

where $1/q = 1/2 - m/d$. For $m = d/2$ (e.g., for the important case of $d = 2$ and $m = 1$), the functions in $H^m(\Omega)$ belong to $L^q(\Omega_b)$, for any finite q $(1 \leq q < \infty)$ and for any bounded and smooth subdomain Ω_b of Ω. The Sobolev embedding (4.25) is derived from an inequality similar to (4.8)–(4.9) that is also invariant by dilatation (dimensionally invariant):

$$\left(\int_{\mathbb{R}^d} |u(\mathbf{x})|^q \, d\mathbf{x} \right)^{1/q} \leq c(m, d, \Omega) \sum_{[\alpha]=m} \left(\int_{\mathbb{R}^d} |D^\alpha u(\mathbf{x})|^2 \, d\mathbf{x} \right)^{1/2}, \tag{4.26}$$

where the constant $c(m, d, \Omega)$ depends on m, d, and the shape of Ω. Note that both sides of this inequality have the dimension of $UL^{d/q} = UL^{d/2-m}$, and the constant $c(m, d, \Omega)$ is a constant that has no dimension.

Remark 4.4 Unlike for (4.8) and (4.9), the constant $c(m, d, \Omega)$ in (4.26) is very difficult to obtain; and the proof of (4.26) is much more involved than that of (4.8) and (4.9). The interested reader can learn more about Sobolev spaces in the books of Adams [1975], Lions and Magenes [1972], and Mazja [1985]; see also a summary in Temam [1997, Chap. II].

Remark 4.5 By interpolation, one can supplement the embedding (4.25) with a large collection of similar inequalities. Indeed, using the Hölder inequality (4.19), it is easy to show that

$$\left(\int_{\mathbb{R}^d} |u(\mathbf{x})|^r \, d\mathbf{x} \right)^{1/r} \leq \left(\int_{\mathbb{R}^d} |u(\mathbf{x})|^2 \, d\mathbf{x} \right)^{\frac{d(q-r)}{r(q-2)}} \left(\int_{\mathbb{R}^d} |u(\mathbf{x})|^q \, d\mathbf{x} \right)^{\frac{d(r-2)}{r(q-2)}} \tag{4.27}$$

when $2 \leq r \leq q$. Hence, with (4.26),

$$\left(\int_{\mathbb{R}^d} |u(\mathbf{x})|^r \, d\mathbf{x} \right)^{1/r}$$

$$\leq c(m, d, \Omega) \left(\int_{\mathbb{R}^d} |u(\mathbf{x})|^2 \, d\mathbf{x} \right)^{\frac{d(q-r)}{r(q-2)}} \left(\sum_{[\alpha]=m} \int_{\mathbb{R}^d} |D^\alpha u(\mathbf{x})|^2 \, d\mathbf{x} \right)^{\frac{dq(r-2)}{2r(q-2)}}. \tag{4.28}$$

Compact Mappings, the Rellich Lemma, and Compact Sobolev Embeddings

We conclude this section with a more technical *nonintuitive* concept: compact embeddings and the Rellich lemma, showing that certain Sobolev embeddings are compact. The Sobolev embedding (4.25) is continuous, which is equivalent to saying that the following inequality holds (cf. (4.26)):

$$\|u\|_{L^q(\Omega)} \leq c\|u\|_{H^m(\Omega)}$$

for some constant c independent of u. This means that the identity operator is continuous from $H^m(\Omega)$ into $L^q(\Omega)$. This holds under the condition that $1/q = 1/2 - m/d$, for $1 \leq m < d/2$ or in the critical case $m = d/2$ as just explained. When the domain Ω is bounded, $L^q(\Omega)$ is included in $L^p(\Omega)$ for any $p < q$. For simplicity, we consider only the case $m = 1$. Then, if Ω is bounded and smooth, we have the embedding

$$H^1(\Omega) \subset L^p(\Omega)$$

for $1/p > 1/2 - 1/d$. The Rellich lemma (see e.g. Adams [1975]) asserts that this embedding is not only continuous but also *compact*. This means that a bounded set in $H^1(\Omega)$ is precompact as a subset of $L^p(\Omega)$. One of the important consequences of this property is that any sequence $\{u_n\}_{n\in\mathbb{N}}$ of functions whose norms in $H^1(\Omega)$ are uniformly bounded contains a subsequence $\{u_{n_j}\}_{j\in\mathbb{N}}$ that converges in the norm of $L^p(\Omega)$ to some element u in $H^1(\Omega)$. The usual notation for compact embedding is

$$H^1(\Omega) \subset\subset L^p(\Omega), \tag{4.29}$$

for $1/p > 1/2 - 1/d$ and for Ω bounded and smooth. This important result is often invoked in this monograph, especially for proving that the Stokes operator has a compact inverse, whence we deduce the existence of an orthonormal basis of eigenfunctions of the Stokes operator (see Section 6 in Chapter II).

Alternatively, saying that the embedding (4.29) is compact means that any sequence weakly convergent in $H^1(\Omega)$ is strongly convergent in $L^p(\Omega)$ for such values of p, that is, convergent in the $L^p(\Omega)$ norm. See Appendix A.1 in Chapter II for some explanations on weak convergences.

II

Elements of the Mathematical Theory of the Navier–Stokes Equations

Introduction

The purpose of this chapter is to recall some elements of the classical mathematical theory of the Navier–Stokes equations (NSE). We try also to explain the physical background of this theory for the physics-oriented reader.

As they stand, the Navier–Stokes equations are presumed to embody all of the physics inherent in the given incompressible, viscous fluid flow. Unfortunately, this does not automatically guarantee that the solutions to those equations satisfy the given physics. In fact, it is not even guaranteed a priori that a satisfactory solution exists. This chapter addresses the means for specifying function spaces – that is, the ensembles of functions consistent with the physics of the situation (such as incompressibility, boundedness of energy and enstrophy, as well as the prescribed boundary conditions) – that can serve as solutions to the Navier–Stokes equations. An important point is made that the kinematic pressure, p, is determined uniquely by the velocity field up to an additive constant. Hence, one cannot specify independently the initial boundary conditions for the pressure. This observation leads naturally to a representation of the NSE by an abstract differential equation in a corresponding function space for the velocity field.

Two types of boundary conditions are considered: *no-slip*, which are relevant to flows in domains bounded by solid impermeable walls; and *space-periodic* boundary conditions, which serve to study some idealized flows (including homogeneous flows) far away from real boundaries.

A simplification, without compromising the physics or mathematical rigor, results from the so-called Helmholtz–Leray decomposition, which is a generalization of the decomposition of any vector field into a gradient of a potential plus a solenoidal (divergence-free) component. The generalization consists of taking into account the accompanying boundary conditions. From the mathematical point of view, it is possible that some solutions of the Navier–Stokes equations could be highly irregular, even though no such irregularities are encountered in nature. However, these solutions when smoothed in some sense could represent physical realities. Toward this end, one can introduce the so-called weak solutions: essentially, an inner product of the possibly irregular solutions with some sufficiently regular (nice) functions. This operation is somewhat similar to taking moments of the velocity field with respect to some distribution function, yielding observable averages of flow quantities. However, care must be taken not to push this analogy too far, because weak solutions are not the same as the statistical solutions.

The next major task undertaken in this chapter is the question of existence and uniqueness of solutions to the Navier–Stokes equations. In three dimensions, where all practical fluid flows arise, we find that this is still an open problem, suggesting that the 3-dimensional NSE are not a complete description of fluid flows. This is not too unexpected, because we know (Schlichting [1979, Chap. III]) that these equations have been obtained under the natural but unverified hypothesis that the normal and shearing stresses are strictly linear functions of the rate of deformation. The departure from linearity may be so small that there are no experimentally observable consequences. Yet an absence of the corresponding correction in the equations can be significant enough to reveal mathematical pathologies. It should be remarked here that other (even small) additional terms serve to regularize the NSE in three dimensions (Lions [1969]). We do not address this problem here. On the other hand, the corresponding situation in 2-dimensional flows is well in hand, with existence and uniqueness of both strong and weak solutions proved beyond any challenge.

Another important issue raised in this chapter addresses the problem that arises because not all the terms in the Navier–Stokes equations belong to the same function space. Thus, all the terms linear in the velocity field $\mathbf{u}$ are divergence-free, whereas the inertial term $(\mathbf{u}\cdot\nabla)\mathbf{u}$ is not. In principle, some mathematical operations involving pairs of quantities (say, a and b) require, in concert with our intuitive physical senses, that a and b be of the "same kind" – not "mixing apples and oranges." Mathematically speaking, this can be articulated as requiring that, in general, a and b belong to the same function space. But sometimes, as in the present case, we must deal with pairs that are *not* in the same function space. In order to make the necessary operations (e.g., inner products) meaningful, it is necessary to introduce special rules. The concept of dual spaces, mentioned in the body of this chapter and dealt with at greater length in Appendix A, serves this purpose.

We address the analyticity of solutions of the equations on the real-time axis. Physically, this corresponds to the demand that the solutions do not evidence any singular behavior for all time, which is true for the case of flows in two dimensions. We return then to the dependence of the solutions on the boundary conditions and study the special case of flows driven by a moving boundary. The chapter ends with Appendices A and B, addressing some delicate mathematical questions raised in the body of the chapter.

In short, the mathematical theory presented here serves to ensure that the solutions of the Navier–Stokes equations, when they exist, are consistent with the physics underlying the flow of incompressible, viscous flows. It leads to tools presented further on, tools that allow us to obtain insight into the nature of turbulent flows without the burden of elaborate numerical computations. At the same time, such tools are useful for testing the validity of purely numerical efforts.

We now describe the content of this chapter in words that are mathematically more technical. In Section 1 we introduce the concepts of kinetic energy and enstrophy; in Section 2, we present the corresponding boundary value problems to be considered. We show also that, in the incompressible case, pressure (using the terminology of meteorology) is a diagnostic variable: this means that, at each instant of time, pressure is a function (functional) of the velocity field – including at time $t = 0$. Hence, as we shall see, at $t = 0$ the initial field of velocities is prescribed, but not the initial pressure distribution. In Section 3 we introduce the Helmholtz–Leray decomposition of vector fields, which is a version of the classical Helmholtz decomposition of vector fields adapted to the boundary conditions. In Section 4 we give the weak formulation of the NSE in which the pressure term disappears; this is the starting point of the mathematical theory of these equations as introduced by the French mathematician Jean Leray [1933, 1934a,b]. In Section 5 we introduce the function spaces that we use, and we explain their physical meaning by relating them to the spaces of finite kinetic energy and finite enstrophy. Section 6 is devoted to the study of the Stokes operator, which is associated with the linear part of the Navier–Stokes equations and is a fundamental tool in the mathematical theory of the NSE.

Section 7 recalls the major existence and uniqueness results for the Navier–Stokes equations and introduces the concepts of weak (less regular) and strong (more regular) solutions. In Section 8 we recall the fact that, in dimension 2, the solutions of the NSE are analytic in time – a property with important physical consequences. Section 9 concerns analyticity in space in the 2- and 3-dimensional cases; we also show how this space analyticity implies the exponential decay of the Fourier coefficients of the solutions. Section 10 introduces the function spaces needed for the Navier–Stokes equations on the whole space ($\mathbb{R}^2$ or $\mathbb{R}^3$). Section 11 gives the mathematical framework needed for the treatment of the no-slip case with moving boundaries. We introduce apropriate "background flows" to rewrite the equations with homogeneous boundary conditions and added (linear) terms. In Section 12, we present universal estimates for the energy dissipation in terms of a characteristic velocity and a characteristic length; in Section 13, we introduce the nondimensional Grashof number and discuss nondimensional estimates. Finally, in Appendices A and B, we give some mathematical complements (for the mathematically oriented reader) and prove some of the results alluded to in this chapter.

1 Energy and Enstrophy

We recall from Chapter I the Navier–Stokes equations for a viscous, incompressible, homogeneous flow:

$$\frac{\partial \mathbf{u}}{\partial t} - \nu\Delta\mathbf{u} + (\mathbf{u}\cdot\nabla)\mathbf{u} + \nabla p = \mathbf{f}, \tag{1.1a}$$

$$\nabla\cdot\mathbf{u} = 0. \tag{1.1b}$$

With the density normalized to $\rho = 1$, the kinetic energy of a fluid with velocity field $\mathbf{u} = \mathbf{u}(x)$ and occupying a region Ω is given by

$$e(\mathbf{u}) = \frac{1}{2}\int_\Omega |\mathbf{u}(\mathbf{x})|^2\,d\mathbf{x}. \tag{1.2}$$

Another important quantity is the enstrophy,

$$E(\mathbf{u}) = \sum_{i=1}^{d}\int_\Omega |\nabla u_i(\mathbf{x})|^2\,dx = \sum_{i,j=1}^{d}\int_\Omega \left|\frac{\partial u_i}{\partial x_j}(\mathbf{x})\right|^2\,d\mathbf{x}, \tag{1.3}$$

where $d = 2$ or 3 depending on whether the flow is 2- or 3-dimensional. As will be seen shortly, the significance of enstrophy is that it determines the rate of dissipation of kinetic energy.

If the domain Ω is the whole space ($\mathbb{R}^2$ or $\mathbb{R}^3$) and if the velocity $\mathbf{u}$ decays sufficiently rapidly at infinity, then integration by parts using the divergence-free condition (1.1b) implies the following representation for the enstrophy:

$$E(\mathbf{u}) = \int_\Omega |\boldsymbol{\omega}(\mathbf{x})|^2\,d\mathbf{x}, \tag{1.4}$$

where $\boldsymbol{\omega} = \operatorname{curl}\mathbf{u}$ is the vorticity vector. (In fact, *strophy* comes from Greek, meaning *rotation*.) Relation (1.4) is also valid when Ω is not the whole space, provided the boundary terms (resulting from the integration by parts) vanish. This is true in many interesting cases.

Now assume, for simplicity, that $\Omega = \mathbb{R}^3$ and that $\mathbf{u}$ and p vanish and decay sufficiently rapidly at infinity. Then, we take the scalar product of the momentum equation (1.1a) with $\mathbf{u}$ and integrate over Ω. The first term yields

$$\int \frac{\partial \mathbf{u}}{\partial t}(\mathbf{x},t)\cdot\mathbf{u}(\mathbf{x},t)\,d\mathbf{x} = \frac{1}{2}\frac{d}{dt}\int_\Omega |\mathbf{u}(\mathbf{x},t)|^2\,d\mathbf{x}.$$

The contributions of the inertial term and of the pressure vanish owing to the divergence-free condition:

$$\begin{aligned}\int_\Omega [(\mathbf{u}\cdot\nabla)\mathbf{u}]\cdot\mathbf{u}\,d\mathbf{x} &= \sum_{i=1}^{3}\sum_{j=1}^{3}\int_\Omega u_j\frac{\partial u_i}{\partial x_j}u_i\,d\mathbf{x} = \frac{1}{2}\sum_{i=1}^{3}\sum_{j=1}^{3}\int_\Omega u_j\frac{\partial(u_i^2)}{\partial x_j}\,d\mathbf{x}\\ &= -\frac{1}{2}\sum_{i=1}^{3}\int_\Omega(\nabla\cdot\mathbf{u})u_i^2\,d\mathbf{x} = 0\end{aligned} \tag{1.5}$$

and

$$\int_\Omega \nabla p\cdot\mathbf{u}\,d\mathbf{x} = \sum_{i=1}^{3}\frac{\partial p}{\partial x_i}u_i\,d\mathbf{x} = -\int_\Omega p(\nabla\cdot\mathbf{u})\,d\mathbf{x} = 0. \tag{1.6}$$

Integrating by parts, we see that the viscous term yields

$$\begin{aligned}-\nu\int_\Omega \Delta\mathbf{u}\cdot\mathbf{u}\,d\mathbf{x} &= -\nu\sum_{i,j=1}^{3}\int_\Omega \frac{\partial^2 u_i}{\partial x_j^2}u_i\,d\mathbf{x} = \nu\sum_{i,j=1}^{3}\int_\Omega\left(\frac{\partial u_i}{\partial x_j}\right)^2 d\mathbf{x}\\ &= \nu E(\mathbf{u}).\end{aligned} \tag{1.7}$$

We are then left with the equation of conservation of energy, which gives the rate of decay of the kinetic energy:

$$\frac{d}{dt}e(\mathbf{u}) + \nu E(\mathbf{u}) = \int_\Omega \mathbf{f} \cdot \mathbf{u}\, dx. \tag{1.8}$$

When there are no volume forces (i.e., when $\mathbf{f} = 0$), the conservation of energy equation implies the decay of the kinetic energy by viscous effect at the rate $-\nu E(\mathbf{u})$:

$$\frac{d}{dt}e(\mathbf{u}) = -\nu E(\mathbf{u}). \tag{1.9}$$

The calculations performed in (1.5), (1.6), and (1.7) play an essential role and are repeatedly used in the mathematical theory of the Navier–Stokes equations, even when the domain Ω is not the whole space.

2 Boundary Value Problems

The Navier–Stokes equations (1.1) are inescapably supplemented with initial and boundary conditions that depend on the physical problem under consideration. Throughout this book we mainly consider two distinct types of boundary conditions: the *no-slip* boundary condition (for bounded domains) and the *space-periodic* boundary condition.

The no-slip boundary condition (flow past a rigid boundary) is one of the few that correspond to a physically accessible boundary condition. Another physically accessible boundary condition is the open boundary (i.e., an open surface of a flowing fluid for part or all of the boundary[1]); we will not treat this case specifically, as it can be handled by methods very similar to the no-slip case. The space-periodic case is not a physically achievable one, but it is relevant on the physical side as a model for some flows and is needed in the study of homogeneous turbulence. On the mathematical side, the space-periodic case includes many of the difficulties encountered in the no-slip case. However, the former is simpler to treat because of the absence of boundaries (no boundary terms and often no boundary layers). Furthermore, using the Fourier series as a tool simplifies the analysis and eases visualization of the physical aspects of the flow.

Flows in unbounded domains, or flows in the whole space ($\mathbb{R}^2$ or $\mathbb{R}^3$), will not be emphasized here. This case is often discussed in the literature but is not appropriate for our present studies owing, in particular, to the lack of compactness. This characteristic raises technical difficulties that are related to the feasibility of standard approximation methods and hence to the computability of quantities such as (1.2) and (1.3). Very special assumptions about the flows must be made to obtain sensible results. Thus, we will discuss only the whole-space case (without a decay condition at infinity) as an idealization in the context of homogeneous statistical solutions; some material preliminary to that study appears in Section 10 of this chapter.

[1] In the language of partial differential equations, this corresponds to a Neumann-type boundary condition; as we shall see, the no-slip case corresponds to a Dirichlet-type boundary condition.

Note, too, that in the periodic case we consider two distinct situations: (i) when the average flow (over the space domain) is zero and (ii) when it is not necessarily zero. For that matter, if the volume forces have zero average and the average flow is zero at some instant of time, then the average flow is zero at all times. This is a particular case of periodic flows that makes some calculations easier.

In what follows, we describe more precisely each of the boundary conditions just mentioned. For notational purposes we denote by $|\Omega|$ the area or the volume of a given bounded set Ω in $\mathbb{R}^d$ for $d = 2$ or 3, respectively. The corresponding mathematical settings (e.g., function spaces) appear in Sections 5 and 10.

No-Slip Boundary Condition

This condition corresponds to the case where the fluid fills a smooth, bounded domain Ω with a rigid boundary $\partial\Omega$. Assuming that the motion of the boundary $\partial\Omega$ is prescribed (velocity at the boundary $= \boldsymbol{\varphi}$), the no-slip boundary condition is

$$\mathbf{u} = \boldsymbol{\varphi} \text{ on } \partial\Omega. \tag{2.1}$$

If the boundary is at rest, then

$$\mathbf{u} = 0 \text{ on } \partial\Omega. \tag{2.2}$$

We always assume that the shape and volume of the domain Ω occupied by the fluid are independent of time. In this context, the case where the normal component of $\boldsymbol{\varphi}$ is not zero corresponds to a permeable boundary (as occurs, for instance, in some flow control problems or with flow past a porous wall).

Space-Periodic Case

As we have already pointed out, the space-periodic boundary conditions are clearly not accessible in realistic physical situations, but they are useful for idealizations. They arise in the study of homogeneous turbulence, when it is assumed that walls are far from the region being studied and thus that the wall effects are not important. These boundary conditions do not contain the difficulties related to rigid boundaries (e.g., the presence of boundary layers), but they retain the complexities due to the nonlinear terms (introduced by the kinematics) that characterize the Navier–Stokes equations.

For space-periodic flows, we assume that the fluid fills the entire space $\mathbb{R}^d$ ($d =$ 2, 3) but with the condition that

$$\mathbf{u}, \mathbf{f}, \text{ and } p \text{ are periodic in each direction } 0x_i,\ i = 1, \dots, d, \text{ with corresponding periods } L_i > 0. \tag{2.3}$$

In this case we use Ω to denote the period:

$$\Omega = \left(-\frac{L_1}{2}, \frac{L_1}{2}\right) \times \left(-\frac{L_2}{2}, \frac{L_2}{2}\right) \times \left(-\frac{L_3}{2}, \frac{L_3}{2}\right)$$

in the 3-dimensional case, and

$$\Omega = \left(-\frac{L_1}{2}, \frac{L_1}{2}\right) \times \left(-\frac{L_2}{2}, \frac{L_2}{2}\right)$$

in the 2-dimensional case. A simplified case is when all the periods are the same, $L_i = L$ $(i = 1, \ldots, d)$; then the period is the cube

$$\Omega = Q(L) \equiv \left(-\frac{L}{2}, \frac{L}{2}\right)^d.$$

A particular case of interest arises when the volume forces have zero space average, that is, when

$$\frac{1}{|\Omega|}\int_\Omega \mathbf{f}(\mathbf{x}, t)\, d\mathbf{x} = 0. \tag{2.4}$$

Here $|\Omega|$ denotes the area or the volume of the domain Ω, depending on whether $d = 2$ or 3. Averaging each term in the NSE we notice – using integration by parts, as illustrated in Section 1, and the periodicity conditions – that several terms vanish; we are left with the equation

$$\frac{d}{dt}\int_\Omega \mathbf{u}(\mathbf{x}, t)\, d\mathbf{x} = 0.$$

Hence, if the initial average flow is zero then the average flow is zero for all time:

$$\frac{1}{|\Omega|}\int_\Omega \mathbf{u}(\mathbf{x}, t)\, d\mathbf{x} \equiv 0. \tag{2.5}$$

As noted earlier, it is sometimes useful (and simpler) to assume that the average flow is zero. Moreover, the general case – where the volume forces and the initial condition do not average to zero – can be reduced to this case, as we show henceforth. First, averaging the momentum equation in this case yields

$$\frac{d}{dt}\int_\Omega \mathbf{u}(\mathbf{x}, t)\, d\mathbf{x} = \int_\Omega \mathbf{f}(\mathbf{x}, t)\, d\mathbf{x}. \tag{2.6}$$

Therefore, if we set

$$\mathbf{M}_\mathbf{u}(t) = \frac{1}{|\Omega|}\int_\Omega \mathbf{u}(\mathbf{x}, t)\, dx \tag{2.7}$$

and define $\mathbf{M}_\mathbf{f}(t)$ similarly, we obtain

$$\frac{d\mathbf{M}_\mathbf{u}(t)}{dt} = \mathbf{M}_\mathbf{f}(t). \tag{2.8}$$

Thus, we easily obtain the average $\mathbf{M}_\mathbf{u}(t)$, for all time, directly from the volume forces $\mathbf{f}$. We can then write

$$\mathbf{u} = \mathbf{M}_\mathbf{u} + \tilde{\mathbf{u}}$$

and obtain for $\tilde{\mathbf{u}}$, which has zero average, an equation similar to the momentum equation (1.1a); the only changes are the addition of the advection term $(\mathbf{M}_\mathbf{u} \cdot \nabla)\tilde{\mathbf{u}}$ and the change in the volume force to its perturbation from the average (i.e., $\tilde{\mathbf{f}} = \mathbf{f} - \mathbf{M}_\mathbf{f}$. Then, we can absorb the advection term into the independent variable by introducing the functions

$$\begin{aligned}\tilde{\tilde{\mathbf{u}}}(\mathbf{x},t) &= \tilde{\mathbf{u}}(\mathbf{x}+\mathbf{I}_{\mathbf{u}}(t),t),\\ \tilde{\tilde{p}}(\mathbf{x},t) &= p(\mathbf{x}+\mathbf{I}_{\mathbf{u}}(t),t),\\ \tilde{\tilde{\mathbf{f}}}(\mathbf{x},t) &= \tilde{\mathbf{f}}(\mathbf{x}+\mathbf{I}_{\mathbf{u}}(t),t),\end{aligned} \tag{2.9}$$

where $\mathbf{I}_{\mathbf{u}}(t)$ is a primitive of $\mathbf{M}_{\mathbf{u}}(t)$ – for example,

$$\mathbf{I}_{\mathbf{u}}(t) = \int_0^t \mathbf{M}_{\mathbf{u}}(s)\,ds.$$

Notice that $\tilde{\tilde{\mathbf{u}}}$, $\tilde{\tilde{p}}$, and $\tilde{\tilde{\mathbf{f}}}$ are also periodic with period Ω. The equations for $\tilde{\tilde{\mathbf{u}}}$ and $\tilde{\tilde{p}}$ are the momentum equation (1.1a) and the divergence-free condition (1.1b) – with $\mathbf{u}$, p, and $\mathbf{f}$ replaced by $\tilde{\tilde{\mathbf{u}}}$, $\tilde{\tilde{p}}$, and $\tilde{\tilde{\mathbf{f}}}$ – all functions periodic and with $\tilde{\tilde{\mathbf{u}}}$ and $\tilde{\tilde{\mathbf{f}}}$ having zero space average. We need to solve those equations for $\tilde{\tilde{\mathbf{u}}}$ and $\tilde{\tilde{p}}$, as well as the equation (2.8) for $\mathbf{M}_{\mathbf{u}}(t)$, in order to recover $\mathbf{u}(\mathbf{x},t)$ and $p(\mathbf{x},t)$.

Channel Flows

Another interesting boundary condition, which is not (fully) physically accessible but is often considered in the mathematical and physical literatures, is the channel flow.

The channel occupies the domain of $\mathbb{R}^3$,

$$\Omega = \left(-\frac{L_1}{2},\frac{L_1}{2}\right)\times\left(-\frac{L_2}{2},\frac{L_2}{2}\right)\times\left(-\frac{L_3}{2},\frac{L_3}{2}\right)$$

(in, say, dimension 3). The boundaries x_3 (or z) $= \pm L_3/2$ are rigid ones, with a no-slip condition as in (2.1); that is,

$$\mathbf{u} = \varphi \text{ at } x_3 = \pm\frac{L_3}{2}. \tag{2.10}$$

On the other hand, the flow is periodic in the directions x_1 and x_2:

$\mathbf{u}$ and p are periodic in the directions $0x_1$ and $0x_2$, with corresponding periods L_1 and L_2. (2.11)

Usually, such a flow is sustained by a pressure gradient P in the direction $0x_1$, which amounts to choosing $\mathbf{f} = P\mathbf{e}_1$ ($\mathbf{e}_1 = (1,0,0)$ in $\mathbb{R}^3$); the total (physical) pressure is $p - Px_1$, which is *not* periodic in the direction $0x_1$. Concerning the function $P = P(t)$, we usually consider one of two cases: either the pressure gradient is prescribed (i.e., $P(t)$ is given) or the flux is prescribed in the direction $0x_1$ – namely,

$$\int_{-L_3/2}^{L_3/2}\int_{-L_2/2}^{L_2/2} u_1(x_1,x_2,x_3)\,dx_2\,dx_3 = F$$

is given and $P(t)$ is an auxiliary unknown (note that F is independent of x_1, by incompressibility).

We will not consider this case in detail here. Of course, the mathematical techniques that it requires combine those of the no-slip and space-periodic cases.

Initial Condition

When studying the evolution of a flow, we must also prescribe the initial distribution of velocities. Mathematically, this amounts to a condition of the form

$$\mathbf{u}(\mathbf{x},0) = \mathbf{u}_0(\mathbf{x}), \quad \mathbf{x} \in \Omega, \tag{2.12}$$

where $\mathbf{u}_0$ is given. For consistency, the initial condition must satisfy the appropriate boundary conditions.

Note that, to obtain a well-posed initial boundary value problem for the Navier–Stokes equations, we do not prescribe the boundary value of the pressure or its initial value. As we will see, the pressure is in fact fully and uniquely determined at all times, including $t = 0$, as a function of the velocity field $\mathbf{u}$ (uniquely up to an additive constant, since only the gradient of the pressure is accounted for in the momentum equations). In the terminology used in meteorology, we would say that $\mathbf{u}$ is a prognostic variable and p a diagnostic variable.

Combining the initial condition (2.12) with the previous boundary conditions yields the following (deterministic) initial boundary value problems, which we discuss in the sequel:

$$\begin{aligned}&(\mathcal{P}_{\text{nsp}}), \text{ consisting of (1.1), (2.2), and (2.12);}\\&(\mathcal{P}_{\text{per}}), \text{ consisting of (1.1), (2.3), and (2.12);}\\&(\dot{\mathcal{P}}_{\text{per}}), \text{ consisting of (1.1), (2.3), (2.4), (2.5), and (2.12);}\\&(\mathcal{P}_{\text{ch}}), \text{ consisting of (1.1), (2.10), (2.11), and (2.12).}\end{aligned} \tag{2.13}$$

These problems are well-posed, subject to some restrictions in dimension 3. Problem $(\dot{\mathcal{P}}_{\text{per}})$ is a special case of $(\mathcal{P}_{\text{per}})$, so for many purposes it suffices to consider $(\mathcal{P}_{\text{per}})$. However, very often we consider $(\dot{\mathcal{P}}_{\text{per}})$ for the purpose of simplification. As mentioned earlier, we will not consider the channel flow problem $(\mathcal{P}_{\text{ch}})$ in detail; it can be handled by combining the techniques for the no-slip and periodic cases.

Simplified Problems

It is worth mentioning some simplified problems associated with the Navier–Stokes equations. First, steady-state (or stationary) flows are flows in which the velocity is independent of time:

$$\mathbf{u}(\mathbf{x},t) = \mathbf{u}(\mathbf{x}) \quad \text{for all } t \geq 0. \tag{2.14}$$

In order to maintain such a flow, the volume forces $\mathbf{f}$ (and φ in the nonhomogeneous no-slip case (2.1)) must be independent of time. In this case, (1.1) reduces to

$$-\nu\Delta\mathbf{u} + (\mathbf{u}\cdot\nabla)\mathbf{u} + \nabla p = \mathbf{f}, \tag{2.15a}$$

$$\nabla\cdot\mathbf{u} = 0. \tag{2.15b}$$

Equations (2.15) constitute the *stationary Navier–Stokes equations*; they must be supplemented with a boundary condition for $\mathbf{u}$, such as (2.1), (2.2), (2.3), or (2.10)–(2.11).

In general, solutions of (2.15) are stable only for sufficiently low Reynolds numbers. The linearized versions of the NSE correspond to cases where the flow is so slow and viscosity so large that the quadratic (inertial) terms are negligible as compared to the linear terms. This is known as "creep flow." Here the inertial terms can be suppressed, and so (1.1) reduces to

$$\frac{\partial \mathbf{u}}{\partial t} - \nu\Delta\mathbf{u} + \nabla p = \mathbf{f}, \tag{2.16a}$$

$$\nabla \cdot \mathbf{u} = 0. \tag{2.16b}$$

Equations (2.16) – together with the initial condition (2.12) and one of the boundary conditions (2.1), (2.2), (2.3), or (2.10)–(2.11) – is a well-posed linear initial boundary value problem.

In the context of this work, more interesting is the *Stokes problem,* which is the linearized stationary version of the Navier–Stokes equations:

$$-\nu\Delta\mathbf{u} + \nabla p = \mathbf{f}, \tag{2.17a}$$

$$\nabla \cdot \mathbf{u} = 0, \tag{2.17b}$$

supplemented by any one of the boundary conditions (2.1), (2.2), (2.3), or (2.10)–(2.11). Flows appearing in rheology and materials processing are governed by these equations. More significantly for us, the Stokes problem arises constantly in the theoretical (mathematical) study of the NSE because the eigenfunctions appearing in (2.17) span the function spaces in which the solutions of the full Navier–Stokes equations reside. We will encounter them in the following sections and chapters.

A Boundary Value Problem for the Pressure

We mentioned earlier that, in the incompressible case, the kinematic pressure field is fully determined at each instant of time by the velocity field **u**. More precisely: at each instant of time, the pressure can be expressed in terms of the velocity field at that time via the solution of a suitable boundary value problem. In particular, at time $t = 0$, by prescribing the initial spatial distribution of velocities $\mathbf{u}_0$ as in (2.12), we automatically prescribe (to within an additive constant) the initial pressure distribution $p(\mathbf{x}, 0)$. The pressure at any given point in space is in fact determined by the velocity field everywhere. This is a consequence of the incompressibility assumption. The sound speed becomes infinite and velocity fluctuations everywhere are coupled instantaneously. Mathematically, this can be seen by taking the divergence of the momentum equation, ending up with a Laplace equation for the pressure.

Indeed, consider a solution **u** of the full Navier–Stokes equations for one of the problems $(\mathcal{P}_{\text{nsp}})$ or $(\mathcal{P}_{\text{per}})$. The other initial and boundary value problems can be handled similarly. We take the divergence at each side of the momentum equation (1.1a), which yields, term by term:

$$\operatorname{div} \frac{\partial \mathbf{u}}{\partial t} = 0,$$

$$\operatorname{div} \Delta \mathbf{u} = 0,$$

$$\operatorname{div} \nabla p = \Delta p,$$

$$\operatorname{div}\{(\mathbf{u} \cdot \nabla)\mathbf{u}\} = \sum_{i,j=1}^{d} \frac{\partial}{\partial x_i}\left(u_j \frac{\partial u_i}{\partial x_j}\right)$$

$$= \sum_{i,j=1}^{d} \frac{\partial u_j}{\partial x_i} \frac{\partial u_i}{\partial x_j} \quad \text{(by (1.1b))}.$$

Hence, we find the following equation for the pressure:

$$\Delta p = \operatorname{div} \mathbf{f} - \sum_{i,j=1}^{d} \frac{\partial u_j}{\partial x_i} \frac{\partial u_i}{\partial x_j}. \tag{2.18}$$

Therefore, in the space-periodic case (2.3), the pressure p is fully defined in terms of the velocity field $\mathbf{u}$ and the volume forces $\mathbf{f}$ by equation (2.18) and the periodicity condition. We obtain p as a quadratic function of $\mathbf{u}$: $p = \psi(\mathbf{u})$.

In the no-slip case (2.2), we obtain a boundary condition for the pressure p by taking the scalar product of each term in the momentum equation (1.1a) with the vector $\mathbf{n}$ normal to the smooth boundary $\partial\Omega$. Thus, using the no-slip condition (2.2), we have

$$\frac{\partial \mathbf{u}}{\partial t} \cdot \mathbf{n} = \frac{\partial}{\partial t}(\mathbf{u} \cdot \mathbf{n}) = 0,$$

$$\nabla p \cdot \mathbf{n} = \frac{\partial p}{\partial \mathbf{n}},$$

$$[(\mathbf{u} \cdot \nabla)\mathbf{u}] \cdot \mathbf{n} = \sum_{i,j=1}^{d} u_j \frac{\partial u_i}{\partial x_j} n_i = 0.$$

Hence, there remains

$$\frac{\partial p}{\partial \mathbf{n}} = [\mathbf{f} + \nu \Delta \mathbf{u}] \cdot \mathbf{n}. \tag{2.19}$$

As may be easily verified, the right-hand sides of (2.18) and (2.19) satisfy the consistency condition

$$\int_{\Omega} \Delta p \, d\mathbf{x} = \int_{\partial\Omega} \frac{\partial p}{\partial \mathbf{n}} \, dS(\mathbf{x}).$$

We conclude that p can be expressed in terms of $\mathbf{u}$ by solving the Neumann problem (2.18)–(2.19). Just as in the periodic case, in the no-slip case we obtain p as a quadratic function of $\mathbf{u}$,

$$p = \psi(\mathbf{u}). \tag{2.20}$$

Note that (2.20) holds at each instant of time. In particular, the initial distribution of pressure $p(\mathbf{x}, 0) = p_0(\mathbf{x})$ is defined by

$$p_0 = \psi(\mathbf{u}_0). \tag{2.21}$$

In all cases, p is defined up to an additive constant.

An Evolution Equation for the Velocity Field **u**

Clearly we can infer from the preceding remarks, particulary (2.20), that the full Navier–Stokes equations amount to an evolution equation for the velocity field **u**. That evolution equation is a functional equation and no longer a partial differential equation. The functional equation obtained by means of (2.20) is not the one commonly used in the mathematical theory that emerges naturally with the weak formulation of the NSE (see Section 4). However, it is instructive at this point to give this slightly different form of the evolution equation for **u**.

With (2.20) in mind, the sum of the inertial term and the pressure gradient is found to be a quadratic function of **u**, which we may denote by

$$\mathcal{B}(\mathbf{u}) \equiv (\mathbf{u}\cdot\nabla)\mathbf{u} + \nabla\psi(\mathbf{u}).$$

Then, the momentum equation (1.1a) can be written solely in terms of **u**:

$$\frac{\partial \mathbf{u}}{\partial t} - \nu\Delta\mathbf{u} + \mathcal{B}(\mathbf{u}) = \mathbf{f}. \tag{2.22}$$

As just mentioned, this equation is not the usual form of the Navier–Stokes equations, but it puts in evidence the fact that those equations may be regarded as an evolution equation for **u**. A more common form of the evolution equation for **u** is based on the Helmholtz–Leray decomposition of vector fields, to which we now turn.

3 Helmholtz–Leray Decomposition of Vector Fields

The Helmholtz decomposition resolves a vector field **u** in $\mathbb{R}^d$ $(d = 2, 3)$ into the sum of a gradient and a curl vector. There is an appropriate generalization, which we will call the Helmholtz–Leray decomposition, that is valid for vector fields defined on a bounded set, taking into account the boundary conditions of the problem. Regardless of these boundary conditions, for a given vector field **w** we seek a decomposition of the form

$$\mathbf{w} = \nabla q + \mathbf{v}, \quad \text{with } \operatorname{div}\mathbf{v} = 0. \tag{3.1}$$

Thus, at least locally, **v** is a curl vector, say $\mathbf{v} = \operatorname{curl}\boldsymbol{\zeta}$. The assumption that $\operatorname{div}\mathbf{v} = 0$ implies

$$\Delta q = \operatorname{div}\mathbf{w}. \tag{3.2}$$

Equation (3.2) is supplemented by boundary conditions that depend on those of the original vector field **w**. Using (3.2) and the boundary conditions, one can derive q from **w**; then, from (3.1), one obtains **v**.

In the space-periodic case, **w** is periodic (in the sense of (2.3)), so we require **v** to be periodic as well. Hence, we impose periodic boundary conditions on q, which together with equation (3.2) determine q uniquely in terms of $\mathbf{w}$ (up to an additive constant). We then obtain $\mathbf{v}$ from (3.1), so that $\mathbf{v}$ is also periodic and divergence-free. Since $\operatorname{div} \mathbf{v} = 0$ it follows that $\mathbf{v}$ is indeed a curl vector, say $\mathbf{v} = \operatorname{curl} \boldsymbol{\zeta}$, with $\boldsymbol{\zeta}$ periodic.

In the no-slip case, $\mathbf{w}$ vanishes at the boundary (as in (2.2)) and we require only that

$$\mathbf{v} \cdot \mathbf{n} = 0 \text{ on } \partial\Omega. \tag{3.3}$$

This implies that $\nabla q \cdot \mathbf{n} = \mathbf{w} \cdot \mathbf{n}$; that is,

$$\frac{\partial q}{\partial \mathbf{n}} = \mathbf{w} \cdot \mathbf{n} \text{ on } \partial\Omega. \tag{3.4}$$

We conclude that q is solution of the Neumann problem (3.2), (3.4). The necessary consistency condition

$$\int_{\Omega} \operatorname{div} \mathbf{w}\, d\mathbf{x} = \int_{\partial\Omega} \mathbf{w} \cdot \mathbf{n}\, dS(\mathbf{x})$$

follows from the divergence theorem. Thus, q is uniquely defined up to an additive constant and $\mathbf{v}$ is equally well-defined.

Since $\operatorname{div} \mathbf{v} = 0$, it follows that $\mathbf{v}$ is the curl of a single-valued function $\boldsymbol{\zeta}$ defined locally. If the boundary $\partial\Omega$ of Ω is connected – that is, if Ω has no holes (i.e., it is a simply connected set), then the conditions

$$\operatorname{div} \mathbf{v} = 0 \text{ in } \Omega \quad \text{and} \quad \mathbf{v} \cdot \mathbf{n} = 0 \text{ on } \partial\Omega \tag{3.5}$$

imply that $\boldsymbol{\zeta}$ is a single-valued function in the whole domain Ω, with $\mathbf{v} = \operatorname{curl} \boldsymbol{\zeta}$, as in the usual Helmholtz decomposition. When the boundary is not connected, some delicate mathematical (topological) issues appear: $\mathbf{v}$ is the sum of the curl of a single-valued function $\boldsymbol{\zeta}$ and of N multivalued functions $\boldsymbol{\zeta}_1, \dots, \boldsymbol{\zeta}_N$, where N is the number of holes in the domain (see Temam [1979, Apx. I]).

We observe that, contrary to the usual Helmoltz decomposition, the Helmholtz–Leray decomposition of $\mathbf{w}$ is unique (up to an additive constant for q). Indeed if $q = q_1 - q_2$ and $\mathbf{v} = \mathbf{v}_1 - \mathbf{v}_2$, where $(q_1, \mathbf{v}_1)$ and $(q_2, \mathbf{v}_2)$ correspond to two such decompositions, then

$$\nabla q + \mathbf{w} = 0;$$

hence,

$$\Delta q = 0.$$

In the space-periodic case this implies that q is constant. In the no-slip case, (3.4) yields

$$\frac{\partial q}{\partial \mathbf{n}} = \mathbf{w} \cdot \mathbf{n} = 0 \text{ on } \partial\Omega,$$

so that q is also constant.

We infer from the preceding remarks that the map $\mathbf{w} \mapsto \mathbf{v}$ is well-defined. We denote this map by

$$P_L \colon \mathbf{w} \mapsto \mathbf{v}(\mathbf{w}). \tag{3.6}$$

This map is a projector; that is, if $\mathbf{w}$ is already divergence-free then $P_L\mathbf{w} = \mathbf{w}$. We will call it the *Leray projector* (for the corresponding boundary conditions).

The Evolution Equation for the Velocity Field

As indicated earlier, the Navier–Stokes equations are equivalent to a functional evolution equation for $\mathbf{u}$ that is not a partial differential equation. We may obtain an appropriate evolution equation for $\mathbf{u}$ using the Leray projector P_L. We apply P_L to both sides of the momentum equation (1.1a). If $\mathbf{u}$ represents an incompressible flow, then the divergence-free condition (1.1b) holds and it is easy to see that

$$P_L\mathbf{u} = \mathbf{u}, \quad P_L\frac{\partial \mathbf{u}}{\partial t} = \frac{\partial \mathbf{u}}{\partial t}, \quad \text{and} \quad P_L\nabla p = 0.$$

Therefore, we find that

$$\frac{d\mathbf{u}}{dt} + \nu A\mathbf{u} + B(\mathbf{u}) = P_L\mathbf{f}, \tag{3.7}$$

where we have written

$$A\mathbf{u} = -P_L\Delta\mathbf{u}, \quad B(\mathbf{u}) = B(\mathbf{u},\mathbf{u}), \quad B(\mathbf{u},\mathbf{v}) = P_L((\mathbf{u}\cdot\nabla)\mathbf{v}). \tag{3.8}$$

The operator A is the *Stokes operator* that we introduced earlier. In the space-periodic case,

$$A\mathbf{u} = -P_L\Delta\mathbf{u} = -\Delta\mathbf{u},$$

so that (3.7) is essentially the same as (2.22). However, in the no-slip case,

$$A\mathbf{u} = -P_L\Delta\mathbf{u} \neq -\Delta\mathbf{u},$$

and (3.7) and (2.22) are slightly different. More precisely,

$$\mathcal{B}(\mathbf{u}) = (I - P_L)(\nu\Delta\mathbf{u} + \mathbf{f}) + P_L((\mathbf{u}\cdot\nabla)\mathbf{u}),$$

where I denotes the identity operator, while

$$B(\mathbf{u}) = P_L((\mathbf{u}\cdot\nabla)\mathbf{u}).$$

The form (3.7) of the NSE was first derived by Leray [1933], using a slightly different presentation based on the so-called weak formulation, which we present in the following section.

For notational simplicity, we may write the functional evolution equation (3.7) as an ordinary differential equation (in infinite dimension). In general, we assume for simplicity that $\mathbf{f}$ belongs to H, so that $P_L\mathbf{f} = \mathbf{f}$; this can always be done, the term $(I - P_L)\mathbf{f}$ being added to the pressure. Then we write

$$\mathbf{u}' = \mathbf{F}(t,\mathbf{u}), \tag{3.9}$$

where

$$\mathbf{u}' = \mathbf{u}_t = \frac{\partial \mathbf{u}}{\partial t}, \qquad \mathbf{F}(t,\mathbf{u}) = \mathbf{f}(t) - \nu A\mathbf{u} - B(\mathbf{u}). \tag{3.10}$$

In case the forcing term $\mathbf{f}$ is time-independent, we say in the dynamical context that the system is autonomous and simply write

$$\mathbf{u}' = \mathbf{F}(\mathbf{u}), \tag{3.11}$$

where

$$\mathbf{F}(\mathbf{u}) = \mathbf{f} - \nu A\mathbf{u} - B(\mathbf{u}). \tag{3.12}$$

4 Weak Formulation of the Navier–Stokes Equations

The weak formulation of the NSE is obtained by multiplying equation (1.1a) by a "test" function $\mathbf{v}$ and integrating the result over Ω. The space V of test functions to be considered is defined more precisely in the following section; loosely speaking, a test function $\mathbf{v}$ is assumed to be divergence-free and to satisfy the same boundary conditions as $\mathbf{u}$.

For either no-slip or periodic boundary conditions, consider at each instant of time the vector field

$$\mathbf{x} \in \Omega \mapsto \mathbf{u}(\mathbf{x}, t).$$

For simplicity, we will denote this vector field at time t by $\mathbf{u}(\cdot, t)$ or simply by $\mathbf{u}(t)$. Then, let $\mathbf{v} = \mathbf{v}(\mathbf{x})$ be a test function belonging to V. In order to obtain the weak formulation of the Navier–Stokes equations, we take the inner product of the momentum equation (1.1a) with the test function $\mathbf{v}(\mathbf{x})$ and integrate the result over Ω. Using integration by parts when necessary, we can rewrite the following terms:

$$\int_\Omega \frac{\partial \mathbf{u}}{\partial t}(\mathbf{x}, t) \cdot \mathbf{v}(\mathbf{x})\, d\mathbf{x} = \frac{d}{dt}\int_\Omega \mathbf{u}(\mathbf{x}, t) \cdot \mathbf{v}(\mathbf{x})\, d\mathbf{x};$$

$$\begin{aligned} -\int_\Omega \Delta\mathbf{u}(\mathbf{x}, t) \cdot \mathbf{v}(\mathbf{x})\, d\mathbf{x} &= -\sum_{i,j=1}^{d} \int_{\partial\Omega} \frac{\partial u_i}{\partial \mathbf{n}}(\mathbf{x}, t) \cdot v_i(\mathbf{x})\, dS(\mathbf{x}) \\ &\quad + \sum_{i,j=1}^{d} \int_\Omega \frac{\partial u_i}{\partial x_j}(\mathbf{x}, t) \frac{\partial v_i}{\partial x_j}(\mathbf{x})\, d\mathbf{x} \\ &= \text{(owing to the boundary conditions on } \mathbf{v}) \\ &= \sum_{i,j=1}^{d} \int_\Omega \frac{\partial u_i}{\partial x_j}(\mathbf{x}, t) \frac{\partial v_i}{\partial x_j}(\mathbf{x})\, d\mathbf{x}. \end{aligned}$$

Then, by virtue of the boundary and the divergence-free conditions on the test function $\mathbf{v}$, we have

$$\int_\Omega \nabla p(\mathbf{x}, t) \cdot \mathbf{v}(\mathbf{x})\, d\mathbf{x} = \int_{\partial\Omega} p(\mathbf{x}, t)\mathbf{v}(\mathbf{x}) \cdot \mathbf{n}(\mathbf{x})\, dS(\mathbf{x}) - \int_\Omega p(\mathbf{x}, t) \operatorname{div} \mathbf{v}(\mathbf{x})\, d\mathbf{x} = 0.$$

Hence, we find that

$$\frac{d}{dt}\int_\Omega \mathbf{u}(\mathbf{x},t)\cdot\mathbf{v}(\mathbf{x})\,d\mathbf{x} + \nu\sum_{i,j=1}^{d}\int_\Omega \frac{\partial u_i}{\partial x_j}(\mathbf{x},t)\frac{\partial v_i}{\partial x_j}(\mathbf{x})\,d\mathbf{x}$$
$$+\sum_{i,j=1}^{d}\int_\Omega u_i(\mathbf{x},t)\frac{\partial u_j}{\partial x_i}(\mathbf{x},t)v_j(\mathbf{x})\,d\mathbf{x} = \int_\Omega \mathbf{f}(\mathbf{x},t)\cdot\mathbf{v}(\mathbf{x})\,d\mathbf{x},$$

which holds for every test function $\mathbf{v}$ of the type indicated. In order to simplify the notation, a more compact form is used by dropping the dummy variable $\mathbf{x}$:

$$\frac{d}{dt}\int_\Omega \mathbf{u}(t)\cdot\mathbf{v}\,d\mathbf{x} + \nu\sum_{i,j=1}^{d}\int_\Omega \frac{\partial u_i}{\partial x_j}(t)\frac{\partial v_i}{\partial x_j}\,d\mathbf{x}$$
$$+\sum_{i,j=1}^{d}\int_\Omega u_i(t)\frac{\partial u_j}{\partial x_i}(t)v_j\,d\mathbf{x} = \int_\Omega \mathbf{f}(t)\cdot\mathbf{v}\,d\mathbf{x}. \tag{4.1}$$

We simplify further by introducing the following classical notation. For every pair of vector fields $\boldsymbol{\varphi}$ and $\boldsymbol{\psi}$ defined on Ω,

$$(\boldsymbol{\varphi},\boldsymbol{\psi}) = \int_\Omega \boldsymbol{\varphi}(\mathbf{x})\cdot\boldsymbol{\psi}(\mathbf{x})\,d\mathbf{x},$$

$$((\boldsymbol{\varphi},\boldsymbol{\psi})) = \int_\Omega \sum_{i=1}^{d}\frac{\partial\boldsymbol{\varphi}(\mathbf{x})}{\partial x_i}\cdot\frac{\partial\boldsymbol{\psi}(\mathbf{x})}{\partial x_i}\,d\mathbf{x},$$

and, if $\boldsymbol{\theta}$ is a third vector field,

$$b(\boldsymbol{\varphi},\boldsymbol{\psi},\boldsymbol{\theta}) = \sum_{i,j=1}^{d}\int_\Omega \varphi_i(\mathbf{x})\frac{\partial\psi_j(\mathbf{x})}{\partial x_i}\theta_j(\mathbf{x})\,d\mathbf{x}. \tag{4.2}$$

Then, the weak equation (4.1) can be expressed as follows:

The function $t\mapsto\mathbf{u}(t)$ takes its values in V and satisfies

$$\frac{d}{dt}(\mathbf{u}(t),\mathbf{v}) + \nu((\mathbf{u}(t),\mathbf{v})) + b(\mathbf{u}(t),\mathbf{u}(t),\mathbf{v}) = (\mathbf{f}(t),\mathbf{v}) \tag{4.3}$$

for every test function $\mathbf{v}\in V$.

Equation (4.3) is supplemented with the initial condition

$$\mathbf{u}(0) = \mathbf{u}_0. \tag{4.4}$$

This is the weak formulation of the Navier–Stokes equations that goes back to the pioneering works of Leray [1933, 1934a,b]. It plays an essential role in the mathematical theory of those equations.

Energy Equation

In the derivation of the energy equation in Section 1, we assumed that the fluid fills the whole space and is at rest at infinity. An extension of this energy equation can be obtained for flows corresponding to the initial and boundary value problems ($\mathcal{P}_{\mathrm{nsp}}$), ($\mathcal{P}_{\mathrm{per}}$), ($\dot{\mathcal{P}}_{\mathrm{per}}$), and ($\mathcal{P}_{\mathrm{ch}}$). This can be obtained using the same idea as in Section 1: using the weak formulation (4.1) of the NSE, we replace $\mathbf{v}$ by $\mathbf{u}(t)$ itself. Then (see (1.5) and Appendix A.2)

$$b(\mathbf{u},\mathbf{u},\mathbf{u}) = \int_\Omega [(\mathbf{u}\cdot\nabla)\mathbf{u}]\cdot\mathbf{u}\,d\mathbf{x} = 0, \tag{4.5}$$

leading to

$$\frac{1}{2}\frac{d}{dt}\int_\Omega |\mathbf{u}(\mathbf{x},t)|^2\,d\mathbf{x} + \nu\sum_{i,j=1}^{d}\int_\Omega\left|\frac{\partial u_i}{\partial x_j}(\mathbf{x},t)\right|^2 d\mathbf{x} = \int_\Omega \mathbf{f}(\mathbf{x},t)\cdot\mathbf{u}(\mathbf{x},t)\,d\mathbf{x}. \tag{4.6}$$

Alternatively, for every vector field φ we set

$$|\varphi| = (\varphi,\varphi)^{1/2} \quad \text{and} \quad \|\varphi\| = ((\varphi,\varphi))^{1/2}, \tag{4.7}$$

so that

$$e(\varphi) = \tfrac{1}{2}|\varphi|^2 \quad \text{and} \quad E(\varphi) = \|\varphi\|^2.$$

We then rewrite (4.6) as

$$\frac{1}{2}\frac{d}{dt}|\mathbf{u}(t)|^2 + \nu\|\mathbf{u}(t)\|^2 = (\mathbf{f}(t),\mathbf{u}(t)) \tag{4.8}$$

or, equivalently, as

$$\frac{d}{dt}e(\mathbf{u}(t)) = -\nu E(\mathbf{u}(t)) + (\mathbf{f}(t),\mathbf{u}(t)). \tag{4.9}$$

The rate of change of the kinetic energy, $(d/dt)e(\mathbf{u}(t))$, is the difference between the power supplied by the external volume forces, $(\mathbf{f}(t),\mathbf{u}(t))$, and the energy dissipation rate by viscosity, $\nu E(\mathbf{u}(t))$. Note that, in reality, almost all of the energy dissipation takes place in the dissipation range; the inertial range is nearly dissipationless. The eddies in the inertial range "dissipate" or disappear principally by breaking up into smaller eddies.

5 Function Spaces

In this section we introduce most of the function spaces (i.e., various ensembles of suitable restricted functions) appropriate for use in mathematical treatments of the Navier–Stokes equations. Thus, in general, the term "(a given) function space" may be regarded as shorthand for a set of functions satisfying some prescribed restrictions (boundary conditions, integrability, boundedness, etc.). Many such restrictions reflect the physics of the case under consideration. We consider here only the no-slip and periodic boundary cases. The function spaces associated with the whole-space case are given in Section 10.

There are two fundamental spaces, denoted H and V, for each choice of boundary conditions. They are natural spaces that take into account the boundary conditions, the incompressibility condition, and the physical quantities $e(\mathbf{u})$ and $E(\mathbf{u})$ (resp., the kinetic energy and the enstrophy). The space H is the space of incompressible vector fields with finite kinetic energy and with the appropriate boundary conditions required by each initial and boundary value problem, and V is the space of incompressible vector fields with finite enstrophy and also with appropriate boundary conditions. We discuss each boundary condition in turn. In general, we consider a bounded domain Ω in $\mathbb{R}^d$ with $d = 2$ or 3, and the starting point is the space $L^2(\Omega)^d$ of square integrable vector fields from Ω into $\mathbb{R}^d$. As previously remarked, this is the space of finite kinetic energy vector fields. This space is endowed with the inner product

$$(\mathbf{u}, \mathbf{v}) = \int_\Omega \mathbf{u}(\mathbf{x}) \cdot \mathbf{v}(\mathbf{x})\, d\mathbf{x} \tag{5.1}$$

and the associated norm

$$|\mathbf{u}| = (\mathbf{u}, \mathbf{u})^{1/2} = \left\{ \int_\Omega |\mathbf{u}(\mathbf{x})|^2\, d\mathbf{x} \right\}^{1/2}. \tag{5.2}$$

Notice that this norm is associated with the energy per unit mass through the relation $|\mathbf{u}|^2 = 2e(\mathbf{u})$. The space $L^2(\Omega)^d$ endowed with this inner product and the associated norm is a Hilbert space.

Another important space, which is associated with the notion of enstrophy, is the Sobolev space $H^1(\Omega)^d$. It consists of the space of vector fields on Ω that are square integrable (finite kinetic energy) and whose gradient is square integrable (finite enstrophy) (see Section 4 in Chapter I). The associated inner product and norm are

$$((\mathbf{u}, \mathbf{v}))_1 = \frac{1}{L^2} \int_\Omega \mathbf{u}(\mathbf{x}) \cdot \mathbf{v}(\mathbf{x})\, d\mathbf{x} + \int_\Omega \sum_{i=1}^{d} \frac{\partial \mathbf{u}}{\partial x_i} \cdot \frac{\partial \mathbf{v}}{\partial x_i}\, d\mathbf{x} \tag{5.3}$$

and

$$\|\mathbf{u}\|_1 = ((\mathbf{u}, \mathbf{u}))_1^{1/2}, \tag{5.4}$$

where L is a typical length – for example, the diameter of Ω (see Section I.4, before (4.6)).

It is useful to distinguish the term related to the enstrophy in the inner product and in the norm in $H^1(\Omega)$. We define

$$((\mathbf{u}, \mathbf{v})) = \int_\Omega \sum_{i=1}^{d} \frac{\partial \mathbf{u}}{\partial x_i} \cdot \frac{\partial \mathbf{v}}{\partial x_i}\, d\mathbf{x} \tag{5.5}$$

and

$$\|\mathbf{u}\| = ((\mathbf{u}, \mathbf{u}))^{1/2} = \left\{ \int_\Omega \sum_{i=1}^{d} \left| \frac{\partial \mathbf{u}}{\partial x_i} \right|^2 d\mathbf{x} \right\}^{1/2}, \tag{5.6}$$

so that

$$((\mathbf{u}, \mathbf{v}))_1 = \frac{1}{L^2}(\mathbf{u}, \mathbf{v}) + ((\mathbf{u}, \mathbf{v}))$$

and

$$\|\mathbf{u}\|_1^2 = \frac{1}{L^2}|\mathbf{u}|^2 + \|\mathbf{u}\|^2.$$

The quantity $\|\cdot\|$ is exactly the square root of the enstrophy (see (1.3)).

A mathematically rigorous and physically intuitive definition of the spaces V and H is as follows:

> V is made up of all the limit points (in the distributional sense) of all the possible sequences of smooth vector fields $\mathbf{u}_m$ which are divergence-free, which satisfy the boundary conditions of the problem, and whose enstrophy remains bounded, that is, $E(\mathbf{u}_m) \leq \text{const.} < \infty$. (5.7)

The space H is defined in a similar way, replacing the boundedness of the enstrophy by the boundedness of the kinetic energy, $e(\mathbf{u}_m) \leq \text{const.} < \infty$. A precise description of these spaces in each case is given in the sequel.

The spaces H and V that we obtain for the problems $(\mathcal{P}_{\text{nsp}})$, $(\mathcal{P}_{\text{per}})$, and $(\dot{\mathcal{P}}_{\text{per}})$ are denoted respectively by H_{nsp} and V_{nsp}, H_{per} and V_{per}, and $\dot{H}_{\text{per}}$ and $\dot{V}_{\text{per}}$. We use those specific notations whenever necessary; otherwise, when the context is clear or the situation applies to any of the cases, we just consider H and V.

Loosely speaking, V is the subspace of $H^1(\Omega)^d$ consisting of vector fields that are divergence-free and satisfy the boundary conditions of the problem (and with space average on Ω equal to zero for $(\dot{\mathcal{P}}_{\text{per}})$). The space H included in $L^2(\Omega)^d$ is similar, but only the boundary conditions on the normal component of the velocity field at the boundary is retained; more specific details will be given shortly.

We now discuss in turn the spaces associated to each boundary condition.

No-Slip Boundary Conditions

In the no-slip case, the domain Ω is assumed to be bounded and to have a smooth boundary. More precisely, we assume that

> Ω is open, bounded, and connected, with a C^2 boundary $\partial\Omega$ and such that Ω is on only one side of $\partial\Omega$. (5.8)

By a C^2 boundary we mean that the boundary can be represented locally as the graph of a C^2 function (i.e., a twice differentiable function).

Based on (5.7), one can prove that[2]

$$H_{\text{nsp}} = \{\mathbf{u} \in L^2(\Omega)^d;\ \nabla \cdot \mathbf{u} = 0,\ \mathbf{u} \cdot \mathbf{n}|_{\partial\Omega} = 0\}. \tag{5.9}$$

Similarly,

$$V_{\text{nsp}} = \{\mathbf{u} \in H^1(\Omega)^d;\ \nabla \cdot \mathbf{u} = 0,\ \mathbf{u}|_{\partial\Omega} = 0\}. \tag{5.10}$$

We endow H_{nsp} with the norm $|\cdot|$ and the associated inner product $(\cdot,\cdot)$, inherited from $L^2(\Omega)^d$. For V_{nsp}, in view of the Poincaré inequality (see (4.11) and (4.12) in Chapter I), it suffices to take into account the part of the norm of $H^1(\Omega)^d$ (defined

[2] Note that the full no-slip boundary condition $\mathbf{u} = 0$ on $\partial\Omega$ is included in the definition of V, whereas for H the space contains only the condition $\mathbf{u} \cdot \mathbf{n} = 0$ on $\partial\Omega$. This is a mathematical technicality that we cannot control.

in (5.4)) that is connected to the enstrophy (see (5.6)). Hence, we endow V_{nsp} with the norm $\|\cdot\|$ and the associated inner product $((\cdot,\cdot))$. The Poincaré inequality in this context reads

$$|\mathbf{u}|^2 \leq \frac{1}{\lambda_1}\|\mathbf{u}\|^2 \quad \text{for all } \mathbf{u} \in V_{\text{nsp}}, \tag{5.11}$$

where λ_1 is defined to be the best (smallest) constant for which this inequality holds. This constant is actually the lowest eigenvalue of the corresponding Stokes equation (see Section 6).

In the mathematical literature, we usually define V_{nsp} and H_{nsp} as the closure in $H^1(\Omega)^d$ and in $L^2(\Omega)^d$ (respectively) of the space

$$\mathcal{V}_{\text{nsp}} = \{\mathbf{u} \in \mathcal{C}_c^\infty(\Omega)^d;\ \nabla \cdot \mathbf{u} = 0\}, \tag{5.12}$$

where $\mathcal{C}_c^\infty(\Omega)^d$ denotes the space of infinitely differentiable vector fields with compact support in Ω. The space $\mathcal{V}_{\text{nsp}}$ resembles the space of test functions in the theory of distributions (Schwartz [1950/51]). In fact, Leray [1933, 1934a,b] introduced it before the theory of distributions and the Sobolev spaces had even been developed. It is easily checked that this definition coincides with (5.7).

We do not discuss the case where Ω is only piecewise smooth (or even a Lipschitz domain, i.e., one whose boundary is locally the graph of a Lipschitz function). This occurs, for instance, with squares in dimension 2 or with wedges; in this case singularities occur in the corners, and handling them properly would necessitate lengthy developments obscuring the intrinsic turbulence issues. For the singularities occurring for flows in domains with corners, see for example Grisvard [1985, 1992], Kellogg and Osborn [1976], and Serre [1983].

Periodic Boundary Conditions

Recall that for periodic boundary conditions the domain Ω is of the form

$$\Omega = \prod_{i=1}^{d}\left(-\frac{L_i}{2}, \frac{L_i}{2}\right).$$

For the sake of the boundary conditions, we first define the spaces $L^2_{\text{per}}(\Omega)^d$ and $H^1_{\text{per}}(\Omega)^d$. They are the spaces of vector fields $\mathbf{u} = \mathbf{u}(\mathbf{x})$ which are defined for all $\mathbf{x} \in \mathbb{R}^d$, which are Ω-periodic in the sense that they are L_i-periodic in each direction $0x_i$ $(i = 1, \ldots, d)$, and which belong respectively to $L^2(\mathcal{O})^d$ and $H^1(\mathcal{O})^d$ for every bounded open set $\mathcal{O} \subset \mathbb{R}^d$.

Based on (5.7), one can prove that

$$V_{\text{per}} = \{\mathbf{v} \in H^1_{\text{per}}(\Omega)^d;\ \nabla \cdot \mathbf{u} = 0\} \tag{5.13}$$

and, similarly, that

$$H_{\text{per}} = \{\mathbf{u} \in L^2_{\text{per}}(\Omega)^d;\ \nabla \cdot \mathbf{u} = 0\}. \tag{5.14}$$

The Poincaré inequality (see Section I.4) is not valid in this case. Hence, we endow V_{per} with the full norm $\|\cdot\|_1$ and the corresponding inner product, using for L, say,

$L = \min\{L_1, L_2, L_3\}$. The space H_{per} is endowed with the norm $|\cdot|$ and the associated inner product $(\cdot,\cdot)$ inherited from $L^2(\Omega)^d$. In this case, we have

$$|\mathbf{u}|^2 \le L^2 \|\mathbf{u}\|_1^2 \quad \text{for all } \mathbf{u} \in V_{\text{per}}, \tag{5.15}$$

In the mathematical literature, we usually define V_{per} and H_{per} as the closure in $H^1(\Omega)^d$ and in $L^2(\Omega)^d$ (respectively) of the space

$$\mathcal{V}_{\text{per}} = \{\mathbf{u} \in C^\infty_{\text{per}}(\Omega)^d;\ \nabla \cdot \mathbf{u} = 0\}, \tag{5.16}$$

where $C^\infty_{\text{per}}(\Omega)^d$ is the space of Ω-periodic, C^∞ vector fields defined on $\mathbb{R}^d$.

A characterization of the spaces H_{per} and V_{per} in terms of Fourier series is given later in this section.

Periodic Boundary Conditions with Zero Space Average

In the case of zero space average, one can prove, in view of definition (5.7), that

$$\dot{V}_{\text{per}} = \left\{\mathbf{v} \in H^1_{\text{per}}(\Omega)^d;\ \int_\Omega \mathbf{u}(\mathbf{x})\, d\mathbf{x} = 0,\ \nabla \cdot \mathbf{u} = 0\right\} \tag{5.17}$$

and, similarly, that

$$\dot{H}_{\text{per}} = \left\{\mathbf{u} \in L^2_{\text{per}}(\Omega)^d;\ \int_\Omega \mathbf{u}(\mathbf{x})\, d\mathbf{x} = 0,\ \nabla \cdot \mathbf{u} = 0\right\}. \tag{5.18}$$

We endow $\dot{H}_{\text{per}}$ with the norm $|\cdot|$ and the associated inner product $(\cdot,\cdot)$ inherited from $L^2(\Omega)^d$. For $\dot{V}_{\text{per}}$ we recall that, for periodic functions with vanishing space average, the Poincaré inequality holds (see Section I.4). Hence, it suffices to endow $\dot{V}_{\text{per}}$ with the norm $\|\cdot\|$ and the associated inner product $((\cdot,\cdot))$. The Poincaré inequality in this context reads

$$|\mathbf{u}|^2 \le \frac{1}{\lambda_1} \|\mathbf{u}\|^2 \quad \text{for all } \mathbf{u} \in \dot{V}_{\text{per}}, \tag{5.19}$$

where λ_1 is defined to be the best (smallest) constant for which this inequality holds. As in the no-slip case, this constant is actually the lowest eigenvalue ($\lambda_1 \sim 1/L^2$) of the corresponding Stokes operator (see Section 6).

A characterization of those spaces in term of Fourier series will be given shortly.

In the mathematical literature, we usually define $\dot{V}_{\text{per}}$ and $\dot{H}_{\text{per}}$ as the closure in $H^1(\Omega)^d$ and in $L^2(\Omega)^d$ (respectively) of the space

$$\dot{\mathcal{V}}_{\text{per}} = \left\{\mathbf{u} \in C^\infty_{\text{per}}(\Omega)^d;\ \int_\Omega \mathbf{u}(\mathbf{x})\, d\mathbf{x} = 0,\ \nabla \cdot \mathbf{u} = 0\right\}. \tag{5.20}$$

Fourier Characterization of the Function Spaces for Periodic Flows

As noted earlier, one of the advantages attendant to space-periodic flows is the possibility of using Fourier series. We now show how the function spaces V and H can be characterized in terms of Fourier series. Then, we introduce further useful spaces.

Similar spaces for the no-slip case are introduced in Section 6, using more abstract entities (eigenfunctions of the Stokes operator).

Consider a bounded domain Ω in $\mathbb{R}^d$, with $d = 2$ or 3, of the form

$$\Omega = \left(-\frac{L_1}{2}, \frac{L_1}{2}\right) \times \left(-\frac{L_2}{2}, \frac{L_2}{2}\right) \times \left(-\frac{L_3}{2}, \frac{L_3}{2}\right)$$

in the 3-dimensional case and

$$\Omega = \left(-\frac{L_1}{2}, \frac{L_1}{2}\right) \times \left(-\frac{L_2}{2}, \frac{L_2}{2}\right)$$

in the 2-dimensional case. For each index $\mathbf{k} = (k_1, \ldots, k_d)$ in $\mathbb{Z}^d$ (d-dimensional vectors with unsigned integer components, hence with no physical dimension) we can associate wavenumbers $k_1/L_1, \ldots, k_d/L_d$. For notational simplicity, we set

$$\frac{\mathbf{k}}{\mathbf{L}} = \left(\frac{k_1}{L_1}, \ldots, \frac{k_d}{L_d}\right). \tag{5.21}$$

We work with complex representation, for which we take $i = \sqrt{-1}$. Then, a square integrable vector field $\mathbf{u} = \mathbf{u}(\mathbf{x})$ on Ω can be represented by the expansion

$$\mathbf{u}(x) = \sum_{\mathbf{k}\in\mathbb{Z}^d} \hat{\mathbf{u}}_{\mathbf{k}} e^{2\pi i \frac{\mathbf{k}}{\mathbf{L}}\cdot\mathbf{x}}, \tag{5.22}$$

where the amplitudes $\hat{\mathbf{u}}_{\mathbf{k}}$ of each set of frequencies $\mathbf{k}$ belong to $\mathbb{C}^d$. The convergence of this expansion is in the L^2 norm. Parseval's identity reads

$$|\mathbf{u}|^2 = |\Omega| \sum_{k\in\mathbb{Z}}^{d} |\hat{\mathbf{u}}_{\mathbf{k}}|^2, \tag{5.23}$$

where $|\Omega|$ is the volume of Ω in the 3-dimensional case (viz., $L_1L_2L_3$) or the area (L_1L_2) in the 2-dimensional case. Since the vector field $\mathbf{u} = \mathbf{u}(\mathbf{x})$ is real-valued, we have

$$\hat{\mathbf{u}}_{-\mathbf{k}} = \overline{\hat{\mathbf{u}}_{\mathbf{k}}}$$

for every $\mathbf{k}$, where $\overline{\hat{\mathbf{u}}_{\mathbf{k}}}$ denotes the complex conjugate of $\hat{\mathbf{u}}_{\mathbf{k}}$. In the Fourier space the divergence-free condition reads

$$\hat{\mathbf{u}}_{\mathbf{k}} \cdot \frac{\mathbf{k}}{\mathbf{L}} = 0 \quad \forall \mathbf{k}.$$

Therefore, the space H_{per} can be represented as

$$H_{\text{per}} = \left\{ \mathbf{u} = \sum_{\mathbf{k}\in\mathbb{Z}^d} \hat{\mathbf{u}}_{\mathbf{k}} e^{2\pi i \frac{\mathbf{k}}{\mathbf{L}}\cdot\mathbf{x}};\ \hat{\mathbf{u}}_{-\mathbf{k}} = \overline{\hat{\mathbf{u}}_{\mathbf{k}}},\ \frac{\mathbf{k}}{\mathbf{L}} \cdot \hat{\mathbf{u}}_{\mathbf{k}} = 0,\ \sum_{\mathbf{k}\in\mathbb{Z}^d} |\hat{\mathbf{u}}_{\mathbf{k}}|^2 < \infty \right\}. \tag{5.24}$$

If the vector field $\mathbf{u} = \mathbf{u}(\mathbf{x})$ has finite enstrophy or (in other words) belongs to $H^1_{\text{per}}(\Omega)^d$, then we can write its enstrophy in terms of its Fourier coefficients:

$$E(\mathbf{u}) = \|\mathbf{u}\|^2 = 2\pi|\Omega| \sum_{\mathbf{k}\in\mathbb{Z}^d} \left|\frac{\mathbf{k}}{\mathbf{L}}\right|^2 |\hat{\mathbf{u}}_{\mathbf{k}}|^2. \tag{5.25}$$

The H^1 norm becomes

$$\|\mathbf{u}\|_1^2 = \frac{1}{L^2}|\mathbf{u}|^2 + \|\mathbf{u}\|^2 = |\Omega| \sum_{\mathbf{k}\in\mathbb{Z}^d} \left(\frac{1}{L^2} + 2\pi \left| \frac{\mathbf{k}}{\mathbf{L}} \right|^2 \right) |\hat{\mathbf{u}}_{\mathbf{k}}|^2, \tag{5.26}$$

where $L = \min\{L_1, \ldots, L_d\}$. Therefore, we can rewrite V_{per} in the form

$$V_{\mathrm{per}} = \left\{ \mathbf{u} = \sum_{\mathbf{k}\in\mathbb{Z}^d} \hat{\mathbf{u}}_{\mathbf{k}} e^{2\pi i \frac{\mathbf{k}}{\mathbf{L}}\cdot\mathbf{x}}; \quad \hat{\mathbf{u}}_{-\mathbf{k}} = \overline{\hat{\mathbf{u}}_{\mathbf{k}}},\ \frac{\mathbf{k}}{\mathbf{L}}\cdot\hat{\mathbf{u}}_{\mathbf{k}} = 0,\ \sum_{\mathbf{k}\in\mathbb{Z}^d} \left(\frac{1}{L^2} + 2\pi \left| \frac{\mathbf{k}}{\mathbf{L}} \right|^2 \right) |\hat{\mathbf{u}}_{\mathbf{k}}|^2 < \infty \right\}. \tag{5.27}$$

For the vanishing space average case, we have the additional condition

$$\mathbf{u}_0 = \int_\Omega \mathbf{u}(\mathbf{x})\, d\mathbf{x} = 0. \tag{5.28}$$

Hence, we obtain the characterizations

$$\dot{H}_{\mathrm{per}} = \left\{ \mathbf{u} = \sum_{\mathbf{k}\in\mathbb{Z}^d\setminus\{0\}} \hat{\mathbf{u}}_{\mathbf{k}} e^{2\pi i \frac{\mathbf{k}}{\mathbf{L}}\cdot\mathbf{x}}; \quad \hat{\mathbf{u}}_{-\mathbf{k}} = \overline{\hat{\mathbf{u}}_{\mathbf{k}}},\ \frac{\mathbf{k}}{\mathbf{L}}\cdot\hat{\mathbf{u}}_{\mathbf{k}} = 0,\ \sum_{\mathbf{k}\in\mathbb{Z}^d\setminus\{0\}} |\hat{\mathbf{u}}_k|^2 < \infty \right\} \tag{5.29}$$

and

$$\dot{V}_{\mathrm{per}} = \left\{ \mathbf{u} = \sum_{\mathbf{k}\in\mathbb{Z}^d\setminus\{0\}} \hat{\mathbf{u}}_{\mathbf{k}} e^{2\pi i \frac{\mathbf{k}}{\mathbf{L}}\cdot\mathbf{x}}; \quad \hat{\mathbf{u}}_{-\mathbf{k}} = \overline{\hat{\mathbf{u}}_{\mathbf{k}}},\ \frac{\mathbf{k}}{\mathbf{L}}\cdot\hat{\mathbf{u}}_{\mathbf{k}} = 0,\ \sum_{\mathbf{k}\in\mathbb{Z}^d\setminus\{0\}} \left| \frac{\mathbf{k}}{\mathbf{L}} \right|^2 |\hat{\mathbf{u}}_{\mathbf{k}}|^2 < \infty \right\}. \tag{5.30}$$

There are many relevant spaces other than H and V that one can introduce, such as spaces between V and H, spaces included in V, and superspaces of H. For the sake of simplicity, we restrict ourselves to the space-periodic case with vanishing space average (see Section 6 for the analog in the no-slip case).

Indeed, for all $s \in \mathbb{R}$ we may consider the space

$$V_s = \left\{ \mathbf{u} = \sum_{\mathbf{k}\in\mathbb{Z}^d\setminus\{0\}} \hat{\mathbf{u}}_{\mathbf{k}} e^{2\pi i \frac{\mathbf{k}}{\mathbf{L}}\cdot\mathbf{x}}; \quad \hat{\mathbf{u}}_{-\mathbf{k}} = \overline{\hat{\mathbf{u}}_{\mathbf{k}}},\ \frac{\mathbf{k}}{\mathbf{L}}\cdot\hat{\mathbf{u}}_{\mathbf{k}} = 0,\ \sum_{\mathbf{k}\in\mathbb{Z}^d\setminus\{0\}} \left| \frac{\mathbf{k}}{\mathbf{L}} \right|^{2s} |\hat{\mathbf{u}}_{\mathbf{k}}|^2 < \infty \right\}. \tag{5.31}$$

Note that $V_{s_1} \subset V_{s_2}$ for $s_1 \geq s_2$, and that $V_1 = V$ $(= \dot{V}_{\mathrm{per}})$ and $V_0 = H$ $(= \dot{H}_{\mathrm{per}})$. Hence, $V \subset V_s \subset H$ for $0 \leq s \leq 1$, $V_s \subset V$ for $s \geq 1$, and $V_s \supset H$ for $s < 0$. It can be shown that V_s is a Hilbert space for the norm

$$\|u\|_{V_s} = \left(\sum_{\mathbf{k}\in\mathbb{Z}^d\setminus\{0\}} \left| \frac{\mathbf{k}}{\mathbf{L}} \right|^{2s} |\hat{\mathbf{u}}_{\mathbf{k}}|^2 \right)^{1/2}.$$

Of particular interest are the spaces V_2 and V_{-1}. The space V_2 is the domain of the Stokes operator $A \doteq -\Delta$ in H, as explained in the following section. The space V_{-1} is the dual space of V, usually denoted V' ($= \dot{V}'_{\text{per}}$, here); this is the space of linear continuous forms on V.[3] More generally, for all $s \geq 0$, V_{-s} is the dual of V_s.

It is natural at this point to introduce also the powers A^r of the Stokes operator in the periodic case. As we said, A is just the mapping

$$\mathbf{u} = \sum_{\mathbf{k}\in\mathbb{Z}^d\setminus\{0\}} \hat{\mathbf{u}}_{\mathbf{k}} e^{2\pi i \frac{\mathbf{k}}{\mathbf{L}}\cdot\mathbf{x}} \mapsto A\mathbf{u} = \sum_{\mathbf{k}\in\mathbb{Z}^d\setminus\{0\}} \left|\frac{\mathbf{k}}{\mathbf{L}}\right|^2 \hat{\mathbf{u}}_{\mathbf{k}} e^{2\pi i \frac{\mathbf{k}}{\mathbf{L}}\cdot\mathbf{x}}.$$

Similarly, we define A^r as the operator

$$\mathbf{u} = \sum_{\mathbf{k}\in\mathbb{Z}^d\setminus\{0\}} \hat{\mathbf{u}}_{\mathbf{k}} e^{2\pi i \frac{\mathbf{k}}{\mathbf{L}}\cdot\mathbf{x}} \mapsto A^r\mathbf{u} = \sum_{\mathbf{k}\in\mathbb{Z}^d\setminus\{0\}} \left|\frac{\mathbf{k}}{\mathbf{L}}\right|^{2r} \hat{\mathbf{u}}_{\mathbf{k}} e^{2\pi i \frac{\mathbf{k}}{\mathbf{L}}\cdot\mathbf{x}}.$$

It is straightforward to see that A^r maps V_{2s} continuously onto V_{2s-2r} ($s, r \in \mathbb{R}$). In particular, for $s \geq 0$, we have $A^{2s}V_{2s} = H$ and so $V_{2s} = D(A^s)$ is the domain of the (unbounded) operator A^s in H.

The similar spaces and concepts for the no-slip case will be introduced in the next section. In that case, the elements $\exp(2\pi i\mathbf{x}\cdot\mathbf{k}/\mathbf{L})$ are replaced by the eigenfunctions of the Stokes operator $A = -P_L\Delta$.

Space–Time Function Spaces

It is useful to regard a function $\mathbf{u} = \mathbf{u}(\mathbf{x}, t)$ as a time-dependent function $\mathbf{u} = \mathbf{u}(t)$ with values in one of the function spaces defined previously. There are several such function spaces that one may consider. For a function space X, we denote by $L^p(0, T; X)$ the space of functions from $[0, T]$ into X whose norm in X to the pth power is integrable over $[0, T]$. Its norm is denoted by

$$\|\mathbf{u}\|_{L^p(0,T;X)} = \left(\int_0^T |\mathbf{u}(t)|_X^p \, dt\right)^{1/p}. \tag{5.32}$$

We also have the space of essentially bounded functions, denoted by $L^\infty(0, T; X)$ and normed with

$$\|\mathbf{u}\|_{L^\infty(0,T;X)} = \operatorname*{sup.ess.}_{t\in[0,T]} |\mathbf{u}(t)|_X. \tag{5.33}$$

where sup.ess. denotes the essential supremum as in the case of real-valued functions. This means that $|\mathbf{u}(t)|_X$ is bounded by a constant for almost every t in $[0, T]$, and the smallest such constant is precisely the $L^\infty(0, T; X)$ norm of $\mathbf{u}$.

The space of functions that are continuous from $[0, T]$ into X is denoted by $\mathcal{C}([0, T]; X)$, and its norm is that of the maximum on $[0, T]$. The most common spaces are $L^2(0, T; X)$ and $\mathcal{C}([0, T]; X)$, with $X = H$, V, and $D(A)$.

[3] The less mathematically oriented reader may refer to Appendix A.1 in this chapter for basic details.

The duality between V_s and V_{-s}, presented earlier in this section, can be extended to time-dependent functions. The dual space to $L^p(0, T; V_s)$ is $L^{p'}(0, T; V_{-s})$ for $1 \leq p < \infty$ and $s \in \mathbb{R}$, where p' is the conjugate of p (i.e., $1/p + 1/p' = 1$).

6 The Stokes Operator

In this section, some functional settings described in Section 5 in the context of Fourier series in the space-periodic case are extended to the no-slip case. We return also to the space-periodic case to provide some additional information. Of necessity, the framework is more abstract than in Section 5.

The Stokes operator was introduced in Section 3. From the mathematical and physical points of view it is associated with the linear part of the Navier–Stokes equations and, as such, plays an important role in the study of the full, nonlinear equations. Here, we aim to give a short account of the main relevant properties of the Stokes operator.

The Stokes operator was formally defined by $A\mathbf{u} = -P_L\Delta\mathbf{u}$, where P_L is the Helmholtz–Leray projector and Δ is the Laplacian. For a rigorous definition we must also define the domain $D(A)$ of A, that is, the space of functions in H for which $A\mathbf{u}$ makes sense. One can show that

$$A\mathbf{u} = -P_L\Delta\mathbf{u} \quad \text{for } \mathbf{u} \in D(A) = V \cap H^2(\Omega)^d, \tag{6.1}$$

where $d = 2$ or 3 is the space dimension; Ω is bounded in $\mathbb{R}^d$ and, in the no-slip case, is smooth. In what follows, we consider separately each choice of boundary conditions.

The Stokes Operator in the No-Slip Case

In the no-slip case we assume that the boundary of Ω is smooth (in the sense of (5.8)). In this case, a difficult-to-prove regularity result asserts that

$$A\mathbf{u} = -P_L\Delta\mathbf{u} \quad \text{for } \mathbf{u} \in D(A) = V_{\text{nsp}} \cap H^2(\Omega)^d, \tag{6.2}$$

and A is one-to-one from $D(A)$ onto H.[4] Note that, since a vector field $\mathbf{u}$ in $D(A)$ belongs to $H^2(\Omega)^d$, the Laplacian $\Delta\mathbf{u}$ makes sense and is square integrable, so that the Helmholtz–Leray projector P_L can be applied to it to yield a vector field $A\mathbf{u}$ in H. It can be shown that, in fact, the Stokes operator maps $D(A)$ onto H. Hence, the inverse A^{-1} is well-defined and takes H onto $D(A)$. Because Ω is bounded and smooth, we have by Rellich's theorem (see e.g. (4.29) in Chapter I and Adams [1975]) that $H^2(\Omega)^d$ is compactly embedded into $L^2(\Omega)^d$. It follows also that $D(A)$ is compactly embedded into H (and V is also compactly embedded into H).

Integration by parts serves to verify that the Stokes operator is symmetric;

[4] See Constantin and Foias [1988], Temam [2001], or the original articles by Agmon, Douglis, and Nirenberg [1959, 1964], Cattabriga [1961], Ghidaglia [1984], and Solonnikov [1964].

$$(A\mathbf{u}, \mathbf{v}) = (\mathbf{u}, A\mathbf{v}) \quad \text{for all } \mathbf{u}, \mathbf{v} \text{ in } D(A).$$

It turns out that A^{-1} is also self-adjoint. From the elementary spectral theory of compact self-adjoint operators in a Hilbert space (see e.g. Courant and Hilbert [1953]), we can infer the existence of an orthonormal basis $\{\mathbf{w}_m\}_{m\in\mathbb{N}}$ in H and a sequence of real eigenvalues $\{\sigma_m\}_{m\in\mathbb{N}}$ accumulating at zero, so that

$$A^{-1}\mathbf{w}_m = \sigma_m \mathbf{w}_m, \quad m = 1, 2, \ldots.$$

Setting $\lambda_m = 1/\sigma_m$, we see that

$$A\mathbf{w}_m = \lambda_m \mathbf{w}_m, \quad m = 1, 2, \ldots.$$

Since A is a positive definite operator – that is, since

$$(A\mathbf{u}, \mathbf{u}) = \|\mathbf{u}\|^2 > 0 \quad \text{for all } \mathbf{u} \neq 0 \text{ in } D(A)$$

(which follows from integration by parts) – we see that each λ_m is positive. Moreover, we can order them such that

$$0 < \lambda_1 \leq \lambda_2 \leq \cdots \leq \lambda_m \leq \cdots, \quad \lambda_m \to +\infty \quad \text{as } m \to +\infty.$$

The first eigenvalue, λ_1, is exactly the best constant for the Poincaré inequality introduced in (5.11) (see Courant and Hilbert [1953]).

Because $\{\mathbf{w}_m\}_{m\in\mathbb{N}}$ is an orthonormal basis in H, we can expand each vector field $\mathbf{u}$ in terms of its projection onto each eigenspace:

$$\mathbf{u} = \sum_{m=1}^{\infty} (\mathbf{u}, \mathbf{w}_m)\mathbf{w}_m \quad \text{for } \mathbf{u} \text{ in } H. \tag{6.3}$$

For notational simplicity, we set $\hat{u}_m = (\mathbf{u}, \mathbf{w}_m)$, so that

$$\mathbf{u} = \sum_{m=1}^{\infty} \hat{u}_m \mathbf{w}_m \quad \text{for } \mathbf{u} \text{ in } H. \tag{6.4}$$

The following relation is known as the Parseval identity:

$$|\mathbf{u}|^2 = \sum_{m=1}^{\infty} |\hat{u}_m|^2. \tag{6.5}$$

Similarly,

$$\|\mathbf{u}\|^2 = \sum_{m=1}^{\infty} \lambda_m |\hat{u}_m|^2 \tag{6.6}$$

and

$$|A\mathbf{u}|^2 = \sum_{m=1}^{\infty} \lambda_m^2 |\hat{u}_m|^2. \tag{6.7}$$

More generally, we can define the spaces V_{2s}, for all $s \geq 0$, by setting

$$\mathbf{u} \in V_{2s} \iff \sum_{m=1}^{\infty} \lambda_m^{2s} |\hat{u}_m|^2 < \infty. \tag{6.8}$$

The powers A^s ($s \geq 0$) of A are defined by

$$A^s \mathbf{u} = \sum_{m=1}^{\infty} \lambda_m^s \hat{u}_m \mathbf{w}_m. \tag{6.9}$$

The domain of A^s in H is $D(A^s) = V_{2s}$. Moreover, A^r maps $D(A^s) = V_{2s}$ into $D(A^{s-r}) = V_{2(s-r)}$.

The domain $D(A^s) = V_{2s}$ is endowed with the inner product

$$(\mathbf{u}, \mathbf{v})_{D(A^s)} = (A^s \mathbf{u}, A^s \mathbf{v}) = \sum_{m=1}^{\infty} \lambda_m^{2s} \hat{u}_m \hat{v}_m, \tag{6.10}$$

where $\hat{v}_m = (\mathbf{v}, \mathbf{w}_m)$, and with the norm

$$|\mathbf{u}|_{D(A^s)} = |A^s \mathbf{u}| = \sum_{m=1}^{\infty} \lambda_m^{2s} |\hat{u}_m|^2. \tag{6.11}$$

Notice that, for $s = 1/2$, we recover the space V; that is,

$$\begin{aligned}
|\mathbf{u}|^2_{D(A^{1/2})} &= \sum_{m=1}^{\infty} \lambda_m |\hat{u}_m|^2 = \sum_{m=1}^{\infty} (\mathbf{u}, \lambda_m \mathbf{w}_m)(\mathbf{u}, \mathbf{w}_m) \\
&= \sum_{m=1}^{\infty} (\mathbf{u}, A\mathbf{w}_m)(\mathbf{u}, \mathbf{w}_m) = \sum_{m=1}^{\infty} (A\mathbf{u}, \mathbf{w}_m)(\mathbf{u}, \mathbf{w}_m) \\
&= (A\mathbf{u}, \mathbf{u}) = ((\mathbf{u}, \mathbf{u})) = \|\mathbf{u}\|^2.
\end{aligned}$$

Hence, $D(A^{1/2}) = V$.

We can also define negative powers of A. Indeed, for $s > 0$ we define $D(A^{-s})$ to be the completed space of H (it is a space larger than H) for the norm

$$|\mathbf{u}|_{D(A^{-s})} = \sum_{m=1}^{\infty} \lambda_m^{-2s} |(\mathbf{u}, \mathbf{w}_m)|^2. \tag{6.12}$$

We define also

$$A^{-s} \mathbf{u} = \sum_{m=1}^{\infty} \lambda_m^{-s} (\mathbf{u}, \mathbf{w}_m) \mathbf{w}_m \quad \text{for } \mathbf{u} \in D(A^{-s}). \tag{6.13}$$

One can show also that, for $s > 0$, the space $D(A^{-s})$ is identical with (isomorphic to) the dual space $D(A^s)'$ of $D(A^s)$ and that (see (A.5))

$$D(A^s) \subset H \subset D(A^{-s}).$$

When $s = 1/2$, we recover V' (i.e., $D(A^{-1/2}) = V'$).

From the expression (6.11) and using Hölder's inequality, it is straightforward to deduce the so-called interpolation inequalities, which relate the norms associated with three different powers of A:

$$|A^s \mathbf{u}| \leq |A^{s_1} \mathbf{u}|^{\theta} |A^{s_2} \mathbf{u}|^{1-\theta} \tag{6.14}$$

for any real $s_1 \leq s \leq s_2$, where θ is given by

$$s = s_1\theta + s_2(1-\theta). \tag{6.15}$$

Further regularity results for the Stokes operator can be obtained as long as the domain Ω is regular enough. More precisely, assume that the domain Ω is of class $\mathcal{C}^{m+2}$ (in the sense that the boundary is locally the graph of a $\mathcal{C}^{m+2}$ function), where $m \in \mathbb{N}$. Then, if $\mathbf{f}$ belongs to $H \cap H^m(\Omega)^d$, the solution $\mathbf{u}$ of the Stokes problem (see (2.17)) $A\mathbf{u} = \mathbf{f}$ belongs to $H \cap H^{m+2}(\Omega)$. Moreover, the corresponding pressure p belongs to $H^{m+1}(\Omega)^d$.

The Stokes Operator in the Space-Periodic Case with Vanishing Space Average

The case of periodic boundary conditions with vanishing space average is almost identical to the no-slip case. The only difference is the definition of the Stokes operator A, which now reads

$$A\mathbf{u} = -P_L\Delta\mathbf{u} = -\Delta\mathbf{u} \quad \text{for } \mathbf{u} \in D(A) = \dot{V}_{\text{per}} \cap H^2_{\text{per}}(\Omega)^d. \tag{6.16}$$

Then, as before, we have that A is a positive self-adjoint operator with compact inverse and possesses a sequence $\{\lambda_m\}_{m\in\mathbb{N}}$ of positive eigenvalues associated with an orthonormal basis $\{\mathbf{w}_m\}_{m\in\mathbb{N}}$. The positive and negative powers of A can be defined similarly, and we have the identifications $D(A^{1/2}) = \dot{V}_{\text{per}}$ and $D(A^{-1/2}) = \dot{V}'_{\text{per}}$.

In the periodic case, the eigenfunctions $\mathbf{w}_m$ can be found explicitly in view of the Fourier expansion (5.22). Given that $\hat{\mathbf{u}}_{-\mathbf{k}} = \overline{\hat{\mathbf{u}}_{\mathbf{k}}}$ and that $\hat{\mathbf{u}}_{\mathbf{k}} \cdot \mathbf{k} = 0$, one can rewrite that expansion in terms of the vector fields as

$$\mathbf{w}_{\mathbf{k}} = \mathbf{a}_{\mathbf{k}} e^{2\pi i \frac{\mathbf{k}}{\mathbf{L}}\cdot\mathbf{x}} + \bar{\mathbf{a}}_{\mathbf{k}} e^{-2\pi i \frac{\mathbf{k}}{\mathbf{L}}\cdot\mathbf{x}}, \tag{6.17}$$

where for each $\mathbf{k}$ the $\mathbf{a}_{\mathbf{k}}$ are $d-1$ independent vectors in $\mathbb{C}^d$ such that $\mathbf{a}_{\mathbf{k}} \cdot \mathbf{k} = 0$ and with $\mathbf{a}_{-\mathbf{k}} = \bar{\mathbf{a}}_{\mathbf{k}}$. Notice that, owing to the condition $\hat{\mathbf{u}}_{-\mathbf{k}} = \overline{\hat{\mathbf{u}}_{\mathbf{k}}}$, we need only consider "half" of the $\mathbf{k}$s in $\mathbb{Z}^d \setminus \{0\}$ in the expansion in terms of $\mathbf{w}_{\mathbf{k}}$.

The eigenvalues are

$$\lambda_{\mathbf{k}} = 4\pi^2 \left|\frac{\mathbf{k}}{\mathbf{L}}\right|^2. \tag{6.18}$$

They can be ordered in nondecreasing order so that, for each $\lambda_{\mathbf{k}}$ ($\mathbf{k} \in \mathbb{Z}^d \setminus \{0\}$), we have a corresponding eigenvalue λ_m for an appropriate $m \in \mathbb{N}$, with $\lambda_{m+1} \geq \lambda_m$; the corresponding eigenfunction is $\mathbf{w}_m = \mathbf{w}_{\mathbf{k}}$.

The Stokes Operator in the General Periodic Case

When the space average is not necessarily zero, the definition of the Stokes operator remains essentially the same:

$$A\mathbf{u} = -P_L\Delta\mathbf{u} = -\Delta\mathbf{u} \quad \text{for } \mathbf{u} \in D(A) = V_{\text{per}} \cap H^2_{\text{per}}(\Omega)^d. \tag{6.19}$$

The difference in this case is that the Stokes operator is no longer positive definite. Indeed, since nonzero constants belong to the domain $D(A)$ of A, it follows that A is no longer one-to-one, and it is not invertible. However, we can consider the operator $\tilde{A}$ defined by

$$\tilde{A}\mathbf{u} = \frac{1}{L^2}\mathbf{u} + A\mathbf{u} \quad \text{for } \mathbf{u} \in D(\tilde{A}) \equiv D(A), \tag{6.20}$$

where, as before, $L = \min\{L_1, \ldots, L_d\}$. On integration by parts it follows that

$$(\tilde{A}\mathbf{u}, \mathbf{v}) = \frac{1}{L^2}(\mathbf{u}, \mathbf{v}) + ((\mathbf{u}, \mathbf{v})) = ((\mathbf{u}, \mathbf{v}))_1$$

for all $\mathbf{u}$ in $D(A)$ and all $\mathbf{v}$ in V. Hence, $\tilde{A}$ is a positive self-adjoint operator with compact inverse. Therefore, $\tilde{A}$ possesses a sequence of positive eigenvalues $\{\tilde{\lambda}_m\}_{m\in\mathbb{N}}$ associated with an orthonormal basis $\{\mathbf{w}_m\}_{m\in\mathbb{N}}$. We can now recover the eigenvalues $\{\lambda_m\}_{m\in\mathbb{N}}$ of the Stokes operator A, which are related to those of $\tilde{A}$ by $\lambda_m = \tilde{\lambda}_m - 1/L^2$. Then,

$$0 = \lambda_1 \le \lambda_2 \le \cdots \le \lambda_m \le \cdots, \quad \lambda_m \to +\infty \text{ as } m \to +\infty.$$

The eigenvalues and eigenfunctions can also be given explicitly, as in the case of vanishing space average. They are actually the same, except that now we include the case $\mathbf{k} = 0$, which is associated with the eigenvalue $\lambda_1 = 0$ and with a d-dimensional eigenspace.

Alternative (Abstract) Definition of the Stokes Operator

An alternative definition of the Stokes operator can be given using the notion of duality introduced in Appendix A.1 of this chapter. Note that, for each vector field $\mathbf{u}$ in V, the map

$$\mathbf{v} \mapsto ((\mathbf{u}, \mathbf{v}))$$

defines a linear functional in V. It is continuous, since the Cauchy–Schwarz inequality implies that

$$((\mathbf{u}, \mathbf{v})) \le \|\mathbf{u}\|\|\mathbf{v}\| \le C_{\mathbf{u}}\|\mathbf{v}\|$$

for some constant $C_{\mathbf{u}}$ and for all $\mathbf{v}$ in V. Therefore, this linear functional belongs to V' and can be represented by an element ℓ in V'. Each $\mathbf{u}$ determines uniquely an element $\ell(\mathbf{u})$ in V' in this fashion. The map $\mathbf{u} \mapsto \ell(\mathbf{u})$ is linear, so we denote it by $A\mathbf{u}$. Clearly, by definition,

$$A\colon V \to V', \qquad (A\mathbf{u}, \mathbf{v}) = ((\mathbf{u}, \mathbf{v})) \quad \text{for all } \mathbf{u}, \mathbf{v} \text{ in } V. \tag{6.21}$$

For a vector field $\mathbf{u}$ smooth enough, it follows easily from integration by parts that, for all $\mathbf{v}$ in V,

$$(A\mathbf{u}, \mathbf{v}) = ((\mathbf{u}, \mathbf{v})) = -(\Delta\mathbf{u}, \mathbf{v}) = (-P_L\Delta\mathbf{u}, \mathbf{v}), \tag{6.22}$$

thus establishing the connection with the previous definition of the Stokes operator. The map A is actually an isomorphism between V and its dual V'. Note that, by

(6.22), $A = -P_L\Delta$, so that $-A$ *is not, in general, the Laplace operator, as it is in the space-periodic case.* By the Riesz representation theorem (see Appendix A.1), A is one-to-one from V onto V' with

$$\|A\mathbf{u}\|_{V'} = \|\mathbf{u}\| \qquad \text{for all } \mathbf{u} \text{ in } V. \tag{6.23}$$

The domain $D(A)$ of A in H in this context can be naturally defined as

$$D(A) = \{\mathbf{u} \in V;\ A\mathbf{u} \in H\}. \tag{6.24}$$

In the no-slip case, if the domain Ω is not regular enough (a case we do not otherwise consider) then one would not recover the characterization (6.2) for the domain of A, and the two definitions of the Stokes operator are not known to be identical.

Asymptotic Behavior of the Eigenvalues of the Stokes Operator

From the physical point of view, the quantities λ_m are the eigenvalues for linear (small or infinitesimal) self-oscillations of the fluid contained in Ω; of course, the corresponding $\mathbf{w}_m$ are the associated vector fields.

Concerning the behavior of the λ_m as m goes to infinity: it has been proved, interestingly enough, that the behavior is the same as that of the eigenvalues of the Laplace operator. See, for example, Courant and Hilbert [1953] for the classical results on the eigenvalues of the Laplace operator; the results concerning the eigenvalues of the Stokes operator are due to Métivier [1978] for the no-slip case.

In dimension $d = 2$ or 3, and in either the no-slip case or the periodic case with vanishing space average, the asymptotic behavior of the eigenvalues is given by

$$\lambda_m \sim \lambda_1 m^{2/d}. \tag{6.25}$$

More precisely, we know that the limit

$$\lim_{m\to\infty} \frac{\lambda_m}{\lambda_1 m^{2/d}} > 0 \tag{6.26}$$

exists and depends only on the dimension d and the shape of the domain Ω (see e.g. Métivier [1978] for details of the no-slip case on smooth domains, Ilyin [1996] for the no-slip case on nonsmooth domains, and Constantin and Foias [1988] for the space-periodic case).

Galerkin (Spectral) Projectors

Associated with the Stokes operator are the Galerkin, or spectral, projectors. They are important in the mathematical theory of the Navier–Stokes equations and will play a fundamental role in the proofs of most of the results in Chapters IV and V. They are also well known for their use in the numerical approximation of the NSE via spectral methods, especially in the periodic case. Here, we simply recall their definition and some of their properties.

For every positive integer m, the Galerkin projector P_m is defined as the orthogonal projector of H onto the space spanned by the first m eigenvectors $\mathbf{w}_1, \mathbf{w}_2, \dots, \mathbf{w}_m$ of the Stokes operator. More explicitly, since $\{\mathbf{w}_m\}_m$ is an orthonormal basis for H, any element $\mathbf{u}$ in H can be expressed as

$$\mathbf{u} = \sum_{m=1}^{\infty} \hat{u}_m \mathbf{w}_m. \tag{6.27}$$

The Galerkin projector P_m is obtained by truncating the highest modes:

$$P_m \mathbf{u} = \sum_{k=1}^{m} \hat{u}_k \mathbf{w}_k. \tag{6.28}$$

By definition, P_m commutes with A, and hence it commutes with every power of A. Therefore, P_m is also an orthogonal projector in each of the spaces $D(A^s)$ for s real and in particular in $V = D(A^{1/2})$ and $V' = D(A^{-1/2})$. The basis $\{\mathbf{w}_m\}_m$ is actually an orthogonal basis (but no longer orthonormal) in each space $D(A^s)$. The following inequalities are straightforward:

$$|\mathbf{u} - P_m \mathbf{u}|^2 \leq \frac{1}{\lambda_m} \|\mathbf{u} - P_m \mathbf{u}\|^2 \quad \text{for all } \mathbf{u} \text{ in } V, \tag{6.29}$$

$$\|P_m \mathbf{u}\|^2 \leq \lambda_m |\mathbf{u}|^2 \quad \text{for all } \mathbf{u} \text{ in } H. \tag{6.30}$$

The following inequality, which is needed in the general periodic case (nonvanishing space average), follows directly from (6.30):

$$\|P_m \mathbf{u}\|_1^2 \leq \left(\frac{1}{L^2} + \lambda_m \right) |\mathbf{u}|^2 \quad \text{for all } \mathbf{u} \text{ in } H. \tag{6.31}$$

7 Existence and Uniqueness of Solutions: The Main Results

In this section we describe the main existence and uniqueness results for the Navier–Stokes equations in dimensions 2 and 3. These results involve two types of solutions, weak and strong; they are related to the possible appearance of singularities in the sense that the magnitude of the vorticity vector may become infinite at some points in space and time. The strong solutions are those for which the vorticity vector (or its square, the enstrophy) is finite at all times, whereas for weak solutions the enstrophy may become infinite at some instants of time. The conjecture of Leray [1933, 1934a,b] concerning turbulence – as yet, neither proved nor disproved – is that the vorticity vector could indeed become infinite. However, we note in passing that no flow experiment performed to date has revealed such a singular behavior.

In the 2-dimensional case, the mathematical theory is fairly complete. The weak solutions turn out to be more regular and are, in fact, strong solutions. Moreover, the solutions are unique for a given initial condition and exist for all time.

In the 3-dimensional case, the mathematical theory is not yet complete. It is known that the weak solutions exist for all time, but it is not known whether they are unique.

On the other hand, strong solutions are unique and can be shown to exist on a certain finite time interval, but it is not known whether they exist for all time.

The results of existence, uniqueness, and regularity that we shall present here are classical ones. They can be found in many references on the mathematical theory of the Navier–Stokes equations (e.g., Constantin and Foias [1988], Ladyzhenskaya [1963], Lions [1969], Temam [1979, 1983]).[5]

The existence of solutions is generally proved by constructing approximate solutions[6] and passing to the limit as the approximation parameter tends to zero (in the Galerkin approximation, this limit is obtained as the number, say m, of modes in the Galerkin projector increases to infinity). For uniqueness we work, as usual, with the equation satisfied by the difference of two solutions satisfying the same initial and boundary conditions. Regularity results are obtained by means of techniques suitable for partial differential equations.

We recall the weak formulation of the Navier–Stokes equations as given in Section 4:

find a function $t \mapsto \mathbf{u}(t)$ taking its values in V and satisfying

$$\frac{d}{dt}(\mathbf{u}(t), \mathbf{v}) + \nu((\mathbf{u}(t), \mathbf{v})) + b(\mathbf{u}(t), \mathbf{u}(t), \mathbf{v}) = (\mathbf{f}(t), \mathbf{v}) \tag{7.1}$$

for every test function $\mathbf{v}$ in V, with $\mathbf{u}(0) = \mathbf{u}_0$.

The results presented next are valid for any of the initial boundary value problems $(\mathcal{P}_{\text{nsp}})$, $(\mathcal{P}_{\text{per}})$, $(\dot{\mathcal{P}}_{\text{per}})$, and $(\mathcal{P}_{\text{ch}})$ (see (2.13)). Therefore, we denote simply by H and V the corresponding spaces of finite kinetic energy and finite enstrophy.

Existence and Uniqueness in Dimension 3

In the 3-dimensional case, we have the following result concerning the existence of weak solutions.

Theorem 7.1 (Existence of Weak Solutions in Three Dimensions) *Assume that* $\mathbf{u}_0$, $\mathbf{f}$, *and* $T > 0$ *are given and satisfy*

$$\mathbf{u}_0 \in H, \qquad \mathbf{f} \in L^2(0, T; H). \tag{7.2}$$

[5] We refrain from proving these classical results here, as they can be found in the books already cited and in other references. In our text and in the appendices we give some indications of the proofs that we need in the rest of the book. References to the original articles can be found in the cited works – in particular, in Constantin and Foias [1988] and Temam [2001, Apx. III]. Briefly, the results presented here are primarily due to Leary [1933, 1934a,b], Hopf [1951], Lions and Prodi [1959], and Ladyzhenskaya [1967].

[6] Approximate solutions are typically obtained by using the Galerkin approximation. This is the equation obtained by applying the Galerkin projector (6.28) to the Navier–Stokes equations, which leads to a finite-dimensional equation with a bilinear nonlinearity. Approximate solutions can be obtained also by finite differences in time and space (see e.g. Temam [2001]).

Then there exists at least one solution $\mathbf{u} = (u_1, u_2, u_3)$ *of* (7.1) *such that*

$$u_i, \frac{\partial u_i}{\partial x_j} \in L^2(\Omega \times (0, T)), \quad i, j = 1, 2, 3, \tag{7.3}$$

and $\mathbf{u}$ *is weakly continuous from* $[0, T]$ *into* H *– that is, for every* $\mathbf{v} \in H$*, the function*

$$t \mapsto (\mathbf{u}(t), \mathbf{v}) = \int_{\Omega} \mathbf{u}(\mathbf{x}, t) \cdot \mathbf{v}(\mathbf{x})\, d\mathbf{x} \tag{7.4}$$

is continuous. Moreover, the following energy inequality *holds*:

$$\int_0^T \{-\tfrac{1}{2}|\mathbf{u}(t)|^2 \psi'(t) + \nu \|\mathbf{u}(t)\|^2 \psi(t)\}\, dt \\ \leq \tfrac{1}{2}|\mathbf{u}(0)|^2 \psi(0) + \int_0^T (\mathbf{f}(t), \mathbf{u}(t))\psi(t)\, dt \tag{7.5}$$

for all nonnegative real-valued C^1 *functions* ψ *on* $[0, T]$ *such that* $\psi(T) = 0$.

Recall that, in Section 1 (see (1.8)), we formally derived the energy equation

$$\frac{1}{2}\frac{d}{dt}|\mathbf{u}(t)|^2 + \nu \|\mathbf{u}(t)\|^2 = (\mathbf{f}(t), \mathbf{u}(t)). \tag{7.6}$$

However, it turns out that, for weak solutions in dimension 3, this energy equation is not necessarily true (more precisely, it is not known whether the equation is true or not). Again, we note that measurements in all flows of real fluids in three dimensions satisfy (7.6), in concert with the basic conservation laws of physics. We can only assert the existence of weak solutions satisfying the energy inequality (7.5). Note that (7.5) implies the following differential inequality (in the distribution sense) on $(0, T)$:

$$\frac{1}{2}\frac{d}{dt}|\mathbf{u}(t)|^2 + \nu \|\mathbf{u}(t)\|^2 \leq (\mathbf{f}(t), \mathbf{u}(t)). \tag{7.7}$$

Since the real-valued function $t \mapsto |\mathbf{u}(t)|^2$ is not known to be differentiable, we can only assert that (7.7) is valid in the distribution sense. In turn, (7.7) – together with a deep result in distribution theory (see Schwartz [1950/51]) – implies that $d(|\mathbf{u}(t)|^2)/dt$ is a bounded (signed) measure (of the form $\alpha\mu_1 - \beta\mu_2$, where μ_1, μ_2 are probability measures and $\alpha, \beta \geq 0$).

It may happen that weak solutions exists that do not satisfy (7.5), but their existence has not been proved and we *will not* consider such solutions: a weak solution for us means a solution with the properties given in Theorem 7.1, including the energy inequality. Other consequences of (7.5), besides (7.7), will be discussed in Appendices A and B (see (A.21), (A.40), and (B.4)).

We do not know whether the weak solutions given in Theorem 7.1 are unique. In fact, the issue of whether the equality in (7.6) holds for all weak solutions lies at the heart of the uniqueness problem. For strong solutions, we have the following result.

Theorem 7.2 (Local Existence and Uniqueness of Strong Solutions in Three Dimensions) *Assume that* $\mathbf{u}_0$, $\mathbf{f}$, *and* $T > 0$ *are given and satisfy*

$$\mathbf{u}_0 \in V, \qquad \mathbf{f} \in L^2(0, T; H). \tag{7.8}$$

Then there exists T_* $(0 < T_* \leq T)$, depending on the data (namely, Ω, ν, $\mathbf{f}$, $\mathbf{u}_0$, and T) such that on $[0, T_*)$ there exists a unique solution $\mathbf{u} = (u_1, u_2, u_3)$ of (7.1) *satisfying*

$$u_i, \frac{\partial u_i}{\partial t}, \frac{\partial u_i}{\partial x_j}, \frac{\partial^2 u_i}{\partial x_j \partial x_k} \in L^2(\Omega \times (0, T)), \quad i, j, k = 1, 2, 3, \tag{7.9}$$

and $\mathbf{u}$ is a continuous function from $[0, T_*)$ into V.

Moreover, the strong solutions are unique in the sense that there is no other strong solution in the sense of (7.8) *and* (7.9) *and no other weak solution on $[0, T^*)$ in the sense of Theorem 7.1.*

Later we present further properties of the solutions in dimension 3. They are more technical and will be needed for some of the proofs in Chapter IV. However, we first survey the main results in dimension 2.

Existence and Uniqueness in Dimension 2

In dimension 2, the weak solutions exist and are unique.

Theorem 7.3 (Existence and Uniqueness of Weak Solutions in Two Dimensions) *Assume that $\mathbf{u}_0$, $\mathbf{f}$, and $T > 0$ are given and satisfy*

$$\mathbf{u}_0 \in H, \qquad \mathbf{f} \in L^2(0, T; H). \tag{7.10}$$

Then there exists a unique solution $\mathbf{u} = (u_1, u_2)$ of (7.1) *such that*

$$u_i, \frac{\partial u_i}{\partial x_j} \in L^2(\Omega \times (0, T)), \quad i, j = 1, 2, \tag{7.11}$$

and $\mathbf{u}$ is continuous from $[0, T]$ into H. Moreover, the following energy equation holds on $[0, T]$:

$$\frac{1}{2}\frac{d}{dt}|\mathbf{u}(t)|^2 + \nu\|\mathbf{u}(t)\|^2 = (\mathbf{f}(t), \mathbf{u}(t)). \tag{7.12}$$

Hence, another difference with the 3-dimensional case is that here the weak solutions are continuous as functions with values in H (i.e., for the kinetic energy norm), and the energy equation is satisfied.

As for strong solutions, we have the following result.

Theorem 7.4 (Existence and Uniqueness of Strong Solutions in Two Dimensions) *Assume that $\mathbf{u}_0$, $\mathbf{f}$, and $T > 0$ are given and satisfy*

$$\mathbf{u}_0 \in V, \qquad \mathbf{f} \in L^2(0, T; H). \tag{7.13}$$

Then there exists a unique solution $\mathbf{u} = (u_1, u_2)$ of (7.1) *satisfying*

$$u_i, \frac{\partial u_i}{\partial t}, \frac{\partial u_i}{\partial x_j}, \frac{\partial^2 u_i}{\partial x_j \partial x_k} \in L^2(\Omega \times (0, T)), \quad i, j, k = 1, 2, \tag{7.14}$$

and $\mathbf{u}$ is a continuous function from $[0, T]$ into V.

As we said, Theorem 7.3 asserts that the weak solutions are continuous in H, that is,

$$\mathbf{u} \in \mathcal{C}([0,T]; H). \tag{7.15}$$

However, it is clear from (7.11) that the Navier–Stokes equations possess a regularization effect; namely, one starts with an initial condition $\mathbf{u}_0$ in H and obtains a weak solution that belongs to $L^2(0,T;V)$. This implies that almost everywhere in $(0,T]$ the solution $\mathbf{u}(t)$ at time t belongs to the space V. Further regularity can then be obtained from Theorem 7.4. Indeed, there exist t_0 arbitrarily close to 0 such that $\mathbf{u}(t_0)$ belongs to V. Hence, the solution is actually strong on $[t_0,T]$; hence, it is strong on $(0,T]$ and is continuous from $(0,T]$ into V, that is,

$$\mathbf{u} \in \mathcal{C}((0,T], V). \tag{7.16}$$

Furthermore, from (7.14), we see that

$$\mathbf{u} \in L^2(0,T; D(A)). \tag{7.17}$$

Further regularity of the solutions can be obtained provided the data are sufficiently regular. For instance, for a positive integer m, denote by H^m any one of the Sobolev spaces $H^m(\Omega)^d$, $H^m_{\text{per}}(\Omega)^d$, or $\dot{H}^m_{\text{per}}(\Omega)^d$ (depending on the boundary conditions) and assume, in the no-slip case, that Ω has a $\mathcal{C}^{m+2}$ boundary $\partial\Omega$ (i.e., locally the graph of a $\mathcal{C}^{m+2}$ function, with Ω only on one side of $\partial\Omega$). Then, for $m \geq 2$, if the initial data $\mathbf{u}_0$ belongs to $V \cap H^m$ and $\mathbf{f}$ is given in $L^2(0,T; H \cap H^{m-1})$, then the solution $\mathbf{u} = \mathbf{u}(t)$ belongs to $L^2(0,T; V \cap H^{m+1})$ and is continuous from $(0,T]$ into $H \cap H^m$.[7] One can also consider the case $m = \infty$, with H^m replaced by $\mathcal{C}^\infty$.

There is a large collection of theoretical results besides the ones we have just presented.[8] One may study solutions in L^p spaces, $p \neq 2$, or in Besov or other spaces. One may also study analytic solutions, either in space or time (or both), which are questions alluded to in Sections 8 and 9.

Remark 7.1 For the proof of Theorem 7.4 (and of (7.16) and (7.17)), we use an equation for conservation of enstrophy that is obtained by replacing $\mathbf{v}$ by $A\mathbf{u}$ in (7.1), instead of replacing $\mathbf{v}$ by $\mathbf{u}$ as in the energy equation; alternatively, we take the scalar product of the abstract form (A.22) of the Navier–Stokes equation (see Appendix A to this chapter) with $A\mathbf{u}$. The resulting enstrophy equation is (A.55). In the 2-dimensional space-periodic case, the orthogonality property (A.62) holds,

$$b(\mathbf{v},\mathbf{v},A\mathbf{v}) = 0 \quad \forall \mathbf{v} \in V,$$

and the enstrophy equation (A.55) becomes simpler and reduces to (A.65). Extensive use of the enstrophy equations (A.55) and (A.65) will be made in subsequent

[7] That the solutions are continuous from $(0,T]$ to $H \cap H^m$ and not from $[0,T]$ to $H \cap H^m$ relates to a difficult problem in the theory of evolution PDEs. The continuity holds from $[0,T]$ to $H \cap H^m$ if certain *compatibility conditions* in the data are satisfied. This problem is not specific to the Navier–Stokes equations; it appears for all initial boundary value problems. See a detailed account of this problem in general and for the NSE in Temam [1982].

[8] See a long (but still incomplete) list of such results, with corresponding references, in Temam [2000].

chapters. Note that in the no-slip case the orthogonality property (A.62) is no longer valid, although we do have

$$b(\mathbf{v}, \mathbf{v}, \Delta\mathbf{v}) = 0 \quad \forall \mathbf{v} \in V.$$

Observe that $A\mathbf{v}$ and $-\Delta\mathbf{v}$ are not equal in the no-slip case (whereas they are equal in the space-periodic case) and that $A\mathbf{v} + \Delta\mathbf{v}$ is a gradient, by definition of A (see (6.1) and (6.2)).

Remark 7.2 In Theorems 7.3 and 7.4, the solutions depend continuously on the data. In particular, for $T > 0$ fixed and finite, the mapping

$$(\nu, \mathbf{u}_0, \mathbf{f}) \mapsto \mathbf{u}$$

is continuous for $\nu \in (0, \infty)$ and for $\mathbf{u}_0$ and $\mathbf{f}$ in the function spaces mentioned in the corresponding theorems.

In the 2-dimensional case, thanks to the uniqueness of the solutions we may define the solution operator that takes the initial condition at a given time to the solution at some later time. Indeed, assuming that the forcing term $\mathbf{f}$ is given in $L^2_{\text{loc}}(0, \infty; H)$ (as defined in (A.14) in Appendix A), then for any initial condition $\mathbf{u}(t_0) = \mathbf{u}_0$ at a given time $t_0 \geq 0$, Theorem 7.3 implies the existence of a unique solution $\mathbf{u} = \mathbf{u}(t)$ that is defined for $t \geq t_0$. The solution operator is defined by

$$S(t; t_0)\mathbf{u}_0 = \mathbf{u}(t) \tag{7.18}$$

and is a continuous map from H into itself. In case the forcing term $\mathbf{f}$ is time-independent, it follows that $S(t; t_0) = S(t - t_0; 0)$, so it suffices to consider $t_0 = 0$. We may then denote the solution operator by

$$S(t)\mathbf{u}_0 = \mathbf{u}(t), \tag{7.19}$$

where $\mathbf{u}(t)$ is the unique solution of (7.1), for $t \geq 0$, such that $\mathbf{u}(0) = \mathbf{u}_0$.

Further Properties of the Solutions in Dimension 3

With respect to the possible occurrence of singularities, we can learn more by examining the structure of the weak solutions of the Navier–Stokes equations. Note first that inequality (7.5) implies in particular that, for all t and for almost all t_0 such that $0 \leq t_0 \leq t \leq T$, the following integral inequality holds:

$$\frac{1}{2}|\mathbf{u}(t)|^2 + \nu \int_{t_0}^{t} \|\mathbf{u}(s)\|^2 \, ds \leq \frac{1}{2}|\mathbf{u}(t_0)|^2 + \int_{t_0}^{t} (\mathbf{f}(s), \mathbf{u}(s)) \, ds; \tag{7.20}$$

see Appendix A.2 for details. The points t_0 for which this inequality holds are the so-called Lebesgue points[9] of the scalar function $t \mapsto |\mathbf{u}(t)|^2$, and those include

[9] We say that t_0 is a *Lebesgue point* of an integrable function $t \mapsto g(t)$ if

$$\frac{1}{h}\int_{t_0}^{t_0+h} g(t)\, dt \to g(t_0) \quad \text{as } h \to 0.$$

The set of points that are not Lebesgue points of g has measure zero.

$t_0 = 0$. There might be other solutions in the weak sense for which the energy inequality (7.5) is not satisfied; however, here, a weak solution with a given initial condition is to be understood as one for which all the properties stated in Theorem 7.1 hold.

Recall the weak continuity

$$\mathbf{u} \in \mathcal{C}([0, T]; H_w), \tag{7.21}$$

where now we have used the notation given in Section 5 (see (A.17)). From this weak continuity, we deduce that $t \to |\mathbf{u}(t)|$ is lower semi-continuous; that is, for every t in $[0, T]$,

$$|\mathbf{u}(t)| \leq \liminf_{s \to t} |\mathbf{u}(s)|^2. \tag{7.22}$$

In particular, we see that $\mathbf{u}(t)$ makes sense for every t in $[0, T]$. Now, from the energy inequality (7.20), we see that a weak solution $\mathbf{u} = \mathbf{u}(t)$ is continuous from the right at all Lebesgue points t_0 of $|\mathbf{u}(\cdot)|^2$; that is,

$$|\mathbf{u}(t)| \to |\mathbf{u}(t_0)| \quad \text{as } t \to t_0, \quad t \geq t_0. \tag{7.23}$$

It follows from (7.20) that[10]

$$\mathbf{u} \in L^\infty(0, T; H) \cap L^2(0, T; V). \tag{7.24}$$

Hence, the solution $\mathbf{u} = \mathbf{u}(t)$ belongs to V for almost every t. Then, recall that Theorem 7.2 assures the existence of a unique (local) strong solution starting with an initial condition in V (unique even among the weak solutions). This means that the weak solution $\mathbf{u} = \mathbf{u}(t)$ must become regular during "short" intervals of time throughout its existence. With that in mind, we consider the following sets:

$$\Sigma = \{t \in [0, T);\ \mathbf{u}(t) \in V\}, \tag{7.25a}$$

$$\Sigma^c = \{t \in [0, T);\ \mathbf{u}(t) \notin V\}, \tag{7.25b}$$

$$\sigma = \{t \in (0, T);\ \mathbf{u} \in \mathcal{C}((t - \tau, t + \tau); V) \text{ for some } \tau > 0\}. \tag{7.25c}$$

By definition, σ is clearly open; hence, it can be written as a countable union of disjoint open intervals – say,

$$\sigma = \bigcup_{j \in \mathbb{N}} I_j, \quad I_j = (\alpha_j, \beta_j). \tag{7.26}$$

By Theorem 7.8, for each $t \in \Sigma$, the solution is strong on some interval $[t, t + \tau)$. Therefore, any point in $\Sigma \setminus \sigma$ (i.e., in Σ but not in σ) is a left endpoint of one of the subintervals I_j. Thus, Σ is of the form

$$\Sigma = \bigcup_{j \in \mathbb{N}} [\alpha_j, \beta_j). \tag{7.27}$$

Moreover, $\Sigma \setminus \sigma$ is countable since it is the set $\{\alpha_j\}_j$. Thus, the Lebesgue measure of σ is the same as that of Σ. On the other hand, since the solution $\mathbf{u}$ belongs to

[10] The space $L^\infty(0, T; H)$ is defined in (A.12); see Appendix A.

$L^2(0, T; V)$, it belongs to V almost everywhere, so that the Lebesgue measure of Σ^c is zero. Thus, σ and Σ have full measure on $[0, T]$. The mappings $t \mapsto |\mathbf{u}(t)|^2$ and $t \mapsto \|\mathbf{u}(t)\|^2$ are continuous on σ (i.e., they are continuous at any point in σ); they are also continuous from the right at the points of $\Sigma \setminus \sigma$.

We can define for any t_1 and t_2, with $t_1 < t_2$, a map $S_V(t_2; t_1)$ on the set

$$D_{S_V(t_2;t_1)} = \{u_0 \in V;\ \text{there exists a strong solution } \mathbf{u}(t) \text{ on } [t_1, t_2] \text{ such that } \mathbf{u}(t_1) = \mathbf{u}_0\} \tag{7.28}$$

by setting

$$S_V(t_2; t_1)\mathbf{u}_0 = \mathbf{u}(t_2). \tag{7.29}$$

One can verify that, for each $t_2 > t_1$, the set $D(S_V(t_2; t_1))$ is an open subset of V. Moreover, $S_V(t_2; t_1)$ is a continuous map from $D_{S_V(t_2;t_1)}$ – endowed with the topology (norm) of V – into V.

In case the forcing term $\mathbf{f}$ is time-independent, it suffices to consider $t_1 = 0$; we set

$$S_V(t_2)\mathbf{u}_0 = \mathbf{u}(t_2) \tag{7.30}$$

on

$$D_{S_V(t_2)} = \{u_0 \in V;\ \text{there exists a strong solution } \mathbf{u}(t) \text{ on } [0, t_2] \text{ such that } \mathbf{u}(0) = \mathbf{u}_0\}. \tag{7.31}$$

Then one can prove the following:

If $u_0 \in V$ and

$$t_2 \le c\nu^3 \min\left\{\frac{\nu^4\lambda_1^2}{|f|^4}, \frac{1}{\|u_0\|^4}\right\}, \tag{7.32}$$

then $u_0 \in D_{S_V(t_2)}$.

Further results are given in Appendix A.2.

8 Analyticity in Time

In the previous section we considered weak and strong solutions of the Navier–Stokes equations and saw that their time derivative is integrable as a function in time, with values in some appropriate spaces. We now show that the strong solutions are actually analytic in time as a function with values in the domain $D(A)$ of the Stokes operator. This is obtained in both the 2- and 3-dimensional cases. Notice that, in the 2-dimensional case, this result applies also to the weak solutions because they are actually strong solutions. We follow the original work of Foias and Temam [1979]. We assume for simplicity that the forcing term $\mathbf{f}$ is time-independent, and we remark that $\mathbf{f}$ is only assumed to belong to H; the case where $\mathbf{f}$ depends on time and $\mathbf{f}$ belongs to $L^\infty(0, T; H)$ can be treated in exactly the same way, provided $\mathbf{f}$ is itself analytic in time in the region of analyticity of $\mathbf{u}$ that we derive in the proof.

A previous analyticity result is due to Masuda [1967], who assumed that $\mathbf{f}$ is time-dependent and analytic both in space and time and then showed that the solution is also analytic both in space and time. Analyticity in time of the solutions of the NSE has also been previously proven by Iooss [1969]. Analyticity in space will be considered in the next section, following the work of Foias and Temam [1989].

We turn our attention now to the time analyticity of the solutions. The result in Foias and Temam [1979] considers also the case of nonhomogeneous no-slip boundary conditions. For simplicity, however, here we consider only the homogeneous no-slip case, as well as the periodic case. We treat both the 2- and 3-dimensional cases at the same time, but the analyticity in the 2-dimensional case is further extended globally in time to an open neighborhood of the positive real time axis. We also present more explicit estimates in the 2-dimensional periodic case.

Time Analyticity in the 3-Dimensional Case

The idea is to start from the Galerkin approximation, for which analyticity in time is trivial because it is a finite-dimensional system with a polynomial nonlinearity. The crucial part, then, is to obtain suitable a priori estimates for the solution in a complex time region that is independent of the Galerkin approximation. Those estimates will allow us to pass to the limit and obtain a time-analytic solution for the Navier–Stokes equations. The passage to the limit is based on classical theorems concerning convergence of analytic functions (see e.g. Dunford and Schwartz [1958]). In contrast with the study of weak solutions, no compactness theorem is needed to pass to the limit for the Galerkin approximation in this case, because the theory of analytic functions provides for the uniform convergence of the functions as well as of their derivatives. Hence, the crucial point is the derivation of the a priori estimates in the complex plane. For the sake of simplicity, we will formally derive them directly for the solution of the Navier–Stokes equations.

In order to extend the solutions of the NSE to complex times, we complexify the spaces H, V, V', and $D(A)$. The elements of the complexified version of a real vector space are of the form $\mathbf{u} = \mathbf{u}_1 + i\mathbf{u}_2$, where $\mathbf{u}_1$ and $\mathbf{u}_2$ belong to the corresponding real space and $i = \sqrt{-1}$. The complexified spaces are denoted by $H_{\mathbb{C}}$, $V_{\mathbb{C}}$, $V'_{\mathbb{C}}$, and $D(A)_{\mathbb{C}}$. The inner product in the complexified space $H_{\mathbb{C}}$ takes the form

$$(\mathbf{u},\mathbf{v})_{\mathbb{C}} = (\mathbf{u}_1 + i\mathbf{u}_2, \mathbf{v}_1 + i\mathbf{v}_2) = (\mathbf{u}_1,\mathbf{v}_1) + (\mathbf{u}_2,\mathbf{v}_2) + i[(\mathbf{u}_2,\mathbf{v}_1) - (\mathbf{u}_1,\mathbf{v}_2)],$$

and similarly for $V_{\mathbb{C}}$ and the other spaces.

We must also extend the linear operator A and the bilinear operator B to the complexified spaces; they take the form

$$A_{\mathbb{C}}\mathbf{u} = A\mathbf{u}_1 + iA\mathbf{u}_2$$

and

$$B(\mathbf{u},\mathbf{v})_{\mathbb{C}} = B(\mathbf{u}_1,\mathbf{v}_1) + B(\mathbf{u}_2,\mathbf{v}_2) + i[B(\mathbf{u}_1,\mathbf{v}_2) + B(\mathbf{u}_2,\mathbf{v}_1)].$$

The operator $A_{\mathbb{C}}$ is a linear self-adjoint operator in $H_{\mathbb{C}}$, with $D(A_{\mathbb{C}}) = D(A)_{\mathbb{C}}$. The operator $B_{\mathbb{C}}\colon V_{\mathbb{C}} \times V_{\mathbb{C}} \to V'_{\mathbb{C}}$ is still a bilinear operator. The compexified trilinear operator is defined via

$$b(\mathbf{u}, \mathbf{v}, \mathbf{w})_{\mathbb{C}} = (B(\mathbf{u}, \mathbf{v})_{\mathbb{C}}, \mathbf{w})_{\mathbb{C}}.$$

Note that $b(\cdot, \cdot, \cdot)_{\mathbb{C}}$ is no longer antisymmetric in the last two variables; because of that, the important orthogonality property (4.5) (see also (A.33)) is lost in the complexified case.

The Navier–Stokes equations can be extended to complex times $\zeta \in \mathbb{C}$ as

$$\frac{d\mathbf{u}}{d\zeta} + \nu A_{\mathbb{C}}\mathbf{u} + B(\mathbf{u})_{\mathbb{C}} = \mathbf{f}, \tag{8.1}$$

where now we have $\mathbf{u} = \mathbf{u}(\zeta)$. Given $\mathbf{u}_0 \in V$ (or even H, for that matter), the Galerkin approximation of (8.1) has an analytic solution with $\mathbf{u}(0) = \mathbf{u}_0$ that is defined for ζ in a complex neighborhood of the origin. For this local analytic solution, the a priori estimates that we are about to derive formally can be obtained rigorously.

For notational simplicity, in the following computations we drop the subscript $\mathbb{C}$ for the inner products, norms, and operators just defined. Only for the functional spaces do we keep, for clarity, the subscript $\mathbb{C}$.

We fix $\theta \in (-\pi/2, \pi/2)$ and consider the time $\zeta = se^{i\theta}$ for $s > 0$. We want to compute the derivative

$$\begin{aligned}
\frac{1}{2}\frac{d}{ds}\|\mathbf{u}(se^{i\theta})\|^2 &= \frac{1}{2}\frac{d}{ds}(\mathbf{u}(se^{i\theta}), A\mathbf{u}(se^{i\theta})) \\
&= \frac{1}{2}\left(e^{i\theta}\frac{d\mathbf{u}}{d\zeta}, A\mathbf{u}\right) + \frac{1}{2}\left(\mathbf{u}, e^{i\theta}A\frac{d\mathbf{u}}{d\zeta}\right) \\
&= \operatorname{Re} e^{i\theta}\left(\frac{d\mathbf{u}}{d\zeta}, A\mathbf{u}\right),
\end{aligned}$$

where in the last step we used the fact that the real Stokes operator is self-adjoint and thus the complexified Stokes operator is hermitian. Then, using (8.1) we find

$$\frac{1}{2}\frac{d}{ds}\|\mathbf{u}(se^{i\theta})\|^2 + \nu\cos\theta|A\mathbf{u}(se^{i\theta})|^2 + \operatorname{Re} e^{i\theta} b(\mathbf{u}, \mathbf{u}, A\mathbf{u}) = \operatorname{Re} e^{i\theta}(\mathbf{f}, A\mathbf{u}). \tag{8.2}$$

From the definition of the complexified trilinear operator b, it is clear that each estimate for the real operator leads to a similar estimate in the complexified case, but with a larger multiplying constant. From the estimate (A.26b), for instance, the following estimate holds (can be proved) in the complexified case:

$$|b(\mathbf{u}, \mathbf{u}, A\mathbf{u})| \le c_1\|\mathbf{u}\|^{3/2}|A\mathbf{u}|^{3/2} \tag{8.3}$$

for a constant c_1, which may be larger than the one for the real case.[11] An application of Young's inequality to (8.3) yields

[11] Actually, we prove (8.3) and other similar inequalities by simply writing $\mathbf{u} = \mathbf{u}_1 + i\mathbf{u}_2$, expanding by linearity, and using the inequalities for real-valued vector fields.

$$|b(\mathbf{u}, \mathbf{u}, A\mathbf{u})| \leq \frac{\nu \cos\theta}{4}|A\mathbf{u}|^2 + \frac{c_2}{\nu^3 \cos^3\theta}\|\mathbf{u}\|^6 \tag{8.4}$$

for another constant c_2. For the forcing term, we use

$$|(\mathbf{f}, A\mathbf{u})| \leq \frac{\nu \cos\theta}{4}|A\mathbf{u}|^2 + \frac{1}{\nu \cos\theta}|\mathbf{f}|^2. \tag{8.5}$$

Hence, we deduce the inequality

$$\frac{d}{ds}\|\mathbf{u}(se^{i\theta})\|^2 + \nu \cos\theta |A\mathbf{u}(se^{i\theta})|^2 \leq \frac{2}{\nu\cos\theta}|\mathbf{f}|^2 + \frac{2c_2}{\nu^3\cos^3\theta}\|\mathbf{u}\|^6. \tag{8.6}$$

Note that on a time ray of fixed angle θ, the role of the viscosity is played by $\nu\cos\theta$, which decreases with $|\theta|$.

We show in Appendix B.2 how one can use (8.6) to derive a priori estimates on $\mathbf{u}$. Then, by using the Cauchy integral formula in complex analysis, we also derive a priori estimates on the time derivative of $\mathbf{u}$. We consider the open set

$$\begin{aligned} \Delta^0(\|\mathbf{u}_0\|) &= \Delta^0(\|\mathbf{u}_0\|, |\mathbf{f}|, \nu, \Omega) \\ &= \Bigg\{\zeta = se^{i\theta};\ |\theta| < \frac{\pi}{2},\ 0 < s\left(\frac{2}{\nu\cos\theta}|\mathbf{f}|^2 + \frac{2c_2}{\nu^3\cos^3\theta}\right) \\ &\qquad < \frac{1}{2(\nu^2\lambda_1^{1/2} + \|\mathbf{u}_0\|^2)^2}\Bigg\}. \end{aligned} \tag{8.7}$$

We show in Appendix B.2 that this set is a domain of analyticity of the solution $\mathbf{u} = \mathbf{u}(\zeta)$. The origin $\zeta = 0$ belongs to the closure of $\Delta^0(\|\mathbf{u}_0\|)$, and $\Delta^0(\|\mathbf{u}_0\|)$ is a neighborhood of the real time interval where the solution is strong. Moreover, inside this domain we have

$$\|\mathbf{u}(\zeta)\|^2 \leq \sqrt{2}(\nu^2\lambda_1^{1/2} + \|\mathbf{u}_0\|^2) \quad \text{for } \zeta \in \Delta^0(\|\mathbf{u}_0\|, |\mathbf{f}|, \nu, \Omega). \tag{8.8}$$

We may now obtain a priori estimates for the time derivative by using the Cauchy formula

$$\frac{d^k\mathbf{u}(\zeta)}{dt^k} = \frac{k!}{2\pi i}\int_\Gamma \frac{\mathbf{u}(z)}{(z-\zeta)^{k+1}}\,dz, \tag{8.9}$$

where Γ is a circle inside $\Delta^0(\|\mathbf{u}_0\|)$ and ζ is inside the circle. Whence,

$$\left\|\frac{d^k\mathbf{u}(\zeta)}{dt^k}\right\| = \frac{k!}{r_\Gamma^k}2^{1/4}\sqrt{\nu^2\lambda_1^{1/2} + \|\mathbf{u}_0\|^2}, \quad k = 0, 1, \ldots, \tag{8.10}$$

where r_Γ is the radius of the circle Γ. From the inequality (A.26b) in the real case, we deduce the following inequality in the complexified case:

$$|B(\mathbf{u})| \leq c_1\|\mathbf{u}\|^{3/2}|A\mathbf{u}|^{1/2}, \tag{8.11}$$

where c_1 is as in (8.3). Then – from the equation (8.1), the estimate (8.10), and the inequality (8.11) – one can see that

$$|A\mathbf{u}(\zeta)| \leq \frac{1}{d_K}C_2(\|\mathbf{u}_0\|, |\mathbf{f}|, \nu, \Omega) \quad \text{for } \zeta \in K, \tag{8.12}$$

where K is a compact subset of $\Delta^0(\|\mathbf{u}_0\|)$ and d_K is the distance between K and the boundary $\partial\Delta^0(\|\mathbf{u}_0\|)$ of the domain of analyticity $\Delta^0(\|\mathbf{u}_0\|)$.

Finally, applying the Stokes operator A to the Cauchy formula (8.9) and now using the estimate (8.12), we deduce the inequalities

$$\left|A\frac{d^k\mathbf{u}(\zeta)}{dt^k}\right| \leq \frac{k!}{d_K^{k+1}}C_2(\|\mathbf{u}_0\|, |\mathbf{f}|, \nu, \Omega) \quad \text{for } \zeta \in K,\ k = 0, 1, 2, \ldots, \tag{8.13}$$

for any compact subset K of $\Delta^0(\|\mathbf{u}_0\|)$.

As mentioned before, these estimates are obtained first for the Galerkin approximation of the solutions. Then, we pass to the limit to find that the strong solutions are analytic in time from $\Delta^0(\|\mathbf{u}_0\|)$ into $D(A)$ and satisfy the same estimates as their Galerkin approximation.

Global Analyticity in the 2-Dimensional Case

In the 2-dimensional case, we know from (A.57) that the solution $\mathbf{u} = \mathbf{u}(t)$ with initial condition $\mathbf{u}_0$ in H belongs to V for all $t > 0$, and it satisfies a uniform bound of the form

$$\|\mathbf{u}(t)\| \leq C(\varepsilon, |\mathbf{u}_0|, |\mathbf{f}|, \nu, \Omega), \quad t \geq \varepsilon > 0, \tag{8.14}$$

for all $\varepsilon > 0$. This dependence is indeed on the H norm of the initial condition $\mathbf{u}_0$. But the bound may blow up as ε decreases to zero, because the initial condition is only assumed to belong to H and hence may not belong to V.

We thus apply our previous argument starting at any time $t_0 > 0$, and we obtain the analyticity in the domain

$$t_0 + \Delta^0(C(t_0/2, |\mathbf{u}_0|, |\mathbf{f}|, \nu, \Omega)) \subset t_0 + \Delta^0(\|\mathbf{u}(t_0)\|).$$

By taking the union for all $t_0 > 0$ of the domains in the LHS of this inclusion, we obtain the analyticity in an open, pencil-like domain

$$\Delta^+(|\mathbf{u}_0|) = \Delta^+(|\mathbf{u}_0|, |\mathbf{f}|, \nu, \Omega) = \bigcup_{t_0>0}\{t_0 + \Delta^0(C(t_0/2, |\mathbf{u}_0|, |\mathbf{f}|, \nu, \Omega))\}; \tag{8.15}$$

this is a neighborhood of the positive real axis and has $\zeta = 0$ at its boundary. Moreover, our previous estimates extend to all of $\Delta^+(|\mathbf{u}_0|)$ in the sense that

$$\|\mathbf{u}(\zeta)\| \leq C_1(\varepsilon, |\mathbf{u}_0|, |\mathbf{f}|, \nu, \Omega) \quad \text{for } \zeta \in \Delta^+(|\mathbf{u}_0|, |\mathbf{f}|, \nu, \Omega),\ \mathrm{Re}\,\zeta \geq \varepsilon > 0, \tag{8.16}$$

for all $\varepsilon > 0$, and

$$\left|A\frac{d^k\mathbf{u}(\zeta)}{dt^k}\right| \leq \frac{k!}{d_K^{k+1}}C_2'(|\mathbf{u}_0|, |\mathbf{f}|, \nu, \Omega) \quad \text{for } \zeta \in K,\ k = 0, 1, 2, \ldots, \tag{8.17}$$

for any closed (not necessarily compact) set K in $\Delta^+(|\mathbf{u}_0|)$ with positive distance d_K from the boundary of $\Delta^+(|\mathbf{u}_0|)$.

If the initial condition $\mathbf{u}_0$ belongs to V, then the bound in (8.16) holds (with different constants) on the whole domain up to $\zeta = 0$. In this case, the construction of the domain $\Delta^+(\|\mathbf{u}_0\|)$ is different but similar to the one just described; it depends now on the enstrophy $\|\mathbf{u}_0\|$ of the initial condition $\mathbf{u}_0$. In this case, we have

$$\|\mathbf{u}(\zeta)\| \le C_1(\|\mathbf{u}_0\|, |\mathbf{f}|, \nu, \Omega) \quad \text{for } \zeta \in \Delta^+(\|\mathbf{u}_0\|, |\mathbf{f}|, \nu, \Omega) \tag{8.18}$$

and

$$\left|A\frac{d^k\mathbf{u}(\zeta)}{dt^k}\right| \le \frac{k!}{d_K^{k+1}} C_2'(\|\mathbf{u}_0\|, |\mathbf{f}|, \nu, \Omega) \quad \text{for } \zeta \in K,\ k = 0, 1, 2, \ldots, \tag{8.19}$$

for any closed (not necessarily compact) set K in $\Delta^+(\|\mathbf{u}_0\|)$ with positive distance d_K from the boundary of $\Delta^+(\|\mathbf{u}_0\|)$.

Improvements in the 2-Dimensional Periodic Case

In the 2-dimensional periodic case, the enstrophy of the solutions admits a simple, explicit bound (in terms of the data $\mathbf{f}$, ν, Ω) that is uniform in time. Indeed, for the solution with initial condition $\mathbf{u}_0$ in H, it follows from (A.66) that

$$\|\mathbf{u}(t)\|^2 \le \frac{2}{\nu^2\lambda_1}|\mathbf{f}|^2 \quad \text{for } t \ge T(|\mathbf{u}_0|, |\mathbf{f}|, \nu, \Omega). \tag{8.20}$$

With this estimate in time, one can obtain explicit bounds on the constants appearing in (8.18) and (8.19), as well as on the width (in the imaginary direction) of the domain of analyticity $\Delta^+(\|\mathbf{u}_0\|)$. The computations in this case are similar to those presented in the next section, where the time analyticity is obtained for the solution not with values just in $D(A)$ but with values in Gevrey-type spaces. Here, we restrict ourselves to presenting the final result.

One can check that the domain of analyticity can be taken as

$$\Delta^+(\|\mathbf{u}_0\|) = \Delta^+(\|\mathbf{u}_0\|, |\mathbf{f}|, \nu, \Omega) = \{\zeta \in \mathbb{C};\ |\operatorname{Im}\zeta| \le \min\{\operatorname{Re}\zeta, \delta_0\}\}, \tag{8.21}$$

where δ_0 is the (largest) width of the pencil-like domain Δ^+, estimated by

$$\delta_0 \ge \left[c_3\nu\lambda_1\left(1+\frac{|\mathbf{f}|^2}{\nu^4\lambda_1^2}\right)\log\left(c_4\left(1+\frac{|\mathbf{f}|^2}{\nu^4\lambda_1^2}\right)\right)\right]^{-1}, \tag{8.22}$$

where c_3 and c_4 depend only on the shape of the domain Ω. In the closure of this domain, the following bound holds:

$$\|\mathbf{u}(\zeta)\|^2 \le c_5\nu^2\lambda_1 + 2\|\mathbf{u}_0\|^2 + \frac{2}{\nu^2\lambda_1}|\mathbf{f}|^2 \quad \text{for } \zeta \in \overline{\Delta^+(\|\mathbf{u}_0\|, |\mathbf{f}|, \nu, \Omega)}, \tag{8.23}$$

where c_5 depends only on the shape of the domain Ω. Similar bounds for the time derivatives of the solutions can also be obtained from the Cauchy formula (8.9).

9 Gevrey Class Regularity and the Decay of the Fourier Coefficients

We have seen already that the Navier–Stokes equations in space dimensions $d = 2$ and 3 possess some regularization properties in the sense that, for positive times, the solutions become more regular than the initial condition (see e.g. Theorems 7.1, 7.2,

7.3, and 7.4). In this section we present a stronger regularization property: we show in the periodic case that, for an initial condition in the space V, the corresponding strong solution becomes analytic in both space and time. This is proven for two and three space dimensions with periodic boundary conditions and zero space average.[12] In the 3-dimensional case, this result is local in time. In the 2-dimensional case, this regularization is global in the sense that it extends to a neighborhood of the positive real time axis. The analyticity in space is obtained with the help of the so-called Gevrey spaces, and it is actually shown that the solutions are analytic in time as functions with values in the Gevrey class of analytic functions in space. For the analyticity in space, it is necessary to assume that the forcing term **f** is also analytic, and we assume (for simplicity) that **f** is time-independent; the case where **f** depends on time can be treated in exactly the same way, provided **f** is itself analytic in time in the region of analyticity of **u** that we derive in the proof. We follow here the work of Foias and Temam [1989]. See the introduction of Section 8 for further remarks on the analyticity of the solutions in time and in space.

After establishing the space analyticity of the solutions in the 2-dimensional periodic case, we derive, as a consequence, the exponential decay of the Fourier coefficients with respect to their Fourier mode.

Gevrey Spaces

In Section 6 we defined, for each $s \in \mathbb{R}$, the domain $D(A^s)$ of the power A^s of the Stokes operator A. Similarly, for each $\sigma, s > 0$, the Gevrey space $D(\exp(\sigma A^s))$ is defined as the domain of the exponential of σA^s. We can give a precise characterization of this space by means of Fourier series as follows. In Section 5, we saw that a vector field $\mathbf{u} \in H = \dot{H}_{\text{per}}$ is characterized in terms of Fourier series as a function

$$\mathbf{u} = \sum_{\mathbf{k}\in\mathbb{Z}^d} \hat{\mathbf{u}}_{\mathbf{k}} e^{2\pi i \frac{\mathbf{k}}{\mathbf{L}}\cdot\mathbf{x}}, \quad \mathbf{u}_{\mathbf{k}} \in \mathbb{C}^d,\ \hat{\mathbf{u}}_{-\mathbf{k}} = \overline{\hat{\mathbf{u}}_{\mathbf{k}}}, \tag{9.1}$$

such that

$$\frac{\mathbf{k}}{\mathbf{L}} \cdot \hat{\mathbf{u}}_{\mathbf{k}} = 0 \quad \text{for all } \mathbf{k} \in \mathbb{Z}^d \tag{9.2}$$

$$|\Omega| \sum_{\mathbf{k}\in\mathbb{Z}^d} |\hat{\mathbf{u}}_{\mathbf{k}}|^2 = |\mathbf{u}|^2 < \infty. \tag{9.3}$$

For the Gevrey spaces, we can define the operator $\exp(\sigma A^s)$ in Fourier space by

$$\exp(\sigma A^s)\mathbf{u} = \sum_{\mathbf{k}\in\mathbb{Z}^d} \exp\left(\sigma\left(2\pi\frac{\mathbf{k}}{\mathbf{L}}\right)^{2s}\right)\hat{\mathbf{u}}_{\mathbf{k}} \exp\left(2\pi i \frac{\mathbf{k}}{\mathbf{L}} \cdot \mathbf{x}\right), \tag{9.4}$$

where the power $(\mathbf{k}/\mathbf{L})^{2s}$ is to be taken componentwise. The domain $D(\exp(\sigma A^s))$ is defined as usual by

[12] So far, the method used in the space-periodic case does not extend to other boundary conditions – such as, for example, the no-slip case.

$$D(e^{\sigma A^s}) = \{\mathbf{u} \in H;\, e^{\sigma A^s}\mathbf{u} \in H\}. \tag{9.5}$$

Therefore, a vector field $\mathbf{u} \in D(\exp(\sigma A^s))$ can be characterized in terms of Fourier series representation (9.1) by the divergence-free condition (9.2) and by the condition that the Fourier coefficients decay exponentially fast in the sense that

$$|\Omega| \sum_{\mathbf{k}\in\mathbb{Z}^d} e^{2\sigma\left|2\pi\frac{\mathbf{k}}{\mathrm{L}}\right|^{2s}} |\hat{\mathbf{u}}_{\mathbf{k}}|^2 = |e^{\sigma A^s}\mathbf{u}|^2 < \infty. \tag{9.6}$$

The norm in the space $D(\exp(\sigma A^s))$ is given by

$$|\mathbf{u}|_{D(e^{\sigma A^s})} = |e^{\sigma A^s}\mathbf{u}| \quad \text{for } \mathbf{u} \in D(e^{\sigma A^s}). \tag{9.7}$$

The space $D(\exp(\sigma A^s))$ is actually a Hilbert space, and the associated inner product is given by

$$(\mathbf{u},\mathbf{v})_{D(e^{\sigma A^s})} = (e^{\sigma A^s}\mathbf{u},\, e^{\sigma A^s}\mathbf{v}) \quad \text{for } \mathbf{u},\mathbf{v} \in D(e^{\sigma A^s}). \tag{9.8}$$

In what follows, we will be mostly concerned with the case $s = 1/2$. Another Gevrey-type space that we will consider is $D(A^{1/2}\exp(\sigma A^{1/2}))$, which is also a Hilbert space; its inner product is given by

$$(\mathbf{u},\mathbf{v})_{D(A^{1/2}e^{\sigma A^{1/2}})} = (A^{1/2}e^{\sigma A^{1/2}}\mathbf{u},\, A^{1/2}e^{\sigma A^{1/2}}\mathbf{v}) = ((e^{\sigma A^{1/2}}\mathbf{u},\, e^{\sigma A^{1/2}}\mathbf{v})) \tag{9.9}$$

for $\mathbf{u},\mathbf{v} \in D(A^{1/2}\exp(\sigma A^{1/2}))$; the associated norm is given by

$$\begin{aligned} |\mathbf{u}|^2_{D(A^{1/2}e^{\sigma A^{1/2}})} &= |A^{1/2}e^{\sigma A^{1/2}}\mathbf{u}|^2 = \|e^{\sigma A^s}\mathbf{u}\|^2 \\ &= 2\pi|\Omega| \sum_{\mathbf{k}\in\mathbb{Z}^d} \left|\frac{\mathbf{k}}{\mathbf{L}}\right|^2 e^{4\pi\sigma\left|\frac{\mathbf{k}}{\mathrm{L}}\right|} |\hat{\mathbf{u}}_{\mathbf{k}}|^2 \end{aligned} \tag{9.10}$$

for $\mathbf{u} \in D(A^{1/2}\exp(\sigma A^{1/2}))$.

For notational simplicity, in this section we set

$$\begin{aligned} |\cdot|_\sigma &= |\cdot|_{D(e^{\sigma A^{1/2}})}, & (\cdot,\cdot)_\sigma &= (\cdot,\cdot)_{D(e^{\sigma A^{1/2}})}, \\ \|\cdot\|_\sigma &= |\cdot|_{D(A^{1/2}e^{\sigma A^{1/2}})}, & (\cdot,\cdot)_\sigma &= (\cdot,\cdot)_{D(A^{1/2}e^{\sigma A^{1/2}})}. \end{aligned} \tag{9.11}$$

Estimates for the Nonlinear Term in the Periodic Case

In the 3- or 2-dimensional periodic cases, each estimate for the nonlinear inertial term in the spaces H, V, and $D(A)$ can be properly generalized to Gevrey-type spaces. We present here two estimates that will be used in the sequel; their proofs will be given in Appendix B.3.

In the 3-dimensional case we know that, for every $\mathbf{u},\mathbf{v},\mathbf{w}$ in $D(A\exp(\sigma A^{1/2})$ with $\sigma > 0$, the bilinear term $B(\mathbf{u},\mathbf{v})$ satisfies the inequality

$$\begin{aligned} &|(e^{\sigma A^{1/2}}B(\mathbf{u},\mathbf{v}),\, e^{\sigma A^{1/2}}A\mathbf{w})| \\ &\quad \le c_1|A^{1/2}e^{\sigma A^{1/2}}\mathbf{u}|^{1/2}|Ae^{\sigma A^{1/2}}\mathbf{u}|^{1/2}|A^{1/2}e^{\sigma A^{1/2}}\mathbf{v}||Ae^{\sigma A^{1/2}}\mathbf{w}|^{1/2}, \end{aligned} \tag{9.12}$$

where c_1 depends only on the shape of the domain Ω (i.e., on the ratios of the periods). The inequality (9.12) implies that the bilinear term $B(\mathbf{u},\mathbf{v})$ belongs to $D(\exp(\sigma A^{1/2})$ in this case.

The inequality (9.12) is also valid in the 2-dimensional case. However, in order to obtain a better estimate for the width of the region of time analyticity, we use the inequality that derives from inequality (A.50) of Brézis and Gallouet [1980] (see (A.51a)). For $\mathbf{u},\mathbf{v},\mathbf{w}$ in $D(A\exp(\sigma A^{1/2})$ with $\sigma > 0$, the bilinear term $B(\mathbf{u},\mathbf{v})$ satisfies the inequality

$$|(e^{\sigma A^{1/2}}B(\mathbf{u},\mathbf{v}), e^{\sigma A^{1/2}}A\mathbf{w})|$$
$$\leq c_2|A^{1/2}e^{\sigma A^{1/2}}\mathbf{u}||A^{1/2}e^{\sigma A^{1/2}}\mathbf{v}||Ae^{\sigma A^{1/2}}\mathbf{w}|\left(1+\log\frac{|Ae^{\sigma A^{1/2}}\mathbf{u}|^2}{\lambda_1|A^{1/2}e^{\sigma A^{1/2}}\mathbf{u}|^2}\right)^{1/2}, \tag{9.13}$$

where c_2 depends only on the shape of the domain Ω.

Analyticity in the 3-Dimensional Periodic Case

As in the time analyticity discussed in Section 8, the idea is to start from the Galerkin approximation (for which analyticity in time is trivial, since it is a finite-dimensional system with a polynomial nonlinearity) and then obtain suitable a priori estimates for the solution in a complex time region that is independent of the Galerkin approximation. Those estimates will allow us to pass to the limit and obtain a time-analytic solution for the Navier–Stokes equations. The passage to the limit is based on classical theorems concerning convergence of analytic functions (see e.g. Dunford and Schwartz [1958]). For the sake of simplicity, however, we will formally derive the a priori estimates directly for the solution of the Navier–Stokes equations. We complexify the spaces and the operators as in Section 8, but we keep the same notation as that used for the real case.

Because we now want to establish the analyticity in time of the solutions as functions with values in Gevrey spaces, we must assume that the forcing term $\mathbf{f}$ itself belongs to a Gevrey space. Hence, we assume that

$$\mathbf{f} \in D(e^{\sigma_1 A^{1/2}}) \tag{9.14}$$

for some $\sigma_1 > 0$.

The NSE can be written for complex times $\zeta \in \mathbb{C}$ as

$$\frac{d\mathbf{u}}{d\zeta} + \nu A\mathbf{u} + B(\mathbf{u}) = \mathbf{f}, \tag{9.15}$$

where $\mathbf{u} = \mathbf{u}(\zeta)$. Given $\mathbf{u}_0 \in V$ (or H), the Galerkin approximation of (9.15) has an analytic solution with $\mathbf{u}(0) = \mathbf{u}_0$ that is defined for ζ in a complex neighborhood of the origin. For this local analytic solution, the a priori estimates that we shall derive formally can be rigorously justified.

We fix $\theta \in (-\pi/2, \pi/2)$ and consider the time $\zeta = se^{i\theta}$ for $s > 0$. Later, we will restrict θ to the interval $(-\pi/4, \pi/4)$. At the initial time $\zeta = 0$, the solution $\mathbf{u}(0) = \mathbf{u}_0$ belongs only to V but not necessarily to any Gevrey-type space. We will show that, as the real part of the complexified time increases, the solution becomes Gevrey-regular and the width of the space analyticity (measured by the parameter σ) also increases. To account for this increase in the width of analyticity, we consider the parameter σ as a function $\sigma = \varphi(s\cos\theta)$ of the real part of the complexified time. We will show that, for an appropriate function $\varphi = \varphi(s\cos\theta)$, we have the regularization

$$\mathbf{u}(se^{i\theta}) \in D(A^{1/2}e^{\varphi(s\cos\theta)A^{1/2}})$$

for s positive and small enough. For simplicity we assume, in view of the assumption (9.14), that $\varphi \leq \sigma_1$ and, moreover, that the bound $\varphi' \leq \nu\lambda_1^{1/2}$ holds.[13]

We show in Appendix B.4 that, as long as

$$\theta \in \left[-\frac{\pi}{4}, \frac{\pi}{4}\right], \quad 0 \leq s \leq T_0(\|\mathbf{u}_0\|), \tag{9.16}$$

where

$$T_0(\|\mathbf{u}_0\|) = T_0(\|\mathbf{u}_0\|, |\mathbf{f}|_{\sigma_1}, \nu, \Omega) = \frac{3\nu^3}{8a(\nu^2\lambda_1^{1/2} + \|\mathbf{u}_0\|^2)^2} \tag{9.17}$$

and

$$a = 1 + 4\sqrt{2}c_1 + \frac{3\sqrt{2}}{\nu^4\lambda_1^{3/2}}|\mathbf{f}|_{\sigma_1}^2, \tag{9.18}$$

we obtain

$$\|\mathbf{u}(se^{i\theta})\|_{\varphi(s\cos\theta)}^2 \leq \nu^2\lambda_1^{1/2} + 2\|\mathbf{u}_0\|^2. \tag{9.19}$$

In view of the assumptions $\varphi' \leq \nu\lambda_1^{1/2}$ and $\varphi \leq \sigma_1$, we can consider $\varphi(\xi) = \min\{\nu\lambda_1^{1/2}\xi, \sigma_1\}$ for all $\xi \geq 0$. This choice of φ is not continuously differentiable but can be approximated by continuously differentiable functions (satisfying the required assumptions), so that (9.19) holds for this choice of φ. Then we define the region

$$\begin{aligned}\Delta^0_{\sigma_1}(\|\mathbf{u}_0\|) &= \Delta^0_{\sigma_1}(\|\mathbf{u}_0\|, |\mathbf{f}|_{\sigma_1}, \nu, \Omega)\\ &= \Big\{\zeta = se^{i\theta};\ |\theta| < \frac{\pi}{4},\\ &\qquad 0 < s < T_0(\|\mathbf{u}_0\|, |\mathbf{f}|_{\sigma_1}, \nu, \Omega),\ \nu\lambda_1^{1/2}s|\sin\theta| < \sigma_1\Big\}.\end{aligned} \tag{9.20}$$

This set is a domain of analyticity of the solution $\mathbf{u} = \mathbf{u}(\zeta)$ of the complexified Navier–Stokes equations. The origin $\zeta = 0$ belongs to the closure of $\Delta^0_{\sigma_1}(\|\mathbf{u}_0\|)$, and $\Delta^0_{\sigma_1}(\|\mathbf{u}_0\|)$ is a neighborhood of the real time interval where the solution of the (real) NSE is strong. Moreover, for the closure of this domain we have

$$|\mathbf{u}(\zeta)|^2_{D(A^{1/2}e^{\varphi(s\cos\theta)A^{1/2}})} \leq \nu^2\lambda_1^{1/2} + 2\|\mathbf{u}_0\|^2 \quad \text{for } \zeta \in \overline{\Delta^0_{\sigma_1}}(\|\mathbf{u}_0\|, |\mathbf{f}|_{\sigma_1}, \nu, \Omega). \tag{9.21}$$

As mentioned before, these estimates are obtained first for the Galerkin approximation of the solutions. Then, we pass to the limit to find that the strong solutions are

[13] Note that for the exponential term $\varphi A^{1/2}$ to be nondimensional, we expect φ to have the dimension of length; since $s\cos\theta$ has the dimension of time, φ' is expected to have the dimension of length over time, which is the dimension of $\nu\lambda_1^{1/2}$.

analytic in time from $\Delta^0_{\sigma_1}(\|\mathbf{u}_0\|)$ into $D(A^{1/2}\exp(\varphi(s\cos\theta)A^{1/2}))$, where $\varphi(\xi) = \min\{\nu\lambda_1^{1/2}\xi, \sigma_1\}$, and that the strong solutions satisfy the same estimates as their Galerkin approximation.

In summary, we have shown that if $\mathbf{u}_0 \in V$ and if $\mathbf{f} \in D(\exp(\sigma_1 A^{1/2}))$ with $\sigma_1 > 0$, then there exists $T_0(\|\mathbf{u}_0\|, |\mathbf{f}|_{\sigma_1}, \nu, \Omega)$, given by (9.17) and (9.18), such that: (i) the corresponding strong solution with $\mathbf{u}(0) = \mathbf{u}_0$ of the Navier–Stokes equations can be extended to a solution of the complexified NSE (9.15) that is analytic from $\Delta^0_{\sigma_1}(\|\mathbf{u}_0\|, |\mathbf{f}|_{\sigma_1}, \nu, \Omega)$ into $D(A^{1/2}\exp(\varphi(s\cos\theta)A^{1/2}))$, where $\varphi(\xi) = \min\{\nu\lambda_1^{1/2}\xi, \sigma_1\}$; and (ii) on the closure of this domain of analyticity, the estimate (9.21) holds.

Analyticity in the 2-Dimensional Periodic Case

In the 2-dimensional case, owing to the uniform bound on the enstrophy of the strong solutions, the domain of analyticity can be extended to a neighborhood of the whole positive real axis. Moreover, we can use more favorable estimates for the nonlinear term to obtain a wider width of analyticity. The estimate we use for the nonlinear term is given by (9.13). Then, as shown in Appendix B.4, as long as

$$\theta \in \left[-\frac{\pi}{4}, \frac{\pi}{4}\right], \quad 0 \le s \le T_0(\|\mathbf{u}_0\|), \tag{9.22}$$

where

$$\begin{aligned} T_0(\|\mathbf{u}_0\|) &= T_0(\|\mathbf{u}_0\|, |\mathbf{f}|_{\sigma_1}, \nu, \Omega) \\ &= \left[c_8\nu\lambda_1\left(1+\frac{|\mathbf{f}|_{\sigma_1}}{\nu^2\lambda_1}+\frac{\|\mathbf{u}_0\|^2}{\nu^2\lambda_1}\right)\log c_9\left(1+\frac{|\mathbf{f}|_{\sigma_1}}{\nu^2\lambda_1}+\frac{\|\mathbf{u}_0\|^2}{\nu^2\lambda_1}\right)\right]^{-1} \end{aligned} \tag{9.23}$$

(here c_8 and c_9 are constants depending only on the shape of the domain Ω), the following estimate holds:

$$\|\mathbf{u}(se^{i\theta})\|^2_{\varphi(s\cos\theta)} \le c_7\lambda_1\nu^2 + 2\|\mathbf{u}_0\|^2, \tag{9.24}$$

where c_7 depends only on the shape of the domain. As before, we can choose $\varphi(\xi) = \min\{\nu\lambda_1^{1/2}\xi, \sigma_1\}$ for $\xi \ge 0$. Then we define the region

$$\begin{aligned} \Delta^0_{\sigma_1}(\|\mathbf{u}_0\|) &= \Delta^0_{\sigma_1}(\|\mathbf{u}_0\|, |\mathbf{f}|_{\sigma_1}, \nu, \Omega) \\ &= \Big\{\zeta = se^{i\theta};\ |\theta| < \frac{\pi}{4}, \\ &\qquad 0 < s < T_0(\|\mathbf{u}_0\|, |\mathbf{f}|_{\sigma_1}, \nu, \Omega),\ \nu\lambda_1^{1/2}s|\sin\theta| < \sigma_1\Big\}. \end{aligned} \tag{9.25}$$

This set is a domain of analyticity of the solution $\mathbf{u} = \mathbf{u}(\zeta)$ of the complexified Navier–Stokes equations. The origin $\zeta = 0$ belongs to the closure of $\Delta^0_{\sigma_1}(\|\mathbf{u}_0\|)$. Moreover, on the closure of this domain we have

$$|\mathbf{u}(\zeta)|^2_{D(A^{1/2}e^{\varphi(s\cos\theta)A^{1/2}})} \le c_7\lambda_1\nu^2 + 2\|\mathbf{u}_0\|^2 \quad \text{for } \zeta \in \overline{\Delta^0_{\sigma_1}}(\|\mathbf{u}_0\|, |\mathbf{f}|_{\sigma_1}, \nu, \Omega). \tag{9.26}$$

In the 2-dimensional case, the strong solutions exist for all positive time and their enstrophy is uniformly bounded. Hence, the domain of analyticity of the solutions can be extended to a neighborhood of the positive real axis. Indeed, we know from (A.66) that, for each $t \geq 0$,

$$\|\mathbf{u}(t)\|^2 \leq \|\mathbf{u}_0\|^2 + \frac{1}{\nu^2\lambda_1}|\mathbf{f}|^2. \tag{9.27}$$

Then, repeating our previous argument with the initial condition $\mathbf{u}(t_0)$ at time $t_0 \geq 0$, we obtain the analyticity of the solution on the domain

$$t_0 + \Delta^0_{\sigma_1}((\|\mathbf{u}_0\|^2 + |\mathbf{f}|^2/(\nu^2\lambda_1))^{1/2}) \subset t_0 + \Delta^0_{\sigma_1}(\|\mathbf{u}(t_0)\|).$$

By taking the union for all $t_0 > 0$ of the domains in the LHS of this expression, we obtain the analyticity in an open, pencil-like domain

$$\Delta^+_{\sigma_1}(\|\mathbf{u}_0\|) = \bigcup_{t_0>0}\{t_0 + \Delta^0((\|\mathbf{u}_0\|^2 + |\mathbf{f}|^2/(\nu^2\lambda_1))^{1/2})\}; \tag{9.28}$$

this is a neighborhood of the positive real axis and has $\zeta = 0$ on its boundary. Moreover, our estimates extend to all of $\Delta^+_{\sigma_1}(|\mathbf{u}_0|)$ in the sense that

$$|\mathbf{u}(\zeta)|^2_{D(A^{1/2}e^{\varphi(s\cos\theta)A^{1/2}})} \leq c_7\lambda_1\nu^2 + 2\|\mathbf{u}_0\|^2 + \frac{2}{\nu^2\lambda_1}|\mathbf{f}|^2$$
$$\text{for } \zeta = se^{i\theta} \in \overline{\Delta^+_{\sigma_1}}(\|\mathbf{u}_0\|, |\mathbf{f}|_{\sigma_1}, \nu, \Omega). \tag{9.29}$$

From (9.23), (9.25), and (9.27), we can write the domain of analyticity as

$$\Delta^+_{\sigma_1}(\|\mathbf{u}_0\|) = \Delta^+_{\sigma_1}(\|\mathbf{u}_0\|, |\mathbf{f}|_{\sigma_1}, \nu, \Omega) = \{\zeta \in \mathbb{C};\ |\operatorname{Im}\zeta| \leq \min\{\operatorname{Re}\zeta, \delta_0\}\}, \tag{9.30}$$

where δ_0 is the (largest) width of the pencil-like domain $\Delta^+_{\sigma_1}$, estimated by

$$\delta_0 \geq \min\left\{\frac{\sigma_1}{\nu\lambda_1^{1/2}}, \left[c_{10}\nu\lambda_1\left(1 + \frac{|\mathbf{f}|^2_{\sigma_1}}{\nu^4\lambda_1^2}\right)\log\left(c_{11}\left(1 + \frac{|\mathbf{f}|^2_{\sigma_1}}{\nu^4\lambda_1^2}\right)\right)\right]^{-1}\right\}, \tag{9.31}$$

where c_{10} and c_{11} depend only on the shape of the domain Ω.

Hence, we have shown that if $\mathbf{u}_0 \in V$ and if $\mathbf{f} \in D(\exp(\sigma_1 A^{1/2}))$ with $\sigma_1 > 0$ then there exists $T_0(\|\mathbf{u}_0\|, |\mathbf{f}|_{\sigma_1}, \nu, \Omega)$, given by (9.23), such that: (i) the corresponding strong solution with $\mathbf{u}(0) = \mathbf{u}_0$ of the NSE can be extended to a solution of the complexified Navier–Stokes equations (9.15) that is analytic from $\Delta^+_{\sigma_1}(\|\mathbf{u}_0\|, |\mathbf{f}|_{\sigma_1}, \nu, \Omega)$ into $D(A^{1/2}\exp(\varphi(s\cos\theta)A^{1/2}))$, where $\varphi(\xi) = \min\{\nu\lambda_1^{1/2}\xi, \sigma_1\}$; and (ii) on the closure of this domain of analyticity, the estimate (9.26) holds.

Finally, if the initial condition $\mathbf{u}_0$ is known to be only in H then we can use the fact that, in the 2-dimensional case, the weak solutions are actually strong solutions, and a uniform bound on the enstrophy of the solution holds for time that is bounded away from the initial time. Then, we obtain the analyticity of the solution in a pencil-like domain that is a neighborhood of the positive real axis. There is also a uniform bound depending on $|\mathbf{u}_0|$ that holds on every subset of the domain of analyticity that is bounded away from zero.

Exponential Decrease of the Fourier Coefficients

An immediate consequence of the space analyticity of NSE solutions in the 2-dimensional periodic case just derived is the exponential decrease of the Fourier coefficients of each solution with respect to the wavenumber. We have proven that, for a forcing term $\mathbf{f}$ in the Gevrey space $D(\exp(\sigma_1 A^{1/2}))$ with $\sigma_1 > 0$, and for an initial velocity field $\mathbf{u}_0$ in V (or even in H), the corresponding flow $\mathbf{u} = \mathbf{u}(t)$ is analytic in both space and time. Moreover, after some short transient time when the radius of analyticity of the solution $\mathbf{u}(t)$ increases, we find $\mathbf{u}(t)$ in the Gevrey space $D(\exp(\delta_0 A^{1/2}))$ with δ_0 as in (9.31). According to (9.29), the norm of $\mathbf{u}(t)$ in this space is bounded uniformly in time:

$$|\mathbf{u}(t)|^2_{D(A^{1/2}e^{\sigma A^{1/2}})} \leq c_7\lambda_1\nu^2 + 2\|\mathbf{u}_0\|^2 + \frac{2}{\nu^2\lambda_1}|\mathbf{f}|^2 \quad \text{for } t \geq \delta_0, \tag{9.32}$$

where δ_0 is the width of analyticity of the pencil-like domain $\Delta^+_{\sigma_1}$ as given by (9.31). From the Fourier series characterization (9.10) of the space $D(\exp(\delta_0 A^{1/2}))$, we obtain

$$|\mathbf{u}(t)|^2_{D(A^{1/2}e^{\delta_0 A^{1/2}})} = 2\pi|\Omega| \sum_{\mathbf{k}\in\mathbb{Z}^d} \left|\frac{\mathbf{k}}{\mathbf{L}}\right|^2 e^{4\pi\delta_0\left|\frac{\mathbf{k}}{\mathbf{L}}\right|} |\hat{\mathbf{u}}_{\mathbf{k}}(t)|^2 \leq M^2, \tag{9.33}$$

where M^2 is the bound on the RHS of (9.32). Whence, it is straightforward to deduce the following crude bound:

$$|\hat{\mathbf{u}}_{\mathbf{k}}|^2 \leq \frac{M}{\sqrt{2\pi|\Omega|}} \left|\frac{\mathbf{k}}{\mathbf{L}}\right| e^{-2\pi\delta_0\left|\frac{\mathbf{k}}{\mathbf{L}}\right|}. \tag{9.34}$$

We thus conclude that each Fourier coefficient $\hat{\mathbf{u}}_{\mathbf{k}}$ decreases exponentially with respect to its wavenumber $2\pi|\mathbf{k}|/|\mathbf{L}|$ – uniformly with respect to time and uniformly with respect to the initial enstrophy $\|\mathbf{u}_0\|$.

Note that if the forcing term is a trigonometric polynomial, as in the Kolmogorov theory of turbulence, then $\mathbf{f}$ belongs to the Gevrey space $D(\exp(\sigma_1 A^{1/2}))$ for any $\sigma_1 > 0$. Therefore, the exponential rate of decrease in (9.34) holds for δ_0 given by (9.31).

If the initial condition $\mathbf{u}_0$ belongs only to the space H of finite kinetic energy, then (owing to the regularization of the solutions in the 2-dimensional case) the solution $\mathbf{u}(t)$ belongs to V and is uniformly bounded in V after a transient time $t \geq t_0 > 0$, with t_0 depending uniformly on $|\mathbf{u}_0|$. We can then proceed as before to obtain an exponential decrease of the Fourier coefficients of $\mathbf{u}(t)$, uniformly with respect to the initial kinetic energy $|\mathbf{u}_0|^2/2$.

Remark 9.1 A similar result based on different considerations was obtained in Foias, Manley, and Sirovich [1989a].

Remark 9.2 A method inspired by the Gevrey-class method was introduced in Grujic and Kukavica [1998, 1999]. They provided analogous results in terms of L^p-norms of initial data and also dealt with the issue of estimating the radius of spatial analyticity of solutions in the case of Dirichlet boundary conditions.

10 Function Spaces for the Whole-Space Case

As mentioned previously, we deal with the whole-space case only in the context of homogeneous statistical solutions, which are the evolution of homogeneous probability distributions of an ensemble of flows (see Chapters IV and V); no individual, deterministic solution is considered. For comparison with results of individual experiments, moments of such distributions are used. The domain filled by the fluid is $\Omega = \mathbb{R}^d$ with $d = 2$ or 3, depending on the space dimension. Like the periodic case, the whole-space case arises in the study of homogeneous turbulence. For this study, we do not ask the total kinetic energy

$$e(\mathbf{u}) = \frac{1}{2}\int_\Omega |\mathbf{u}(\mathbf{x})|^2\,dx$$

to be finite; otherwise the very assumption of homogeneity would be contradicted, since the flow would have to vanish at infinity. Naturally, in this case we will consider only local energies. In particular we will derive, in the space homogeneous case, a suitable energy equation for the local average kinetic energy

$$e(\mathbf{u}) = \frac{1}{2}\frac{1}{|Q(L)|}\int_{Q(L)} |\mathbf{u}(\mathbf{x})|^2\,d\mathbf{x}.$$

Thanks to the homogeneity assumption, the local average kinetic energy is, in fact, independent of the subdomain considered.

We introduce now the function spaces that will be useful for flows on the whole space $\mathbb{R}^d$ $(d = 2, 3)$. For notational purposes we consider the scalar product and norm in the spaces $L^2(\mathbb{R}^d)^d$ and $H^1(\mathbb{R}^d)^d$ given by

$$\begin{aligned}(\mathbf{u},\mathbf{v}) &= \int_{\mathbb{R}^d} \mathbf{u}(\mathbf{x})\cdot\mathbf{v}(\mathbf{x})\,dx,\\ ((\mathbf{u},\mathbf{v}))_1 &= \frac{1}{L^2}(\mathbf{u},\mathbf{v})_0 + \sum_{i=1}^{d}(\partial_{x_i}\mathbf{u}, \partial_{x_i}\mathbf{v})_0\end{aligned} \tag{10.1}$$

and

$$|\mathbf{u}| = (\mathbf{u},\mathbf{u})_0^{1/2}, \qquad \|\mathbf{u}\|_1 = ((\mathbf{u},\mathbf{u}))_1^{1/2}. \tag{10.2}$$

Here, L is a typical length scale that could be associated, by dimensional analysis, with the kinematic viscosity and a typical velocity or with the force driving the flow, whatever the force (since there is no typical length in the geometry of the problem).[14] As in the other cases, we distinguish the term associated with the enstrophy:

$$((\mathbf{u},\mathbf{v})) = \sum_{i=1}^{d}(\partial_{x_i}\mathbf{u}, \partial_{x_i}\mathbf{v})_0, \qquad \|\mathbf{u}\| = ((\mathbf{u},\mathbf{u}))^{1/2}. \tag{10.3}$$

[14] The outer length of order $(\nu^3/\varepsilon)^{1/4}\,\mathrm{Re}^{3/4} \simeq L_{\mathrm{Taylor}}\,\mathrm{Re}^{1/2}$ can serve as such an L (Foias and Temam [1983], Foias, Manley, and Temam [1986]).

However, as we have just mentioned, we omit flows decaying sufficiently fast at infinity for the total energy and the total enstrophy to be finite; such a situation is not consistent with the homogeneity assumption. Therefore, we shall consider spaces of functions with only finite local kinetic energy and finite local enstrophy. For that purpose it is appropriate to define the space $L^1_{\text{loc}}(\mathbb{R}^d)$ (resp., $L^2_{\text{loc}}(\mathbb{R}^d)$) of real-valued functions on $\mathbb{R}^d$ that are locally (square) integrable – that is, of real-valued functions on $\mathbb{R}^d$ that belong to $L^1(Q)$ (resp., $L^2(Q)$) when restricted to any bounded measurable set Q in $\mathbb{R}^d$. Similarly, the space $H^1_{\text{loc}}(\mathbb{R}^d)$ is the space of real-valued functions on $\mathbb{R}^d$ that belong to $H^1(Q)$ when restricted to a bounded open set Q. These spaces can obviously be generalized to vector-valued spaces $L^2_{\text{loc}}(\mathbb{R}^d)^d$ and $H^1_{\text{loc}}(\mathbb{R}^d)^d$.

We can now consider divergence-free vector fields with finite local energy and finite local enstrophy by setting

$$H_{\text{loc}} = \{\mathbf{u} \in L^2_{\text{loc}}(\mathbb{R}^d)^d;\ \nabla \cdot \mathbf{u} = 0\} \tag{10.4}$$

and

$$V_{\text{loc}} = \{\mathbf{u} \in H^1_{\text{loc}}(\mathbb{R}^d)^d;\ \nabla \cdot \mathbf{u} = 0\}. \tag{10.5}$$

It is appropriate to define measures of the local average energy and the local average enstrophy. For that purpose, we define the following quantities for a given bounded set Q in $\mathbb{R}^d$, with volume or area denoted by $|Q|$:

$$(\mathbf{u}, \mathbf{v})_Q = \frac{1}{|Q|} \int_Q \mathbf{u}(\mathbf{x}) \cdot \mathbf{v}(\mathbf{x})\, d\mathbf{x}, \tag{10.6}$$

$$((\mathbf{u}, \mathbf{v}))_Q = \sum_{i=1}^{d} (\partial_{x_i}\mathbf{u}, \partial_{x_i}\mathbf{v})_Q, \tag{10.7}$$

$$((\mathbf{u}, \mathbf{v}))_{1,Q} = (\mathbf{u}, \mathbf{v})_Q + ((\mathbf{u}, \mathbf{v}))_Q, \tag{10.8}$$

and

$$|\mathbf{u}|_Q = (\mathbf{u}, \mathbf{u})_Q^{1/2}, \quad \|\mathbf{u}\|_Q = ((\mathbf{u}, \mathbf{u}))_Q^{1/2}, \quad |\mathbf{u}|_{1,Q} = (\mathbf{u}, \mathbf{u})_{1,Q}^{1/2}. \tag{10.9}$$

By setting $Q(L) = (-L/2, L/2)^d$ for $L > 0$, we endow H_{loc} and V_{loc} with the family of seminorms[15] $|\cdot|_{Q(L)}$ and $\|\cdot\|_{1,Q(L)}$ for $L \in \mathbb{N}$, respectively, which makes them complete metric spaces when endowed with the following distance between two vector fields $\mathbf{u}$ and $\mathbf{v}$:

$$d_{H_{\text{loc}}}(\mathbf{u}, \mathbf{v}) = \sum_{L\in\mathbb{N}} \frac{|\mathbf{u}-\mathbf{v}|_{Q(L)}}{2^L(1 + |\mathbf{u}-\mathbf{v}|_{Q(L)})}, \qquad d_{V_{\text{loc}}}(\mathbf{u}, \mathbf{v}) = \sum_{L\in\mathbb{N}} \frac{\|\mathbf{u}-\mathbf{v}\|_{Q(L)}}{2^L(1 + \|\mathbf{u}-\mathbf{v}\|_{Q(L)})}.$$

We need dual spaces in this case, also. Dual spaces are needed to handle the inertial term of the Navier–Stokes equations in order to make sense out of $(\mathbf{u} \cdot \nabla)\mathbf{u}$ when $\mathbf{u}$ is only in V (V_{loc}, in the present case). In the whole-space case, the appropriate duality is associated with the space V_c of vector fields in V_{loc} that have compact

[15] A seminorm has the same properties as a norm, except that $|\mathbf{u}|_{Q(L)} = 0$ does not imply $\mathbf{u} = 0$. For example, $\left|\int_\Omega u\, dx\right|$ for the functions u defined and integrable on a set Ω is a seminorm.

support on $\mathbb{R}^d$. Indeed, if $\mathbf{w}$ belongs to V_c and if $\mathbf{u}$ and $\mathbf{v}$ belong to V_{loc} then, for L sufficiently large that $Q(L)$ contains the support of $\mathbf{w}$, it follows from (A.26f) that

$$|b(\mathbf{u},\mathbf{v},\mathbf{w})| \le c(L)|\mathbf{u}|_{Q(L)}^{1/4}\|\mathbf{u}\|_{Q(L)}^{3/4}|\mathbf{v}|_{Q(L)}^{1/4}\|\mathbf{v}\|_{Q(L)}^{3/4}\|\mathbf{w}\|_{Q(L)}. \tag{10.10}$$

The constant, $c(L)$, now depends on L because the norms involve averages and thus have powers of L in the denominator. Notice that the trilinear term $b(\mathbf{u},\mathbf{v},\mathbf{w})$ is now defined formally as an integral over the whole space, but it agrees with that on (A.26f) because $\mathbf{w}$ has compact support. Estimate (10.10) shows that $b(\cdot,\cdot,\cdot)$ extends to a continuous operator from $V_{\text{loc}} \times V_{\text{loc}} \times V_c$ into $\mathbb{R}$. Therefore, it is natural to consider the dual space V_c' of V_c given by the continuous linear functionals defined on V_c. The topology in V_c is such that $\mathbf{v}_n$ converges to $\mathbf{v}$ in V_c if and only if there exists $L > 0$ sufficiently large such that (a) $Q(L)$ contains the compact support of $\mathbf{v}$ and $\mathbf{v}_n$ for all n in $\mathbb{N}$ and (b)

$$\|\mathbf{v}_n - \mathbf{v}\|_{Q(L)} \to 0$$

as n goes to infinity. With such topology, V_c is a linear subspace of V_{loc}.

For a given linear functional ℓ in V_c', we can denote its value $\ell(\mathbf{v})$ at an element $\mathbf{v}$ in V_c by the product

$$\ell(\mathbf{v}) = (\ell, \mathbf{v}). \tag{10.11}$$

As in the bounded cases, we can identify vector fields in H_{loc} as linear functionals in H_{loc} via the L^2 inner product. Then we have the continuous injections

$$V_c \subset V_{\text{loc}} \subset H_{\text{loc}} \subset V_c'. \tag{10.12}$$

The estimate (10.10) shows that $B(\mathbf{u},\mathbf{v})$ can be extended to be a continuous bilinear operator on $V_{\text{loc}} \times V_{\text{loc}}$ with values on the dual space V_c' of V_c:

$$B(\cdot,\cdot)\colon V_{\text{loc}} \times V_{\text{loc}} \to V_c'. \tag{10.13}$$

11 The No-Slip Case with Moving Boundaries

As mentioned in Section 2, the no-slip case can be considered with nonvanishing boundary conditions. They may represent either permeable or flexible (deformable) boundaries – when the normal component of the nonvanishing boundary condition is not zero – or moving boundaries: moving objects embedded in a fluid, a container moving with fluid inside, or moving boundaries surrounding a fixed domain (as with, e.g., Couette–Taylor flows). In the case of moving boundaries, special geometries must be considered in order for the physical domain occupied by the fluid to remain the same (e.g., a rotating sphere or cylinder and sliding channels). Otherwise, a more elaborate mathematical modeling would be necessary, but this case is not addressed here.

In the remainder of this monograph (except for this section and the next), we will not discuss nonvanishing boundary conditions in general, but the reader should keep in mind that several of the results presented in this monograph can be extended to the case where the boundary is not at rest. Several such extensions are in fact available in the literature, and we cite them where appropriate. In this section, we shall describe only the mathematical framework used to handle this case, which essentially amounts to removing a suitable background flow from the solution, thus reducing the problem to an evolution equation with added linear terms and vanishing boundary conditions.

We follow the framework presented by Foias and Temam [1979]. The domain Ω occupied by the fluid is assumed to be an open, bounded, connected set in $\mathbb{R}^d$ ($d = 2, 3$); its boundary $\partial\Omega$ is assumed to be smooth in the sense that it is locally the graph of a C^2 function with Ω located on one side of the boundary. It is also assumed that $\partial\Omega$ has a finite number of connected components $\Gamma_1, \ldots, \Gamma_k$, $k \in \mathbb{N}$.

The nonhomogeneous boundary condition reads

$$\mathbf{u} = \boldsymbol{\varphi} \text{ on } \partial\Omega. \tag{11.1}$$

We assume that the function $\boldsymbol{\varphi}$ is time-independent and belongs to $H^{3/2}(\partial\Omega)^d$, which is the space of the trace of functions[16] in $H^2(\Omega)$. Moreover, we assume that the average of the normal component of $\boldsymbol{\varphi}$ over each connected component of the boundary vanishes:

$$\int_{\Gamma_j} \boldsymbol{\varphi} \cdot \mathbf{n}\, d\Gamma = 0, \quad j = 1, \ldots, k, \tag{11.2}$$

where $\mathbf{n}$ is the outward unit normal on $\partial\Omega$. It is clear that, for the sake of conservation of mass, the average of the normal component of $\boldsymbol{\varphi}$ over all $\partial\Omega$ should be zero; hence, the sum of the quantities in (11.2) should vanish. From the mathematical point of view, this zero mass flux follows from the divergence theorem and the incompressibility condition. Hence, (11.2) is more than that required by incompressibility alone; it is needed for our construction of the background flow (defined shortly).

The space of functions $\boldsymbol{\varphi}$ in $H^{3/2}(\partial\Omega)^d$ satisfying (11.2) will be denoted by $\dot{H}^{3/2}(\partial\Omega)^d$. From Foias and Temam [1978] and Lions [1969] (see also Hopf [1951]) we see that, given $\delta > 0$, there exists for any $\boldsymbol{\varphi}$ in $\dot{H}^{3/2}(\partial\Omega)^d$ a vector field $\boldsymbol{\Phi}_\delta$ in $H^2(\Omega)^d$ such that

$$\nabla \cdot \boldsymbol{\Phi}_\delta = 0 \text{ in } \Omega, \qquad \boldsymbol{\Phi}_\delta = \boldsymbol{\varphi} \text{ on } \partial\Omega, \tag{11.3}$$

and

$$|b(\mathbf{v}, \boldsymbol{\Phi}_\delta, \mathbf{v})| \le \delta \|\boldsymbol{\varphi}\|_{\dot{H}^{3/2}(\partial\Omega)^d} \|\mathbf{v}\|^2 \tag{11.4}$$

for all $\mathbf{v}$ in V. The map $\boldsymbol{\varphi} \mapsto \Lambda_\delta \boldsymbol{\varphi} \equiv \boldsymbol{\Phi}_\delta$ defines a linear continuous operator Λ_δ from $\dot{H}^{3/2}(\partial\Omega)^d$ into $H^2(\Omega)^d$.

The importance of the operator Λ_δ that takes $\boldsymbol{\varphi}$ into $\boldsymbol{\Phi}_\delta$ is twofold. First, it allows us to transform the problem into one with vanishing boundary conditions (see (11.8)), which is more convenient for a number of purposes. Second, it guarantees that the

[16] For any function $\boldsymbol{\Phi}$ in $H^2(\Omega)^d$, we can define the *trace* of $\boldsymbol{\Phi}$ on $\partial\Omega$, denoted $\boldsymbol{\Phi}|_{\partial\Omega}$; it coincides with the restriction of $\boldsymbol{\Phi}$ to the boundary $\partial\Omega$ when $\boldsymbol{\Phi}$ is a smooth function. The trace $\boldsymbol{\Phi}|_{\partial\Omega}$ belongs to $H^{3/2}(\partial\Omega)^d$, and the mapping $\boldsymbol{\Phi} \to$ trace $\boldsymbol{\Phi}|_{\partial\Omega}$ is a bounded linear operator from $H^2(\Omega)^d$ *onto* $H^{3/2}(\partial\Omega)^d$.

linear part of the transformed system of equations is positive (see (11.13)), which is essential for the derivation of certain important energy estimates (see (11.14)). A concrete example of such an operator is given by (12.29).

Let a boundary condition $\boldsymbol{\varphi}$ be given in $\dot{H}^{3/2}(\partial\Omega)^d$. For each δ, the vector field $\boldsymbol{\Phi}_\delta$ is divergence-free and is equal to $\boldsymbol{\varphi}$ on the boundary of Ω. Then, we may look for a solution of the nonhomogeneous Navier–Stokes equations in the form

$$\mathbf{u} = \boldsymbol{\Phi}_\delta + \bar{\mathbf{u}}, \tag{11.5}$$

with $\bar{\mathbf{u}}$ vanishing on the boundary of the domain Ω. In this context, the term $\boldsymbol{\Phi}_\delta$ is usually called "background flow" in the literature, although it is not actually a flow, in general: it is not a stationary or time-dependent solution of the Navier–Stokes equations. By (11.3), the vector field $\bar{\mathbf{u}}$ must also be divergence-free. Substituting (11.5) in the NSE, we find the following equation for $\bar{\mathbf{u}}$:

$$\frac{\partial \bar{\mathbf{u}}}{\partial t} - \nu\Delta\bar{\mathbf{u}} + (\bar{\mathbf{u}}\cdot\nabla)\bar{\mathbf{u}} + (\boldsymbol{\Phi}_\delta\cdot\nabla)\bar{\mathbf{u}} + (\bar{\mathbf{u}}\cdot\nabla)\boldsymbol{\Phi}_\delta + \nabla p = \bar{\mathbf{f}}, \tag{11.6a}$$

$$\nabla\cdot\bar{\mathbf{u}} = 0, \tag{11.6b}$$

where

$$\bar{\mathbf{f}} = \mathbf{f} + \nu\Delta\boldsymbol{\Phi}_\delta - (\boldsymbol{\Phi}_\delta\cdot\nabla)\boldsymbol{\Phi}_\delta. \tag{11.7}$$

The boundary condition for $\bar{\mathbf{u}}$ reduces to the homogeneous no-slip condition:

$$\bar{\mathbf{u}} = 0 \text{ on } \partial\Omega, \tag{11.8}$$

and the initial condition reads

$$\bar{\mathbf{u}}(0) = \bar{\mathbf{u}}_0 \equiv \mathbf{u}_0 - \boldsymbol{\Phi}_\delta, \tag{11.9}$$

where $\mathbf{u}_0$ is the initial condition for the evolution of the velocity field $\mathbf{u} = \mathbf{u}(t)$.

Because of the homogeneous boundary condition (11.8) and the divergence-free condition (11.6b), we expect the vector field $\bar{\mathbf{u}}$ to belong to H or V. Hence, we can rewrite equations (11.6) in a functional form in V' similar to the homogeneous case:

$$\frac{d\bar{\mathbf{u}}}{dt} + \nu A\bar{\mathbf{u}} + B(\bar{\mathbf{u}}) + B(\bar{\mathbf{u}}, \boldsymbol{\Phi}_\delta) + B(\boldsymbol{\Phi}_\delta, \bar{\mathbf{u}}) = P_L\bar{\mathbf{f}}, \tag{11.10}$$

with the initial condition

$$\bar{\mathbf{u}}(0) = \bar{\mathbf{u}}_0. \tag{11.11}$$

The distinction from the homogeneous case (boundary at rest) is the presence of additional terms in the linear part of the equation:

$$\nu A\bar{\mathbf{u}} + B(\bar{\mathbf{u}}, \boldsymbol{\Phi}_\delta) + B(\boldsymbol{\Phi}_\delta, \bar{\mathbf{u}}). \tag{11.12}$$

The estimate (11.4) is then crucial to show that, for small enough δ, the linear term (11.12) is V-elliptic, that is,

$$(\nu A\bar{\mathbf{u}} + B(\bar{\mathbf{u}}, \boldsymbol{\Phi}_\delta) + B(\boldsymbol{\Phi}_\delta, \bar{\mathbf{u}}), \bar{\mathbf{u}}) \geq \frac{\nu}{2}\|\bar{\mathbf{u}}\|^2 \tag{11.13}$$

for all $\bar{\mathbf{u}}$ in V. Note that the third term in the LHS of (11.13) actually vanishes owing to the orthogonality property of the trilinear term (see (A.33)).

Moreover, since $\mathbf{\Phi}_\delta$ belongs to $H^2(\Omega)^d$, one can check from (11.7) that $\bar{\mathbf{f}}$ belongs to $L^2(\Omega)^2$. Hence $P_L\bar{\mathbf{f}}$ does indeed make sense and belongs to H.

From (11.10) and (11.13), it follows (upon taking the inner product of (11.10) in H with $\bar{\mathbf{u}}$) that

$$\frac{1}{2}\frac{d}{dt}|\bar{\mathbf{u}}|^2 + \frac{\nu}{2}\|\bar{\mathbf{u}}\|^2 \le |\bar{\mathbf{f}}||\bar{\mathbf{u}}|.$$

Whence,

$$|\bar{\mathbf{u}}(t)|^2 \le |\bar{\mathbf{u}}_0|^2 e^{-\nu\lambda_1 t} + \frac{1}{\nu^2\lambda_1}|\bar{\mathbf{f}}|^2. \tag{11.14}$$

This expression provides the basic a priori estimates for the construction of the solutions. Note that this estimate relies on the fact that $\mathbf{\Phi}_\delta$ (nonunique) can be constructed so that it satisfies (11.13).

With the estimate (11.13) in mind, one can obtain the appropriate existence and uniqueness results similar to the homogeneous case, as well as the results on time and space analyticity. Similar long-time estimates (derived in the Appendix A for the homogeneous case) can also be obtained.

Other explicit ways to construct a background flow (extension of the boundary data inside the domain as divergence-free vector fields, not necessarily solutions of the NSE) have been proposed and used in the literature for specific purposes. Miranville [1993] proposed such a construction to improve the estimate of the attractor dimension for channel flows. Constantin and Doering [1992] proposed a crude construction of the background flow that was sufficient for their objective (see Section 12). Finally, Wang (see Temam and Wang [1995] and Doering and Wang [1998]) proposed the construction of a background flow with components oscillating in the direction perpendicular to the boundary – with the aims of assessing boundary layer problems and extending the work of Constantin and Doering (see Section 12).

12 Dissipation Rate of Flows

An important quantity in the statistical study of turbulence is the average rate of energy dissipation ϵ, defined heuristically as

$$\epsilon = \nu\langle(\nabla\mathbf{u})^2\rangle, \tag{12.1}$$

where $\langle\cdot\rangle$ denotes a suitable ensemble average. According to Kolmogorov [1941a], a universal bound on ϵ can be derived heuristically in the following way: If

$$e = \tfrac{1}{2}\langle\mathbf{u}^2\rangle \tag{12.2}$$

represents the average energy per unit mass in a turbulent flow with a rate of energy dissipation ϵ, then $t_\epsilon = e/\epsilon$ should represent a characteristic time for the dissipation of energy; the characteristic mean velocity should be $U = \sqrt{2e}$. The corresponding length $\ell = Ut_\epsilon$ can be viewed as the average distance traveled by the turbulent eddies until they dissipate. Thus, we find that the average rate of energy dissipation should be of the order

$$\epsilon \sim \frac{U^2}{t_\epsilon} = \frac{U^3}{\ell}. \tag{12.3}$$

In a series of articles, Constantin and Doering [1992, 1994, 1995] derived rigorous bounds in the form of (12.3), in a suitable sense, directly from the Navier–Stokes equations. The bounds were derived for shear flows and channel flows and were later extended to more general boundary-driven flows by Doering and Wang [1998]. For the sake of illustration, we consider here the bound in the case of 3-dimensional shear flows, following Constantin and Doering [1992]. We also present the bounds obtained in the periodic case by Foias [1997].

Bounds on the Energy Dissipation for a 3-Dimensional Shear Flow

Our analysis is based on the 3-dimensional Navier–Stokes equations:

$$\frac{\partial \mathbf{u}}{\partial t} - \nu\Delta\mathbf{u} + (\mathbf{u}\cdot\nabla)\mathbf{u} + \nabla p = 0, \tag{12.4a}$$

$$\nabla\cdot\mathbf{u} = 0. \tag{12.4b}$$

The fluid is contained between rigid parallel plates located at $z = 0$ and $z = h$. We write $\mathbf{x} = (x, y, z)$ and $\mathbf{u} = (u_1, u_2, u_3)$. We impose L-periodic boundary conditions in the x and y directions, so that the domain is of the form $\Omega = (0, L)\times(0, L)\times(0, h)$.

The fluid is driven by the boundary at $z = 0$ moving in the x direction at a speed U, so that

$$\mathbf{u}(x, y, 0, t) = U\mathbf{e}_1, \qquad \mathbf{u}(x, y, h, t) = 0, \tag{12.5}$$

where $\mathbf{e}_1$ is the unit vector in the x direction.

The ensemble average in the heuristic definition (12.1) of the energy dissipation rate is interpreted rigorously as

$$\langle g\rangle = \limsup_{T\to\infty} \frac{1}{T}\int_0^T g(t)\,dt, \tag{12.6}$$

which is well-defined for bounded functions $g = g(t)$. Then, the rate of energy dissipation is interpreted as

$$\epsilon = \frac{\nu}{|\Omega|}\langle\|\mathbf{u}\|^2\rangle, \tag{12.7}$$

where $|\Omega| = hL^2$ is the volume of the domain Ω. The aim is to establish rigorous bounds on ϵ in terms of ν, h, L, and U.

For the laminar flow we have

$$\mathbf{u}_{\text{laminar}} = U\frac{h-z}{h}\mathbf{e}_1, \tag{12.8}$$

and the dissipation rate is given by

$$\epsilon_{\text{laminar}} = \nu\frac{U^2}{h^2}. \tag{12.9}$$

This is generally regarded as a lower bound on the dissipation rate of turbulent flows, but some authors believe that certain quantities increasing with turbulence can also reach values below those of laminar flows.

For nonexplicit solutions we must find means to estimate the total enstrophy. By taking the inner product of the momentum equation (12.4a) with $\mathbf{u}$, integrating on Ω, and using integration by parts (as done in Section 1), we find that

$$\frac{1}{2}\frac{d}{dt}|\mathbf{u}(t)|^2 + \nu\|\mathbf{u}(t)\|^2 = -\nu U \iint_{(0,L)^2} \frac{\partial u_1}{\partial z}(x, y, 0, t)\, dx\, dy. \tag{12.10}$$

Assuming that $|\mathbf{u}(t)|$ is bounded uniformly in time, the time average of this equation yields

$$\nu\langle\|\mathbf{u}\|^2\rangle = \nu U \int_{(0,L)^2} \left\langle -\frac{\partial u_1}{\partial z}(x, y, 0)\right\rangle dx\, dy. \tag{12.11}$$

However, we have no means of bounding the term on the RHS of (12.11) in terms of the data of the problem. To circumvent this difficulty, the velocity field is decomposed into two parts in a way akin to what has been done for the well-posedness of flows with moving boundaries, as described in Section 11: one of the terms in the decomposition satisfies the boundary conditions of the problem, while the other vanishes at both boundaries $z = 0$ and h. More precisely, we write

$$\mathbf{u}(x, y, z, t) = \phi(z)\mathbf{e}_1 + \mathbf{v}(x, y, z, t), \tag{12.12}$$

where

$$\phi(h) = U, \qquad \phi(0) = 0, \tag{12.13}$$

and

$$\mathbf{v}(x, y, 0, t) = 0, \qquad \mathbf{v}(x, y, h, t) = 0. \tag{12.14}$$

It follows from (12.12) that $\mathbf{v}$ must be divergence-free and periodic with period L in the directions x and y. Substituting for $\mathbf{u}$ in the Navier–Stokes equations (12.4), we find an evolution equation for the field $\mathbf{v} = (v_1, v_2, v_3)$:

$$\frac{\partial \mathbf{v}}{\partial t} - \nu\Delta\mathbf{v} + (\mathbf{v}\cdot\nabla)\mathbf{v} + \phi\frac{\partial \mathbf{v}}{\partial x} + v_3\phi'\mathbf{e}_1 + \nabla p = \nu\phi''\mathbf{e}_1, \tag{12.15a}$$

$$\nabla\cdot\mathbf{v} = 0. \tag{12.15b}$$

The terms ϕ' and ϕ'' represent the first- and second-order derivatives of ϕ.

From the relation (12.12), we find that

$$\|\mathbf{u}\|^2 = \|\mathbf{v}\|^2 + 2\int_\Omega \phi'\frac{\partial v_1}{\partial z}\, d\mathbf{x} + L^2\int_0^h |\phi'|^2\, dz. \tag{12.16}$$

Thus, the dissipation rate ϵ can be expressed in terms of an ensemble average involving the vector field $\mathbf{v}$:

$$\epsilon = \frac{\nu}{hL^2}\left\langle \|\mathbf{v}\|^2 + 2\int_\Omega \phi'\frac{\partial v_1}{\partial z}\, d\mathbf{x}\right\rangle + \frac{\nu}{h}\int_0^h |\phi'|^2\, dz. \tag{12.17}$$

To estimate the ensemble average in (12.17), we use the energy equation for $\mathbf{v}$, which is obtained by multiplying the momentum equation (12.15a) by $\mathbf{v}$ and integrating over the domain Ω. Using integration by parts (as illustrated in Section 1), we find

$$\frac{1}{2}\frac{d}{dt}|\mathbf{v}(t)|^2 + \nu\|\mathbf{v}(t)\|^2 + \nu\int_\Omega \phi'\frac{\partial v_1}{\partial z}\,d\mathbf{x} = -\int_\Omega \phi' v_1 v_3\,d\mathbf{x}. \tag{12.18}$$

The idea is to choose ϕ appropriately so that the term that is quadratic in $\mathbf{v}$ in the right-hand side of (12.18) is dominated by the dissipative term on the left-hand side. This is accomplished by choosing ϕ such that its first-order derivative is small in some suitable sense. Because of the different boundary conditions at $z = 0$ and $z = h$, we cannot simply choose ϕ to be a constant. The solution is to choose ϕ constant throughout most of the domain, except on a small layer near the boundary.

We split the quadratic term into two parts: one with the integration in z between 0 and $h/2$, and the other with the integration between $z = h/2$ and $z = h$. For the first part, we use the fundamental theorem of calculus and the fact that $\mathbf{v}$ vanishes at the boundary $z = 0$ to write

$$v_j(x, y, z, t) = \int_0^z \frac{\partial v_j}{\partial \zeta}(x, y, \zeta, t)\,d\zeta$$

for $j = 1, 3$. This yields, upon using the Cauchy–Schwarz inequality, the following estimate for the integral over x and y of this part of the quadratic term:

$$\begin{aligned}
&\left|\iint_{(0,L)\times(0,L)} v_1(x, y, z, t) v_3(x, y, z, t)\,dx\,dy\right| \\
&\le z \iint_{(0,L)\times(0,L)} \left(\int_0^z \left|\frac{\partial v_1}{\partial \zeta}(x, y, \zeta, t)\right|^2 d\zeta\right)^{1/2} \\
&\qquad\qquad \left(\int_0^z \left|\frac{\partial v_3}{\partial \zeta}(x, y, \zeta, t)\right|^2 d\zeta\right)^{1/2} dx\,dy \\
&\le \frac{z}{2}\iint_{(0,L)\times(0,L)} \int_0^{h/2} \left(\left|\frac{\partial v_1}{\partial \zeta}(x, y, \zeta, t)\right|^2 + \left|\frac{\partial v_3}{\partial \zeta}(x, y, \zeta, t)\right|^2\right) dx\,dy\,dz.
\end{aligned} \tag{12.19}$$

Similarly, for the second part we write

$$v_j(x, y, z, t) = -\int_z^h \frac{\partial v_j}{\partial \zeta}(x, y, \zeta, t)\,d\zeta,$$

and we obtain

$$\begin{aligned}
&\left|\iint_{(0,L)\times(0,L)} v_1(x, y, z, t) v_3(x, y, z, t)\,dx\,dy\right| \\
&\le \frac{h - z}{2}\iint_{(0,L)\times(0,L)} \int_{h/2}^{h} \left(\left|\frac{\partial v_1}{\partial \zeta}(x, y, \zeta, t)\right|^2 + \left|\frac{\partial v_3}{\partial \zeta}(x, y, \zeta, t)\right|^2\right) dx\,dy\,dz.
\end{aligned} \tag{12.20}$$

Combining (12.19) and (12.20), we bound the quadratic term on the RHS of (12.18) as

$$-\int_\Omega \phi' v_1 v_3\,d\mathbf{x} \le \frac{1}{2}\max\left\{\int_0^{h/2} z|\phi'(z)|\,dz,\ \int_{h/2}^{h} (h - z)|\phi'(z)|\,dz\right\}\|\mathbf{v}\|^2. \tag{12.21}$$

Then, we look for ϕ such that ϕ' is small in the sense that

$$\max\left\{\int_0^{h/2} z|\phi'(z)|\,dz,\ \int_{h/2}^{h}(h-z)|\phi'(z)|\,dz\right\} \leq \nu. \tag{12.22}$$

For such ϕ, using (12.21) and (12.22) we obtain the following estimate from (12.18):

$$\frac{d}{dt}|\mathbf{v}(t)|^2 + \nu\|\mathbf{v}(t)\|^2 + 2\nu\int_\Omega \phi'\frac{\partial v_1}{\partial z}\,d\mathbf{x} \leq 0. \tag{12.23}$$

From this estimate, if the kinetic energy of $\mathbf{v}$ is uniformly bounded in time then we can take the time average of (12.23) to find that

$$\left\langle \|\mathbf{v}\|^2 + 2\int_\Omega \phi'\frac{\partial v_1}{\partial z}\,d\mathbf{x}\right\rangle \leq 0. \tag{12.24}$$

Using this estimate in (12.17) yields the bound

$$\epsilon \leq \frac{\nu}{h}\int_0^h |\phi'|^2\,dz. \tag{12.25}$$

We must still show that the kinetic energy of $\mathbf{v}$ is bounded and find ϕ such that (12.22) holds. Then we will be able to obtain a more explicit bound in (12.25).

To show that the kinetic energy of $\mathbf{v}$ is bounded, we estimate the term linear in $\mathbf{v}$ in the LHS of (12.18); we use simply the Cauchy–Schwarz and Young inequalities:

$$-\nu\int_\Omega \phi'\frac{\partial v_1}{\partial z}\,d\mathbf{x} \leq \nu\left(L^2\int_0^h |\phi'(z)|^2\,dz\right)^{1/2}\|\mathbf{v}\| \tag{12.26}$$

$$\leq \nu L^2\int_0^h |\phi'(z)|^2\,dz + \frac{\nu}{4}\|\mathbf{v}\|^2. \tag{12.27}$$

Inserting (12.27) into (12.23), we find that

$$\frac{d}{dt}|\mathbf{v}(t)|^2 + \frac{\nu}{2}\|\mathbf{v}(t)\|^2 \leq 2\nu L^2\int_0^h |\phi'(z)|^2\,dz. \tag{12.28}$$

This implies that the kinetic energy is uniformly bounded in time.

Hence, it remains only to find ϕ satisfying (12.22) and the boundary condition. The goal is to find such a ϕ that yields a bound in (12.25) as sharp as possible (i.e., as small as possible). Constantin and Doering [1992] chose ϕ of the form

$$\phi(z) = \begin{cases} \frac{U}{2\delta}(2\delta - z) & \text{if } 0 \leq z \leq \delta, \\ \frac{U}{2} & \text{if } \delta \leq z \leq h-\delta, \\ \frac{U}{2\delta}(h-z) & \text{if } h-\delta \leq z \leq h, \end{cases} \tag{12.29}$$

where $0 < \delta \ll h$ is the thickness of the boundary layer for ϕ as explained earlier. Notice that, with this choice, the "background flow" $\phi\mathbf{e}_1$ is not a real flow because it is not a solution of the Navier–Stokes equations. Observe also that this is an example of the lifting operator Λ_δ described in Section 11.

For this choice of ϕ, the condition (12.22) becomes

$$\frac{U\delta}{4} \leq \nu. \tag{12.30}$$

Hence, we let

$$\delta = \frac{4\nu}{U}. \tag{12.31}$$

For this choice of ϕ and δ, we find

$$\int_0^h |\phi'(z)|^2 \, dz = \frac{U^2}{2\delta} = \frac{U^3}{8\nu}. \tag{12.32}$$

Inserting this estimate into (12.25) yields the following bound on the energy dissipation rate:

$$\epsilon \leq \frac{U^3}{8h}. \tag{12.33}$$

This is a rigorous bound of the order of that predicted by Kolmogorov's theory.

Bounds on the Energy Dissipation for Periodic Flows

Consider now the fully periodic case – that is, when the flow is periodic in all directions. For simplicity, we assume the periods are the same in all directions; hence $\Omega = (-L/2, L/2)^d$, where $d = 2$ or 3, depending on the space dimension. We assume that $\mathbf{f}$ belongs to $V = D(A^{1/2})$. For the characteristic speed, we take the quantity

$$U = \frac{1}{L^{d/2}} \limsup_{T\to\infty} \left(\frac{1}{T} \int_0^T |\mathbf{u}(t)|^2 \, dt \right)^{1/2}. \tag{12.34}$$

We also consider a length scale associated with the forcing term $\mathbf{f}$:

$$L_f = \left(\frac{|A^{-1/2}\mathbf{f}|}{|A^{1/2}\mathbf{f}|} \right)^{1/2}. \tag{12.35}$$

We will show that

$$\epsilon = \frac{\nu}{L^d} \langle \|\mathbf{u}\|^2 \rangle \leq c_0 \left(\frac{L}{L_f} \right)^{\frac{d+2}{2}} \frac{U^3}{L}, \tag{12.36}$$

provided the Reynolds number is sufficiently large, that is, if

$$\mathrm{Re} = \frac{UL}{\nu} \geq \frac{1}{c_1}, \tag{12.37}$$

where c_1 is a numerical constant independent of the data of the problem.

In the space-periodic case, we have the energy inequality (see (7.7)):

$$\frac{1}{2} \frac{d}{dt} |\mathbf{u}(t)|^2 + \nu \|\mathbf{u}(t)\|^2 \leq (\mathbf{f}, \mathbf{u}(t)). \tag{12.38}$$

As shown in Appendix A, the kinetic energy in this case is uniformly bounded in time. Hence, we can take the time average of (12.38) to find that

$$\nu \langle \|\mathbf{u}\|^2 \rangle \leq \langle (\mathbf{f}, \mathbf{u}) \rangle \leq |\mathbf{f}| \langle |\mathbf{u}| \rangle \leq |\mathbf{f}| \langle |\mathbf{u}|^2 \rangle^{1/2}. \tag{12.39}$$

Upon dividing by $L^{d/2}$, we obtain

$$\epsilon \leq \frac{|\mathbf{f}|U}{L^{d/2}}. \tag{12.40}$$

We now look for an estimate for $|\mathbf{f}|$ in terms of the length parameters L and L_f and of the average velocity U. We take the inner product in H of the momentum equation (1.1a) with $A^{-1}\mathbf{f}$ to find

$$\frac{1}{2}\frac{d}{dt}(\mathbf{u}, A^{-1}\mathbf{f}) + \nu(\mathbf{u}, A^{-1}\mathbf{f}) + b(\mathbf{u}, \mathbf{u}, A^{-1}\mathbf{f}) = (\mathbf{f}, A^{-1}\mathbf{f}). \tag{12.41}$$

The RHS of (12.41) is equal to $|A^{-1/2}\mathbf{f}|^2$. Hence, we take the time average of (12.41) – and use that the kinetic energy of $\mathbf{u}$ is uniformly bounded in time – to obtain

$$|A^{-1/2}\mathbf{f}|^2 \leq \nu\langle(\mathbf{u}, \mathbf{f})\rangle + \langle b(\mathbf{u}, \mathbf{u}, A^{-1}\mathbf{f})\rangle. \tag{12.42}$$

We bound the first term in the RHS of (12.42) as in (12.39). For the second term, we use the skew symmetry of the trilinear term together with Hölder's inequality:

$$|b(\mathbf{u}, \mathbf{u}, A^{-1}\mathbf{f})| = |b(\mathbf{u}, A^{-1}\mathbf{f}, \mathbf{u})| \leq c_2|\mathbf{u}|^2\|A^{-1/2}\mathbf{f}\|_{L^\infty(\Omega)}.$$

Using now Agmon's inequality (A.29) or (A.48), we have

$$|b(\mathbf{u}, \mathbf{u}, A^{-1}\mathbf{f})| \leq c_1|\mathbf{u}|^2|A^{(d-3)/2}\mathbf{f}|^{1/2}|A^{1/2}\mathbf{f}|^{1/2}. \tag{12.43}$$

Then, (12.42) yields

$$|A^{-1/2}\mathbf{f}|^2 \leq \nu L^{d/2}|\mathbf{f}|U + c_1 L^d|A^{(d-3)/2}\mathbf{f}|^{1/2}|A^{1/2}\mathbf{f}|^{1/2}U^2. \tag{12.44}$$

Multiplying (12.44) by $|\mathbf{f}|$ and dividing the result by $|A^{-1/2}\mathbf{f}|^2$ implies that

$$|\mathbf{f}| \leq \nu L^{d/2}\frac{|\mathbf{f}|^2}{|A^{-1/2}\mathbf{f}|^2}U + c_1 L^d\frac{|\mathbf{f}||A^{1/2}\mathbf{f}|^{1/2}|A^{(d-3)/2}\mathbf{f}|^{1/2}}{|A^{-1/2}\mathbf{f}|^2}U^2. \tag{12.45}$$

Inserting this into (12.40) yields

$$\epsilon \leq \nu\frac{|\mathbf{f}|^2}{|A^{-1/2}\mathbf{f}|^2}U^2 + c_1 L^{d/2}\frac{|\mathbf{f}||A^{1/2}\mathbf{f}|^{1/2}|A^{(d-3)/2}\mathbf{f}|^{1/2}}{|A^{-1/2}\mathbf{f}|^2}U^3. \tag{12.46}$$

If the condition (12.37) on the Reynolds number holds, we obtain

$$\epsilon \leq c_1 L\frac{|\mathbf{f}|^2}{|A^{-1/2}\mathbf{f}|^2}U^3 + c_1 L^{d/2}\frac{|\mathbf{f}||A^{1/2}\mathbf{f}|^{1/2}|A^{(d-3)/2}\mathbf{f}|^{1/2}}{|A^{-1/2}\mathbf{f}|^2}U^3. \tag{12.47}$$

The interpolation inequalities (6.14) derived in Section 6 give

$$\begin{aligned} |\mathbf{f}| &\leq |A^{-1/2}\mathbf{f}|^{1/2}|A^{1/2}\mathbf{f}|^{1/2}, \\ |A^{(d-3)/2}\mathbf{f}| &\leq |A^{-1/2}\mathbf{f}|^{(4-d)/2}|A^{1/2}\mathbf{f}|^{(d-2)/2}. \end{aligned} \tag{12.48}$$

Hence, from (12.47),

$$\epsilon \leq c_1\left(L\frac{|A^{1/2}\mathbf{f}|}{|A^{-1/2}\mathbf{f}|} + L^{d/2}\frac{|A^{-1/2}\mathbf{f}|^{(d+2)/4}}{|A^{-1/2}\mathbf{f}|^{(d+2)/4}}\right)U^3. \tag{12.49}$$

From the definition (12.35) of the length scale L_f, we can rewrite (12.49) as

$$\epsilon \leq c_1 \left(\frac{L}{L_f}\right)^{\frac{d+2}{2}} \left(1 + \left(\frac{L_f}{L}\right)^{\frac{d-2}{2}}\right) \frac{U^3}{L}. \tag{12.50}$$

Since the first eigenvalue of the Stokes operator in the space-periodic case is $4\pi^2/L^2$ (see (6.18)), one can check that

$$L_f \leq \frac{L}{2\pi}. \tag{12.51}$$

Therefore, (12.50) yields

$$\epsilon \leq c_0 \left(\frac{L}{L_f}\right)^{\frac{d+2}{2}} \frac{U^3}{L}, \tag{12.52}$$

where c_0 is a numerical constant. Hence, (12.36) holds.

The relation (12.52) is not exactly the same as that derived by Kolmogorov [1941a], owing to the additional term involving the length scale L_f. However, one must bear in mind that, under Kolmogorov's derivation, the forcing term was assumed to be a trigonometric polynomial containing only lower-order modes. More precisely, $\mathbf{f}$ is of the form

$$\mathbf{f}(x) = \sum_{\substack{\mathbf{k} \in \mathbb{Z}^d \setminus \{0\} \\ \frac{2\pi}{L}|\mathbf{k}| \leq \kappa_0}} \hat{\mathbf{f}}_{\mathbf{k}} e^{i\frac{2\pi}{L}\mathbf{k}\cdot\mathbf{x}} \tag{12.53}$$

for some $\kappa_0 > 0$, which represents the highest possible wavenumber in the Fourier representation of $\mathbf{f}$. In this case, one can check that

$$L_f \geq \frac{1}{\kappa_0}. \tag{12.54}$$

Therefore, (12.52) implies

$$\epsilon \leq c_0 (L\kappa_0)^{\frac{d+2}{2}} \frac{U^3}{L}. \tag{12.55}$$

For $\kappa_0 \sim 2\pi/L$, we recover rigorously Kolmogorov's estimate (12.3).

13 Nondimensional Estimates and the Grashof Number

As mentioned in Section I.1, the variables associated with fluid flows, as in many other physical systems, are not completely independent; rescaling two or more variables appropriately leads to the same fluid flow behavior. In the case of flows of incompressible viscous fluids, multiplying the velocity field by a given factor and dividing the characteristic length scale of the physical domain occupied by the fluid by the same factor leaves the Reynolds number unchanged; hence, the resulting flows are essentially the same. This fact, known as *Reynolds similarity,* makes wind tunnel simulations appropriate. With this in mind, it is useful to consider nondimensional quantities that allow us to relate different systems. One important nondimensional quantity is the Reynolds number just mentioned. Another nondimensional quantity that we use often is the so-called Grashof number, which is proportional to the forcing term $\mathbf{f}$. The Grashof number is denoted by G; it is defined in terms of the forcing term **f**, the viscosity ν, and some other parameter with the dimension of length. The usual candidate for the length parameter is derived from the first eigenvalue of the Stokes operator, λ_1, which has the dimension of L^{-2}. For dimensional reasons, the explicit form of G depends on the spatial dimension, so we consider the two cases separately. This Grashof number was introduced in Foias et al. [1983b] by analogy with the Grashof number used in thermohydraulics, where the force **f** is the buoyancy force.

The 2-Dimensional Case

We assume first that **f** is time-independent and that it belongs to the space H. The forcing term has the dimension of acceleration – that is, L/T^2, where T is a time scale and L is a length scale. Hence, the norm $|\mathbf{f}|$ in H of **f** has the dimension of L^2/T^2. Since the kinematic viscosity ν has the dimension of L^2/T, we find that $|\mathbf{f}|/\nu^2$ has the dimension of L^{-2}. Then, since λ_1 (the first eigenvalue of the Stokes operator) has the dimension of L^{-2}, we see that $|\mathbf{f}|/(\nu^2\lambda_1)$ is nondimensional. Hence, in the 2-dimensional case, we define the Grashof number by

$$\mathrm{G} = \frac{|\mathbf{f}|}{\nu^2\lambda_1}. \tag{13.1}$$

Several estimates obtained for the solutions of the Navier–Stokes equations in the previous sections and in the appendices can now be expressed in terms of the Grashof number. For instance, in the 2-dimensional case, with periodic boundary condition and vanishing space average or with a no-slip boundary condition, the energy equation (7.7) implies the following asymptotic bounds, as shown in Appendix A (see (A.53) and (A.54)):

$$\limsup_{t\to\infty} |\mathbf{u}(t)|^2 \leq \nu^2\mathrm{G}^2; \tag{13.2}$$

$$\limsup_{T\to\infty} \frac{1}{T}\int_0^T \|\mathbf{u}(s)\|^2\, ds \leq \nu^2\lambda_1\mathrm{G}^2; \tag{13.3}$$

and, for any $T > 0$,

$$\limsup_{t\to\infty} \frac{1}{T}\int_t^{t+T} \|\mathbf{u}(s)\|^2\, ds \leq \left(\frac{\nu}{T} + \nu^2\lambda_1\right)\mathrm{G}^2. \tag{13.4}$$

Note that the left-hand sides of these expressions are not nondimensional; the quantities multiplying the Grashof number on the corresponding right-hand sides make the expressions dimensionally consistent. All the limits in (13.2)–(13.4) are uniform with respect to initial conditions bounded in H.

In the 2-dimensional periodic case we have, owing to the orthogonality property (see (A.62)),

$$b(\mathbf{u}, \mathbf{u}, A\mathbf{u}) = 0.$$

We also obtain (see (A.66) and (A.67)):

$$\limsup_{t\to\infty} \|\mathbf{u}(t)\|^2 \le \nu^2\lambda_1 \mathrm{G}^2; \tag{13.5}$$

$$\limsup_{T\to\infty} \frac{1}{T}\int_0^T |A\mathbf{u}(s)|^2\,ds \le \nu^2\lambda_1^2\mathrm{G}^2; \tag{13.6}$$

and, for any $T > 0$,

$$\limsup_{t\to\infty} \frac{1}{T}\int_t^{t+T} |A\mathbf{u}(s)|^2\,ds \le \left(\frac{\nu\lambda_1}{T} + \nu^2\lambda_1^2\right)\mathrm{G}^2. \tag{13.7}$$

As before, the limits are uniform with respect to initial conditions bounded in H (not necessarily V; this is due to the regularization property of the NSE).

In the 2-dimensional periodic case, for $\mathbf{f}$ still in H, we can also obtain a simple and explicit bound for the width δ_0 of the pencil-like domain Δ^+ of time analyticity of the solutions (see (8.22)) derived in Section 8:

$$\delta_0 \ge \frac{1}{c_3\nu\lambda_1(1+\mathrm{G}^2)\log(c_4(1+\mathrm{G}^2))}, \tag{13.8}$$

where c_3 and c_4 depend only on the shape of the domain Ω (i.e., on the ratio L_2/L_1). Moreover, if $\mathbf{f}$ belongs to the Gevrey space $D(\exp(\sigma_1 A^{1/2}))$ for some $\sigma_1 > 0$, then the same bound holds (with possibly different constants c_3 and c_4) as obtained in Section 9 (see (9.31)) for the width of the domain $\Delta^+_{\sigma_1}$ of time analyticity with values in Gevrey space.

If the forcing term $\mathbf{f} = \mathbf{f}(t)$ is time-dependent and belongs to $L^\infty(0,\infty; H)$, then the bounds (13.2)–(13.7) can also be obtained by defining the Grashof number as[17]

$$\mathrm{G} = \frac{1}{\nu^2\lambda_1}\limsup_{t\to\infty}|\mathbf{f}(t)|. \tag{13.9}$$

In the case of estimates (13.2), (13.3), and (13.4), we can also consider forces only in V'. Here the Grashof number is proportional to $\|\mathbf{f}\|_{V'}$, and the powers of ν and λ_1 must be modified accordingly.

The 3-Dimensional Case

In the 3-dimensional case, the quantity $|\mathbf{f}|$ has the dimension of $L^{5/2}/T^2$. Hence, we must take different powers of ν and λ_1 in the definition of the Grashof number. Since ν has the dimension of L^2/T, we see that $|f|/\nu^2$ has the dimension of $L^{-3/2}$. Then, since λ_1 has the dimension of L^{-2}, we consider the power $\lambda_1^{3/4}$, which has the dimension of $L^{-3/2}$. Hence, we define the Grashof number in the 3-dimensional case by

$$\mathrm{G} = \frac{|\mathbf{f}|}{\nu^2\lambda_1^{3/4}}. \tag{13.10}$$

[17] Since in this case $\mathbf{f}$ may only be bounded almost everywhere, the limit is taken in the sense of the essential supremum. We recall that the essential suppremum of a function g is the smallest M such that the set of points where $|g|$ exceeds M has measure zero. The limit in (13.9) is to be understood as the smallest number M such that, for any $\varepsilon > 0$, there exists a $T > 0$ such that the set of points t in $[T,\infty)$ where $|\mathbf{f}(t)|$ exceeds $M+\varepsilon$ has measure zero.

With this definition of the Grashof number, the estimates corresponding to (13.2), (13.3), and (13.4) now read as follows:

$$\limsup_{t\to\infty} |\mathbf{u}(t)|^2 \le \frac{\nu^2}{\lambda_1^{1/2}} \mathrm{G}^2; \tag{13.11}$$

$$\limsup_{T\to\infty} \frac{1}{T}\int_0^T \|\mathbf{u}(s)\|^2\,ds \le \nu^2 \lambda_1^{1/2} \mathrm{G}^2; \tag{13.12}$$

and, for any $T > 0$,

$$\limsup_{t\to\infty} \frac{1}{T}\int_t^{t+T} \|\mathbf{u}(s)\|^2\,ds \le \frac{1}{\lambda_1^{1/2}}\left(\frac{\nu}{T} + \nu^2\lambda_1\right)\mathrm{G}^2. \tag{13.13}$$

The lack of the orthogonality property (A.62) means that other estimates in the 3-dimensional case corresponding to those just obtained for the 2-dimensional case are exponential in G.

Appendix A Mathematical Complements

A.1 Function Spaces

The aim in this section is to recall in more detail some of the function spaces that we use throughout the book and that are based on the spaces H and V of finite kinetic energy (L^2) and of finite enstrophy (H^1) (see Section 5). We start with the concept of duality in function spaces, then we consider the weak topology in H, and finally we examine space–time function spaces.

Dual Spaces – Riesz Representation Theorem

For every Hilbert space V associated with every choice of boundary conditions, we can define a corresponding dual space V' consisting of all the linear functionals defined on V. We may do this in a way akin to the theory of distributions or generalized functions. The distributions are continuous linear functionals acting on the space of compactly supported infinitely differentiable functions; the square integrable functions are identified as linear functionals through the L^2 inner product. Here, we proceed in a similar fashion. Consider the set of all continuous linear functionals

$$\mathbf{v} \mapsto \ell(\mathbf{v})$$

from V into $\mathbb{R}$; continuity here means that there exists a constant C, depending on the functional ℓ, such that

$$|\ell(\mathbf{v})| \le C\|\mathbf{v}\|_V$$

for all vector fields $\mathbf{v}$ in V. We denote by V' the set of all continuous linear functionals on V. This is a vector space that we endow with the norm

$$\|\ell\|_{V'} = \sup_{\substack{\mathbf{v}\in V\\ \mathbf{v}\neq 0}} \frac{|\ell(\mathbf{v})|}{\|\mathbf{v}\|_V}. \tag{A.1}$$

The Riesz representation theorem is a simple yet powerful result. It states that any Hilbert space V is isomorphic to its dual; that is, there exists a one-to-one linear mapping Λ from V onto its dual V'. Furthermore, for every ℓ in V' and $\mathbf{u}$ in V such that $\Lambda\mathbf{u} = \ell$, we have

$$(\Lambda\mathbf{u}, \mathbf{v})_{V',V} = (\mathbf{u}, \mathbf{v})_V, \quad \text{for every } \mathbf{v} \text{ in } V,$$

where $(\cdot, \cdot)_{V',V}$ denotes the pairing (duality) between V' and V.

In our case, this operator Λ is, in general, the same as the Stokes operator A.

The Pair V, H

The pair of Hilbert spaces V, H appears in the mathematical theory of the evolutionary Navier–Stokes equations: V is a subspace of the Sobolev space $H^1(\Omega)^d$ (the suitable space of vector fields with finite enstrophy), and H is the closure of V in $L^2(\Omega)^d$, the space of square integrable vector fields. Hence,

$$V \subset H,$$

V is dense in H, and the injection (embedding) of V in H is continuous.

If the injection of V in H is compact, we have seen in Section 6 how one can define an unbounded operator A in H (same as the Λ just described) whose inverse A^{-1} is compact in H. We can also define all the powers A^s of A, and then V, H, and V' are the same as (isomorphic to) the domains of $A^{1/2}$, I, and $A^{-1/2}$, respectively; in particular, we obtain the triplet V, H, V' as in (A.5). Here, we want to introduce and describe the triplet V, H, V' by a different method – one that extends to the case where the injection of V in H is not compact.

One can identify the space H with a subspace of V'. Indeed, given a vector field $\mathbf{u}$ in H, we can consider $\mathbf{u}$ as a linear functional in V' given by

$$\mathbf{v} \mapsto (\mathbf{u}, \mathbf{v}). \tag{A.2}$$

It is continuous on V because

$$|(\mathbf{u}, \mathbf{v})| \leq |\mathbf{u}|_H |\mathbf{v}|_H \leq c_1 |\mathbf{u}| \|\mathbf{v}\|_V \leq C_2 \|\mathbf{v}\|_V$$

for suitable (dimensional) constants $c_1, C_2 > 0$, with c_1 independent of $\mathbf{u}$ and C_2 depending on $\mathbf{u}$; here we have used the usual Cauchy–Schwarz inequality and the fact that the norm on V, $\|\cdot\|_V$, dominates (in the sense of (5.11) and (5.19)) the norm on H, $|\cdot|_H$. More specifically, in the periodic case with nonvanishing space average, it follows from inequality (5.15) and the definition (A.1) of the dual norm that

$$\|\mathbf{u}\|_{V'} \leq L|\mathbf{u}|_H, \tag{A.3}$$

where $L = \min\{L_1, \ldots, L_d\}$. In the no-slip case and in the periodic case with vanishing space average, it now follows from inequalities (5.11) and (5.19) that

$$\|\mathbf{u}\|_{V'} \leq \frac{1}{\lambda_1^{1/2}} |\mathbf{u}|_H. \tag{A.4}$$

Since V is dense in H and the injection mapping from V into H is continuous, we infer by duality that its adjoint from H' into V' is continuous with a dense image. Identifying further H with its dual H', we infer that

$$V \subset H \subset V', \tag{A.5}$$

where each space is dense in the following one and the embeddings are continuous. With this identification in mind, we can consider elements in V' also as vector fields and denote them by the usual letters $\mathbf{u}$ and $\mathbf{v}$, though strictly speaking an element in V' is not a function on Ω and does not make sense pointwise. We will denote by $(\ell, \mathbf{v})_{V',V}$, or more simply $(\ell, \mathbf{v})$, the action $\ell(\mathbf{v})$ of an element ℓ of V' on an element $\mathbf{v}$ of V. This notation is consistent with (A.2) when $\ell = \mathbf{u} \in H$. If ℓ belongs to V' but not to H, then the product $(\ell, \mathbf{v})$ is called the *duality product* between V' and V, and it is simply the application of the linear functional $\mathbf{u}$ to $\mathbf{v}$; in general, it does not make sense as an integral of the product of two functions. However, when $\mathbf{u}$ is more regular (i.e., when $\mathbf{u} \in H$), this application becomes exactly the L^2 inner product between $\mathbf{u}$ and $\mathbf{v}$. The duality product can be thought of as a generalization of the L^2 product to less regular vector fields $\mathbf{u}$ (belonging to V' but not to H) at the expense of $\mathbf{v}$ being more regular (belonging to V). It is similar to the situation in the theory of distributions, where, for instance, the Dirac delta is not a function in the classical sense. The dual norm of an element $\mathbf{u}$ in V' can be rewritten as

$$\|\mathbf{u}\|_{V'} = \sup_{\substack{\mathbf{v} \in V \\ \mathbf{v} \neq 0}} \frac{|(\mathbf{u}, \mathbf{v})|}{\|\mathbf{v}\|_V}, \tag{A.6}$$

and the following generalized form of the Cauchy–Schwarz inequality holds:

$$(\mathbf{u}, \mathbf{v}) \leq \|\mathbf{u}\|_{V'} \|\mathbf{v}\|_V \quad \text{for all } \mathbf{u} \text{ in } V' \text{ and all } \mathbf{v} \text{ in } V. \tag{A.7}$$

For each choice of the boundary conditions we obtain, then, the dual spaces V'_{nsp}, V'_{per}, and $\dot{V}'_{\text{per}}$.

The space V' is naturally present in the mathematical theory of the Navier–Stokes equations. The reason is that, for a flow $\mathbf{u}$ in the natural space V of finite enstrophy, the inertial term $(\mathbf{u} \cdot \nabla)\mathbf{u}$ does not, in general, belong to H; it belongs instead to the larger space V'.

We indicated in Section 6 that usually we can identify the spaces V and V' with the domain of powers of the Stokes operators, $D(A^{1/2})$ and $D(A^{-1/2})$, respectively. The only exception is the periodic case with nonvanishing space average, where some adjustment is needed because A is positive but not positive definite.

Weak Topology in Normed Spaces

In a Hilbert space X, one can define a topology (concept of convergence) weaker than the usual topology corresponding to the norm $\|\cdot\|_X$. We say that a sequence $\{\mathbf{u}_n\}_n$ converges weakly to $\mathbf{u}$ in X if, for all $\mathbf{v}$ in X,

$$(\mathbf{u}_n, \mathbf{v})_X \to (\mathbf{u}, \mathbf{v})_X \quad \text{as } n \to \infty.$$

In order to emphasize the difference, we say sometimes that $\mathbf{u}_n$ converges strongly to $\mathbf{u}$ in X if $\mathbf{u}_n$ converges to $\mathbf{u}$ in the usual sense – that is, if $\|\mathbf{u}_n - \mathbf{u}\|_X \to 0$ as $n \to \infty$.

It is easy to see that $\{\mathbf{u}_n\}_n$ converges strongly to $\mathbf{u}$ in X if and only if $\mathbf{u}_n$ converges weakly to $\mathbf{u}$ in X and $\|\mathbf{u}_n\|_X$ converges to $\|\mathbf{u}\|_X$.

In order to visualize the concept of weak convergence, one can think of the functions $u_n = \sin nx$ in $X = L^2(0,1)$; it is easy to see that u_n goes to 0 weakly in $L^2(0,1)$ but not strongly.

More generally, if X and X' are two normed spaces in duality (X' dual of X in the aforementioned sense) then one can define two different topologies on X: (i) that of the norm, $\mathbf{u}_n \to \mathbf{u}$ if and only if $\|\mathbf{u}_n - \mathbf{u}\| \to 0$ as $n \to \infty$; and (ii) the weak topology denoted $\sigma(X, X')$, for which $\mathbf{u}_n \to \mathbf{u}$ if and only if, for all $\mathbf{v}$ in X, $(\mathbf{u}_n, \mathbf{v})_{X,X'} \to (\mathbf{u}, \mathbf{v})_{X,X'}$ as $n \to \infty$.

Of particular interest to us in this chapter is the weak topology on H; we denote by H_w the space H endowed with this weak topology. This space is important because, in general, the continuity of the weak solutions of the 3-dimensional NSE (see (A.18)) is known to be achieved only in this weak sense (i.e., in the space (A.17)).

One must keep in mind that, unlike the spaces V and H (whose norms are embedded in the physics of the Navier–Stokes equations as enstrophy and energy), the weak topology on H has been introduced as a mere technicality commanded by our present limitations in the understanding of the theory of these equations. One may surmise (or hope) that better theorems will be derived, ones for which such a concept is no longer needed.

Space–Time Function Spaces

It is customary and useful to regard a function $\mathbf{u} = \mathbf{u}(\mathbf{x}, t)$ as a time-dependent function $\mathbf{u} = \mathbf{u}(t)$ with values in appropriate function spaces. There are several function spaces that one may consider; moreover, the function $t \mapsto \mathbf{u}(t)$ might be continuous, continuously differentiable, integrable, square integrable, and so on, leading to several types of space–time function spaces. In order to set down the notation we consider, for instance, a function space X, which might be H, or V, or the domain $V_{2s} = D(A^s)$ of some power $s \in \mathbb{R}$ of the Stokes operator A.

The space of continuous functions on an interval I and with values in X is denoted by

$$\mathcal{C}(I, X). \tag{A.8}$$

Typically, the interval I will be $I = [0, T]$ or $(0, T)$ but, more generally, it can be any interval with endpoints a and b, $-\infty \leq a < b \leq +\infty$, and it can be either open or closed at either a or b. The norm in $\mathcal{C}(I, X)$ is taken to be the maximum of the X-norm over I, unless otherwise stated.

One can also work with spaces of integrable functions. For functions defined on an interval I with endpoints a and b ($-\infty \leq a < b \leq +\infty$), one may consider the space

$$L^p(a, b; X) \quad \text{or} \quad L^p(I; X) \tag{A.9}$$

of functions whose X-norm to the pth power is integrable over I, with $p \geq 1$.[18] In this case the functions are defined almost everywhere (i.e., up to a set of measure zero), so it does not make any difference whether I is open or closed. Hence, it has been the convention to denote this space by $L^p(a, b; X)$ instead of $L^p((a, b), X)$ or $L^p([a, b]; X)$. The norm in this space is

$$\|\mathbf{u}\|_{L^p(a,b;X)} = \left(\int_a^b |\mathbf{u}(t)|_X^p \, dt \right)^{1/p}, \tag{A.10}$$

where we denote the norm of X by $|\cdot|_X$. The Cauchy–Schwarz and Hölder inequalities hold on those spaces exactly as in the real-valued case, as long as $p \geq 1$. The spaces $L^p(a, b; X)$ are Banach spaces, that is, complete normed spaces.

As in the real-valued case, when $p = 2$ we obtain a Hilbert space $L^2(a, b; X)$ with inner product

$$(\mathbf{u}, \mathbf{v})_{L^2(a,b;X)} = \int_a^b (\mathbf{u}(t), \mathbf{v}(t))_X \, dt; \tag{A.11}$$

again, we have denoted the inner product in X by $(\cdot, \cdot)_X$.

For $p = \infty$, we denote by

$$L^\infty(a, b; X) \quad \text{or} \quad L^\infty(I; X) \tag{A.12}$$

the space of essentially bounded functions from I into X (i.e., bounded uniformly except possibly on a set of measure zero). We norm this space with the norm

$$\|\mathbf{u}\|_{L^\infty(a,b;X)} = \underset{t \in [a,b]}{\text{sup.ess.}} |\mathbf{u}(t)|_X; \tag{A.13}$$

here, "sup.ess." denotes the essential supremum (as in the case of real-valued functions). Hence, $|\mathbf{u}(t)|_X$ is bounded by its $L^\infty(a, b; X)$-norm except possibly on a subset of $[a, b]$ of measure zero.

For all these L^p spaces, if the interval I is infinite then there is a corresponding "local" space

$$L^p_{\text{loc}}(a, b; X) \quad \text{or} \quad L^p_{\text{loc}}(I; X), \tag{A.14}$$

which is the space of functions that are locally in L^p in the sense that they belong to $L^p(J; X)$ for every bounded subinterval J of I. This space has a metric (but not a norm) under which it is complete, but we will not work with it explicitly. A nonzero constant function, for instance, belongs to $L^p_{\text{loc}}(\mathbb{R}; X)$ for all $p \geq 1$, but not to $L^p(\mathbb{R}; X)$. Typically, the solutions of the Navier–Stokes equations belong to $L^2_{\text{loc}}(0, \infty; V)$ but not to $L^2(0, \infty; V)$, since they may not decay to zero as $t \to \infty$.

The duality product between V and V' carries over to space–time function spaces. The duality product between $L^2(a, b; V)$ and $L^2(a, b; V')$ reads

$$\int_a^b \langle \mathbf{u}(t), \mathbf{v}(t) \rangle \, dt, \tag{A.15}$$

and the Cauchy–Schwarz inequality in this space takes the form

[18] A case with $p < 1$ is discussed later (see (A.43)), but the space is no longer a normed space.

$$\left|\int_a^b (\mathbf{u}(t),\mathbf{v}(t))\,dt\right| \le \left(\int_a^b |\mathbf{u}(t)|_{V'}^2\,dt\right)^{1/2}\left(\int_a^b |\mathbf{v}(t)|_V^2\,dt\right)^{1/2}. \quad \text{(A.16)}$$

Similarly, for any $s \in \mathbb{R}$ and any $1 \le p < \infty$, $L^{p'}(a,b;V_{-s})$ is the dual space of $L^p(a,b;V_s)$, where $1/p + 1/p' = 1$.

Another useful space is that of continuous functions with values in $X = H_w$, where we recall that H_w is the space H endowed with its weak topology. This leads to the space

$$\mathcal{C}(I;H_w), \quad \text{(A.17)}$$

which is the space of weakly continuous functions from I into H. The functions on this space are weakly continuous in the sense that, for each vector field $\mathbf{v}$ in H, the function

$$t \mapsto (\mathbf{u}(t),\mathbf{v})_H \quad \text{(A.18)}$$

is continuous from the interval I into $\mathbb{R}$. This is a space occurring naturally when we work with the weak solutions of the Navier–Stokes equations. Typically, a weak solution of the NSE in dimension 3 belongs to $\mathcal{C}([0,\infty);H_w)$.

A.2 Weak and Strong Solutions of the NSE in Dimension 3

Here we present further results concerning the solutions of the NSE in dimension 3. These results will be needed for some of the proofs in Chapter IV.

The Energy Inequality

Consider a weak solution $\mathbf{u} = \mathbf{u}(t)$ on the interval $[0,T]$ in time. The forcing term $\mathbf{f} = \mathbf{f}(t)$ is assumed to belong to $L^2(0,T;H)$.

As stated in Theorem 7.1, this solution satisfies the energy inequality (7.5). This implies in particular that, for every nonnegative test function $\varphi \in \mathcal{C}_c^\infty(0,T)$ (i.e., φ is an infinitely differentiable, nonnegative, real-valued function with compact support on $(0,T)$),

$$-\frac{1}{2}\int_0^T |\mathbf{u}(s)|^2\varphi'(s)\,ds + \nu\int_0^T \varphi(s)\|\mathbf{u}(s)\|^2\,ds \le \int_0^T \varphi(s)(\mathbf{f}(s),\mathbf{u}(s))\,ds. \quad \text{(A.19)}$$

This means that the weak solution satisfies the following differential inequality in the distribution sense on $[0,T]$:

$$\frac{1}{2}\frac{d}{dt}|\mathbf{u}(t)|^2 + \nu\|\mathbf{u}(t)\|^2 \le (\mathbf{f}(t),\mathbf{u}(t)). \quad \text{(A.20)}$$

The energy inequality (7.5) is obtained by passing to the limit (as the number of modes goes to infinity) in the corresponding equation for the solution of the Galerkin approximation of the Navier–Stokes equations, which is obtained by applying the Galerkin projector (6.28) to the system of equations. As shown in Appendix B.1, the inequality (A.19) implies that, for every Lebesgue point[19] t_0 of $t \mapsto |\mathbf{u}(t)|^2$ (hence, for almost every t_0; see footnote 9) and every t with $0 \le t_0 \le t \le T$,

[19] The Lebesgue points of a function were defined in footnote 9.

$$\frac{1}{2}|\mathbf{u}(t)|^2 + \nu \int_{t_0}^{t} \|\mathbf{u}(s)\|^2\, ds \le \frac{1}{2}|\mathbf{u}(t_0)|^2 + \int_{t_0}^{t} (\mathbf{f}(s), \mathbf{u}(s))\, ds. \tag{A.21}$$

The energy inequality (7.5) also implies that $t_0 = 0$ belongs to the Lebesgue set of $t \mapsto |\mathbf{u}(t)|^2$, so that (A.21) holds, in particular, for $t_0 = 0$. Conversely, it can be shown that a weak solution which satisfies the differential inequality (A.20) in the distribution sense on $[0, T]$ and for which $t_0 = 0$ belongs to the Lebesgue set of $t \mapsto |\mathbf{u}(t)|^2$ must also satisfy the energy inequality (7.5) and hence be a weak solution in the sense of Theorem 7.1. As mentioned earlier, there may be solutions that do not satisfy the energy inequality (A.21), but, for us, a weak solution always means a solution which is weakly continuous, which belongs to $L^2(\Omega \times [0, T])$ together with its first-order derivatives in space, and for which the energy inequality holds in the distribution sense, with $t_0 = 0$ being a Lebesgue point of $|\mathbf{u}(\cdot)|^2$.

More generally, we consider weak solutions on an arbitrary time interval I in $\mathbb{R}$, finite or not, and open or closed at either endpoint. Then, a weak solution on I is a solution which is weakly continuous in I, which for every bounded interval $J \subset I$ belongs to $L^2(\Omega \times J)$ together with its first-order derivatives in space, and for which the energy inequality (A.21) holds (in the distribution sense) in I. Moreover, if the interval I is closed at the left endpoint then we require that the left endpoint be a Lebesgue point of $|\mathbf{u}(\cdot)|^2$.

The Abstract Functional Equation and Nonlinear Inequalities

Recall the functional form of the Navier–Stokes equations given in (3.7):

$$\frac{d\mathbf{u}}{dt} + \nu A\mathbf{u} + B(\mathbf{u}) = \mathbf{f}. \tag{A.22}$$

We have dropped the Helmholtz–Leray projector P_L on $\mathbf{f}$ because we already assume that $\mathbf{f} = \mathbf{f}(t)$ takes its values in H. As in (3.9), we may write this equation in the concise form

$$\mathbf{u}' = \mathbf{F}(t, \mathbf{u}), \tag{A.23}$$

with

$$\mathbf{u}' = \mathbf{u}_t = \frac{\partial \mathbf{u}}{\partial t}, \qquad \mathbf{F}(t, \mathbf{u}) = \mathbf{f}(t) - \nu A\mathbf{u} - B(\mathbf{u}). \tag{A.24}$$

The operator $B(\cdot) = B(\cdot, \cdot)$ was defined in (3.8) by $B(\mathbf{u}, \mathbf{v}) = P_L((\mathbf{u} \cdot \nabla)\mathbf{v})$. This requires $(\mathbf{u} \cdot \nabla)\mathbf{v}$ to be in L^2. Hence, $B(\mathbf{u}, \mathbf{v})$ is a priori not defined for $\mathbf{u}$ and $\mathbf{v}$ in V. More regularity is needed. One may take, for instance, either $\mathbf{u}$ or $\mathbf{v}$ in the domain $D(A)$ of the Stokes operator. In fact, consider the formal relation

$$(B(\mathbf{u}, \mathbf{v}), \mathbf{w}) = b(\mathbf{u}, \mathbf{v}, \mathbf{w}), \tag{A.25}$$

where the trilinear operator $b(\cdot, \cdot, \cdot)$ was defined in (4.2). The trilinear form $b(\mathbf{u}, \mathbf{v}, \mathbf{w})$ is defined and in three dimensions the following inequalities hold, for some appropriate constant c_1, in the following cases:

$$|b(\mathbf{u},\mathbf{v},\mathbf{w})| \le c_1|\mathbf{u}|^{1/4}\|\mathbf{u}\|^{3/4}\|\mathbf{v}\|^{1/4}|A\mathbf{v}|^{3/4}|\mathbf{w}|, \quad \mathbf{u}\in V,\ \mathbf{v}\in D(A),\ \mathbf{w}\in H; \tag{A.26a}$$

$$|b(\mathbf{u},\mathbf{v},\mathbf{w})| \le c_1\|\mathbf{u}\|^{1/2}|A\mathbf{u}|^{1/2}\|\mathbf{v}\||\mathbf{w}|, \quad \mathbf{u}\in D(A),\ \mathbf{v}\in V,\ \mathbf{w}\in H; \tag{A.26b}$$

$$|b(\mathbf{u},\mathbf{v},\mathbf{w})| \le c_1|\mathbf{u}|^{1/4}|A\mathbf{u}|^{3/4}\|\mathbf{v}\||\mathbf{w}|, \quad \mathbf{u}\in D(A),\ \mathbf{v}\in V,\ \mathbf{w}\in H; \tag{A.26c}$$

$$|b(\mathbf{u},\mathbf{v},\mathbf{w})| \le c_1|\mathbf{u}|\|\mathbf{v}\||\mathbf{w}|^{1/4}|A\mathbf{w}|^{3/4}, \quad \mathbf{u}\in H,\ \mathbf{v}\in V,\ \mathbf{w}\in D(A); \tag{A.26d}$$

$$|b(\mathbf{u},\mathbf{v},\mathbf{w})| \le c_1|\mathbf{u}|^{1/4}\|\mathbf{u}\|^{3/4}\|\mathbf{v}\||\mathbf{w}|^{1/4}\|\mathbf{w}\|^{3/4}, \quad \mathbf{u}\in V,\ \mathbf{v}\in V,\ \mathbf{w}\in V; \tag{A.26e}$$

$$|b(\mathbf{u},\mathbf{v},\mathbf{w})| \le c_1|\mathbf{u}|^{1/4}\|\mathbf{u}\|^{3/4}|\mathbf{v}|^{1/4}\|\mathbf{v}\|^{3/4}\|\mathbf{w}\|, \quad \mathbf{u}\in V,\ \mathbf{v}\in V,\ \mathbf{w}\in V. \tag{A.26f}$$

These inequalities can be obtained by applying Hölder's inequality, the Ladyzhenskaya inequality (see Section I.4)

$$|\mathbf{u}|_{L^4} \le c_2|\mathbf{u}|^{1/4}\|\mathbf{u}\|^{3/4}, \quad \mathbf{u}\in V, \tag{A.27}$$

and Agmon's inequalities:

$$|\mathbf{u}|_{L^\infty} \le c_2|\mathbf{u}|^{1/4}|A\mathbf{u}|^{3/4}, \quad \mathbf{u}\in D(A); \tag{A.28}$$

$$|\mathbf{u}|_{L^\infty} \le c_2\|\mathbf{u}\|^{1/2}|A\mathbf{u}|^{1/2}, \quad \mathbf{u}\in D(A). \tag{A.29}$$

Here c_2 is some appropriate constant (see e.g. Constantin and Foias [1988] or Temam [1979] for details).

One can see from (A.26a) and (A.26c) that $B(\mathbf{u},\mathbf{v})$ is indeed well-defined (with values in H) for $\mathbf{u}$ in V and $\mathbf{v}$ in $D(A)$ or vice versa. On the other hand, the inequality (A.26f) can be used to extend the definition of $B(\cdot,\cdot)$ to less regular spaces. In fact, for $\mathbf{u}$ and $\mathbf{v}$ in V we see from (A.26f) that $B(\mathbf{u},\mathbf{v})$ belongs to the dual space $V' = D(A^{-1/2})$, with

$$\|B(\mathbf{u},\mathbf{v})\|_{V'} \le c_1|\mathbf{u}|^{1/4}\|\mathbf{u}\|^{3/4}|\mathbf{v}|^{1/4}\|\mathbf{v}\|^{3/4}, \quad \mathbf{u},\mathbf{v}\in V. \tag{A.30}$$

This is actually an extension of the operator $B(\cdot,\cdot)$ as defined in (3.8), but for notational simplicity we still denote this extension by $B(\cdot,\cdot)$. In this case $B(\mathbf{u},\mathbf{v})$ belongs to V', so the trilinear operation $b(\mathbf{u},\mathbf{v},\mathbf{w})$ is now to be understood in the duality sense

$$b(\mathbf{u},\mathbf{v},\mathbf{w}) = (B(\mathbf{u},\mathbf{v}),\mathbf{w}), \quad \mathbf{u},\mathbf{v},\mathbf{w}\in V. \tag{A.31}$$

A useful inequality follows from (A.30) when $\mathbf{v} = \mathbf{u}$:

$$\|B(\mathbf{u})\|_{V'} \le c_1|\mathbf{u}|^{1/2}\|\mathbf{u}\|^{3/2}, \quad \mathbf{u}\in V. \tag{A.32}$$

Moreover, the orthogonality property of the inertial term illustrated in (1.5) is fundamental and extends to

$$b(\mathbf{u},\mathbf{v},\mathbf{v}) = (B(\mathbf{u},\mathbf{v}),\mathbf{v}) = 0, \quad \mathbf{u},\mathbf{v}\in V, \tag{A.33}$$

for all types of boundary conditions considered here. The orthogonality property (A.33) implies, in particular, the skew symmetry of the trilinear operator with respect to the last two arguments:

$$b(\mathbf{u},\mathbf{v},\mathbf{w}) \ (= (B(\mathbf{u},\mathbf{v}),\mathbf{w})) = -b(\mathbf{u},\mathbf{w},\mathbf{v}) \ (= -(B(\mathbf{u},\mathbf{w}),\mathbf{v})),$$
$$\mathbf{u},\mathbf{v},\mathbf{w} \in V. \quad \text{(A.34)}$$

The proof of (A.33) follows in the same manner as that of (1.5).

A Priori Estimates

We now return to the weak solutions of the Navier–Stokes equations. If $\mathbf{u} = \mathbf{u}(t)$ is a weak solution on a time interval $[0, T]$, then we know from (7.24) that $\mathbf{u}$ belongs to $L^\infty(0, T; H)$ and to $L^2(0, T; V)$. Therefore, one can check from (A.24), (A.32), and (6.23) that

$$t \mapsto \mathbf{F}(t, \mathbf{u}(t)) = \mathbf{f}(t) - A\mathbf{u}(t) - B(\mathbf{u}(t)) \in L^{4/3}(0, T; V'). \quad \text{(A.35)}$$

Hence, the functional equation holds in the dual space V' with

$$\mathbf{u}' \in L^{4/3}(0, T; V'). \quad \text{(A.36)}$$

Before we proceed, let us observe the following. With (A.36), we can now assert that (A.22) holds for almost every t in $[0, T]$, and we can take the scalar product of (A.22) with $\mathbf{u}(t)$, for almost every t, in the duality (pairing) between V and V'. We obtain

$$(\mathbf{u}'(t), \mathbf{u}(t)) + \nu\|\mathbf{u}(t)\|^2 + b(\mathbf{u}(t), \mathbf{u}(t), \mathbf{u}(t)) = (\mathbf{f}(t), \mathbf{u}(t)).$$

In view of (1.5) or (A.33), there remains

$$(\mathbf{u}'(t), \mathbf{u}(t)) + \nu\|\mathbf{u}(t)\|^2 = (\mathbf{f}(t), \mathbf{u}(t)). \quad \text{(A.37)}$$

This relation is similar but not identical to the energy inequality (7.7). Namely, since the function $t \mapsto (\mathbf{u}'(t), \mathbf{u}(t))$ is not known to be integrable (we know only that $\mathbf{u}$ belongs to $L^2(0, T; V)$ and $\mathbf{u}'$ belongs to $L^{4/3}(0, T; V')$), we cannot assert that

$$(\mathbf{u}', \mathbf{u})_{V',V} = \frac{1}{2}\frac{d}{dt}|\mathbf{u}|^2$$

in the distributional sense.

We now proceed as follows. From the energy inequality (A.21), we can derive a priori estimates for the weak solutions. Hereafter for simplicity we consider only the no-slip and zero-average periodic cases, for which the Poincaré inequalities hold (see (4.12), (4.13), and (4.14) in Chapter I). Then, using the Cauchy–Schwarz and Young inequalities on (A.21), we find that

$$\int_{t_0}^{t} \|\mathbf{u}(s)\|^2\, ds \le \frac{1}{\nu}|\mathbf{u}(t_0)|^2 + \frac{1}{\nu^2\lambda_1}\int_{t_0}^{t} |\mathbf{f}(s)|^2\, ds \quad \text{(A.38)}$$

for almost all t_0 (including $t_0 = 0$) and all t with $0 \le t_0 \le t \le T$. The following a priori estimate can also be obtained, but its proof is more involved and is left to Appendix B.1:

$$|\mathbf{u}(t)|^2 \leq |\mathbf{u}(t_0)|^2 e^{-\nu\lambda_1(t-t_0)} + \frac{1}{\nu\lambda_1}\int_{t_0}^{t} e^{-\nu\lambda_1(t-s)}|\mathbf{f}(s)|^2\,ds \tag{A.39}$$

for almost all t_0 (including $t_0 = 0$) and all t with $0 \leq t_0 \leq t \leq T$.

In the particular case where $\mathbf{f}$ belongs to $L^\infty(0,T;H)$ (e.g., when $\mathbf{f}$ is time-independent), the estimates (A.38) and (A.39) imply that

$$\int_{t_0}^{t} \|\mathbf{u}(s)\|^2\,ds \leq \frac{1}{\nu}|\mathbf{u}(t_0)|^2 + \frac{1}{\nu^2\lambda_1}\|\mathbf{f}\|^2_{L^\infty(t_0,t;H)}(t-t_0) \tag{A.40}$$

and

$$|\mathbf{u}(t)|^2 \leq |\mathbf{u}(t_0)|^2 e^{-\nu\lambda_1(t-t_0)} + \frac{1}{\nu^2\lambda_1^2}\|\mathbf{f}\|^2_{L^\infty(t_0,t;H)}(1-e^{-\nu\lambda_1(t-t_0)}). \tag{A.41}$$

For the weak solution given in Theorem 7.1, it follows in particular that

$$|\mathbf{u}(t)|^2 \leq |\mathbf{u}_0|^2 e^{-\nu\lambda_1 t} + \frac{1}{\nu^2\lambda_1^2}\|\mathbf{f}\|^2_{L^\infty(0;T;H)}(1-e^{-\nu\lambda_1 t}) \tag{A.42}$$

for all $0 \leq t \leq T$.

If the forcing term $\mathbf{f} = \mathbf{f}(t)$ belongs to $L^\infty(0,\infty;H)$, then any weak solution $\mathbf{u} = \mathbf{u}(t)$ defined for all $t \geq 0$ satisfies also the following a priori estimate:

$$\int_0^T |A\mathbf{u}(t)|^{2/3}\,dt \leq C_3\left(\frac{1}{\nu\lambda_1} + T\right) \quad \forall T \geq 0 \tag{A.43}$$

for some constant $C_3 \geq 0$ independent of T. This estimate is particularly useful for obtaining further regularity for the support of the stationary statistical solutions when the forcing term is time-independent. The proof of this estimate is more involved and is given in Appendix B.1; note that (A.43) implies that $\mathbf{u}(t)$ belongs to $D(A)$ for almost every t, even for weak solutions.

A.3 Weak and Strong Solutions of the NSE in Dimension 2

We now consider the 2-dimensional case.

A Priori Estimates Using the Energy Equation

Recall from Section 1 that the energy equation (7.12) is obtained by (a) taking the inner product of the momentum equation with the solution $\mathbf{u}$ and (b) using the orthogonality property $b(\mathbf{u},\mathbf{u},\mathbf{u}) = 0$ of the inertial term. As in the 3-dimensional case, this orthogonality property actually holds in a more general sense. Namely,

$$b(\mathbf{u},\mathbf{v},\mathbf{v}) = 0 \tag{A.44}$$

for any pair of vector fields $\mathbf{u}$ and $\mathbf{v}$ in V. This implies, in particular, the skew symmetry of the trilinear operator with respect to the last two arguments:

$$b(\mathbf{u},\mathbf{v},\mathbf{w}) = -b(\mathbf{u},\mathbf{w},\mathbf{v}), \quad \mathbf{u},\mathbf{v},\mathbf{w} \in V. \tag{A.45}$$

In the 2-dimensional case, the trilinear term also satisfies the following inequalities for some appropriate constant c_1:

$$|b(\mathbf{u},\mathbf{v},\mathbf{w})| \le c_1|\mathbf{u}|^{1/2}\|\mathbf{u}\|^{1/2}\|\mathbf{v}\|^{1/2}|A\mathbf{v}|^{1/2}|\mathbf{w}|, \quad \mathbf{u}\in V,\ \mathbf{v}\in D(A),\ \mathbf{w}\in H; \tag{A.46a}$$

$$|b(\mathbf{u},\mathbf{v},\mathbf{w})| \le c_1|\mathbf{u}|^{1/2}|A\mathbf{u}|^{1/2}\|\mathbf{v}\||\mathbf{w}|, \quad \mathbf{u}\in D(A),\ \mathbf{v}\in V,\ \mathbf{w}\in H; \tag{A.46b}$$

$$|b(\mathbf{u},\mathbf{v},\mathbf{w})| \le c_1|\mathbf{u}|\|\mathbf{v}\||\mathbf{w}|^{1/2}|A\mathbf{w}|^{1/2}, \quad \mathbf{u}\in H,\ \mathbf{v}\in V,\ \mathbf{w}\in D(A); \tag{A.46c}$$

$$|b(\mathbf{u},\mathbf{v},\mathbf{w})| \le c_1|\mathbf{u}|^{1/2}\|\mathbf{u}\|^{1/2}\|\mathbf{v}\||\mathbf{w}|^{1/2}\|\mathbf{w}\|^{1/2}, \quad \mathbf{u}\in V,\ \mathbf{v}\in V,\ \mathbf{w}\in V; \tag{A.46d}$$

$$|b(\mathbf{u},\mathbf{v},\mathbf{w})| \le c_1|\mathbf{u}|^{1/2}\|\mathbf{u}\|^{1/2}|\mathbf{v}|^{1/2}\|\mathbf{v}\|^{1/2}\|\mathbf{w}\|, \quad \mathbf{u}\in V,\ \mathbf{v}\in V,\ \mathbf{w}\in V. \tag{A.46e}$$

These inequalities can be obtained by applying Hölder's inequality, the Ladyzhenskaya inequality (see Section I.4)

$$\|\mathbf{u}\|_{L^4(\Omega)} \le c_2|\mathbf{u}|^{1/2}\|\mathbf{u}\|^{1/2}, \quad \mathbf{u}\in V, \tag{A.47}$$

and Agmon's inequality

$$\|\mathbf{u}\|_{L^\infty(\Omega)} \le c_2|\mathbf{u}|^{1/2}|A\mathbf{u}|^{1/2}, \quad \mathbf{u}\in D(A), \tag{A.48}$$

where c_2 is some appropriate constant. We also have the interpolation inequality

$$\|\mathbf{u}\| \le c_2|\mathbf{u}|^{1/2}|A\mathbf{u}|^{1/2}, \quad \mathbf{u}\in D(A). \tag{A.49}$$

Brézis and Gallouet [1980] derived the useful inequality

$$\|\mathbf{u}\|_{L^\infty(\Omega)} \le c_2\|\mathbf{u}\|\left(1+\log\frac{|A\mathbf{u}|^2}{\lambda_1\|\mathbf{u}\|^2}\right)^{1/2}, \quad \mathbf{u}\in D(A), \tag{A.50}$$

which implies the following inequalities for the trilinear term:

$$|b(\mathbf{u},\mathbf{v},\mathbf{w})| \le c_1\|\mathbf{u}\|\|\mathbf{v}\|\left(1+\log\frac{|A\mathbf{u}|^2}{\lambda_1\|\mathbf{u}\|^2}\right)^{1/2}|\mathbf{w}|, \quad \mathbf{u}\in D(A),\ \mathbf{v}\in V,\ \mathbf{w}\in H; \tag{A.51a}$$

$$|b(\mathbf{u},\mathbf{v},\mathbf{w})| \le c_1|\mathbf{u}||A\mathbf{v}|\left(1+\log\frac{|A^{3/2}\mathbf{v}|^2}{\lambda_1|A\mathbf{v}|^2}\right)^{1/2}|\mathbf{w}|, \quad \mathbf{u}\in H,\ \mathbf{v}\in D(A),\ \mathbf{w}\in H. \tag{A.51b}$$

In the 2-dimensional case, the energy equation (7.12) in Theorem 7.3 means that a weak solution $\mathbf{u}=\mathbf{u}(t)$ on $[0, T]$ satisfies the integral equation

$$\frac{1}{2}|\mathbf{u}(t)|^2 + \nu\int_{t_0}^{t}\|\mathbf{u}(s)\|^2\,ds = \frac{1}{2}|\mathbf{u}(t_0)|^2 + \int_{t_0}^{t}(\mathbf{f}(s),\mathbf{u}(s))\,ds \tag{A.52}$$

for all $0\le t_0\le t\le T$. In this case, it is straightforward to obtain the analog of the a priori estimates (A.38) and (A.39) as

$$|\mathbf{u}(t)|^2 \le |\mathbf{u}(t_0)|^2 e^{-\nu\lambda_1(t-t_0)} + \frac{1}{\nu^2\lambda_1^2}\|\mathbf{f}\|^2_{L^\infty(t_0,t;\,H)}(1-e^{-\nu\lambda_1(t-t_0)}) \tag{A.53}$$

and

$$\int_{t_0}^{t} \|\mathbf{u}(s)\|^2 \, ds \le \frac{1}{\nu}|\mathbf{u}(t_0)|^2 + \frac{1}{\nu^2\lambda_1}\|\mathbf{f}\|^2_{L^\infty(t_0,t;\,H)}(t-t_0) \tag{A.54}$$

for all $0 \le t_0 \le t \le T$.

A Priori Estimates Using the Enstrophy Equation

Further estimates can be derived from the enstrophy equation. This equation is obtained by taking the inner product of the momentum equation with the vector field $A\mathbf{u}$. Upon doing so, we find that

$$\frac{1}{2}\frac{d}{dt}\|\mathbf{u}\|^2 + \nu|A\mathbf{u}|^2 + b(\mathbf{u},\mathbf{u},A\mathbf{u}) = (\mathbf{f},A\mathbf{u}). \tag{A.55}$$

From this equation, one can obtain the following estimates:

$$\|\mathbf{u}(t)\| \le C(\|\mathbf{f}\|_{L^\infty(0,\infty;\,H)},\nu,\Omega), \quad t \ge T(|\mathbf{u}_0|,\|\mathbf{f}\|_{L^\infty(0,\infty;\,H)},\nu,\Omega), \tag{A.56}$$

$$\|\mathbf{u}(t)\| \le C(\varepsilon,|\mathbf{u}_0|,\|\mathbf{f}\|_{L^\infty(\varepsilon,\infty;\,H)},\nu,\Omega), \quad t \ge \varepsilon > 0, \tag{A.57}$$

$$\|\mathbf{u}(t)\| \le C(\|\mathbf{u}_0\|,\|\mathbf{f}\|_{L^\infty(0,\infty;\,H)},\nu,\Omega), \quad t \ge 0; \tag{A.58}$$

also,

$$\frac{1}{T}\int_t^{t+T} |A\mathbf{u}(s)|^2 \, ds \le C(\|\mathbf{f}\|_{L^\infty(t,t+T;\,H)},\nu,\Omega), \quad t \ge T(|\mathbf{u}_0|,|\mathbf{f}|,\nu,\Omega). \tag{A.59}$$

One can also show that

$$|A\mathbf{u}(t)| \le C(\|\mathbf{f}\|_{L^\infty(0,\infty;\,H)},\nu,\Omega), \quad t \ge T(|\mathbf{u}_0|,\|\mathbf{f}\|_{L^\infty(0,\infty;\,H)},\nu,\Omega), \tag{A.60}$$

$$|A\mathbf{u}(t)| \le C(\varepsilon,|\mathbf{u}_0|,\|\mathbf{f}\|_{L^\infty(\varepsilon,\infty;\,H)},\nu,\Omega), \quad t \ge \varepsilon > 0. \tag{A.61}$$

Improvements in the 2-Dimensional Periodic Case

In the 2-dimensional periodic case with a vanishing space average, the inertial term satisfies a further orthogonality property that enables us to obtain better a priori estimates. The orthogonality property is that

$$b(\mathbf{u},\mathbf{u},A\mathbf{u}) = 0 \tag{A.62}$$

for all vector fields in $D(A)$. A useful consequence of this property can be obtained by applying (A.62) to $\mathbf{u}+\mathbf{v}$ and to $\mathbf{u}-\mathbf{v}$ and then subtracting the results (or by differentiating (A.62) with respect to $\mathbf{u}$). This gives us the identity

$$b(\mathbf{v},\mathbf{u},A\mathbf{u}) + b(\mathbf{u},\mathbf{v},A\mathbf{u}) + b(\mathbf{u},\mathbf{u},A\mathbf{v}) = 0, \tag{A.63}$$

which holds for all vector fields $\mathbf{u}$ and $\mathbf{v}$ in $D(A)$. Similarly, differentiating (A.63) with respect to $\mathbf{u}$ yields the identity

$$\begin{aligned} b(\mathbf{v},\mathbf{w},A\mathbf{u}) &+ b(\mathbf{v},\mathbf{u},A\mathbf{w}) + b(\mathbf{w},\mathbf{v},A\mathbf{u}) \\ &+ b(\mathbf{u},\mathbf{v},A\mathbf{w}) + b(\mathbf{w},\mathbf{u},A\mathbf{v}) + b(\mathbf{u},\mathbf{w},A\mathbf{v}) = 0, \end{aligned} \tag{A.64}$$

which is valid for all vector fields $\mathbf{u},\mathbf{v},\mathbf{w}$ in $D(A)$.

Now, with (A.62) in mind, the enstrophy equation (A.55) in the 2-dimensional periodic case reduces to

$$\frac{1}{2}\frac{d}{dt}\|\mathbf{u}(t)\|^2 + \nu|A\mathbf{u}(t)|^2 = (\mathbf{f}(t), A\mathbf{u}(t)). \tag{A.65}$$

This provides us with an estimate for the enstrophy which is similar to that for the kinetic energy – namely,

$$\|\mathbf{u}(t)\|^2 \leq \|\mathbf{u}(t_0)\|^2 e^{-\nu\lambda_1(t-t_0)} + \frac{1}{\nu^2\lambda_1}\|\mathbf{f}\|^2_{L^\infty(t_0,t;H)}(1 - e^{-\nu\lambda_1(t-t_0)}) \tag{A.66}$$

for all $0 \leq t_0 \leq t \leq T$. We also obtain the estimate

$$\int_{t_0}^{t} |A\mathbf{u}(s)|^2\, ds \leq \frac{1}{\nu}\|\mathbf{u}(t_0)\|^2 + \frac{1}{\nu^2}\|\mathbf{f}\|^2_{L^\infty(t_0,t;H)}(t - t_0) \tag{A.67}$$

for all $0 \leq t_0 \leq t \leq T$.

Appendix B Proofs of Technical Results in Chapter II

B.1 Energy Equation and A Priori Estimates

We now present the proofs of some of the a priori estimates mentioned in Section 7.

Proof of the Integral Form (A.19) of the Energy Inequality

Assume that the forcing term $\mathbf{f} = \mathbf{f}(t)$ belongs to $L^2(0, T; H)$ and let $\mathbf{u} = \mathbf{u}(t)$ be a weak solution defined on $[0, T]$. As stated in Theorem 7.1, this solution satisfies the energy inequality (7.5). In particular this implies that, for every nonnegative test function $\varphi \in \mathcal{C}_c^\infty(0, T)$ (i.e., φ is an infinitely differentiable, nonnegative, real-valued function with compact support on $(0, T)$), the following inequality holds:

$$-\frac{1}{2}\int_0^T |\mathbf{u}(s)|^2\varphi'(s)\, ds + \nu\int_0^T \varphi(s)\|\mathbf{u}(s)\|^2\, ds \leq \int_0^T \varphi(s)(\mathbf{f}(s), \mathbf{u}(s))\, ds. \tag{B.1}$$

It is also valid for any continuous nonnegative function φ that vanishes at $t = 0$ and T and whose derivative φ' is integrable; this can be proven by approximating φ by a sequence $\{\varphi_n\}_n$ of functions in $\mathcal{C}_c^\infty(0, T)$ and then passing to the limit $n \to \infty$ in the relations (B.1) for φ_n.

Let ψ be a nonnegative function in $\mathcal{C}^1([0, T])$. For given t_0 and t with $0 \leq t_0 < t \leq T$, consider $m \in \mathbb{N}$ such that $t_0 + 1/m < t - 1/m$. For any such m, let θ_m be the continuous piecewise linear function that equals 1 on $[t_0 + 1/m, t - 1/m]$, 0 on $[0, t_0]$ and $[t, T]$, and is linear on $[t_0, t_0 + 1/m]$ and $[t - 1/m, t]$. Then (B.1) holds for $\varphi = \theta_m\psi$, and we have $\varphi' = \theta_m'\psi + \theta_m\psi'$. The relation (B.1) for $\varphi = \theta_m\psi$ reads

$$-\frac{1}{2}\int_{t_0}^{t_0+\frac{1}{m}} |\mathbf{u}(s)|^2\theta_m'(s)\psi(s)\,ds - \frac{1}{2}\int_{t-\frac{1}{m}}^{t} |\mathbf{u}(s)|^2\theta_m'(s)\psi(s)\,ds$$

$$-\frac{1}{2}\int_{t_0}^{t_0+\frac{1}{m}} |\mathbf{u}(s)|^2\theta_m(s)\psi'(s)\,ds - \frac{1}{2}\int_{t-\frac{1}{m}}^{t} |\mathbf{u}(s)|^2\theta_m(s)\psi'(s)\,ds$$

$$+\nu\int_0^T \varphi(s)\|\mathbf{u}(s)\|^2\,ds \le \int_0^T \varphi(s)(\mathbf{f}(s),\mathbf{u}(s))\,ds. \quad \text{(B.2)}$$

Using the Lebesgue differentiation theorem (See Appendix A.5 in Chapter IV), we can pass to the limit as m goes to infinity in (B.2). From the choice of θ_m, and since ψ is continuous and $|\mathbf{u}(\cdot)|^2$ is integrable (it is essentially bounded) on $(0, T)$, the first two terms in the LHS of (B.2) converge to

$$-\tfrac{1}{2}|\mathbf{u}(t_0)|^2\psi(t_0) + \tfrac{1}{2}|\mathbf{u}(t)|^2\psi(t),$$

provided t_0 and t are Lebesgue points[20] of $|\mathbf{u}(\cdot)|^2$. The next two terms in the LHS of (B.2) vanish at the limit. Then we find that, for any pair of Lebesgue points t_0 and t of $|\mathbf{u}(\cdot)|^2$ (hence, almost everywhere) with $0 \le t_0 < t \le T$,

$$\frac{1}{2}|\mathbf{u}(t)|^2\psi(t) + \nu\int_{t_0}^{t} \psi(s)\|\mathbf{u}(s)\|^2\,ds$$
$$\le \frac{1}{2}|\mathbf{u}(t_0)|^2\psi(t_0) + \int_{t_0}^{t} \psi(s)(\mathbf{f}(s),\mathbf{u}(s))\,ds. \quad \text{(B.3)}$$

The next step is to obtain (B.3) for every t in $(t_0, T]$. For that purpose, fix an arbitrary $t > t_0$, with t_0 a Lebesgue point of $|\mathbf{u}(\cdot)|^2$ in $[0, T)$. Let t_j be a sequence of Lebesgue points of $|\mathbf{u}(\cdot)|^2$ converging to t. Hence, (B.3) holds with t replaced by t_j. From the weak continuity in H of the weak solution, we find

$$|\mathbf{u}(t)|^2 \le \liminf_{j\to\infty} |\mathbf{u}(t_j)|^2.$$

Hence, by passing to the limit $j \to \infty$ in (B.3) (with t_j instead of t), we obtain that (B.3) holds for any t and any Lebesgue point t_0 of $|\mathbf{u}(\cdot)|^2$, as well as for any nonnegative function ψ in $\mathcal{C}^1([0, T])$. Thus we have established that, for any nonnegative $\mathcal{C}^1$ function ψ on $[0, T]$, any t in $[0, T]$, and any Lebesgue point t_0 of $|\mathbf{u}(\cdot)|^2$ (hence, for almost every t_0) and with $0 \le t_0 < t \le T$, the following inequality holds:

$$\frac{1}{2}|\mathbf{u}(t)|^2\psi(t) + \nu\int_{t_0}^{t} \psi(s)\|\mathbf{u}(s)\|^2\,ds$$
$$\le \frac{1}{2}|\mathbf{u}(t_0)|^2\psi(t_0) + \int_{t_0}^{t} \psi(s)(\mathbf{f}(s),\mathbf{u}(s))\,ds. \quad \text{(B.4)}$$

In particular, by choosing ψ identically equal to 1, we prove the energy inequality (A.19).

Finally, we want to establish that $t_0 = 0$ is a Lebesgue point of $t \mapsto |\mathbf{u}(t)|^2$ (i.e., that (B.4) holds also for $t_0 = 0$). For that purpose we work directly with (7.5). First

[20] The Lebesgue points of a function are defined in footnote 9.

notice that (7.5) holds for any continuous, nonnegative function ψ that vanishes at $t = T$ and whose derivative ψ' is integrable; this can be proven by approximating ψ by suitable functions as we did for (B.1). Then, for any τ such that $0 < \tau < T$, take ψ to be identically equal to zero on $[\tau, T]$, linear on $[0, \tau]$, and such that $\psi(0) = 1$. Then we obtain

$$\int_0^\tau \left\{ \frac{1}{2\tau} |\mathbf{u}(t)|^2 + \nu \|\mathbf{u}(t)\|^2 \psi(t) \right\} dt \leq \frac{1}{2} |\mathbf{u}(0)|^2 + \int_0^\tau (\mathbf{f}(t), \mathbf{u}(t)) \psi(t)\, dt. \quad \text{(B.5)}$$

Let τ go to zero to find that

$$\limsup_{\tau \to 0+} \frac{1}{\tau} \int_0^\tau |\mathbf{u}(t)|^2\, dt \leq |\mathbf{u}(0)|^2. \quad \text{(B.6)}$$

From the lower semicontinuity of $t \mapsto |\mathbf{u}(t)|^2$, which follows from the weak continuity in H of the weak solutions, one can show the opposite inequality in (B.6) – and with lim inf instead of lim sup. Hence, $t_0 = 0$ is a Lebesgue point of $t \mapsto |\mathbf{u}(t)|^2$.

Proof of the A Priori Estimate (A.39)

We start from the energy inequality (B.4). We bound the term with the volume forces using the Poincaré inequality:

$$(\mathbf{f}, \mathbf{u}) \leq |\mathbf{f}||\mathbf{u}| \leq \frac{1}{\lambda_1^{1/2}} |\mathbf{f}| \|\mathbf{u}\| \leq \frac{1}{2\nu\lambda_1} |\mathbf{f}|^2 + \frac{\nu}{2} \|\mathbf{u}\|^2.$$

Hence, we obtain

$$|\mathbf{u}(t)|^2 \psi(t) + \nu \int_{t_0}^t \psi(s) \|\mathbf{u}(s)\|^2\, ds \leq |\mathbf{u}(t_0)|^2 \psi(t_0) + \frac{1}{\nu\lambda_1} \int_{t_0}^t \psi(s) |\mathbf{f}(s)|^2\, ds.$$

Choosing $\psi(t) = \exp(\nu\lambda_1 t)$ and neglecting the second term in the left-hand side yields

$$|\mathbf{u}(t)|^2 e^{\nu\lambda_1 t} \leq |\mathbf{u}(t_0)|^2 e^{\nu\lambda_1 t_0} + \frac{1}{\nu\lambda_1} \int_{t_0}^t e^{\nu\lambda_1 s} |\mathbf{f}(s)|^2\, ds,$$

which implies (A.39).

Proof of the A Priori Estimate (A.43)

Assume that $\mathbf{f} = \mathbf{f}(t)$ belongs to $L^\infty(0, \infty; H)$ and let $\mathbf{u} = \mathbf{u}(t)$ be a weak solution defined for all $t \geq 0$. We know (see Section 7) that there exists a relatively open set $\sigma \subset [0, \infty)$ of total measure in $[0, \infty)$ such that $\mathbf{u}$ is a strong solution on the intervals of σ. For each $T > 0$ consider the open set $\sigma_T = \sigma \cap (0, T)$. Let $\{(\alpha_i, \beta_i)\}_i$ be the connected components of σ_T, so that $\sigma_T = \bigcup (\alpha_i, \beta_i)$ and $\mathbf{u}$ is a strong solution on each (α_i, β_i). We also know that each $[\alpha_i, \beta_i)$ is a maximal interval of regularity of the weak solution, where "regularity" is used in the sense of being a strong solution on that interval. Note that the family $\{(\alpha_i, \beta_i)\}$ depends on T, but for simplicity we omit this fact from the notation. Then, on each subinterval (α_i, β_i) we are allowed to take the inner product of the momentum equation with $A\mathbf{u}$, which implies (using (A.26) and Young's inequality) that

$$\frac{1}{2}\frac{d}{dt}\|\mathbf{u}\|^2 + \nu|A\mathbf{u}|^2 \le |\mathbf{f}(t)|^2 + C_0\|\mathbf{u}\|^6 \tag{B.7}$$

on (α_i, β_i) for each i and for some constant $C_0 \ge 0$ that is independent of i. From (B.7) and the assumption that $\mathbf{f}$ belongs to $L^\infty(0,\infty;H)$, we have

$$\frac{1}{2}\frac{d}{dt}\|\mathbf{u}\|^2 + \nu|A\mathbf{u}|^2 \le C_1(\nu^2\lambda_1^{1/2} + \|\mathbf{u}\|^2)^3 \tag{B.8}$$

for another constant $C_1 \ge 0$. Note that $\nu^2\lambda_1^{1/2}$ has the same physical dimension as $\|\mathbf{u}\|^2$. Divide (B.8) by $(\nu^2\lambda_1^{1/2} + \|\mathbf{u}\|^2)^2$ to find that

$$\frac{1}{(1+\|\mathbf{u}\|^2)^2}\frac{d}{dt}\|\mathbf{u}\|^2 + 2\frac{\nu|A\mathbf{u}|^2}{(\nu^2\lambda_1^{1/2} + \|\mathbf{u}\|^2)^2} \le 2C_1(\nu^2\lambda_1^{1/2} + \|\mathbf{u}\|^2);$$

upon integration on (α, β), for $\alpha_i < \alpha < \beta < \beta_i$ we obtain

$$-\frac{1}{\nu^2\lambda_1^{1/2} + \|\mathbf{u}(\beta)\|^2} + \frac{1}{\nu^2\lambda_1^{1/2} + \|\mathbf{u}(\alpha)\|^2} + 2\nu\int_\alpha^\beta \frac{|A\mathbf{u}(t)|^2}{(\nu^2\lambda_1^{1/2} + \|\mathbf{u}(t)\|^2)^2}\,dt$$
$$\le 2C_1\int_\alpha^\beta (\nu^2\lambda_1^{1/2} + \|\mathbf{u}(t)\|^2)\,dt.$$

Because $[\alpha_i, \beta_i)$ is a maximal interval of regularity, it follows that $\|\mathbf{u}(\beta)\|$ goes to infinity as $\beta \nearrow \beta_i$. Hence, we find

$$\nu\int_{\alpha_i}^{\beta_i} \frac{|A\mathbf{u}(t)|^2}{(\nu^2\lambda_1^{1/2} + \|\mathbf{u}(t)\|^2)^2}\,dt \le C_1\int_{\alpha_i}^{\beta_i} (\nu^2\lambda_1^{1/2} + \|\mathbf{u}(t)\|^2)\,dt.$$

Summing up this result in i yields

$$\nu\int_{\sigma_T} \frac{|A\mathbf{u}(t)|^2}{(\nu^2\lambda_1^{1/2} + \|\mathbf{u}(t)\|^2)^2}\,dt \le C_1\int_{\sigma_T} (\nu^2\lambda_1^{1/2} + \|\mathbf{u}(t)\|^2)\,dt.$$

Since σ_T has total measure in $[0, T]$, the preceding inequality means that

$$\nu\int_0^T \frac{|A\mathbf{u}(t)|^2}{(\nu^2\lambda_1^{1/2} + \|\mathbf{u}(t)\|^2)^2}\,dt \le C_1\int_0^T (\nu^2\lambda_1^{1/2} + \|\mathbf{u}(t)\|^2)\,dt. \tag{B.9}$$

We infer from (A.40) that

$$\nu\int_0^T \|\mathbf{u}(t)\|^2\,dt \le |\mathbf{u}_0|^2 + \frac{T}{\nu\lambda_1}|\mathbf{f}|^2 \quad \forall T \ge 0, \tag{B.10}$$

and we deduce from (B.9) that

$$\nu\int_0^T \frac{|A\mathbf{u}(t)|^2}{(\nu^2\lambda_1^{1/2} + \|\mathbf{u}(t)\|^2)^2}\,dt \le C_2\left(\frac{1}{\nu^3\lambda_1^{1/2}}|\mathbf{u}_0|^2 + T\right) \quad \forall T \ge 0 \tag{B.11}$$

for some constant $C_2 \ge 0$ independent of T. Using the Hölder inequality, (B.11), and again (B.10), we then see that

$$\begin{aligned}\int_0^T |A\mathbf{u}(t)|^{2/3}\,dt &= \int_0^T \frac{|A\mathbf{u}(t)|^{2/3}}{(\nu^2\lambda_1^{1/2}+\|\mathbf{u}(t)\|^2)^{2/3}}(\nu^2\lambda_1^{1/2}+\|\mathbf{u}(t)\|^2)^{2/3}\,dt\\ &\le \left(\int_0^T \frac{|A\mathbf{u}(t)|^2}{(\nu^2\lambda_1^{1/2}+\|\mathbf{u}\|^2)^2}\,dt\right)^{1/3}\left(\int_0^T(\nu^2\lambda_1^{1/2}+\|\mathbf{u}(t)\|^2)\,dt\right)^{2/3}\\ &\le C_3\left(\frac{1}{\nu^3\lambda_1^{1/2}}|\mathbf{u}_0|^2+T\right)^{1/3}\left(\frac{1}{\nu^3\lambda_1^{1/2}}|\mathbf{u}_0|^2+T\right)^{2/3},\end{aligned}$$

so that

$$\int_0^T |A\mathbf{u}(t)|^{2/3}\,dt \le C_3\left(\frac{1}{\nu^3\lambda_1^{1/2}}|\mathbf{u}_0|^2+T\right)\quad \forall T\ge 0 \tag{B.12}$$

for some constant $C_3 \ge 0$ independent of T. This finishes the proof of (A.43).

B.2 Time Analyticity

We now prove the estimate (8.8) concerning the time analyticity of the strong solutions of the 3-dimensional Navier–Stokes equations.

The NSE can be extended to complex times $\zeta \in \mathbb{C}$ as

$$\frac{d\mathbf{u}}{d\zeta} + \nu A_{\mathbb{C}}\mathbf{u} + B(\mathbf{u})_{\mathbb{C}} = \mathbf{f}, \tag{B.13}$$

where $\mathbf{u} = \mathbf{u}(\zeta)$. Given $\mathbf{u}_0 \in V$ (or even H, for that matter), the Galerkin approximation of (B.13) has an analytic solution with $\mathbf{u}(0) = \mathbf{u}_0$ that is defined for ζ in a complex neighborhood of the origin. For this local analytic solution, the a priori estimates that we are about to derive formally can be obtained rigorously.

For notational simplicity, in the following computations we drop the subscript $\mathbb{C}$ for the inner products, norms, and operators defined previously. We only keep, for clarity, the subscript $\mathbb{C}$ for the functional spaces.

We fix $\theta \in (-\pi/2, \pi/2)$ and consider the time $\zeta = se^{i\theta}$ for $s > 0$. We want to compute the derivative

$$\begin{aligned}\frac{1}{2}\frac{d}{ds}\|\mathbf{u}(se^{i\theta})\|^2 &= \frac{1}{2}\frac{d}{ds}(\mathbf{u}(se^{i\theta}), A\mathbf{u}(se^{i\theta}))\\ &= \frac{1}{2}\left(e^{i\theta}\frac{d\mathbf{u}}{d\zeta}, A\mathbf{u}\right) + \frac{1}{2}\left(\mathbf{u}, e^{i\theta}A\frac{d\mathbf{u}}{d\zeta}\right)\\ &= \operatorname{Re} e^{i\theta}\left(\frac{d\mathbf{u}}{d\zeta}, A\mathbf{u}\right),\end{aligned}$$

where in the last step we used the fact that the real Stokes operator is self-adjoint and thus the complexified Stokes operator is hermitian. Then, using (B.13) we find

$$\frac{1}{2}\frac{d}{ds}\|\mathbf{u}(se^{i\theta})\|^2 + \nu\cos\theta|A\mathbf{u}(se^{i\theta})|^2 + \operatorname{Re} e^{i\theta} b(\mathbf{u}, \mathbf{u}, A\mathbf{u}) = \operatorname{Re} e^{i\theta}(\mathbf{f}, A\mathbf{u}). \tag{B.14}$$

From the definition of the complexified trilinear operator b, it is clear that each estimate for the real operator leads to a similar estimate in the complexified case, but with a larger multiplying constant. From the estimate (A.26b), for instance, the following estimate holds (can be proved) in the complexified case:

$$|b(\mathbf{u}, \mathbf{u}, A\mathbf{u})| \le c_1 \|\mathbf{u}\|^{3/2} |A\mathbf{u}|^{3/2} \tag{B.15}$$

for a constant c_1, which may be larger than the one for the real case.[21] An application of Young's inequality to (B.15) yields

$$|b(\mathbf{u}, \mathbf{u}, A\mathbf{u})| \le \frac{\nu \cos\theta}{4} |A\mathbf{u}|^2 + \frac{c_2}{\nu^3 \cos^3\theta} \|\mathbf{u}\|^6 \tag{B.16}$$

for another constant c_2. For the forcing term, we use

$$|(\mathbf{f}, A\mathbf{u})| \le \frac{\nu\cos\theta}{4} |A\mathbf{u}|^2 + \frac{1}{\nu\cos\theta} |\mathbf{f}|^2. \tag{B.17}$$

Hence, as in Section 8, we deduce the inequality

$$\frac{d}{ds} \|\mathbf{u}(se^{i\theta})\|^2 + \nu\cos\theta |A\mathbf{u}(se^{i\theta})|^2 \le \frac{2}{\nu\cos\theta} |\mathbf{f}|^2 + \frac{2c_2}{\nu^3\cos^3\theta} \|\mathbf{u}\|^6. \tag{B.18}$$

Note that on a time ray of fixed angle θ, the role of the viscosity is played by $\nu\cos\theta$, which decreases with $|\theta|$.

Setting

$$y(s) = \|\mathbf{u}(se^{i\theta})\|^2,$$

we see that

$$y' \le \frac{2}{\nu\cos\theta} |\mathbf{f}|^2 + \frac{2c_2}{\nu^3\cos^3\theta} y^3. \tag{B.19}$$

This expression is of the form

$$y' \le a + by^3.$$

Divide (B.19) by $(c+y)^3$ with $c = \nu^2 \lambda^{1/2}$ (hence c has the same dimension as y in the 3-dimensional case; in the 2-dimensional case we take $c = \nu^2\lambda_1$). Then integrate between s_0 and s to find

$$-\frac{1}{2}\frac{d}{dt}\frac{1}{(c+y)^2} \le \int_{s_0}^{s} \left(\frac{a}{(c+y)^3} + \frac{by^3}{(c+y)^3} \right) ds \le (a+b)(s-s_0).$$

Thus,

$$c + y(s) \le \frac{c + y(s_0)}{\sqrt{1-(a+b)(s-s_0)(c+y(s_0))^2}}.$$

Hence, as long as

$$s - s_0 \le \frac{1}{2(a+b)(c+y(s_0))^2},$$

it follows that

$$c + y(s) \le \sqrt{2}(c + y(s_0)).$$

[21] Actually, we prove (B.15) and other similar inequalities by simply writing $\mathbf{u} = \mathbf{u}_1 + i\mathbf{u}_2$, expanding by linearity, and using the inequalities for real-valued vector fields.

Therefore, if

$$s\left(\frac{2}{\nu\cos\theta}|\mathbf{f}|^2+\frac{2c_2}{\nu^3\cos^3\theta}\right)\leq\frac{1}{2(c+\|\mathbf{u}_0\|^2)^2} \tag{B.20}$$

then we have

$$\|\mathbf{u}(se^{i\theta})\|^2\leq\sqrt{2}(\nu^2\lambda_1^{1/2}+\|\mathbf{u}_0\|^2). \tag{B.21}$$

Now consider the open set

$$\begin{aligned}\Delta^0(\|\mathbf{u}_0\|)&=\Delta^0(\|\mathbf{u}_0\|,|\mathbf{f}|,\nu,\Omega)\\&=\left\{\zeta=se^{i\theta};\ |\theta|<\frac{\pi}{2},\ 0<s\left(\frac{2}{\nu\cos\theta}|\mathbf{f}|^2+\frac{2c_2}{\nu^3\cos^3\theta}\right)\right.\\&\qquad\left.<\frac{1}{2(\nu^2\lambda_1^{1/2}+\|\mathbf{u}_0\|^2)^2}\right\}.\end{aligned} \tag{B.22}$$

This set is a domain of analyticity of the solution $\mathbf{u}=\mathbf{u}(\zeta)$. The origin $\zeta=0$ belongs to the closure of $\Delta^0(\|\mathbf{u}_0\|)$, and $\Delta^0(\|\mathbf{u}_0\|)$ is a neighborhood of the real time interval where the solution is strong. Inside this domain we have

$$\|\mathbf{u}(\zeta)\|^2\leq\sqrt{2}(\nu^2\lambda_1^{1/2}+\|\mathbf{u}_0\|^2)\quad\text{for }\zeta\in\Delta^0(\|\mathbf{u}_0\|,|\mathbf{f}|,\nu,\Omega). \tag{B.23}$$

B.3 Bilinear Estimates in Gevrey Spaces

We want to prove the estimates (9.12) and (9.13) in Section 9. We borrow the proof from Foias and Temam [1989]. We write (see Section 5)

$$\mathbf{u}(x)=\sum_{\mathbf{k}\in\mathbb{Z}^d}\hat{\mathbf{u}}_{\mathbf{k}}e^{2\pi i\frac{\mathbf{k}}{\mathbf{L}}\cdot\mathbf{x}},$$
$$\mathbf{u}^*(x)=\sum_{\mathbf{k}\in\mathbb{Z}^d}\hat{\mathbf{u}}^*_{\mathbf{k}}e^{2\pi i\frac{\mathbf{k}}{\mathbf{L}}\cdot\mathbf{x}},\quad\hat{\mathbf{u}}^*_{\mathbf{k}}=e^{\sigma|\mathbf{k}|}\hat{\mathbf{u}}_{\mathbf{k}}.$$

We use similar notations for $\mathbf{v}$ and $\mathbf{w}$. We have

$$(B(\mathbf{u},\mathbf{v}),\mathbf{w})=|\Omega|i\sum_{\mathbf{j}+\mathbf{k}=\mathbf{l}}(\hat{\mathbf{u}}_{\mathbf{k}}\cdot\mathbf{j})(\hat{\mathbf{v}}_{\mathbf{j}}\cdot\overline{\hat{\mathbf{w}}}_{\mathbf{l}}), \tag{B.24}$$

where $\mathbf{j},\mathbf{k},\mathbf{l}\in\mathbb{Z}^d$. Similarly,

$$\begin{aligned}(e^{\sigma A^{1/2}}B(\mathbf{u},\mathbf{v}),e^{\sigma A^{1/2}}A\mathbf{w})&=|\Omega|i\sum_{\mathbf{j}+\mathbf{k}=\mathbf{l}}(\hat{\mathbf{u}}_{\mathbf{k}}\cdot\mathbf{j})(\hat{\mathbf{v}}_{\mathbf{j}}\cdot\overline{\hat{\mathbf{w}}}_{\mathbf{l}})|\mathbf{l}|^2e^{2\sigma|\mathbf{l}|}\\&=|\Omega|i\sum_{\mathbf{j}+\mathbf{k}=\mathbf{l}}(\hat{\mathbf{u}}^*_{\mathbf{k}}\cdot\mathbf{j})(\hat{\mathbf{v}}^*_{\mathbf{j}}\cdot\overline{\hat{\mathbf{w}}}^*_{\mathbf{l}})|\mathbf{l}|^2e^{2\sigma(|\mathbf{l}|-|\mathbf{k}|-|\mathbf{j}|)}.\end{aligned}$$

Since

$$|\mathbf{l}|-|\mathbf{k}|-|\mathbf{j}|=|\mathbf{k}+\mathbf{j}|-|\mathbf{k}|-|\mathbf{j}|\leq 0,$$

we have

$$|(e^{\sigma A^{1/2}}B(\mathbf{u},\mathbf{v}),e^{\sigma A^{1/2}}A\mathbf{w})|\leq|\Omega|\sum_{\mathbf{j}+\mathbf{k}=\mathbf{l}}|\hat{\mathbf{u}}^*_{\mathbf{k}}||\mathbf{j}||\hat{\mathbf{v}}^*_{\mathbf{j}}||\hat{\mathbf{w}}^*_{\mathbf{l}}||\mathbf{l}|^2. \tag{B.25}$$

The RHS of (B.25) is equal to the integral

$$\int_{\Omega} \xi(\mathbf{x})\psi(\mathbf{x})\theta(\mathbf{x})\, d\mathbf{x},$$

where

$$\xi(\mathbf{x}) = \sum_{\mathbf{k}\in\mathbb{Z}^d} |\hat{\mathbf{u}}^*_{\mathbf{k}}| e^{2\pi i \frac{\mathbf{k}}{L}\cdot\mathbf{x}}, \quad \psi(\mathbf{x}) = \sum_{\mathbf{j}\in\mathbb{Z}^d} |\mathbf{j}||\hat{\mathbf{v}}^*_{\mathbf{j}}| e^{2\pi i \frac{\mathbf{j}}{L}\cdot\mathbf{x}}, \quad \theta(\mathbf{x}) = \sum_{\mathbf{l}\in\mathbb{Z}^d} |\mathbf{l}|^2 |\hat{\mathbf{u}}^*_{\mathbf{l}}| e^{2\pi i \frac{\mathbf{l}}{L}\cdot\mathbf{x}}.$$

This integral can be estimated exactly as $(B(\mathbf{u}^*, \mathbf{v}^*), A\mathbf{w}^*)$. Hence, the standard estimates for the nonlinear term (see Appendix A) allow us to obtain, for instance, (9.12) and (9.13)

B.4 Time Analyticity in Gevrey Spaces

We now prove the estimates (9.21) and (9.26) concerning the time analyticity of the strong solutions of the Navier–Stokes equations as functions with values in Gevrey spaces. Recall that the norms appearing in (9.21) and (9.26) were defined in (9.11).

Analyticity in the 3-Dimensional Case

The forcing term $\mathbf{f}$ is assumed to belong to a Gevrey space:

$$\mathbf{f} \in D(e^{\sigma_1 A^{1/2}}) \tag{B.26}$$

for some $\sigma_1 > 0$.

The Navier–Stokes equations can be written for complex times $\zeta \in \mathbb{C}$ as

$$\frac{d\mathbf{u}}{d\zeta} + \nu A\mathbf{u} + B(\mathbf{u}) = \mathbf{f}, \tag{B.27}$$

where $\mathbf{u} = \mathbf{u}(\zeta)$. Given $\mathbf{u}_0 \in V$ (or even H), the Galerkin approximation of (B.27) has an analytic solution with $\mathbf{u}(0) = \mathbf{u}_0$ that is defined for ζ in a complex neighborhood of the origin. For this local analytic solution, the a priori estimates that we are about to derive formally can be rigorously justified.

We fix $\theta \in (-\pi/2, \pi/2)$ and consider the time $\zeta = se^{i\theta}$ for $s > 0$. Later, we will restrict θ to the interval $(-\pi/4, \pi/4)$. At the initial time $\zeta = 0$, the solution $\mathbf{u}(0) = \mathbf{u}_0$ belongs only to V but not necessarily to any Gevrey-type space. We will show that, as the real part of the complexified time increases, the solution becomes Gevrey-regular and the width of the space analyticity (measured by the parameter σ) also increases. To account for this increase in the width of analyticity, we consider the parameter σ as a function $\sigma = \varphi(s\cos\theta)$ of the real part of the complexified time. We will show that, for an appropriate function $\varphi = \varphi(s\cos\theta)$, we have the regularization

$$\mathbf{u}(se^{i\theta}) \in D(A^{1/2} e^{\varphi(s\cos\theta)A^{1/2}})$$

for s positive and small enough. For simplicity we assume, in view of the assumption (B.26), that $\varphi \le \sigma_1$ and, moreover, that the bound $\varphi' \le \nu\lambda_1^{1/2}$ holds.[22]

[22] See footnote 13.

With all this in mind, we wish to compute the derivative

$$\begin{aligned}\frac{1}{2}\frac{d}{ds}\|\mathbf{u}(se^{i\theta})\|^2_{\varphi(s\cos\theta)}\\ &= \frac{1}{2}\frac{d}{ds}|A^{1/2}e^{\varphi(s\cos\theta)A^{1/2}}\mathbf{u}(se^{i\theta})|^2\\ &= \mathrm{Re}\left(\left(\frac{d}{ds}(e^{\varphi(s\cos\theta)A^{1/2}}\mathbf{u}(se^{i\theta})), e^{\varphi(s\cos\theta)A^{1/2}}\mathbf{u}(se^{i\theta})\right)\right). \qquad \text{(B.28)}\end{aligned}$$

Using

$$\begin{aligned}\frac{d}{ds}(e^{\varphi(s\cos\theta)A^{1/2}}\mathbf{u}(se^{i\theta})) &= \varphi'(s\cos\theta)\cos\theta A^{1/2}e^{\varphi(s\cos\theta)A^{1/2}}\mathbf{u}(se^{i\theta})\\ &\quad + e^{i\theta}e^{\varphi(s\cos\theta)A^{1/2}}\frac{d\mathbf{u}}{d\zeta}(se^{i\theta}), \qquad \text{(B.29)}\end{aligned}$$

we find from (B.28) that

$$\begin{aligned}\frac{1}{2}\frac{d}{ds}\|\mathbf{u}(se^{i\theta})\|^2_{\varphi(s\cos\theta)} &= \varphi'(s\cos\theta)\cos\theta\,\mathrm{Re}\,e^{i\theta}(A\mathbf{u}, A^{1/2}\mathbf{u})_{\varphi(s\cos\theta)}\\ &\quad + \mathrm{Re}\,e^{i\theta}\left(\frac{d\mathbf{u}}{d\zeta}(se^{i\theta}), A\mathbf{u}\right)_{\varphi(s\cos\theta)}. \qquad \text{(B.30)}\end{aligned}$$

Omitting the arguments of $\mathbf{u} = \mathbf{u}(se^{i\theta})$ and $\varphi = \varphi(s\cos\theta)$ for notational simplicity, we obtain via equation (B.27) that

$$\begin{aligned}\frac{1}{2}\frac{d}{ds}\|\mathbf{u}\|^2_\varphi + \nu\cos\theta|A\mathbf{u}|^2_\varphi &= \varphi'(s\cos\theta)\cos\theta(A\mathbf{u}, A^{1/2}\mathbf{u})_\varphi\\ &\quad - \mathrm{Re}\,e^{i\theta}(B(\mathbf{u}), A\mathbf{u})_\varphi + \mathrm{Re}\,e^{i\theta}(\mathbf{f}, A\mathbf{u})_\varphi. \qquad \text{(B.31)}\end{aligned}$$

Using estimate (9.12) for the nonlinear term, using that $\varphi' \le \nu\lambda^{1/2}$, and using the Cauchy–Schwarz and Young inequalities, we have

$$\begin{aligned}&\frac{1}{2}\frac{d}{ds}\|\mathbf{u}\|^2_\varphi + \nu\cos\theta|A\mathbf{u}|^2_\varphi\\ &\le \nu\lambda_1^{1/2}\cos\theta|A\mathbf{u}|_\varphi\|\mathbf{u}\|_\varphi + c_1\|\mathbf{u}\|^{3/2}_\varphi|A\mathbf{u}|^{3/2}_\varphi + |\mathbf{f}|_\varphi|A\mathbf{u}|_\varphi\\ &\le \frac{\nu\cos\theta}{2}|A\mathbf{u}|^2_\varphi + \frac{3}{2}\nu\lambda_1\cos\theta\|\mathbf{u}\|^2_\varphi + \frac{c_3}{\nu^3\cos^3\theta}\|\mathbf{u}\|^6_\varphi + \frac{3}{2\nu\cos\theta}|\mathbf{f}|^2_\varphi\\ &\le \frac{\nu\cos\theta}{2}|A\mathbf{u}|^2_\varphi + \frac{1}{2}\nu^3\lambda_1^{3/2}\cos^3\theta + \frac{c_4}{\nu^3\cos^3\theta}\|\mathbf{u}\|^6_\varphi + \frac{3}{2\nu\cos\theta}|\mathbf{f}|^2_\varphi,\end{aligned}$$

where c_3 and c_4 depend only on the shape of the domain Ω. Thus, using also assumption (B.26) for the forcing term and that $\varphi \le \sigma_1$, it follows that

$$\frac{d}{ds}\|\mathbf{u}\|^2_\varphi + \nu\cos\theta|A\mathbf{u}|^2_\varphi \le \frac{3}{\nu\cos\theta}|\mathbf{f}|^2_{\sigma_1} + \nu^3\lambda_1^{3/2} + \frac{2c_4}{\nu^3\cos^3\theta}\|\mathbf{u}\|^6_\varphi, \qquad \text{(B.32)}$$

where $\mathbf{u} = \mathbf{u}(se^{i\theta})$ and $\varphi = \varphi(s\cos\theta)$. Now, assuming $|\theta| \le \pi/4$ and defining

$$y(s) = 1 + \frac{1}{\nu^2 \lambda_1^{3/2}} \|\mathbf{u}(se^{i\theta})\|^2_{\varphi(s\cos\theta)}, \tag{B.33}$$

we see that

$$\frac{dy}{ds} \le \nu\lambda_1 a y^3, \tag{B.34}$$

where

$$a = 1 + 4\sqrt{2}c_1 + \frac{3\sqrt{2}}{\nu^4 \lambda_1^{3/2}} |\mathbf{f}|^2_{\sigma_1}. \tag{B.35}$$

Notice that a is nondimensional and that $y(s)$ is a nondimensional function of time.

Then, as long as

$$\theta \in \left[-\frac{\pi}{4}, \frac{\pi}{4}\right], \quad 0 \le s \le T_0(\|\mathbf{u}_0\|), \tag{B.36}$$

where

$$T_0(\|\mathbf{u}_0\|) = T_0(\|\mathbf{u}_0\|, |\mathbf{f}|_{\sigma_1}, \nu, \Omega) = \frac{3\nu^3}{8a(\nu^2\lambda_1^{1/2} + \|\mathbf{u}_0\|^2)^2}, \tag{B.37}$$

we obtain $y(s) \le 2y(0)$; that is,

$$\|\mathbf{u}(se^{i\theta})\|^2_{\varphi(s\cos\theta)} \le \nu^2\lambda_1^{1/2} + 2\|\mathbf{u}_0\|^2. \tag{B.38}$$

In view of our assumptions that $\varphi' \le \nu\lambda_1^{1/2}$ and $\varphi \le \sigma_1$, we can consider $\varphi(\xi) = \min\{\nu\lambda_1^{1/2}\xi, \sigma_1\}$ for all $\xi \ge 0$. This choice of φ is not continuously differentiable, but it can be approximated by continuously differentiable functions (satisfying the required assumptions) so that (B.38) holds for this choice of φ. Then we define the region

$$\begin{aligned} \Delta^0_{\sigma_1}(\|\mathbf{u}_0\|) &= \Delta^0_{\sigma_1}(\|\mathbf{u}_0\|, |\mathbf{f}|_{\sigma_1}, \nu, \Omega) \\ &= \Big\{ \zeta = se^{i\theta};\ |\theta| < \frac{\pi}{4}, \\ &\qquad 0 < s < T_0(\|\mathbf{u}_0\|, |\mathbf{f}|_{\sigma_1}, \nu, \Omega),\ \nu\lambda_1^{1/2} s|\sin\theta| < \sigma_1 \Big\}. \end{aligned} \tag{B.39}$$

This set is a domain of analyticity of the solution $\mathbf{u} = \mathbf{u}(\zeta)$ of the complexified Navier–Stokes equations. The origin $\zeta = 0$ belongs to the closure of $\Delta^0_{\sigma_1}(\|\mathbf{u}_0\|)$, and $\Delta^0_{\sigma_1}(\|\mathbf{u}_0\|)$ is a neighborhood of the real time interval where the solution of the (real) NSE is strong. Moreover, for the closure of this domain we have

$$|\mathbf{u}(\zeta)|^2_{D(A^{1/2}e^{\varphi(s\cos\theta)A^{1/2}})} \le \nu^2\lambda_1^{1/2} + 2\|\mathbf{u}_0\|^2 \quad \text{for } \zeta \in \overline{\Delta^0_{\sigma_1}}(\|\mathbf{u}_0\|, |\mathbf{f}|_{\sigma_1}, \nu, \Omega). \tag{B.40}$$

As mentioned before, these estimates are first obtained for the Galerkin approximation of the solutions. Then, we pass to the limit to find that the strong solutions are analytic in time from $\Delta^0_{\sigma_1}(\|\mathbf{u}_0\|)$ into $D(A^{1/2}\exp(\varphi(s\cos\theta)A^{1/2}))$, where $\varphi(\xi) = \min\{\nu\lambda_1^{1/2}\xi, \sigma_1\}$, and that the strong solutions satisfy the same estimates as their Galerkin approximation.

In summary, we have shown (as in Section 9) that if $\mathbf{u}_0 \in V$ and $\mathbf{f} \in D(\exp(\sigma_1 A^{1/2}))$ with $\sigma_1 > 0$ then there exists $T_0(\|\mathbf{u}_0\|, |\mathbf{f}|_{\sigma_1}, \nu, \Omega)$, given by (B.37) and (B.35), such that: (i) the corresponding strong solution with $\mathbf{u}(0) = \mathbf{u}_0$ of the Navier–Stokes equations can be extended to a solution of the complexified NSE (B.27) that is analytic from $\Delta^0_{\sigma_1}(\|\mathbf{u}_0\|, |\mathbf{f}|_{\sigma_1}, \nu, \Omega)$ into $D(A^{1/2}\exp(\varphi(s\cos\theta)A^{1/2}))$, where $\varphi(\xi) = \min\{\nu\lambda_1^{1/2}\xi, \sigma_1\}$; and (ii) on the closure of this domain of analyticity, the estimate (B.40) holds.

Analyticity in the 2-Dimensional Periodic Case

In the 2-dimensional case, owing to the uniform bound on the enstrophy of the strong solutions, the domain of analyticity can be extended to a neighborhood of the whole positive real axis. Moreover, we can use more favorable estimates for the nonlinear term to obtain a wider width of analyticity. The estimate we use for the nonlinear term is given by (9.13), and the calculations are similar to the 3-dimensional case – up to equation (B.31). Then, we use instead (9.13) to find that

$$\begin{aligned}\frac{1}{2}\frac{d}{ds}\|\mathbf{u}\|_\varphi^2 + \nu\cos\theta|A\mathbf{u}|_\varphi^2 &\le \nu\lambda_1^{1/2}\cos\theta|A\mathbf{u}|_\varphi\|\mathbf{u}\|_\varphi + |\mathbf{f}|_\varphi|A\mathbf{u}|_\varphi \\ &\quad + c_2\|\mathbf{u}\|_\varphi^2|A\mathbf{u}|_\varphi\left(1+\log\frac{|A\mathbf{u}|_\varphi^2}{\lambda_1\|\mathbf{u}\|_\varphi^2}\right)^{1/2} \\ &\le \frac{\nu\cos\theta}{2}|A\mathbf{u}|_\varphi^2 + \frac{3}{2}\nu\lambda_1\cos\theta\|\mathbf{u}\|_\varphi^2 + \frac{3}{2\nu\cos\theta}|\mathbf{f}|_\varphi^2 \\ &\quad + \frac{3c_2^2}{2\nu\cos\theta}\|\mathbf{u}\|_\varphi^4\left(1+\log\frac{|A\mathbf{u}|_\varphi^2}{\lambda_1\|\mathbf{u}\|_\varphi^2}\right) \\ &\le \frac{\nu\cos\theta}{2}|A\mathbf{u}|_\varphi^2 + \frac{3}{2\nu\cos\theta}|\mathbf{f}|_\varphi^2 + \frac{1}{2}\nu^3\lambda_1^2\cos^3\theta \\ &\quad + \frac{c_5}{\nu\cos\theta}\|\mathbf{u}\|_\varphi^4\left(1+\log\frac{|A\mathbf{u}|_\varphi^2}{\lambda_1\|\mathbf{u}\|_\varphi^2}\right),\end{aligned}$$

where c_5 depends only on the shape of the domain Ω. Then, using the assumption (B.26) on the forcing term and that $\varphi \le \sigma_1$, we have

$$\begin{aligned}\frac{d}{ds}\|\mathbf{u}\|_\varphi^2 + \nu\cos\theta|A\mathbf{u}|_\varphi^2 &\le \frac{3}{\nu\cos\theta}|\mathbf{f}|_{\sigma_1}^2 + \nu^3\lambda_1^2\cos^3\theta \\ &\quad + \frac{2c_5}{\nu\cos\theta}\|\mathbf{u}\|_\varphi^4\left(1+\log\frac{|A\mathbf{u}|_\varphi^2}{\lambda_1\|\mathbf{u}\|_\varphi^2}\right).\end{aligned} \tag{B.41}$$

We now handle the logarithmic term as follows. First define

$$z = \frac{|A\mathbf{u}|_\varphi^2}{\lambda_1\|\mathbf{u}\|_\varphi^2}. \tag{B.42}$$

Then we can gather a few terms from (B.41) and write

$$-\frac{\nu\cos\theta}{2}|A\mathbf{u}|_\varphi^2+\frac{2c_5}{\nu\cos\theta}\|\mathbf{u}\|_\varphi^4\left(1+\log\frac{|A\mathbf{u}|_\varphi^2}{\lambda_1\|\mathbf{u}\|_\varphi^2}\right)$$

$$=\|\mathbf{u}\|_\varphi^2\left(-\nu\lambda_1 z\cos\theta+\frac{2c_5}{\nu\cos\theta}\|\mathbf{u}\|_\varphi^2(1+\log z)\right). \tag{B.43}$$

The RHS of (B.43) can be written as a function of z in the form

$$-\alpha z+\beta(1+\log z) \tag{B.44}$$

for $z>0$, which has a global maximum at $z=\beta/\alpha$, so that

$$-\alpha z+\beta(1+\log z)\le\beta\log(\beta/\alpha). \tag{B.45}$$

Therefore, we find from (B.43) that

$$-\frac{\nu\cos\theta}{2}|A\mathbf{u}|_\varphi^2+\frac{2c_5}{\nu\cos\theta}\|\mathbf{u}\|_\varphi^4\left(1+\log\frac{|A\mathbf{u}|_\varphi^2}{\lambda_1\|\mathbf{u}\|_\varphi^2}\right)$$

$$\le\frac{2c_5}{\nu\cos\theta}\|\mathbf{u}\|_\varphi^4\log\frac{4c_5\|\mathbf{u}\|^2}{\lambda_1\nu^2\cos^2\theta}. \tag{B.46}$$

Inserting this estimate into (B.41) yields

$$\frac{d}{ds}\|\mathbf{u}\|_\varphi^2+\frac{\nu\cos\theta}{2}|A\mathbf{u}|_\varphi^2\le\frac{3}{\nu\cos\theta}|\mathbf{f}|_{\sigma_1}^2+\nu^3\lambda_1^2\cos^3\theta$$

$$+\frac{2c_5}{\nu\cos\theta}\|\mathbf{u}\|_\varphi^4\log\frac{4c_5\|\mathbf{u}\|_\varphi^2}{\lambda_1\nu^2\cos^2\theta} \tag{B.47}$$

with $\mathbf{u}=\mathbf{u}(se^{i\theta})$ and $\varphi=\varphi(s\cos\theta)$. Defining y by

$$y(s)=e+\frac{1}{\nu^2\lambda_1\cos^2\theta}|\mathbf{f}|_{\sigma_1}^2+\frac{4c_5}{\lambda_1\nu^2\cos^2\theta}\|\mathbf{u}(se^{i\theta})\|_{\varphi(s\cos\theta)}^2, \tag{B.48}$$

we see that

$$\frac{dy}{ds}\le\nu\lambda_1c_6y^2\log y, \tag{B.49}$$

where c_6 is a constant depending only on the shape of the domain Ω. If $y(s)\le 2y(0)$, then

$$\frac{dy}{ds}\le\nu\lambda_1c_6y^2\log 2y(0). \tag{B.50}$$

Hence, as long as

$$0\le s\le\frac{1}{2\nu\lambda_1c_6y(0)\log 2y(0)}, \tag{B.51}$$

we have $y(s)\le 2y(0)$; this implies that

$$\|\mathbf{u}(se^{i\theta})\|_{\varphi(s\cos\theta)}^2\le c_7\lambda_1\nu^2+2\|\mathbf{u}_0\|^2, \tag{B.52}$$

where c_7 depends only on the shape of the domain. Thus, (B.52) holds as long as

$$\theta\in\left[-\frac{\pi}{4},\frac{\pi}{4}\right],\quad 0\le s\le T_0(\|\mathbf{u}_0\|), \tag{B.53}$$

where

$$T_0(\|\mathbf{u}_0\|) = T_0(\|\mathbf{u}_0\|, |\mathbf{f}|_{\sigma_1}, \nu, \Omega)$$
$$= \left[c_8 \nu\lambda_1\left(1+\frac{|\mathbf{f}|_{\sigma_1}}{\nu^2\lambda_1}+\frac{\|\mathbf{u}_0\|^2}{\nu^2\lambda_1}\right)\log c_9\left(1+\frac{|\mathbf{f}|_{\sigma_1}}{\nu^2\lambda_1}+\frac{\|\mathbf{u}_0\|^2}{\nu^2\lambda_1}\right)\right]^{-1} \quad \text{(B.54)}$$

(here c_8 and c_9 are constants depending only on the shape of the domain Ω). As before, we can choose $\varphi(\xi) = \min\{\nu\lambda_1^{1/2}\xi, \sigma_1\}$ for $\xi \geq 0$. Then we define the region

$$\Delta^0_{\sigma_1}(\|\mathbf{u}_0\|) = \Delta^0_{\sigma_1}(\|\mathbf{u}_0\|, |\mathbf{f}|_{\sigma_1}, \nu, \Omega)$$
$$= \Big\{\zeta = se^{i\theta};\ |\theta| < \frac{\pi}{4},$$
$$0 < s < T_0(\|\mathbf{u}_0\|, |\mathbf{f}|_{\sigma_1}, \nu, \Omega),\ \nu\lambda_1^{1/2}s|\sin\theta| < \sigma_1\Big\}. \quad \text{(B.55)}$$

This set is a domain of analyticity of the solution $\mathbf{u} = \mathbf{u}(\zeta)$ of the complexified Navier–Stokes equations. The origin $\zeta = 0$ belongs to the closure of $\Delta^0_{\sigma_1}(\|\mathbf{u}_0\|)$. Moreover, on the closure of this domain we have

$$|\mathbf{u}(\zeta)|^2_{D(A^{1/2}e^{\varphi(s\cos\theta)A^{1/2}})} \leq c_7\lambda_1\nu^2 + 2\|\mathbf{u}_0\|^2 \quad \text{for } \zeta \in \overline{\Delta^0_{\sigma_1}}(\|\mathbf{u}_0\|, |\mathbf{f}|_{\sigma_1}, \nu, \Omega). \quad \text{(B.56)}$$

III

Finite Dimensionality of Flows

Introduction

In principle, the idea that solutions of the Navier–Stokes equations (NSE) might be adequately represented in a finite-dimensional space arose as a result of the realization that the rapidly varying, high-wavenumber components of the turbulent flow decay so rapidly as to leave the energy-carrying (lower-wavenumber) modes unaffected. With the understanding gained from Kolmogorov's [1941a,b] phenomenological theory (see also Section 3), it appeared that, in 3-dimensional turbulent flows, only wavenumbers up to the cutoff value $\kappa_d = (\epsilon/\nu^3)^{1/4}$ need be considered. This is the boundary between the inertial range, which is dominated by the inertial term in the equation, and the dissipation range, which is dominated by the viscous term. As explained by Landau and Lifshitz [1971], the question is then reduced to finding the number of resolution elements needed to describe the velocity field in a volume – say, a cube of length ℓ_0 on each side. Clearly, if the smallest resolved distance is to be $\ell_d = 1/\kappa_d$, then the number of resolution elements is simply $(\ell_0/\ell_d)^3$. On adducing some phenomenological and intuitive arguments, it was argued that this ratio is $\mathrm{Re}^{9/4}$, where Re is the Reynolds number. An alternate way to count the number of active modes is as follows: since these modes are those in the inertial range, their frequency κ satisfies $\kappa_0 < \kappa < \kappa_d$, with $\kappa_0 = 1/\ell_0$; we conclude that, for κ_d/κ_0 large, that number is of the order of $(\kappa_d/\kappa_0)^3 = (\ell_0/\ell_d)^3$.

A similar approach in the case of 2-dimensional turbulent flows leads to the estimated number of the resolution elements as $(\ell_0/\ell_\eta)^2 = (\kappa_\eta/\kappa_0)^2$, where ℓ_η is the so-called Kraichnan cutoff length and $\kappa_\eta = 1/\ell_\eta$. Again, phenomenological and intuitive arguments lead to the conclusion that here this ratio is simply Re.

Most of this chapter is devoted to determining the extent to which these estimates are realistic. Such an exercise is important because the estimates (perhaps we should say "guesstimates") just described are, in practice, very large numbers – sometimes beyond the capabilities of modern computational resources. This barrier to computational fluid dynamics tempts the practitioner to seek low-resolution models of turbulent flows. However, this would be an egregious error, especially if one descends to such models as the Lorenz [1963] three-mode model of convectively driven turbulence. It is now known that the chaotic behavior of that model is an artifact of its severe truncation (see Trève and Manley [1982]), leaving behind essentially a caricature of the true system. The Lorenz system is still extremely instructive because it shows that deceptively simple nonlinearities can lead to chaotic, turbulent-like behavior, but it definitely is not an adequate model of real convective turbulence.

On the mathematical side, the first type of finite dimensionality result appeared in the work of Foias and Prodi [1967]; finite dimensionality was also alluded to by Ladyzhenskaya [1972]. Then, Foias and Temam [1979] showed that the attractors of the Navier–Stokes equations have finite dimension. As we shall explain, this work was followed by subsequent efforts in three different directions: (i) generalizations of the work of Foias and Prodi [1967]; (ii) making connections with the approaches of Kolmogorov and of Landau and Lifshitz; and (iii) exploring some numerical applications. As mentioned in the Preface, we will not explore numerical applications; this chapter is thus devoted to the first two directions.

In Section 1 we study the proposition that NSE solutions in terms of eigenmode expansions can be represented by a large but finite number of so-called determining modes. The case of 2-dimensional flows with different boundary conditions is explored in some detail. It turns out that the upper bounds on the number of determining modes for the no-slip and periodic boundary conditions is somewhat larger than the corresponding estimates arrived at by the nonrigorous arguments mentioned earlier. One may believe that the larger values are overestimates generated by the limitations of the available functional analysis tools; namely, the inequalities we use lead to upper bounds only and so the end results of this (and the following) section are at best sufficient conditions. Nonetheless, evidence exists that such estimates are optimal; in particular, the lower estimates on the dimension of attractors by Babin and Vishik [1983] and Ziane [1997]. The resolution of this apparent contradiction would be that there are indeed many active modes but only a few of them are significant. This is only a speculative thought, and there is still much work to do in reconciling the very large estimates on the numbers of degrees of freedom (from the mathematical theory as well as from the conventional theory of turbulence) with the practical needs of engineers and the practical limitations in computational fluid dynamics.

Section 2 deals with the so-called determining nodes – that is, the question of the necessary set of points in the configuration space at which the evolution of a 2-dimensional turbulent flow must be followed so as to obtain a faithful representation of the flow. Not surprisingly, the bound on the number of these nodes, for the corresponding boundary conditions, is essentially the same as in the case of eigenfunction expansions.

Because of the fundamental difficulties with the solutions of Navier–Stokes equations in three dimensions, no significant effort is expended here on the corresponding numbers of determining modes and nodes. However, for some partial results the reader is referred to the original papers: Constantin, Foias, and Temam [1985b], Constantin, Foias, Manley, and Temam [1985a], and Foias and Temam [1984].

Section 3 takes a completely different approach to the question of finiteness of the function space in which the permanent solutions to the NSE reside: studying the properties of their attractors. These entities contain the solutions for all possible initial conditions after a long time. Clearly, the attractors are invariant in the sense that, if an initial condition happens to be on the attractor, it stays there forever by definition. An important attribute of an attractor is its dimension, that is, the number of orthogonal coordinate axes in the space in which it exists. That space is necessarily a subspace of the function space of the solution. The practical use of this attribute, say $\dim(\mathcal{A})$, lies in its relation to the number of degrees of freedom of the solution (e.g., the number, say n, needed to parameterize the attractor) $n \leq 2\dim(\mathcal{A}) + 1$ (see e.g. Mañe [1981], Foias and Olson [1995], Robinson [1998, 1999], and Friz and Robinson [2000]; see also Eden, Foias, Nicolaenko, and Temam [1994] and Takens [1985]).

Before we proceed with the description of our results, let us recall the concept of dimension of a fractal set. Although many other definitions are available, only two notions of dimension will be considered here: the Hausdorff and the fractal dimensions. The *fractal dimension* of a bounded set X in a metric space H is defined by

$$d_f(X) = \limsup_{\delta \to 0^+} \frac{\ln N_X(\delta)}{\ln(1/\delta)}, \tag{0.1}$$

where $N_X(\delta)$ is the smallest number of balls of radius δ necessary to cover X (which may be infinite, since X is not required to be compact). The definition of Hausdorff dimension is slightly more involved and is done through coverings of X by balls of radii not larger than δ and counting their total volume. More precisely, given $d, \varepsilon > 0$, we first define the quantity

$$\mu_H(X, d, \varepsilon) = \inf \sum_{i \in I} r_i^d, \tag{0.2}$$

where the infimum is for all coverings of X by a family $\{B_i\}_{i\in I}$ of balls of radii $r_i \leq \varepsilon$. Clearly, $\mu_H(X, d, \varepsilon)$ is a nonincreasing function of ε. The quantity $\mu_H(X, d) \in [0, \infty]$ defined by

$$\mu_H(X, d) = \lim_{\varepsilon \to 0} \mu_H(X, d, \varepsilon) = \sup_{\varepsilon > 0} \mu_H(X, d, \varepsilon) \tag{0.3}$$

is the d-dimensional Hausdorff measure of X. It is easy to see that, if $\mu(X, d') < \infty$ for some d', then $\mu_H(X, d) = 0$ for every $d > d'$. Thus, there exists $d_0 \in [0, \infty]$ such that $\mu_H(X, d) = 0$ for $d > d_0$ and $\mu_H(X, d) = \infty$ for $d < d_0$. This number d_0 is called the *Hausdorff dimension* of X and is denoted $d_H(X)$. *The fractal dimension can also be defined similarly to the Hausdorff dimension by taking only coverings of X by balls of radius exactly ε.* The reader is referred to Federer [1969] for the properties of Hausdorff measures and dimension; see also Falconer [1985]. The Hausdorff dimension is always smaller than or equal to the fractal dimension. They are usually different, and it can even happen that the fractal dimension of X is infinite while its Hausdorff dimension is zero.

We start again with flows in two dimensions. Preliminary to the actual estimation of the attractor dimension, it is necessary to discuss the concept of compactness. This may seem to be an exercise in pedantry, but its importance in the present case is shown by the following consideration. One readily appreciates that continuous functions on any finite bounded interval of the real line have the intuitively obvious fundamental property of achieving there their maximum and/or minimum. In more abstract settings, such as the present case of multidimensional function spaces, we need a similar property – that is, compactness – which is necessary for arriving at best approximations, iterated solution methods, and so forth. It must be noted that not all spaces have such a property. Hence we must exercise care, especially since one can invent simple counterexamples – even in the 1-dimensional case, where compactness does not prevail (see Wouk [1979]).

Initially, the dimension of the attractors for the Navier–Stokes equations was calculated by exploiting the available Sobolev inequalities and then employing the relationship between that dimension and the prevailing Lyapounov exponents, as first conjectured by Farmer, Ott, and Yorke [1983]. This effective (but somewhat cumbersome) approach was significantly simplified by the use of the Lieb–Thirring [1976] improvement of the Sobolev inequalities, introduced for the NSE by Temam [1986]. Here one need not deal with the hard-to-obtain Lyapounov exponents; instead, one uses the well-known eigenvalues of the orthonormal functions arising in the Lieb–Thirring inequalities to calculate the attractor dimension. Apart from the simplification, the estimate of the attractor dimension is improved and yields dimensions that are comparable to those estimated in terms of the ratios (arrive at phenomenologically) of the macroscopic lengths to the cutoff lengths.

The last part of this section addresses the dimension of the attractor for flows in three dimensions. Backed by the empirical evidence that – at least at the macroscopic observational level – real turbulent flows evince no singular behavior, it is assumed that there exists a subset of solutions in the relevant function spaces that is regular. For those solutions one can then calculate the dimension of their attractor by methods outlined in the preceding paragraph for the 2-dimensional cases. The result shows that the conventional, phenomenological estimate of $\mathrm{Re}^{9/4}$ is actually an upper bound for the attractor dimension.

Section 4, the final section of this chapter, addresses a means for decreasing to a manageable size the number of modes needed for calculating the solutions to Navier–Stokes equations. Until recently, such efforts were confined to using somewhat controversial "closure approximations." It has now been realized in the context of the general theory of nonlinear partial differential equations that, for some of these equations, the high-wavenumber components of the spectrum could be expressed uniquely in terms of the low-wavenumber components, forming the so-called inertial manifolds. Unfortunately, the available necessary conditions for the existence of such manifolds have not been shown to hold for the NSE; this is related to the property of fluid turbulence that all length scales are present in the turbulence. However, it has been shown that an approximate inertial manifold can be constructed for turbulent solutions (see Foias, Manley, and Temam [1987a, 1988b]). This suffices for constructing an approximate Navier–Stokes equation whose attractor can be made arbitrarily close to the attractor for the exact equation. This implies that the exact solution to the approximate equation (or model) can be made arbitrarily close to the exact (though perhaps uncomputable, in practice) solution. Three practical consequences of the existence of the approximate inertial manifold have been realized: (1) novel time-saving computational algorithms, the nonlinear Galerkin method (Marion and Temam [1989]) and the "incremental unknown" methods that operate in the context of the spectral methods, the finite element methods, and finite differences (see e.g. Dubois, Jauberteau, and Temam [1999], Costa, Dettori, Gottlieb, and Temam [2001], and Faure, Laminie, and Temam [2001]); (2) an effective viscosity accounting for the relegation of the high-wavenumber components to the approximate inertial manifold onto which the evolution of the low-wavenumber components is constrained (Foias, Manley, and Temam [1991]); and (3) the recognition that an n-mode eigenmode expansion of the solution to the approximate Navier–Stokes equation yields an accuracy comparable with a $2n$-mode approximate solution of the exact equation (see García-Archilla, Novo, and Titi [1999]).

Elements of the Mathematical Theory of the NSE

For the reader's convenience, we now recall, from Chapter II, some elements of the mathematical theory of the Navier–Stokes equations that will be needed in this chapter.

We consider the Navier–Stokes equations

$$\mathbf{u}_t - \nu\Delta\mathbf{u} + (\mathbf{u}\cdot\nabla)\mathbf{u} + \nabla p = \mathbf{f}, \tag{0.4a}$$

$$\nabla\cdot\mathbf{u} = 0, \tag{0.4b}$$

in $\Omega \subset \mathbb{R}^d$ ($d = 2$ or 3) with $\nu > 0$ and $\mathbf{f} = \mathbf{f}(\mathbf{x})$. We endow the system with either the no-slip boundary conditions or the space-periodic boundary conditions. In the first case we assume more precisely that

$\Omega \subset \mathbb{R}^d$ is open, bounded, and connected, with a $\mathcal{C}^2$ boundary $\partial\Omega$, (0.5)

and require that

$$\mathbf{u} = 0 \quad \text{on } \partial\Omega. \tag{0.6}$$

In the space-periodic case, we assume that

$$\Omega = \prod_{i=1}^{d}\left(\frac{L_i}{2}, \frac{L_i}{2}\right), \quad L_i > 0,\ i = 1, \ldots, d, \tag{0.7}$$

and require that

$\mathbf{u} = \mathbf{u}(\mathbf{x}, t)$, $p = p(\mathbf{x}, t)$, and $\mathbf{f} = \mathbf{f}(\mathbf{x})$ with $\mathbf{x} = (x_1, \ldots, x_d)$ are L_i-periodic in each variable $x_1, \ldots, x_d$; (0.8)

for simplicity,

$$\int_\Omega \mathbf{f}(\mathbf{x})\, d\mathbf{x} = \int_\Omega \mathbf{u}(\mathbf{x}, t)\, d\mathbf{x} = 0. \tag{0.9}$$

We now recall the mathematical setting of the problem. In the no-slip case, we denote by $\mathcal{V}$ the set of smooth (say, $\mathcal{C}^\infty$) divergence-free vector fields on Ω that are compactly supported in Ω. In the space-periodic case, $\mathcal{V}$ is the space of smooth ($\mathcal{C}^\infty$) divergence-free vector fields on $\mathbb{R}^d$ that are periodic with period Ω. In either case we let H be the closure of $\mathcal{V}$ in $L^2(\Omega)^d$ and let V be the closure of $\mathcal{V}$ in $H^1(\Omega)^d$. In H and V we consider (respectively) the scalar products

$$(\mathbf{u}, \mathbf{v}) = \int_\Omega \mathbf{u}(\mathbf{x}) \cdot \mathbf{v}(\mathbf{x})\, d\mathbf{x}, \qquad ((\mathbf{u}, \mathbf{v})) = \int_\Omega \sum_{i=1}^d \frac{\partial \mathbf{u}}{\partial x_i} \cdot \frac{\partial \mathbf{v}}{\partial x_i}\, d\mathbf{x}$$

and the associated norms, denoted by

$$|\mathbf{u}| = (\mathbf{u}, \mathbf{u})^{1/2} \quad \text{for } \mathbf{u} \in H, \qquad \|\mathbf{v}\| = ((\mathbf{v}, \mathbf{v}))^{1/2} \quad \text{for } \mathbf{v} \in V.$$

We recall that the spaces H and V can be characterized as follows. In the no-slip case,

$$H = \{\mathbf{u} \in L^2(\Omega)^d;\ \nabla \cdot \mathbf{u} = 0,\ \mathbf{u} \cdot \mathbf{n}|_{\partial\Omega} = 0\}$$

and

$$V = \{\mathbf{u} \in H^1(\Omega)^d;\ \nabla \cdot \mathbf{u} = 0,\ \mathbf{u}|_{\partial\Omega} = 0\},$$

where $\mathbf{n}$ denotes the outward unit normal to $\partial\Omega$. We remark that $\mathbf{u} \cdot \mathbf{n}|_{\partial\Omega}$ makes sense when $\mathbf{u} \in L^2(\Omega)^d$ and $\nabla \cdot \mathbf{u} = 0$ in the distribution sense. In the space-periodic case,

$$H = \left\{\mathbf{u} \in L^2_{\text{per}}(\Omega)^d;\ \nabla \cdot \mathbf{u} = 0,\ \int_\Omega \mathbf{u}(\mathbf{x})\, d\mathbf{x} = 0\right\}$$

and

$$V = \left\{\mathbf{v} \in H^1_{\text{per}}(\Omega)^d;\ \nabla \cdot \mathbf{u} = 0,\ \int_\Omega \mathbf{u}(\mathbf{x})\, d\mathbf{x} = 0\right\},$$

where $H^1_{\text{per}}(\Omega)^d$ is the space of $\mathbb{R}^d$-valued functions $\mathbf{u}$ defined on $\mathbb{R}^d$ that are L_i-periodic in each variable x_i ($i = 1, \ldots, d$) and such that $\mathbf{u}|_{\mathcal{O}} \in H^1(\mathcal{O})^d$ for every bounded open set $\mathcal{O}$ in $\mathbb{R}^d$. The functions in $H^1_{\text{per}}(\Omega)^d$ are easily characterized by their Fourier series expansion

$$H^1_{\text{per}}(\Omega)^d = \left\{\mathbf{u} = \sum_{\mathbf{k} \in \mathbb{Z}^d} \hat{\mathbf{u}}_{\mathbf{k}} e^{2\pi i \frac{\mathbf{k}}{\mathbf{L}} \cdot \mathbf{x}};\ \hat{\mathbf{u}}_{-\mathbf{k}} = \bar{\hat{\mathbf{u}}}_{\mathbf{k}},\ \sum_{\mathbf{k} \in \mathbb{Z}^d} \left(\frac{1}{L^2} + 2\pi \left|\frac{\mathbf{k}}{\mathbf{L}}\right|^2\right) |\hat{\mathbf{u}}_{\mathbf{k}}|^2 < \infty\right\},$$

where $L = \min\{L_1, \ldots, L_d\}$ and

$$\frac{\mathbf{k}}{\mathbf{L}} = \left(\frac{k_1}{L_1}, \ldots, \frac{k_d}{L_d}\right).$$

Hence, we also have

$$V = \left\{\mathbf{u} = \sum_{\mathbf{k} \in \mathbb{Z}^d \setminus \{0\}} \hat{\mathbf{u}}_{\mathbf{k}} e^{2\pi i \frac{\mathbf{k}}{\mathbf{L}} \cdot \mathbf{x}};\ \hat{\mathbf{u}}_{-\mathbf{k}} = \bar{\hat{\mathbf{u}}}_{\mathbf{k}},\ \frac{\mathbf{k}}{\mathbf{L}} \cdot \hat{\mathbf{u}}_{\mathbf{k}} = 0,\ \sum_{\mathbf{k} \in \mathbb{Z}^d \setminus \{0\}} \left|\frac{\mathbf{k}}{\mathbf{L}}\right|^2 |\hat{\mathbf{u}}_{\mathbf{k}}|^2 < \infty\right\}$$

as well as

$$H = \left\{\mathbf{u} = \sum_{\mathbf{k} \in \mathbb{Z}^d \setminus \{0\}} \hat{\mathbf{u}}_{\mathbf{k}} e^{2\pi i \frac{\mathbf{k}}{\mathbf{L}} \cdot \mathbf{x}};\ \hat{\mathbf{u}}_{-\mathbf{k}} = \bar{\hat{\mathbf{u}}}_{\mathbf{k}},\ \frac{\mathbf{k}}{\mathbf{L}} \cdot \hat{\mathbf{u}}_{\mathbf{k}} = 0,\ \sum_{\mathbf{k} \in \mathbb{Z}^d \setminus \{0\}} |\hat{\mathbf{u}}_{\mathbf{k}}|^2 < \infty\right\}.$$

The problem posed in the weak formulation of the NSE with either no-slip or space-periodic boundary conditions is as follows. Given $T > 0$, $\mathbf{u}_0$ in H, and $\mathbf{f}$ in $L^2(0, T; H)$, find a function $\mathbf{u}$ in $L^\infty(0, T; H) \cap L^2(0, T; V)$ such that

$$\frac{d}{dt}(\mathbf{u}, \mathbf{v}) + \nu((\mathbf{u}, \mathbf{v})) + b(\mathbf{u}, \mathbf{u}, \mathbf{v}) = (\mathbf{f}, \mathbf{v}) \quad \text{for all } \mathbf{v} \in V \tag{0.10}$$

in the distribution sense on $(0, T)$ and with $\mathbf{u}(0) = \mathbf{u}_0$ in some suitable sense. Here

$$b(\mathbf{u}, \mathbf{v}, \mathbf{w}) = \sum_{i,j=1}^{d} \int_\Omega u_i \frac{\partial v_j}{\partial x_i} w_j \, d\mathbf{x},$$

and if $\mathbf{f}$ is square integrable but not in H then we can replace $\mathbf{f}$ by its Leray projection on H, so that $\mathbf{f}$ is always assumed to be in H.

Note that the pressure term disappears in the weak formulation of the problem. This is because the gradient of a function is orthogonal to the space of solenoidal functions. More precisely, with respect to the inner product in $L^2(\Omega)^d$, we have

$$H^\perp = \{\mathbf{u} \in L^2(\Omega)^d;\ \mathbf{u} = \nabla p,\ p \in H^1(\Omega)^d\}$$

in the no-slip case and

$$H^\perp = \{\mathbf{u} \in L^2(\Omega)^d;\ \mathbf{u} = \nabla p,\ p \in H^1_{\text{per}}(\Omega)^d\}$$

in the space-periodic case. In any case, we have the so-called Leray projector

$$P_L\colon L^2(\Omega)^d \to H,$$

which is the orthogonal projector onto H in $L^2(\Omega)^d$.

The trilinear operator $b = b(\mathbf{u}, \mathbf{v}, \mathbf{w})$ defined previously can be extended to a continuous trilinear operator defined on V. Moreover,

$$b(\mathbf{u}, \mathbf{v}, \mathbf{v}) = 0 \quad \text{for } \mathbf{u}, \mathbf{v} \in V, \tag{0.11}$$

which implies that

$$b(\mathbf{u}, \mathbf{v}, \mathbf{w}) = -b(\mathbf{u}, \mathbf{w}, \mathbf{v}) \quad \text{for } \mathbf{u}, \mathbf{v}, \mathbf{w} \in V. \tag{0.12}$$

An equivalent formulation for the NSE is achieved with the definition of the Stokes operator

$$A\mathbf{u} = -P_L \Delta \mathbf{u} \quad \text{for all } \mathbf{u} \in D(A) = V \cap H^2(\Omega)^d. \tag{0.13}$$

The Stokes operator is a positive self-adjoint operator, so we can work with fractional powers of A. We will consider, as explained in Chapter II, the positive and negative powers of A: A^α with $\alpha \in \mathbb{R}$ and $\alpha \geq 0$ or $\alpha < 0$; we also explained in Chapter II how $D(A^\alpha)$ and $D(A^{-\alpha})$ are paired through the "duality" product (a bilinear form from $D(A^\alpha) \times D(A^{-\alpha})$ into $\mathbb{R}$). In particular, we have $D(A^{1/2}) = V$, and it follows that $A\colon D(A^{1/2}) \to D(A^{-1/2})$ with

$$(A\mathbf{u}, \mathbf{v}) = ((\mathbf{u}, \mathbf{v})) \quad \text{for all } \mathbf{u}, \mathbf{v} \in D(A^{1/2}),$$

where $(\cdot,\cdot)$ is the duality product between $D(A^{-1/2})$ and $D(A^{1/2})$. Moreover, the trilinear operator $b(\mathbf{u},\mathbf{v},\mathbf{w})$ leads to the bilinear operator $B\colon D(A^{1/2})\times D(A^{1/2})\mapsto D(A^{-1/2})$ defined by

$$(B(\mathbf{u},\mathbf{v}),\mathbf{w}) = b(\mathbf{u},\mathbf{v},\mathbf{w}) \quad \text{for all } \mathbf{u},\mathbf{v},\mathbf{w}\in D(A^{1/2}).$$

We then have the following functional formulation of the NSE: Given $T > 0$, $\mathbf{u}_0$ in H, and $\mathbf{f}$ in $L^2(0,T;H)$, find $\mathbf{u}$ in $L^\infty(0,T;H)\cap L^2(0,T;V)$ with $\mathbf{u}'\in L^1(0,T;D(A^{-1/2}))$ for all $T>0$ such that

$$\mathbf{u}' + \nu A\mathbf{u} + B(\mathbf{u}) = \mathbf{f} \tag{0.14}$$

and $\mathbf{u}(0)=\mathbf{u}_0$ in some suitable sense, where $B(\mathbf{u}) = B(\mathbf{u},\mathbf{u})$.

Because the domain Ω is assumed to be bounded, it follows by Rellich's lemma that V is compactly embedded in H, so that A^{-1} is compact as a closed operator in H. Hence, there exists an orthonormal basis $\{\mathbf{w}_m\}_{m=1}^\infty$ in H such that

$$A\mathbf{w}_m = \lambda_m \mathbf{w}_m, \quad m = 1,2,\ldots.$$

Moreover, we have that

$$0<\lambda_1\le\lambda_2\le\cdots\le\lambda_m\le\cdots, \quad \lambda_m\to+\infty \text{ as } m\to+\infty.$$

We also recall the Poincaré inequality

$$|\mathbf{u}|^2 \le \frac{1}{\lambda_1}\|\mathbf{u}\|^2 \quad \forall \mathbf{u}\in V. \tag{0.15}$$

In what follows we denote by P_m the orthogonal projector of H onto the space spanned by $\mathbf{w}_1,\mathbf{w}_2,\ldots,\mathbf{w}_m$. The following inequalities will be useful:

$$|\mathbf{u}-P_m\mathbf{u}|^2 \le \frac{1}{\lambda_m}\|\mathbf{u}-P_m\mathbf{u}\|^2 \quad \forall \mathbf{u}\in V,\ \forall m\in\mathbb{N}; \tag{0.16}$$

$$\|P_m\mathbf{u}\|^2 \le \lambda_m|\mathbf{u}|^2 \quad \forall \mathbf{u}\in H,\ \forall m\in\mathbb{N}. \tag{0.17}$$

In fact, as shown in Chapter II, we need to define solutions of (0.14) that possess further properties; in this way we introduced in Section II.7 the concepts of *weak* and *strong* solutions to the Navier–Stokes equations (0.14). We have seen also that the results of existence and uniqueness of solutions are different in space dimensions 2 and 3 (see Theorems 7.1–7.4 in Chapter II) and that, for technical reasons, the very notion of a weak solution differs from one space dimension to the other. In space dimension 2, the weak solutions of (0.14) are also assumed to satisfy

$$\mathbf{u}'\in L^2(0,T;D(A^{-1/2})) \quad \text{for all } T>0, \tag{0.18}$$

and this suffices for all technical purposes. In space dimension 3, we can only obtain weak solutions of (0.14) that satisfy, instead of (0.18),

$$\mathbf{u}'\in L^{4/3}(0,T;D(A^{-1/2})) \quad \text{for all } T>0. \tag{0.19}$$

Since this information is not sufficient, we also require the 3-dimensional weak solutions to satisfy further properties, which we proved in Chapter II (see Theorem 7.1

and the discussion in the section entitled "Further properties of the solutions in dimension 3"). Particularly important for many purposes are the energy inequalities (7.5) and (7.7) and their consequences (II.A.38) and (II.A.43).

On a more technical side, we will also need from Chapter II (more specifically, Sections 7–9 and Appendix A) several estimates for the inertial term – either through $B(\mathbf{u}, \mathbf{v})$ or through $b(\mathbf{u}, \mathbf{v}, \mathbf{w})$.

1 Determining Modes

As discussed in the previous section, theories based on dimensional analysis and heuristic arguments suggest that the long-time behavior of turbulent flows is determined by a finite number of degrees of freedom. The first rigorous result in this direction, based on the Navier–Stokes equations, was given for 2-dimensional flows by Foias and Prodi [1967]. Loosely speaking, they showed that if a number of Fourier modes of two different solutions of the NSE have the same asymptotic behavior as t goes to infinity, then the remaining infinite number modes also have the same asymptotic behavior. Subsequent efforts have been concerned with estimates on how many low modes are necessary to determine the behavior of the remaining modes; on finding other modes, besides the Fourier modes, that can determine a flow in this sense; and on finding other forms of the finite dimensionality of the flow. As we stated in the Introduction, this chapter is devoted to this topic; we start here with the Fourier determining modes.

To be more precise, let us consider two velocity fields $\mathbf{u} = \mathbf{u}(\mathbf{x}, t)$ and $\mathbf{v} = \mathbf{v}(\mathbf{x}, t)$ satisfying the 2-dimensional Navier–Stokes equations and corresponding to two possibly different forcing terms $\mathbf{f} = \mathbf{f}(\mathbf{x}, t)$ and $\mathbf{g} = \mathbf{g}(\mathbf{x}, t)$. More explicitly, $\mathbf{u}$ and $\mathbf{v}$ satisfy the equations

$$\frac{\partial \mathbf{u}}{\partial t} - \nu\Delta\mathbf{u} + (\mathbf{u} \cdot \nabla)\mathbf{u} + \nabla p = \mathbf{f}, \tag{1.1a}$$

$$\nabla \cdot \mathbf{u} = 0, \tag{1.1b}$$

and

$$\frac{\partial \mathbf{v}}{\partial t} - \nu\Delta\mathbf{v} + (\mathbf{v} \cdot \nabla)\mathbf{v} + \nabla q = \mathbf{g}, \tag{1.2a}$$

$$\nabla \cdot \mathbf{v} = 0, \tag{1.2b}$$

with the corresponding pressure terms $p = p(\mathbf{x}, t)$ and $q = q(\mathbf{x}, t)$. The boundary conditions are the same for both problems: either no-slip on a bounded smooth domain or periodic with vanishing space average (see Section II.2). Clearly, (1.1) and (1.2) share the same Stokes operator.

Recall from Section II.6 the Galerkin projections P_m associated with the first m modes of the Stokes operator. We can expand each solution in the form

$$\mathbf{u}(\mathbf{x}, t) = \sum_{k=1}^{\infty} \hat{u}_k(t)\mathbf{w}_k(\mathbf{x}), \qquad \mathbf{v}(\mathbf{x}, t) = \sum_{k=1}^{\infty} \hat{v}_k(t)\mathbf{w}_k(\mathbf{x}),$$

where the $\mathbf{w}_k$ are eigenfunctions of the Stokes operator. The Galerkin projections correspond to the first (say, m) modes:

$$P_m\mathbf{u}(\mathbf{x},t) = \sum_{k=1}^{m} \hat{u}_k(t)\mathbf{w}_k(\mathbf{x}), \qquad P_m\mathbf{v}(\mathbf{x},t) = \sum_{k=1}^{m} \hat{v}_k(t)\mathbf{w}_k(\mathbf{x}).$$

It is assumed that the forcing terms $\mathbf{f}$ and $\mathbf{g}$ have the same asymptotic behavior for large time, that is,

$$\int_\Omega |\mathbf{f}(\mathbf{x},t) - \mathbf{g}(\mathbf{x},t)|^2\,d\mathbf{x} \to 0 \text{ as } t \to \infty. \tag{1.3}$$

Then, the first m modes associated with P_m are called *determining modes* if the condition

$$\int_\Omega |P_m\mathbf{u}(\mathbf{x},t) - P_m\mathbf{v}(\mathbf{x},t)|^2\,d\mathbf{x} \to 0 \text{ as } t \to \infty \tag{1.4}$$

implies

$$\int_\Omega |\mathbf{u}(\mathbf{x},t) - \mathbf{v}(\mathbf{x},t)|^2\,d\mathbf{x} \to 0 \text{ as } t \to \infty. \tag{1.5}$$

Of course, (1.5) then implies stronger results. The two solutions $\mathbf{u} = \mathbf{u}(t)$ and $\mathbf{v} = \mathbf{v}(t)$ have the same asymptotic behavior in a stronger sense – namely, their difference decays to zero in stronger norms, for example, that associated with the enstrophy:

$$\sum_{j=1}^{2} \int_\Omega |\nabla u_j(\mathbf{x},t) - \nabla v_j(\mathbf{x},t)|^2\,d\mathbf{x} \to 0 \text{ as } t \to \infty, \tag{1.6}$$

or even many stronger norms.

The first explicit estimate of the value of the number m in terms of nondimensional parameters was given by Foias and colleagues [1983b].[1] Such estimates are usually obtained in terms of the Grashof number G, which was introduced in Section II.7. It is a nondimensional number that depends on the viscosity ν, the first eigenvalue λ_1 of the Stokes operator, and the asymptotic strength of the forcing term measured in terms of its L^2-norm, that is,[2]

$$F = \limsup_{t\to\infty} \left(\int_\Omega |\mathbf{f}(\mathbf{x},t)|^2\,d\mathbf{x} \right)^{1/2}. \tag{1.7}$$

Note from (1.3) that F would be the same if defined in terms of the forcing term $\mathbf{g}$ in (1.2).

The Grashof number is defined as

$$\mathrm{G} = \frac{F}{\lambda_1 \nu^2}. \tag{1.8}$$

[1] This was preceded somewhat earlier by a purely physically reasoned argument identifying the connection between (a) the number of degrees of freedom (sufficient number of modes) in terms of the Grashof number and (b) the conservation of energy and momentum in Benard convection (Trève and Manley [1981]).

[2] The limsup in (1.7) is actually to be understood in the sense of the essential supremum, since the functions are only required to be in $L^\infty(0,\infty;H)$; see footnote 17 in Chapter II.

The first estimates given in Foias et al. [1983b] were (i) an upper bound for the smallest number m of determining modes of the order of $\mathrm{G}(1+\log \mathrm{G})^{1/2}$ for the case of periodic boundary conditions and (ii) an upper bound of the order of G^2 for the case of no-slip boundary conditions. Jones and Titi [1993] subsequently improved the upper bound (in the case of space-periodic flows) to be G. We shall prove a result that combines those two works; the proof is based on the following generalization of the classical Gronwall lemma.

Lemma 1.1 *Let $\alpha = \alpha(t)$ and $\beta = \beta(t)$ be locally integrable real-valued functions on $[0, \infty)$ that satisfy the following conditions for some $T > 0$:*

$$\liminf_{t\to\infty} \frac{1}{T}\int_t^{t+T} \alpha(\tau)\,d\tau > 0, \tag{1.9}$$

$$\limsup_{t\to\infty} \frac{1}{T}\int_t^{t+T} \alpha^-(\tau)\,d\tau < \infty, \tag{1.10}$$

$$\lim_{t\to\infty} \frac{1}{T}\int_t^{t+T} \beta^+(\tau)\,d\tau = 0, \tag{1.11}$$

where $\alpha^-(t) = \max\{-\alpha(t), 0\}$ and $\beta^+(t) = \max\{\beta(t), 0\}$. Suppose that $\xi = \xi(t)$ is an absolutely continuous nonnegative function on $[0, \infty)$ that satisfies the following inequality almost everywhere on $[0, \infty)$:

$$\frac{d\xi}{dt} + \alpha\xi \le \beta. \tag{1.12}$$

Then $\xi(t) \to 0$ as $t \to 0$.

The proof of Lemma 1.1 is given in the Appendix to this chapter. A weaker version, with the assumption that $\beta(t) \to 0$ as $t \to \infty$, is due to Foias et al. [1983b]. The improvement in Lemma 1.1 is due to Jones and Titi [1992] and is needed for the periodic case.

We treat each boundary condition separately. We start with the no-slip case, obtaining an upper bound of the order of G^2 for the number of determining modes. In the periodic case, in order to obtain an upper bound of the order of G, we work with the V-norm (the square root of the enstrophy) and exploit the orthogonality property $B(\mathbf{u}, \mathbf{u}, A\mathbf{u}) = 0$ (see (A.62) in Chapter II), which is not valid in the no-slip case; we show that the difference between the velocity fields $\mathbf{u}(t)$ and $\mathbf{v}(t)$ decays to zero in the V-norm as t goes to infinity. We remark that, in view of the upper bounds on the dimension of the global attractor (see Section 3), which are of the order of G in the no-slip case and of $\mathrm{G}^{2/3}(1+\log \mathrm{G})$ in the space-periodic case, there is room for future improvement in the upper bound estimates for the lowest number of determining modes.

Determining Modes in the No-Slip Case

We consider the solutions $\mathbf{u}$ and $\mathbf{v}$ of (1.1) and (1.2), respectively, with no-slip boundary conditions and with the forcing terms $\mathbf{f}$ and $\mathbf{g}$ satisfying (1.3). We assume that

(1.4) holds. We want to show that if m is sufficiently large (of the order of G^2) then (1.5) holds. For this purpose, we write $\mathbf{w} = \mathbf{u} - \mathbf{v}$. From (1.4) we know that

$$|P_m \mathbf{w}(t)| \to 0 \quad \text{as } t \to 0. \tag{1.13}$$

We need to show that $|Q_m \mathbf{w}(t)|$ decays to zero as well, where $Q_m = I - P_m$ is the projection onto the modes higher than m. By writing the functional form of the Navier–Stokes equations for $\mathbf{u}$ and $\mathbf{v}$ (as in Chapter II, equation (3.7)) and subtracting them, we find that

$$\frac{d\mathbf{w}}{dt} + \nu A\mathbf{w} + B(\mathbf{w}, \mathbf{u}) + B(\mathbf{v}, \mathbf{w}) = \mathbf{f}(t) - \mathbf{g}(t). \tag{1.14}$$

Taking the inner product of (1.14) with $Q_m \mathbf{w}$ in H yields

$$\frac{1}{2}\frac{d}{dt}|Q_m \mathbf{w}(t)|^2 + \nu \|Q_m \mathbf{w}\|^2 + b(\mathbf{w}, \mathbf{u}, Q_m \mathbf{w}) + b(\mathbf{v}, \mathbf{w}, Q_m \mathbf{w}) \\ = (\mathbf{f}(t) - \mathbf{g}(t), Q_m \mathbf{w}). \tag{1.15}$$

We want to apply Lemma 1.1 with $\xi(t) = |Q_m \mathbf{w}(t)|^2$. First, we write

$$b(\mathbf{w}, \mathbf{u}, Q_m \mathbf{w}) = b(P_m \mathbf{w}, \mathbf{u}, Q_m \mathbf{w}) + b(Q_m \mathbf{w}, \mathbf{u}, Q_m \mathbf{w}). \tag{1.16}$$

We bound the second term in (1.16) by using the estimate (A.46d) from Chapter II and Young's inequality as follows:

$$|b(Q_m \mathbf{w}, \mathbf{u}, Q_m \mathbf{w})| \le c_1 |Q_m \mathbf{w}| \|Q_m \mathbf{w}\| \|\mathbf{u}\| \le \frac{c_1^2}{2\nu} |Q_m \mathbf{w}|^2 \|\mathbf{u}\|^2 + \frac{\nu}{2} \|Q_m \mathbf{w}\|^2.$$

For the first term in (1.16), we use the estimate (A.46e) from Chapter II:

$$|b(P_m \mathbf{w}, \mathbf{u}, Q_m \mathbf{w})| \le c_1 |P_m \mathbf{w}|^{1/2} \|P_m \mathbf{w}\|^{1/2} |\mathbf{u}|^{1/2} \|\mathbf{u}\|^{1/2} \|Q_m \mathbf{w}\|.$$

Now, for the remaining trilinear term in (1.15), we use the orthogonality property (II.A.44) and write

$$b(\mathbf{v}, \mathbf{w}, Q_m \mathbf{w}) = b(\mathbf{v}, P_m \mathbf{w}, Q_m \mathbf{w});$$

then we estimate this term in a manner similar to the preceeding one:

$$|b(\mathbf{v}, P_m \mathbf{w}, Q_m \mathbf{w})| \le c_1 |\mathbf{v}|^{1/2} \|\mathbf{v}\|^{1/2} |P_m \mathbf{w}|^{1/2} \|P_m \mathbf{w}\|^{1/2} \|Q_m \mathbf{w}\|.$$

For the RHS of (1.15), we use the Cauchy–Schwarz inequality to obtain

$$(\mathbf{f}(t) - \mathbf{g}(t), Q_m \mathbf{w}) \le |\mathbf{f}(t) - \mathbf{g}(t)| |Q_m \mathbf{w}|.$$

Taking these estimates into account, we find from (1.15) that

$$\frac{1}{2}\frac{d}{dt}|Q_m \mathbf{w}(t)|^2 + \frac{\nu}{2}\|Q_m \mathbf{w}\|^2 - \frac{c_1^2}{2\nu}\|\mathbf{u}\|^2 |Q_m \mathbf{w}|^2 \\ \le c_1 |P_m \mathbf{w}|^{1/2} \|P_m \mathbf{w}\|^{1/2} |\mathbf{u}|^{1/2} \|\mathbf{u}\|^{1/2} \|Q_m \mathbf{w}\| \\ + c_1 |\mathbf{v}|^{1/2} \|\mathbf{v}\|^{1/2} |P_m \mathbf{w}|^{1/2} \|P_m \mathbf{w}\|^{1/2} \|Q_m \mathbf{w}\| \\ + |\mathbf{f}(t) - \mathbf{g}(t)| |Q_m \mathbf{w}|. \tag{1.17}$$

Using the inequality $\lambda_{m+1}|Q_m\mathbf{w}|^2 \le \|Q_m\mathbf{w}\|^2$ in the second term of the LHS of (1.17), we obtain a relation for $\xi(t) = |Q_m\mathbf{w}(t)|^2$ of the form

$$\frac{d\xi}{dt} + \alpha\xi \le \beta, \tag{1.18}$$

with

$$\alpha(t) = \nu\lambda_{m+1} - \frac{c_1^2}{\nu}\|\mathbf{u}(t)\|^2 \tag{1.19}$$

and

$$\begin{aligned}\beta(t) &= c_1|P_m\mathbf{w}|^{1/2}\|P_m\mathbf{w}\|^{1/2}|\mathbf{u}|^{1/2}\|\mathbf{u}\|^{1/2}\|Q_m\mathbf{w}\| \\ &\quad + c_1|\mathbf{v}|^{1/2}\|\mathbf{v}\|^{1/2}|P_m\mathbf{w}|^{1/2}\|P_m\mathbf{w}\|^{1/2}\|Q_m\mathbf{w}\| \\ &\quad + |\mathbf{f}(t) - \mathbf{g}(t)||Q_m\mathbf{w}|.\end{aligned} \tag{1.20}$$

Since the solutions $\mathbf{u} = \mathbf{u}(t)$ and $\mathbf{v} = \mathbf{v}(t)$ are bounded uniformly for t bounded away from zero in both the H and V norms (for the initial condition in H, see Section II.7), and since by assumption $\|P_m\mathbf{w}(t)\|$ decays to zero as t goes to infinity, it follows that

$$\beta(t) \to 0 \text{ as } t \to \infty.$$

From (II.A.52) and using that $|\mathbf{u}(t)|$ is uniformly bounded in time, one can deduce the following inequality for sufficiently large T:

$$\frac{1}{T}\int_t^{t+T} \|\mathbf{u}(s)\|^2\, ds \le \frac{2}{\nu^2\lambda_1}\|\mathbf{f}\|^2_{L^\infty(t,t+T;H)}. \tag{1.21}$$

Then, it is straightforward to see that α satisfies the condition (1.10) of Lemma 1.1. For the condition (1.9) we have, using (1.21) and (1.7),

$$\begin{aligned}\liminf_{t\to\infty} \frac{1}{T}\int_t^{t+T} \alpha(\tau)\, d\tau &\ge \nu\lambda_{m+1} - \frac{c_1^2}{\nu}\limsup_{t\to\infty}\frac{1}{T}\int_t^{t+T} \|\mathbf{u}(t)\|^2\, d\tau \\ &\ge \nu\lambda_{m+1} - \frac{2c_1^2F^2}{\nu^3\lambda_1}.\end{aligned}$$

Therefore, if m is sufficiently large that

$$\lambda_{m+1} > \frac{2c_1^2F^2}{\nu^4\lambda_1}, \tag{1.22}$$

then (1.9) is satisfied and we can apply Lemma 1.1 to deduce that $\xi(t) = |Q_m\mathbf{w}(t)|^2$ goes to zero as t goes to infinity. Since, for $m \to \infty$, we have $\lambda_m \sim c'\lambda_1 m$ for some nondimensional constant c' (see (6.25) in Chapter II), we see that (1.22) is satisfied provided m is such that

$$m \ge c\mathrm{G}^2, \tag{1.23}$$

where G is the Grashof number defined in (1.8) and c is some other nondimensional constant depending only on the shape of the domain Ω. We have thus proved the following theorem.

Theorem 1.1 *Suppose that $m \in \mathbb{N}$ is such that*

$$m \geq c\mathrm{G}^2, \tag{1.24}$$

where G *is the Grashof number defined in* (1.8) *and c is a constant depending only on the shape of the domain* Ω. *Then, the first m modes are determining in the sense of* (1.4) *and* (1.5) *for the 2-dimensional Navier–Stokes equations with no-slip boundary conditions.*

Determining Modes in the Space-Periodic Case

We now consider the space-periodic case. Let $\mathbf{u}$ and $\mathbf{v}$ be the solutions of (1.1) and (1.2), respectively, with periodic boundary conditions and with the forcing terms $\mathbf{f}$ and $\mathbf{g}$ satisfying (1.3). We assume that (1.4) holds. We want to show that if m is sufficiently large (of the order of G) then (1.5) holds. As in the no-slip case, we write $\mathbf{w} = \mathbf{u} - \mathbf{v}$ and know from assumption (1.4) that

$$|P_m\mathbf{w}(t)| \to 0 \text{ as } t \to \infty. \tag{1.25}$$

We need to show that $|Q_m\mathbf{w}(t)|$ decays to zero as $t \to \infty$, where $Q_m = I - P_m$ is the projection onto the modes higher than m. We will actually prove this convergence in a stronger norm; namely, we will show that

$$\|Q_m\mathbf{w}(t)\| \to 0 \text{ as } t \to \infty. \tag{1.26}$$

This provides the best available estimate to date (i.e., a sufficient condition) for an upper bound on the number of determining modes in the space-periodic case.

As in the no-slip case, we have

$$\frac{d\mathbf{w}}{dt} + \nu A\mathbf{w} + B(\mathbf{u}, \mathbf{w}) + B(\mathbf{v}, \mathbf{w}) = \mathbf{f}(t) - \mathbf{g}(t). \tag{1.27}$$

Taking the inner product of (1.27) with $AQ_m\mathbf{w}$ in H, we obtain

$$\frac{1}{2}\frac{d}{dt}\|Q_m\mathbf{w}(t)\|^2 + \nu\|Q_mA\mathbf{w}\|^2 + b(\mathbf{w}, \mathbf{u}, AQ_m\mathbf{w}) + b(\mathbf{v}, \mathbf{w}, AQ_m\mathbf{w}) \\ = (\mathbf{f}(t) - \mathbf{g}(t), AQ_m\mathbf{w}). \tag{1.28}$$

We want to apply Lemma 1.1 with $\xi(t) = \|Q_m\mathbf{w}(t)\|^2$. Since $Q_m\mathbf{w} = \mathbf{w} - P_m\mathbf{w}$, we write

$$b(\mathbf{w}, \mathbf{u}, AQ_m\mathbf{w}) + b(\mathbf{v}, \mathbf{w}, AQ_m\mathbf{w}) = b(\mathbf{w}, \mathbf{u}, A\mathbf{w}) + b(\mathbf{v}, \mathbf{w}, A\mathbf{w}) \\ - b(\mathbf{w}, \mathbf{u}, AP_m\mathbf{w}) - b(\mathbf{v}, \mathbf{w}, AP_m\mathbf{w}).$$

Since $\mathbf{v} = \mathbf{u} - \mathbf{w}$ and $b(\mathbf{w}, \mathbf{w}, A\mathbf{w}) = 0$ (see (II.A.62)), we then write

$$b(\mathbf{w}, \mathbf{u}, AQ_m\mathbf{w}) + b(\mathbf{v}, \mathbf{w}, AQ_m\mathbf{w}) = b(\mathbf{w}, \mathbf{u}, A\mathbf{w}) + b(\mathbf{u}, \mathbf{w}, A\mathbf{w}) \\ - b(\mathbf{w}, \mathbf{u}, AP_m\mathbf{w}) - b(\mathbf{v}, \mathbf{w}, AP_m\mathbf{w}).$$

Using now the identity (II.A.63) yields

$$b(\mathbf{w}, \mathbf{u}, AQ_m\mathbf{w}) + b(\mathbf{v}, \mathbf{w}, AQ_m\mathbf{w}) = b(\mathbf{w}, \mathbf{w}, A\mathbf{u}) - b(\mathbf{w}, \mathbf{u}, AP_m\mathbf{w}) - b(\mathbf{v}, \mathbf{w}, AP_m\mathbf{w}).$$

Finally, writing $\mathbf{w} = P_m\mathbf{w} + Q_m\mathbf{w}$, we obtain

$$\begin{aligned} &b(\mathbf{w}, \mathbf{u}, AQ_m\mathbf{w}) + b(\mathbf{v}, \mathbf{w}, AQ_m\mathbf{w}) \\ &\quad = b(Q_m\mathbf{w}, Q_m\mathbf{w}, A\mathbf{u}) + b(P_m\mathbf{w}, Q_m\mathbf{w}, A\mathbf{u}) \\ &\quad\quad + b(Q_m\mathbf{w}, P_m\mathbf{w}, A\mathbf{u}) + b(P_m\mathbf{w}, P_m\mathbf{w}, A\mathbf{u}) \\ &\quad\quad - b(\mathbf{w}, \mathbf{u}, AP_m\mathbf{w}) - b(\mathbf{v}, \mathbf{w}, AP_m\mathbf{w}). \end{aligned} \tag{1.29}$$

We now estimate each term in (1.29), starting with the terms involving $P_m\mathbf{w}$. Using the estimates (II.A.46), we find

$$\begin{aligned} |b(P_m\mathbf{w}, Q_m\mathbf{w}, A\mathbf{u})| &\le c_1|P_m\mathbf{w}|^{1/2}|AP_m\mathbf{w}|^{1/2}\|Q_m\mathbf{w}\||A\mathbf{u}| \\ &\le c_1\lambda_m^{1/2}|P_m\mathbf{w}|^2\|Q_m\mathbf{w}\||A\mathbf{u}|, \\ |b(Q_m\mathbf{w}, P_m\mathbf{w}, A\mathbf{u})| &\le c_1|Q_m\mathbf{w}|^{1/2}|AQ_m\mathbf{w}|^{1/2}\|P_m\mathbf{w}\||A\mathbf{u}| \\ &\le c_1\lambda_m^{1/2}|P_m\mathbf{w}||Q_m\mathbf{w}|^{1/2}|A\mathbf{u}|^{3/2}, \end{aligned}$$

$$|b(P_m\mathbf{w}, P_m\mathbf{w}, A\mathbf{u})| \le c_1|P_m\mathbf{w}|^{1/2}|AP_m\mathbf{w}|^{1/2}\|P_m\mathbf{w}\||A\mathbf{u}| \le c_1\lambda_m|P_m\mathbf{w}|^2|A\mathbf{u}|,$$

$$\begin{aligned} |b(\mathbf{w}, \mathbf{u}, AP_m\mathbf{w})| &\le c_1|\mathbf{w}|^{1/2}\|\mathbf{w}\|^{1/2}\|\mathbf{u}\||AP_m\mathbf{w}|^{1/2}\|AP_m\mathbf{w}\|^{1/2} \\ &\le c_1\lambda_m^{5/4}|\mathbf{w}|^{1/2}\|\mathbf{w}\|^{1/2}\|\mathbf{u}\||P_m\mathbf{w}|, \end{aligned}$$

$$\begin{aligned} |b(\mathbf{v}, \mathbf{w}, AP_m\mathbf{w})| &\le c_1|\mathbf{v}|^{1/2}\|\mathbf{v}\|^{1/2}\|\mathbf{w}\||AP_m\mathbf{w}|^{1/2}\|AP_m\mathbf{w}\|^{1/2} \\ &\le c_1\lambda_m^{5/4}|\mathbf{v}|^{1/2}\|\mathbf{v}\|^{1/2}\|\mathbf{w}\||P_m\mathbf{w}|. \end{aligned}$$

For the remaining term, we again use Young's inequality:

$$\begin{aligned} |b(Q_m\mathbf{w}, Q_m\mathbf{w}, A\mathbf{u})| &\le c_1|Q_m\mathbf{w}|^{1/2}|AQ_m\mathbf{w}|^{1/2}\|Q_m\mathbf{w}\||A\mathbf{u}| \\ &\le c_1\frac{1}{\lambda_{m+1}^{1/2}}|AQ_m\mathbf{w}|\|Q_m\mathbf{w}\||A\mathbf{u}| \\ &\le \frac{c_1^2}{\nu\lambda_{m+1}}|A\mathbf{u}|^2\|Q_m\mathbf{w}\|^2 + \frac{\nu}{4}|AQ_m\mathbf{w}|^2. \end{aligned}$$

Finally, we estimate

$$(\mathbf{f}(t) - \mathbf{g}(t), AQ_m\mathbf{w}) \le |\mathbf{f}(t) - \mathbf{g}(t)||AQ_m\mathbf{w}| \le \frac{1}{\nu}|\mathbf{f}(t) - \mathbf{g}(t)|^2 + \frac{\nu}{4}|AQ_m\mathbf{w}|^2.$$

Taking these estimates into consideration, we find from (1.28) the differential inequality

$$\frac{1}{2}\frac{d}{dt}\|Q_m\mathbf{w}\|^2 + \frac{\nu}{2}|AQ_m\mathbf{w}|^2 - \frac{c_1^2}{\nu\lambda_{m+1}}|A\mathbf{u}|^2\|Q_m\mathbf{w}\|^2 \le \frac{\beta(t)}{2}, \tag{1.30}$$

where $\beta = \beta(t)$; this is similar to (but more involved than) equation (1.20), and it contains all the terms containing $P_m\mathbf{w}$. From (II.A.52) and using that $\|\mathbf{u}(t)\|$ is uniformly bounded for t bounded away from zero, one can deduce the following inequality for sufficiently large T:

$$\frac{1}{T}\int_t^{t+T} |A\mathbf{u}(s)|^2\,ds \le \frac{2}{\nu^2}|\mathbf{f}|^2_{L^\infty(t,t+T;H)}.$$

Thus,

$$\limsup_{t\to\infty}\frac{1}{T}\int_t^{t+T} |A\mathbf{u}(s)|^2\,ds \le \frac{2F^2}{\nu^2}. \tag{1.31}$$

Using (1.31) and the fact that the solutions $\mathbf{u} = \mathbf{u}(t)$ and $\mathbf{v} = \mathbf{v}(t)$ are bounded uniformly for t bounded away from zero in both the H and V norms (for the initial condition in H, see Section II.7), one can verify that $\beta = \beta(t)$ satisfies the condition (1.11) of Lemma 1.1.

Let now

$$\alpha(t) = \nu\lambda_{m+1} - \frac{2c_1^2}{\nu\lambda_{m+1}}|A\mathbf{u}(t)|^2, \tag{1.32}$$

and set $\xi(t) = \|Q_m\mathbf{w}(t)\|^2$. With the inequality $|AQ_m\mathbf{w}|^2 \ge \lambda_{m+1}\|Q_m\mathbf{w}\|^2$, we rewrite (1.30) as

$$\frac{d\xi}{dt} + \alpha\xi \le \beta. \tag{1.33}$$

It remains to check condition (1.9) of Lemma 1.1. Using (1.31), we see that

$$\liminf_{t\to\infty}\frac{1}{T}\int_t^{t+T} \alpha(\tau)\,d\tau = \nu\lambda_{m+1} - \frac{2c_1^2}{\nu\lambda_{m+1}}\frac{2F^2}{\nu^2} > 0,$$

where the strict inequality holds if

$$\lambda_{m+1}^2 > \frac{4c_1^2F^2}{\nu^4\lambda_1}. \tag{1.34}$$

Because $\lambda_m \sim c'\lambda_1 m$ in two dimensions (see (II.6.25)), we see that (1.34) is satisfied provided m is such that

$$m \ge c\mathrm{G}, \tag{1.35}$$

where G is the Grashof number defined in (1.8) and c is a constant depending only on the shape of the domain Ω (here, the ratio between the periods in the two space directions). Therefore, we have proved the following theorem. Note that the weaker dependence on G in the space-periodic case (cf. (1.24)) is attributed to the absence of boundary layers in the former case.

Theorem 1.2 *Suppose that $m \in \mathbb{N}$ is such that*

$$m \ge c\mathrm{G}, \tag{1.36}$$

where G *is the Grashof number defined in* (1.8) *and c is constant depending only on the shape of the domain Ω (i.e., the ratio between the periods in the two space directions). Then, for the 2-dimensional Navier–Stokes equations with periodic boundary conditions and vanishing space average, the first m modes are determining in the sense of* (1.4), (1.5), *and* (1.6).

2 Determining Nodes

The notion of determining modes considered in the previous section is a natural one – when associated with a Fourier decomposition of the flow. However, in many practical situations, the experimental data are collected from measurements at a finite number of points in the physical space, often through devices such as the hot-wire anemometer (see Hinze [1975]) or, in modern technology, laser velocimetry (see Buchahave, George, and Lumley [1979]). In view of those cases, a more appropriate notion (for nodal values) is that of *determining nodes*.

A set of points in the physical space (i.e., in the domain filled by the fluid) is called a set of determining nodes if, whenever the difference between the measurements at those points of the velocity field of any two flows goes to zero as time goes to infinity, then the difference between those velocity fields goes to zero uniformly on the domain. The first result proving the existence of a finite number of determining nodes was given by Foias and Temam [1984]. A number of subsequent works have been devoted to reducing the estimate for the lowest number of determining nodes, particularly in the periodic case. The presentation here relies on the article by Foias and Temam [1984] just mentioned and on a more recent result by Jones and Titi [1993]. The latter article provides an upper bound of the order of G, the Grashof number introduced in Section II.7 (see (1.8)) in the case of 2-dimensional flows with periodic boundary conditions. We will also show the existence of a finite number of determining nodes in the case of no-slip boundary conditions. However, the corresponding estimate is more involved and depends exponentially on the Grashof number; this estimate is probably not physically significant, and we do not attempt to derive an explicit expression of the bound in this case.

As for the number of determining nodes, we observe that – in view of the upper bounds on the dimension of the global attractor (see Section 3), which are of the order of G in the no-slip case and of $G^{2/3}(1 + \log G)$ in the space-periodic case – there is room for future improvement in the upper bound estimates for the lowest number of determining nodes obtained so far.

We consider two velocity fields $\mathbf{u} = \mathbf{u}(\mathbf{x}, t)$ and $\mathbf{v} = \mathbf{v}(\mathbf{x}, t)$ satisfying the 2-dimensional Navier–Stokes equations

$$\frac{\partial \mathbf{u}}{\partial t} - \nu\Delta\mathbf{u} + (\mathbf{u}\cdot\nabla)\mathbf{u} + \nabla p = \mathbf{f}, \tag{2.1a}$$

$$\nabla\cdot\mathbf{u} = 0, \tag{2.1b}$$

and

$$\frac{\partial \mathbf{v}}{\partial t} - \nu\Delta\mathbf{v} + (\mathbf{v}\cdot\nabla)\mathbf{v} + \nabla q = \mathbf{g}, \tag{2.2a}$$

$$\nabla\cdot\mathbf{v} = 0, \tag{2.2b}$$

corresponding to two different forcing terms $\mathbf{f} = \mathbf{f}(\mathbf{x}, t)$ and $\mathbf{g} = \mathbf{g}(\mathbf{x}, t)$; the corresponding pressure terms are $p = p(\mathbf{x}, t)$ and $q = q(\mathbf{x}, t)$. The boundary conditions for both problems are either no-slip on a bounded smooth domain or periodic with vanishing space average (see Section II.2). We assume, as in the case of determining modes, that **f** and **g** have the same time-asymptotic behavior, that is,

$$\int_\Omega |\mathbf{f}(\mathbf{x},t) - \mathbf{g}(\mathbf{x},t)|^2 \, d\mathbf{x} \to 0 \text{ as } t \to \infty. \tag{2.3}$$

We consider a set of N nodes or measurement points, denoted by

$$\mathcal{E} = \{\mathbf{x}^1, \mathbf{x}^2, \ldots, \mathbf{x}^N\}, \tag{2.4}$$

where superscripts are used to avoid confusion with the coordinates of a given point in space. We assume that the points in $\mathcal{E}$ are uniformly distributed within the domain Ω in the sense that Ω can be covered by N identical squares $Q_1, \ldots, Q_N$ such that each square contains one and only one of the given points. Of course, the simplest realization of such a set $\mathcal{E}$ (which may not provide the smallest value of N), is obtained by considering points $\mathbf{x}^j$ on a regular mesh in the space.

The assumption that the measurements of the flows **u** and **v** at the points in $\mathcal{E}$ reveal the same time-asymptotic behavior can be expressed, for instance, by the condition

$$\max_{j=1,\ldots,N} |\mathbf{u}(\mathbf{x}^j,t) - \mathbf{v}(\mathbf{x}^j,t)| \to 0 \text{ as } t \to \infty. \tag{2.5}$$

The set $\mathcal{E}$ is then called a set of *determining nodes* if (2.5) implies

$$\int_\Omega |\mathbf{u}(\mathbf{x},t) - \mathbf{v}(\mathbf{x},t)|^2 \, d\mathbf{x} \to 0 \text{ as } t \to \infty. \tag{2.6}$$

As with determining modes (space-periodic case), we will actually show that, when (2.5) holds, the two solutions $\mathbf{u} = \mathbf{u}(t)$ and $\mathbf{v} = \mathbf{v}(t)$ have the same asymptotic behavior in a stronger sense – namely, their difference decays to zero in the stronger norm that is associated with the enstrophy:

$$\sum_{j=1}^{2} \int_\Omega |\nabla u_j(\mathbf{x},t) - \nabla v_j(\mathbf{x},t)|^2 \, d\mathbf{x} \to 0 \text{ as } t \to \infty. \tag{2.7}$$

In order to measure the difference between the velocity fields throughout the set $\mathcal{E}$, we introduce the following quantity, which is defined for each velocity field **w**:

$$\eta(\mathbf{w}) = \max_{1 \le j \le N} |\mathbf{w}(\mathbf{x}^j)|. \tag{2.8}$$

There are two key ingredients in the proof of the existence of a finite number of nodes. One is Lemma 1.1, already used for determining modes. The other is the following lemma, which is due to Jones and Titi [1993] and is an improvement over a similar result in Foias and Temam [1984]. A proof of this lemma is given in the Appendix.

Lemma 2.1 *Let the domain Ω be covered by N identical squares. Consider the set $\mathcal{E} = \{\mathbf{x}^1, \mathbf{x}^2, \ldots, \mathbf{x}^N\}$ of points in Ω, distributed one in each square. Then, for each vector field* **w** *in $D(A)$, the following inequalities hold:*

$$|\mathbf{w}|^2 \le \frac{c}{\lambda_1}\eta(\mathbf{w})^2 + \frac{c}{\lambda_1^2 N^2}|A\mathbf{w}|^2, \tag{2.9}$$

$$\|\mathbf{w}\|^2 \le cN\eta(\mathbf{w})^2 + \frac{c}{\lambda_1 N}|A\mathbf{w}|^2, \tag{2.10}$$

$$\|\mathbf{w}\|^2_{L^\infty(\Omega)} \le cN\eta(\mathbf{w})^2 + \frac{c}{\lambda_1 N}|A\mathbf{w}|^2, \tag{2.11}$$

where the constant c depends only on the shape of the domain Ω.

We now consider each type of boundary condition separately, starting with the no-slip case.

Determining Nodes in the No-Slip Case

We consider two solutions $\mathbf{u}$ and $\mathbf{v}$ of (2.1) and (2.2), respectively, with no-slip boundary conditions and with the forcing terms $\mathbf{f}$ and $\mathbf{g}$ satisfying (2.3). We assume that (2.5) holds. Let the domain Ω be covered by N identical squares, and let $\mathcal{E} = \{\mathbf{x}^1, \mathbf{x}^2, \dots, \mathbf{x}^N\}$ be a set of points in Ω, distributed one in each square. We want to show that if N is sufficiently large, then (2.6) holds. For that purpose, we write $\mathbf{w} = \mathbf{u} - \mathbf{v}$. From (2.5) we know that

$$\eta(\mathbf{w}(t)) \to 0 \text{ as } t \to 0. \tag{2.12}$$

We need to show that $|\mathbf{w}(t)|$ decays to zero as well. Actually, we will show that

$$\|\mathbf{w}(t)\| \to 0 \text{ as } t \to \infty. \tag{2.13}$$

By writing the functional form of the Navier–Stokes equations (equation (II.3.7)) for $\mathbf{u}$ and $\mathbf{v}$ and then subtracting them, we find

$$\frac{d\mathbf{w}}{dt} + \nu A\mathbf{w} + B(\mathbf{w}, \mathbf{u}) + B(\mathbf{v}, \mathbf{w}) = \mathbf{f}(t) - \mathbf{g}(t). \tag{2.14}$$

Taking the inner product of (2.14) with $A\mathbf{w}$ in H, we obtain

$$\frac{1}{2}\frac{d}{dt}\|\mathbf{w}(t)\|^2 + \nu|A\mathbf{w}|^2 + b(\mathbf{w}, \mathbf{u}, A\mathbf{w}) + b(\mathbf{v}, \mathbf{w}, A\mathbf{w}) = (\mathbf{f}(t) - \mathbf{g}(t), A\mathbf{w}). \tag{2.15}$$

In the no-slip case we do not attempt to derive an upper bound in terms of the Grashof number, so for simplicity we denote by C a generic constant that might depend on Ω, ν, $\mathbf{f}$, $\mathbf{g}$ but not on the initial conditions $\mathbf{u}_0$ and $\mathbf{v}_0$; for clarity, we still keep c to denote constants that might depend only on the shape of the domain. Our only aim here is to show that (a) there exists a finite number of determining nodes and (b) they are independent of the initial conditions.

We estimate the trilinear terms in (2.15) using (II.A.46b) and Young's inequality:

$$|b(\mathbf{w}, \mathbf{u}, A\mathbf{w})| \le c_1|\mathbf{w}|^{1/2}\|\mathbf{u}\||A\mathbf{w}|^{3/2} \le \frac{\nu}{6}|A\mathbf{w}|^2 + \frac{c}{\nu^3}\|\mathbf{u}\|^4|\mathbf{w}|^2, \tag{2.16}$$

$$|b(\mathbf{v}, \mathbf{w}, A\mathbf{w})| \le c_1|\mathbf{v}|^{1/2}|A\mathbf{v}|^{1/2}\|\mathbf{w}\||A\mathbf{w}| \le \frac{\nu}{6}|A\mathbf{w}|^2 + \frac{c}{\nu}|\mathbf{v}||A\mathbf{v}|\|\mathbf{w}\|^2. \tag{2.17}$$

The RHS of (2.15) is estimated as follows:

$$|(\mathbf{f}(t)-\mathbf{g}(t), A\mathbf{w})| \le \frac{\nu}{6}|A\mathbf{w}|^2 + \frac{3}{2\nu}|\mathbf{f}-\mathbf{g}|^2. \tag{2.18}$$

Thus, (2.15) yields

$$\frac{d}{dt}\|\mathbf{w}(t)\|^2 + \nu|A\mathbf{w}|^2 \le \frac{c}{\nu^3}\|\mathbf{u}\|^4|\mathbf{w}|^2 + \frac{c}{\nu}|\mathbf{v}||A\mathbf{v}|\|\mathbf{w}\|^2 + \frac{3}{\nu}|\mathbf{f}-\mathbf{g}|^2. \tag{2.19}$$

From (2.10), we have

$$|A\mathbf{w}|^2 \ge c\lambda_1 N\|\mathbf{w}\|^2 - cN\eta(\mathbf{w})^2. \tag{2.20}$$

Using the Poincaré inequality in the first term on the RHS of (2.19) and using (2.20), we obtain

$$\frac{d}{dt}\|\mathbf{w}(t)\|^2 + \left(c\nu\lambda_1 N - \frac{c}{\lambda_1\nu^3}\|\mathbf{u}\|^2 - \frac{c}{\nu}|\mathbf{w}||A\mathbf{w}|\right)\|\mathbf{w}\|^2$$
$$\le \frac{3}{\nu}|\mathbf{f}-\mathbf{g}|^2 + c\lambda_1^2 N\eta(\mathbf{w})^2. \tag{2.21}$$

This inequality is of the form

$$\frac{d\xi}{dt} + \alpha\xi \le \beta \tag{2.22}$$

for $\xi(t) = \|\mathbf{w}(t)\|^2$. We infer from (2.3) and (2.12) that $\beta(t)$ goes to zero as t goes to infinity. We want to apply Lemma 1.1, and for that purpose we need only check the required conditions on α. They follow, for instance, from the fact that, for large times, the solutions are bounded uniformly in V and the time average of the square of their norm in $D(A)$ is bounded uniformly; both bounds are independent of the initial conditions but depend on the viscosity, the domain, and the forcing terms (see (A.56) and (A.59) in Chapter II). For the first condition (1.10), we find

$$\liminf_{t\to\infty} \frac{1}{T}\int_t^{t+T} \alpha(\tau)\,d\tau \ge c\nu\lambda_1 N - C(\mathbf{f}, \mathbf{g}, \nu, \Omega). \tag{2.23}$$

Therefore, for N sufficiently large, condition (1.10) is also satisfied and it follows from Lemma 1.1 that $\xi(t) = \|\mathbf{w}(t)\|^2$ goes to zero as t goes to infinity. Hence, the set $\mathcal{E}$ is a set of determining nodes. We have thus established the following result.

Theorem 2.1 *Let the domain Ω be covered by N identical squares. Consider a set $\mathcal{E} = \{\mathbf{x}^1, \mathbf{x}^2, \ldots, \mathbf{x}^N\}$ of points in Ω, distributed one in each square. Let* $\mathbf{f}$ *and* $\mathbf{g}$ *be two forcing terms in $L^\infty(0,\infty; H)$ that satisfy* (2.3), *and let*

$$F = \limsup_{t\to\infty}|\mathbf{f}(t)| = \limsup_{t\to\infty}|\mathbf{g}(t)|. \tag{2.24}$$

Then there exists a constant $C = C(F, \nu, \Omega)$ such that, if

$$N \ge C(F, \nu, \Omega), \tag{2.25}$$

then $\mathcal{E}$ is a set of determining nodes in the sense of (2.5), (2.6), *and* (2.7) *for the 2-dimensional Navier–Stokes equations with no-slip boundary conditions.*

Determining Nodes in the Space-Periodic Case

We now consider two solutions $\mathbf{u}$ and $\mathbf{v}$ of (2.1) and (2.2), respectively, with periodic boundary conditions and with the forcing terms $\mathbf{f}$ and $\mathbf{g}$ satisfying (2.3). We assume that (2.5) holds. As before, we let the domain Ω be covered by N identical squares and let $\mathcal{E} = \{\mathbf{x}^1, \mathbf{x}^2, \ldots, \mathbf{x}^N\}$ be a set of points in Ω, distributed one in each square. We want to show that if N is sufficiently large – at least of the order of the Grashof number G defined in (1.8) – then (2.6) and (2.7) hold. For that purpose, we write $\mathbf{w} = \mathbf{u} - \mathbf{v}$. From (2.5) we know that

$$\eta(\mathbf{w}(t)) \to 0 \quad \text{as}\ t \to 0. \tag{2.26}$$

As in the no-slip case, we will show that $\|\mathbf{w}(t)\|$ also decays to zero. We write the functional form of the Navier–Stokes equations for $\mathbf{u}$ and $\mathbf{v}$ and then subtract them to find

$$\frac{d\mathbf{w}}{dt} + \nu A\mathbf{w} + B(\mathbf{w}, \mathbf{u}) + B(\mathbf{v}, \mathbf{w}) = \mathbf{f}(t) - \mathbf{g}(t). \tag{2.27}$$

By taking the inner product of (2.27) with $A\mathbf{w}$ in H, we obtain

$$\frac{1}{2}\frac{d}{dt}\|\mathbf{w}(t)\|^2 + \nu|A\mathbf{w}|^2 + b(\mathbf{w}, \mathbf{u}, A\mathbf{w}) + b(\mathbf{v}, \mathbf{w}, A\mathbf{w}) = (\mathbf{f}(t) - \mathbf{g}(t), A\mathbf{w}). \tag{2.28}$$

Now we want to exploit the orthogonality property that is valid in the space-periodic case,

$$b(\mathbf{w}, \mathbf{w}, A\mathbf{w}) = 0, \tag{2.29}$$

as well as the polar identity obtained by differentiating (2.29) (see (A.62) and (A.63) in Chapter II):

$$b(\mathbf{u}, \mathbf{w}, A\mathbf{w}) + b(\mathbf{w}, \mathbf{u}, A\mathbf{w}) + b(\mathbf{w}, \mathbf{w}, A\mathbf{u}) = 0. \tag{2.30}$$

Then, we rewrite the trilinear terms in (2.28) as

$$\begin{aligned} &b(\mathbf{w}, \mathbf{u}, A\mathbf{w}) + b(\mathbf{v}, \mathbf{w}, A\mathbf{w}) \\ &\quad = b(\mathbf{w}, \mathbf{u}, A\mathbf{w}) + b(\mathbf{u}, \mathbf{w}, A\mathbf{w}) \quad (\text{using } \mathbf{v} = \mathbf{u} - \mathbf{w} \text{ and } (2.29)) \\ &\quad = -b(\mathbf{w}, \mathbf{w}, A\mathbf{u}) \quad (\text{using } (2.30)). \end{aligned}$$

Therefore, (2.28) becomes

$$\frac{1}{2}\frac{d}{dt}\|\mathbf{w}(t)\|^2 + \nu|A\mathbf{w}|^2 = b(\mathbf{w}, \mathbf{w}, A\mathbf{u}) + (\mathbf{f}(t) - \mathbf{g}(t), A\mathbf{w}). \tag{2.31}$$

We estimate the terms in the RHS of (2.31) by means of the inequalities (II.A.46a), (2.9), and the Young inequality:

$$
\begin{aligned}
b(\mathbf{w}, \mathbf{w}, A\mathbf{u}) &+ (\mathbf{f}(t) - \mathbf{g}(t), A\mathbf{w}) \\
&\leq c_1 |\mathbf{w}|^{1/2} \|\mathbf{w}\| |A\mathbf{w}|^{1/2} |A\mathbf{u}| + |\mathbf{f} - \mathbf{g}| |A\mathbf{w}| \\
&\leq \frac{c}{\lambda_1^{1/4}} \eta(\mathbf{w})^{1/2} \|\mathbf{w}\| |A\mathbf{w}|^{1/2} |A\mathbf{u}| + \frac{c}{\lambda_1^{1/2} N^{1/2}} |A\mathbf{w}| \|\mathbf{w}\| |A\mathbf{u}| \\
&\quad + \frac{1}{\nu} |\mathbf{f} - \mathbf{g}|^2 + \frac{\nu}{4} |A\mathbf{w}|^2 \\
&\leq \frac{\nu}{2} |A\mathbf{w}|^2 + \frac{c}{\lambda_1 \nu N} |A\mathbf{u}|^2 \|\mathbf{w}\|^2 \\
&\quad + \frac{c}{\lambda_1^{1/4}} \eta(\mathbf{w})^{1/2} \|\mathbf{w}\| |A\mathbf{w}|^{1/2} |A\mathbf{u}| + \frac{1}{\nu} |\mathbf{f} - \mathbf{g}|^2. \qquad (2.32)
\end{aligned}
$$

Using (2.32), we find from (2.31) that

$$
\begin{aligned}
\frac{d}{dt} \|\mathbf{w}(t)\|^2 + \nu |A\mathbf{w}|^2 &- \frac{c}{\lambda_1 \nu N} |A\mathbf{u}|^2 \|\mathbf{w}\|^2 \\
&\leq \frac{c}{\lambda_1^{1/4}} \eta(\mathbf{w})^{1/2} \|\mathbf{w}\| |A\mathbf{w}|^{1/2} |A\mathbf{u}| + \frac{1}{\nu} |\mathbf{f} - \mathbf{g}|^2. \qquad (2.33)
\end{aligned}
$$

Now we use (2.10) to bound (from below) the second term in the LHS of (2.33):

$$
|A\mathbf{w}|^2 \geq c\lambda_1 N \|\mathbf{w}\|^2 - cN\eta(\mathbf{w})^2. \qquad (2.34)
$$

Hence, (2.33) yields

$$
\frac{d}{dt} \|\mathbf{w}(t)\|^2 + \left(c\lambda_1 N\nu - \frac{c}{\lambda_1 \nu N} |A\mathbf{u}|^2 \right) \|\mathbf{w}\|^2 \leq \beta, \qquad (2.35)
$$

where $\beta = \beta(t)$ contains all the terms involving powers of $\eta(\mathbf{w})$ and $|\mathbf{f} - \mathbf{g}|$. For t bounded away from zero, the time average of the square of the norm of the solutions in $D(A)$ is uniformly bounded (see (A.59) in Chapter II), so one can can verify that the time average of $\beta(t)$ goes to zero as t goes to infinity. We now rewrite (2.35) in the form

$$
\frac{d\xi}{dt} + \alpha\xi \leq \beta. \qquad (2.36)
$$

Again, since (for t bounded away from zero) the time average of the square of the norm of the solutions in $D(A)$ is uniformly bounded, one can check that condition (1.10) on α in Lemma 1.1 is satisfied. Finally, in order to verify condition (1.9), we use estimate (A.67) (which is valid in the periodic case) to find

$$
\liminf_{t \to \infty} \frac{1}{T} \int_t^{t+T} \alpha(\tau)\, d\tau \geq c\lambda_1 N\nu - \frac{c}{\lambda_1 \nu N} \frac{F^2}{\nu^2}. \qquad (2.37)
$$

Thus, (1.9) is satisfied provided

$$
N \geq \frac{cF}{\nu^2 \lambda_1} = cG. \qquad (2.38)
$$

Therefore, for N satisfying (2.38), we can apply Lemma 1.1 to deduce that $\xi(t) = \|\mathbf{w}(t)\|^2$ goes to zero as t goes to infinity. Hence, the set $\mathcal{E}$ is a set of determining nodes, and we have established the following result.

Theorem 2.2 *Let the domain Ω be covered by N identical squares and consider a set $\mathcal{E} = \{\mathbf{x}^1, \mathbf{x}^2, \ldots, \mathbf{x}^N\}$ of points in Ω, distributed one in each square. Let* **f** *and* **g** *be two forcing terms in $L^\infty(0, \infty; H)$ satisfying* (2.3), *and let*

$$F = \limsup_{t\to\infty} |\mathbf{f}(t)| = \limsup_{t\to\infty} |\mathbf{g}(t)|. \tag{2.39}$$

If

$$N \geq c\mathrm{G}, \tag{2.40}$$

where c is a constant that depends only on the shape of Ω (i.e., the ratio between the periods in the two space directions), then the set $\mathcal{E}$ is a set of determining nodes in the sense of (2.5), (2.6), *and* (2.7) *for the 2-dimensional Navier–Stokes equations with periodic boundary conditions.*

3 Attractors and Their Fractal Dimension

The finite-dimensional behavior of a system can also be established by using the concept of a global attractor. It was shown in the 1970s that the Navier–Stokes equations possess a compact attractor of finite dimension – without any restriction in space dimension 2, and with a regularity assumption in space dimension 3.

The global attractor encompasses most of the possible permanent regimes of the flow. It is a subset of the phase space H, which attracts all the possible trajectories of the system. In particular, for a given initial condition, the corresponding flow $\mathbf{u}(\mathbf{x}, t)$ at each time t belongs to the phase space H and, as time evolves, approaches the global attractor in the metric of H. If the fluid is very viscous (ν is sufficiently large) then the flow becomes laminar and the attractor reduces to one point, which is the representation of the only stable laminar flow in the phase space. As the viscosity decreases (more precisely, as the Reynolds number increases), it is conjectured on the basis of experimental data and by comparison with simpler differential equations that the attractor becomes more complicated – possibly a fractal set – and that the flow becomes more complex as the orbit representing the flow in the phase space tends to wander around the attractor.

Establishment of finite Hausdorff and fractal dimensionalities of the global attractor implies, at least in theory, the possible representation (parameterization) of the permanent regimes of the flow in terms of a finite number of parameters. In this section we discuss in more detail the concept of the global attractors, their dimension, and several estimates of their dimension in terms of physical parameters depending on the space dimension and the choice of boundary conditions. A connection with the Kolmogorov–Landau–Lifshitz point of view is also made. We start with the 2-dimensional case; then we discuss the generalization of the relevant concepts to the 3-dimensional case.

Throughout this section, we assume that the force **f** in (1.1a) is time-independent.

3.1 The Global Attractor for the 2-Dimensional Navier–Stokes Equations

The starting point in the theory of global attractors is the formalization of the solution operator as a semigroup acting on a function space. In the case of the 2-dimensional NSE, given a divergence-free vector field $\mathbf{u}_0 = \mathbf{u}_0(\mathbf{x})$ in the space H of finite energy, there exists a unique solution $\mathbf{u}(\mathbf{x}, t)$, $t \geq 0$, with the initial condition $\mathbf{u}(\cdot, 0) = \mathbf{u}_0$ and such that $\mathbf{u}(\cdot, t)$ belongs to H for all $t \geq 0$; this defines a family of operators $\{S(t)\}_{t\geq 0}$ that associates, to each such $\mathbf{u}_0 \in H$, the flow at time $t \geq 0$: $S(t)\mathbf{u}_0 = \mathbf{u}(\cdot, t) \in H$. From the existence and uniqueness properties of the solutions, we also deduce the semigroup property of the family of operators:

$$S(t) \circ S(s) = S(t + s) \quad \forall t, s \geq 0. \tag{3.1}$$

The well-posedness of the system of equations also assures that each $S(t)$ is a continuous operator in H and that each trajectory $t \mapsto S(t)\mathbf{u}_0$ is continuous in H (see Theorem 7.4 in Chapter II). The *global attractor* for $\{S(t)\}_{t\geq 0}$ is a set $\mathcal{A}$ in H with the following properties.

(i) $\mathcal{A}$ is compact in H.
(ii) $\mathcal{A}$ is invariant for the semigroup (i.e., $S(t)\mathcal{A} = \mathcal{A}$ for all $t \geq 0$).
(iii) $\mathcal{A}$ attracts all bounded sets in H; that is, for every bounded set B in H,

$$\operatorname{dist}_H(S(t)B, \mathcal{A}) \to 0 \quad \text{as } t \to \infty.$$

Here, dist_H denotes the distance in H between two subsets.[3] The global attractor, when it exists, is unique.

An important preliminary step in proving the existence of the global attractor is proving the existence of a bounded *absorbing set,* which is a bounded set $\mathcal{B}$ in H with the property that, for each bounded subset B of H, there is a time $T = T(B)$ such that $S(t)B \subset \mathcal{B}$ for all $t \geq T$. The existence of such a set is actually necessary, since any bounded neighborhood of the global attractor must be a bounded absorbing set.[4] In the literature on dynamical systems, some authors refer to dissipative systems that possess an absorbing set as "dissipative." This concept is indeed intimately connected to the physical concepts of dissipativity; usually, proving the existence of an absorbing set depends heavily on the positivity of the viscosity factor, $\nu > 0$.

[3] This is defined by $\operatorname{dist}_H(A, B) = \sup_{a\in A} \inf_{b\in B} |a - b|_H$. Strictly speaking, it is a semidistance between sets and not a distance, because $\operatorname{dist}_H(A, B) = 0$ implies only that $A \subset \bar{B}$. The Hausdorff distance in H between compact sets is $\delta_H(A, B) = \operatorname{dist}_H(A, B) + \operatorname{dist}_H(B, A)$.

[4] Physically, the presence of an absorbing set relates to (assumes) the dissipative nature of the equation.

Another property needed for proving the existence of the global attractor is some kind of compactness of the semigroup. The semigroup $\{S(t)\}_{t\geq 0}$ is called *asymptotically compact* in H if, for any sequence $\{t_n\}_n$ of nonnegative numbers with $t_n \to \infty$ and any sequence of "initial conditions" $\{\mathbf{u}_{0n}\}_n$ bounded in H, the sequence $\{S(t_n)\mathbf{u}_{0n}\}_n$ is precompact in H (i.e., it contains a convergent subsequence). The asymptotic compactness property, together with the existence of a bounded absorbing set $\mathcal{B}$, guarantees the existence of the global attractor $\mathcal{A}$. This set turns out to be the ω-limit set of $\mathcal{B}$, which is defined as follows:

$$\omega(\mathcal{B}) = \left\{\mathbf{u} \in H;\ \mathbf{u} = \lim_{n\to\infty} S(t_n)\mathbf{u}_{0n},\ t_n \to \infty,\ \{\mathbf{u}_{0n}\}_n \subset H \text{ bounded}\right\}. \tag{3.2}$$

One can see from this definition that the concept of asymptotic compactness is a very natural one. Moreover, the existence of the global attractor is equivalent to the existence of a bounded absorbing set and the asymptotic compactness property of the semigroup. The notion of asymptotic compactness is due to Ladyzhenskaya [1991]; see also Abergel [1989, 1990]. A related concept is that of asymptotic smoothness (see Hale [1988]), which requires the existence of a compact attracting set; it is essentially equivalent to the asymptotic compactness property. Another form of compactness is used in Temam [1997].

In the case of the Navier–Stokes equations, a stronger form of compactness is available: the semigroup $\{S(t)\}_{t\geq 0}$ for the NSE is *uniformly compact* in the sense that, for every bounded set B in H,

$$\bigcup_{t\geq t_0} S(t)B \tag{3.3}$$

is precompact in H for any $t_0 > 0$. This follows from the a priori estimate (A.56) in Appendix II.A, which reads, for $\mathbf{f}$ independent of t:

$$\|S(t)\mathbf{u}_0\| \leq C(|\mathbf{f}|, \nu, \Omega), \quad t \geq T(|\mathbf{u}_0|, |\mathbf{f}|, \nu, \Omega). \tag{3.4}$$

The estimate (3.4) implies that the set in (3.3) is a bounded set in the space of finite enstrophy V, which is compactly embedded in H.

The condition of uniform compactness is stronger than that of asymptotic compactness. The difference is that, under uniform compactness, the sequence t_n need not go to infinity for the set $\{S(t_n)\mathbf{u}_{0n}\}_n$ to be precompact in H. The uniform compactness property is common for parabolic systems on bounded domains. This property usually follows from the smoothing effect of the system combined with the compactness of the Sobolev embeddings.[5] The asymptotic compactness property appears in parabolic systems on unbounded domains and hyperbolic systems on arbitrary domains. We refer the reader to Temam [1997] for a thorough discussion of the subject.

The estimate (3.4) implies not only the uniform compactness of the semigroup but also the existence of a compact absorbing set, since the constant is independent of the initial condition (only the time of absorption depends on the initial condition); this

[5] The concept of Sobolev embeddings is introduced in Chapter I, Section 4. Furthermore, as indicated there, such embeddings can be compact.

assures the existence of the global attractor for the Navier–Stokes equations in dimension 2. It also follows that the global attractor is actually a bounded subset of V. The long-term behavior of the solutions to the 2-dimensional Navier–Stokes equations was first investigated by Foias and Prodi [1967]. Ladyzhenskaya [1972] discussed a number of issues related to attractors for these equations. Then Foias and Temam [1979] proved the first result of existence of a global attractor in the 2-dimensional case, and they also proved that its Hausdorff dimension is finite (with a proof that easily extends to the fractal dimension). Another useful regularity property proved by Foias and Temam [1979] is that the global attractor is not only compact in H and bounded in V but is also compact in the domain $D(A)$ of the Stokes operator. This is obtained by using the time analyticity of the solutions of the NSE in dimension 2 (Section 8 in Chapter II).

In Foias and Temam [1979], the domain Ω is assumed to be bounded and of class $\mathcal{C}^2$. On the mathematical side, subsequent efforts were devoted to establishing the existence of the global attractor on unbounded and/or nonsmooth domains: Ladyzhenskaya [1992] proved the existence of the global attractor on bounded, nonsmooth domains; Abergel [1989] and Babin [1992] considered smooth, unbounded, channel-like domains; Ilyin [1996] treated the case of nonsmooth domains with finite measure; and Rosa [1998] considered nonsmooth, unbounded channel-like domains. Another sequence of publications – extending the results of the first two articles on the mathematical and physical sides – is devoted to the question of Hausdorff and fractal dimensions of the global attractor, with discussions of the *physical significance of the estimates.* We now turn our attention to this question; in the course of the discussion, the relevant references will be cited.

Attractor Dimension

The dimension of the global attractor can be estimated in terms of the physical parameters of the problem. For the 2-dimensional Navier–Stokes equations, it is customary to use a single nondimensional constant called the Grashof number, defined as

$$\mathrm{G} = \frac{|f|}{\nu^2 \lambda_1}. \tag{3.5}$$

As noted earlier, the first proof of the finite dimensionality of the global attractor of the 2-dimensional NSE is due to Foias and Temam [1979]. The actual bound there was exponential in G, and subsequent endeavors aimed at improving (reducing) that estimate and making it physically more realistic (see e.g. Babin and Vishik [1983, 1986], Constantin and Foias [1985], Constantin et al. [1985a], and Constantin, Foias, and Temam [1988]).

Two kinds of attractor dimensions are of interest here: the fractal dimension ($\dim_f$) and the Hausdorff dimension ($\dim_H$); the definition of these dimensions was recalled in the Introduction to this chapter. In case the domain Ω is bounded and of class $\mathcal{C}^2$, the Hausdorff and fractal dimensions of the global attractor for the 2-dimensional Navier–Stokes equations were estimated by Temam [1986] as

$$\dim_H(\mathcal{A}) \leq c_1 \mathrm{G} \quad \text{and} \quad \dim_f(\mathcal{A}) \leq c_2 \mathrm{G}, \tag{3.6}$$

respectively, where the constants c_1 and c_2 depend only on the shape of the domain Ω (i.e., they are scale-invariant). The estimate (3.6) was possible owing to the use in Temam [1985, 1986], the first time in this context, of the Lieb–Thirring inequalities, which are collective Sobolev inequalities (see (A.20)). It also relied on the use of global Lyapunov exponents introduced for the NSE by Constantin and Foias [1985], who obtained (3.6) with G replaced by G^2 for general bounded domains.[6] The technique of computing Lyapunov exponents for the estimation of the dimension was later extended to more general equations by Constantin, Foias, and Temam [1988] in what is now known as the CFT framework.

As we just mentioned, the estimate (3.6) was obtained for bounded, smooth domains (of class $\mathcal{C}^2$). An estimate of the same order was obtained by Ilyin [1996] for nonsmooth unbounded domains with finite area (see also Abergel [1989] and Rosa [1998] for unbounded channel-like domains). A further improvement in the periodic case is due to Constantin et al. [1988] and is described later in this section.

From the physical viewpoint, the prevalent phenomenon in space dimension 2 is the cascade of enstrophy, and it would be natural to estimate the dimension N of the attractor in terms of the Kraichnan dissipation length $\ell_\eta = (\nu^3/\eta)^{1/6}$, where η is the enstrophy dissipation rate (see Kraichnan [1967]):

$$N \sim \left(\frac{\ell_0}{\ell_\eta}\right)^2, \tag{3.7}$$

where ℓ_0 is a characteristic macroscopic length of the flow. In the space-periodic case, a rigorous proof of this result was eventually obtained by Constantin et al. [1988] as the conclusion of a work period that extended over more than five years. This result is reported at the end of Section 3.1.

Before that, in Section 3.1 we present the (chronologically earlier) result in space dimension 2 that does not differentiate between (i.e., is valid for both) the space-periodic case and the no-slip case. These results are expressed in terms of a length ℓ_d, which is the 2-dimensional analog of the Kolmogorov dissipation length:

$$N \sim \left(\frac{\ell_0}{\ell_d}\right)^2. \tag{3.8}$$

As discussed (in physical terms) in Section 5 of Chapter IV ("The effects of walls on Kraichnan's dissipation length"), a rather compelling physical argument shows that this length ℓ_d is actually comparable to Kraichnan's length ℓ_η. Therefore, the

[6] And by $\mathrm{G}(\log \mathrm{G})^{1/2}$ in the space-periodic case. A further improvement for the 2-dimensional space-periodic case is described later.

(chronologically earlier) estimate based on ℓ_d, $N \sim (\ell_0/\ell_d)^2$, is actually a good approximation of the physical estimate $(\ell_0/\ell_\eta)^2$ obtained in the space-periodic case. In the no-slip case, the mathematical and physical difficulty[7] is that we do not have a good estimate of the Kraichnan dissipation length ℓ_η; thus we are content with the estimate based on ℓ_d, which we now describe.

Let us recall the definition of ℓ_d and introduce its estimate in terms of the Grashof number G. For heuristical reasons, the quantity ℓ_d is defined through dimensional analysis as $\ell_d = (\nu^3/\epsilon)^{1/4}$, where ϵ is the energy dissipation rate. The energy dissipation is formally defined in turbulence theory as the ensemble average

$$\epsilon = \nu\langle|\nabla\mathbf{u}|^2\rangle; \tag{3.9}$$

here $|\nabla\mathbf{u}|^2$ denotes the modulus of the tensor $\nabla\mathbf{u} = \nabla\mathbf{u}(\mathbf{x}, t)$ and not a functional norm. We may define a rigorous analog of this space-dependent quantity by replacing the ensemble average with the supremum over all initial conditions on the global attractor of an asymptotic time average:

$$\epsilon = \nu\lambda_1 \limsup_{t\to\infty} \sup_{\mathbf{u}_0\in\mathcal{A}} \frac{1}{t}\int_0^t \|\mathbf{u}(s)\|^2\, ds, \tag{3.10}$$

where $\mathbf{u} = \mathbf{u}(s)$ is the solution corresponding to the initial condition $\mathbf{u}_0$.[8] For the macroscopic characteristic length we take, say, $\ell_0 = 1/\lambda_1^{1/2}$. Then, instead of (3.6), one may estimate directly that

$$\dim_H(\mathcal{A}) \le c_1\left(\frac{\ell_0}{\ell_d}\right)^2, \qquad \dim_f(\mathcal{A}) \le c_2\left(\frac{\ell_0}{\ell_d}\right)^2. \tag{3.11}$$

From (A.54) in Chapter II (written with $t_0 = 0$ and $\mathbf{u}(t_0) = \mathbf{u}_0$), we recall the estimate

$$\int_0^t \|\mathbf{u}(s)\|^2\, ds \le \frac{1}{\nu}|\mathbf{u}_0|^2 + \frac{t}{\nu^2\lambda_1}|\mathbf{f}|^2, \tag{3.12}$$

which is valid for all $t > 0$. Therefore, one can bound the energy dissipation as defined in (3.10) as

$$\epsilon \le \frac{1}{\nu}|\mathbf{f}|^2 = \nu^3\lambda_1^2\mathrm{G}^2. \tag{3.13}$$

Hence,

$$\left(\frac{\ell_0}{\ell_d}\right)^2 \le \mathrm{G}, \tag{3.14}$$

and one can see that (3.6) is actually a consequence of (3.11).

[7] On the mathematical side, the difficulty is that the orthogonality property $b(\mathbf{u}, \mathbf{u}, A\mathbf{u}) = 0$, which is valid in the space-periodic case (see (A.62) in Chapter II), is not valid in the no-slip case. This results in a physically unrealistic upper bound estimate of ℓ_0/ℓ_η (and thus of the attractor dimension) as an exponential function of ℓ_0/ℓ_d.

[8] To compare (3.9) and (3.10), note that $\lambda_1\|\mathbf{u}\|^2$ and $\langle\nabla\mathbf{u}\rangle^2$ have the same dimension.

As mentioned before, in the space-periodic case an improved estimate of the dimension of the attractor – using the dissipation of enstrophy η and the corresponding Kraichnan dissipation length ℓ_η – is available and is presented later in this section.

The estimate on the dimension of the attractor is obtained via the CFT framework (Constantin, Foias, and Temam [1985]; see also Temam [1997]). In the CFT framework, one computes the evolution of infinitesimal volume elements in the phase space. If we consider the evolution of an m-dimensional volume element, the volume element may expand in some directions and contract in others. In general, however, the linear part of the equation dominates the nonlinear term in the sense that the "majority" of nearby orbits in the phase space approach each other. Hence, if m is large enough, then the volume element will expand in some directions but contract in most directions – in such a way that its m-dimensional volume decreases in time. The fractal and Hausdorff dimensions are related to how large one needs to take the dimension m of the volume element for its volume to decrease in time regardless of its position in space. We now sketch the principle of the proof.

If we write the Navier–Stokes equations in the short form (see Section II.3)

$$\frac{d\mathbf{u}}{dt} = \mathbf{F}(\mathbf{u}), \qquad \mathbf{u}(0) = \mathbf{u}_0, \tag{3.15}$$

then an infinitesimal box with sides $\boldsymbol{\xi}^1, \ldots, \boldsymbol{\xi}^m \in H$ and a common vertex initially at $\mathbf{u}_0$ evolves in time according to the first variation equation corresponding to the linearization of (3.15) around the actual orbit $\mathbf{u} = \mathbf{u}(t)$:

$$\frac{d\boldsymbol{\xi}^i}{dt} = D\mathbf{F}(\mathbf{u})\boldsymbol{\xi}^i, \qquad \boldsymbol{\xi}^i(0) = \boldsymbol{\xi}_0^i, \tag{3.16}$$

for every $i = 1, \ldots, m$; here $\boldsymbol{\xi}_0^1, \ldots, \boldsymbol{\xi}_0^m$ are the sides of the infinitesimal box at time zero, and $D\mathbf{F}(\mathbf{u})$ is the Fréchet derivative of $\mathbf{F}$ at $\mathbf{u}$ (actually at $\mathbf{u}(t)$, t being dropped). The volume of the box is the m-dimensional determinant of the vectors $\boldsymbol{\xi}^1, \ldots, \boldsymbol{\xi}^m$, denoted by $|\boldsymbol{\xi}^1(t) \wedge \cdots \wedge \boldsymbol{\xi}^m(t)|^2$, and the exponential rate of variation of the volume is given by the following trace:

$$\mathrm{Tr}(D\mathbf{F}(\mathbf{u}) \circ Q_m), \tag{3.17}$$

where $Q_m = Q_m(t) = Q_m(t; \mathbf{u}_0, \boldsymbol{\xi}_0^1, \ldots, \boldsymbol{\xi}_0^m)$ is the orthogonal projector of H onto the space spanned by $\boldsymbol{\xi}^1(t), \ldots, \boldsymbol{\xi}^m(t)$. Indeed, one can show (see e.g. Temam [1997]) that the evolution of the m-dimensional volume element obeys the equation

$$\frac{1}{2}\frac{d}{dt}|\boldsymbol{\xi}^1(t) \wedge \cdots \wedge \boldsymbol{\xi}^m(t)|^2 = \mathrm{Tr}(D\mathbf{F}(\mathbf{u}(t)) \circ Q_m(t))|\boldsymbol{\xi}^1(t) \wedge \cdots \wedge \boldsymbol{\xi}^m(t)|^2. \tag{3.18}$$

If the trace is negative then the volume is decreasing. In order to estimate this rate uniformly in space and asymptotically in time, one considers the following quantity:

$$q_m = \limsup_{t\to\infty} \sup_{\mathbf{u}_0 \in \mathcal{A}} \sup_{\substack{\boldsymbol{\xi}_0^i \in H \\ |\boldsymbol{\xi}_0^i| \le 1 \\ i=1,\ldots,m}} \left\{ \frac{1}{t}\int_0^t \mathrm{Tr}(D\mathbf{F}(S(s)\mathbf{u}_0) \circ Q_m(s; \mathbf{u}_0, \boldsymbol{\xi}_0^1, \ldots, \boldsymbol{\xi}_0^m))\, ds \right\}, \tag{3.19}$$

and we look for m such that

$$q_m < 0. \tag{3.20}$$

For such an m, every m-dimensional infinitesimal volume eventually decays in time. From this and with further analysis one can usually show that the Hausdorff and fractal dimensions of $\mathcal{A}$ are bounded according to

$$\dim_H(\mathcal{A}) \leq m, \qquad \dim_f(\mathcal{A}) \leq Cm, \tag{3.21}$$

where C is an appropriate constant depending on the data. The details of the estimate of the trace (3.17) and of q_m are technical and are given in Appendix A.3.

Lower bounds on the dimension of the attractor are also available. The first result is due to Babin and Vishik [1983]; see also Liu [1993]. The lower bound in Babin and Vishik [1983] matches the upper bound later obtained by Ziane [1997] for a 2-dimensional elongated channel driven by a constant pressure, which shows that, at least in this case, *the upper bound obtained with the CFT theory is sharp.*

Improvements in the 2-Dimensional Space-Periodic Case

In the space-periodic case, estimates such as (3.6) of the dimension of the global attractor can be improved by considering the dimension with respect to the V-norm instead of the H-norm. Notice that the global attractor is a bounded set in $D(A)$ (according to estimates that exploit the time analyticity of the solutions; see Section II.8) and hence is also compact in V. Thus, the Hausdorff and fractal dimensions, whose definitions depend on the metric considered, can also be computed with respect to the V-norm. However, for the global attractor of the Navier–Stokes equations, the dimensions turn out to be the same whether they are considered in the H-norm or in the V-norm. In the space-periodic case, it is (for technical reasons) by working with the V-norm that we obtain the best estimate. In this estimate the Kraichnan dissipation length $\ell_\eta = (\nu^3/\eta)^{1/6}$, which is related to the enstrophy dissipation rate η, arises naturally.

For the estimate of the dimension with respect to the V-norm, one computes the evolution of the finite-dimensional volume elements and then calculates the volume in terms of the V-norm. In this case, the q_m are defined by

$$q_m = \limsup_{t\to\infty} \sup_{\mathbf{u}_0\in\mathcal{A}} \sup_{\substack{\boldsymbol{\xi}_0^i\in V \\ \|\boldsymbol{\xi}_0^i\|\leq 1 \\ i=1,\ldots,m}} \left\{ \frac{1}{t}\int_0^t \mathrm{Tr}(D\mathbf{F}(S(s)\mathbf{u}_0) \circ \tilde{Q}_m(s;\mathbf{u}_0,\boldsymbol{\xi}_0^1,\ldots,\boldsymbol{\xi}_0^m))\,ds \right\}, \tag{3.22}$$

where now we use the orthogonal projector in V onto the space spanned by $\boldsymbol{\xi}^1(t),\ldots,\boldsymbol{\xi}^m(t)$, which we denote by $\tilde{Q}_m = \tilde{Q}(t;\mathbf{u}_0,\boldsymbol{\xi}_0^1,\ldots,\boldsymbol{\xi}_0^m)$. The trace is computed using the inner product in V, and the trilinear term is handled with the help of the polar identity (II.A.63), which is valid only in the periodic case. The Lieb–Thirring inequality, suitably modified, is also used. We thus obtain the following estimates for the dimensions of the global attractor in the periodic case:

$$\dim_H(\mathcal{A}) \le c_6 \left(\frac{\ell_0}{\ell_\eta}\right)^2 \left(1 + \log\left(\frac{\ell_0}{\ell_\eta}\right)\right)^{1/3}, \tag{3.23}$$

$$\dim_f(\mathcal{A}) \le 2c_6 \left(\frac{\ell_0}{\ell_\eta}\right)^2 \left(1 + \log\left(\frac{\ell_0}{\ell_\eta}\right)\right)^{1/3}. \tag{3.24}$$

Here $\ell_0 = 1/\lambda_1^{1/2}$ is a characteristic macroscopic length, and $\ell_\eta = (\nu^3/\eta)^{1/6}$ is the Kraichnan dissipation length. The quantity η is the enstrophy dissipation rate, defined according to

$$\eta = \nu\lambda_1 \sup_{\mathbf{u}_0 \in \mathcal{A}} \limsup_{t\to\infty} \frac{1}{t}\int_0^t |A\mathbf{u}(s)|^2\, ds. \tag{3.25}$$

This definition is the rigorous analog of the definition in the conventional theory of turbulence through ensemble averaging, namely,[9]

$$\eta = \nu\langle|\Delta\mathbf{u}|^2\rangle; \tag{3.26}$$

here $|\Delta\mathbf{u}|$ denotes the modulus of the vector $\Delta\mathbf{u} = \Delta\mathbf{u}(\mathbf{x}, t)$ and not a functional norm. Using the dissipation length ℓ_η, the heuristic estimate of the number of degrees of freedom of a 2-dimensional turbulent flow is

$$N \sim \left(\frac{\ell_0}{\ell_\eta}\right)^2; \tag{3.27}$$

up to the logarithmic term, this is precisely what we obtained in (3.23) and (3.24).

Using the enstrophy equation (II.A.65), one can proceed as in the no-slip case to obtain a bound on the enstrophy dissipation rate:

$$\eta \le \frac{\lambda_1}{\nu}|\mathbf{f}|^2 = \nu^3\lambda_1^3 \mathrm{G}^2; \tag{3.28}$$

whence we deduce that

$$\dim_H(\mathcal{A}) \le c_6 \mathrm{G}^{2/3}\left(1 + \log \mathrm{G}\right)^{1/3}, \tag{3.29}$$

$$\dim_f(\mathcal{A}) \le 2c_6 \mathrm{G}^{2/3}\left(1 + \log \mathrm{G}\right)^{1/3}. \tag{3.30}$$

This improvement, which is consistent with the predictions of the conventional theory of turbulence, is due to Constantin et al. [1988].

An Example of Trivial Attractors for Arbitrarily Large Grashof Numbers in the Space-Periodic Case

The previous results may seem to imply that the dimension of the global attractor increases with the Grashof number G, which is proportional to $|\mathbf{f}|$. This is not quite

[9] To compare (3.25) and (3.26), note that $\lambda_1|A\mathbf{u}|^2$ and $\langle\Delta\mathbf{u}\rangle^2$ have the same dimension.

the case, as we have derived only the upper bounds. The actual dependence of the dimension on the forcing term is more subtle: we may have G arbitrarily large, yet the global attractor may be reduced to a single point and thus have zero dimension. This occurs, for example, in the 2-dimensional space-periodic case when the forcing term **f** is a particular eigenfunction associated with the first (smallest) eigenvalue of the Stokes operator. In this case there is only one fixed point, and this fixed point is asymptotically stable (i.e., it attracts all the other trajectories); the global attractor consists only of this fixed point. This result is due to Marchioro [1986]. In Appendix A.4 we present the proof given in Constantin et al. [1988].

We now consider the 2-dimensional periodic case with the same period L in both directions (i.e., $L_1 = L_2 = L$). We assume that the forcing term **f** has the particular form

$$\mathbf{f}(\mathbf{x}) = \begin{pmatrix} 0 \\ \alpha\sqrt{2} \end{pmatrix} \sin \frac{2\pi x_1}{L}, \tag{3.31}$$

where α is an arbitrary real number and $\mathbf{x} = (x_1, x_2)$. It is straightforward to check that **f** is an eigenvector of the Stokes operator associated with the first eigenvalue $\lambda_1 = 4\pi^2/L^2$:

$$A\mathbf{f} = \lambda_1 \mathbf{f}. \tag{3.32}$$

One can also check directly that

$$B(\mathbf{f}, \mathbf{f}) = (\mathbf{f} \cdot \nabla)\mathbf{f} = 0. \tag{3.33}$$

The result that we shall describe holds not only for (3.31) but also for any force satisfying both (3.32) and (3.33), even if the boundary conditions are different. Since we repeat the eigenvalues according to their multiplicity (see Section 6 in Chapter II), the first four eigenvalues coincide: $\lambda_1 = \lambda_2 = \lambda_3 = \lambda_4$. They correspond to the choices $\mathbf{k} = (1, 0)$ and $(0, 1)$ – with $a_{\mathbf{k}}$ in the 2-dimensional vector space $\mathbb{C}$ (as a vector space over $\mathbb{R}$) – in the Fourier expansion in terms of the basis (6.17) given in Chapter II.

Consider the vector field

$$\bar{\mathbf{u}} = \frac{1}{\nu\lambda_1}\mathbf{f}. \tag{3.34}$$

From (3.32) and (3.33), we find

$$\nu A\bar{\mathbf{u}} = \mathbf{f}, \qquad B(\bar{\mathbf{u}}, \bar{\mathbf{u}}) = 0.$$

Hence, $\bar{\mathbf{u}}$ is a fixed point of the Navier–Stokes equations:

$$\nu A\bar{\mathbf{u}} + B(\bar{\mathbf{u}}, \bar{\mathbf{u}}) = \mathbf{f}.$$

We will show in Appendix A.4 that the global attractor consists only of $\bar{\mathbf{u}}$, that is, $\mathcal{A} = \{\bar{\mathbf{u}}\}$. This is achieved by showing that every solution $\mathbf{u} = \mathbf{u}(t)$ converges to $\bar{\mathbf{u}}$ asymptotically in time:

$$\lim_{t\to\infty} |\mathbf{u}(t) - \bar{\mathbf{u}}| = 0. \tag{3.35}$$

Since this holds for the solution $\mathbf{u} = \mathbf{u}(t)$ with an arbitrary initial condition $\mathbf{u}_0$ in H, we deduce that the global attractor consists of exactly $\{\bar{\mathbf{u}}\}$. Recall that this was obtained for an arbitrary forcing term $\mathbf{f}$ of the form (3.31). For such a force, the corresponding Grashof number reads

$$G = \frac{|\mathbf{f}|}{\nu^2 \lambda_1} = \frac{L|\alpha|}{\nu^2 \lambda_1}.$$

Since α is arbitrary, the Grashof number can be arbitrarily large, but the attractor is always $\mathcal{A} = \{\bar{\mathbf{u}}\}$.

3.2 The 3-Dimensional Navier–Stokes Equations

The main difficulties in the 3-dimensional case are due again to the lack of well-posedness of the system of equations. In particular, we cannot speak of a semigroup in the phase space H, and the notion of global attractor must be adapted in a suitable way. In what follows we consider two related notions.

First, we consider invariant sets bounded in V and, as predicted heuristically in the conventional theory of turbulence, obtain estimates for their fractal and Hausdorff dimensions of the same order as the number of degrees of freedom of a turbulent flow. Note that one of the properties of the global attractor for an arbitrary semigroup is invariance, which is naturally associated with permanent regimes. The global attractor is the maximal bounded invariant set (for the inclusion relation), but within the global attractor there might be orbits or sets of orbits that are more regular and still invariant. In the 2-dimensional NSE, for instance, the global attractor is actually compact in $D(A)$ (for forces in H; see e.g. Foias and Temam [1979]), and the global attractor itself (or parts of it) might be even more regular. Likewise, in the 3-dimensional case there might be several subsets that consist of global orbits belonging to V or to a more regular space. Because of the local uniqueness of the solutions that start with an initial condition in the space V, it makes sense to speak of invariant sets that are bounded in V. A semigroup can be defined on those subsets, and their fractal and Hausdorff dimensions can be estimated; our first aim in this section is to present such estimates.

Second, we present the notion of a weak global attractor. It is essentially a global attractor with respect to the weak topology: it is weakly compact, invariant in a suitable sense, and attracts all the orbits in the weak topology; it encompasses most of the asymptotic regimes of the system. This weak global attractor will be studied in Chapter IV in connection with the stationary statistical solutions of the 3-dimensional Navier–Stokes equations.

The Hausdorff and Fractal Dimensions of Invariant Sets Bounded in V

The technique of estimating the fractal and Hausdorff dimensions through the linearized equations (in order to estimate the decay of volume elements) requires us to consider only sets that are bounded in V and invariant. The invariance here is to be understood in the following sense: A set X in V is *invariant* if, for any initial condition $\mathbf{u}_0$ in X, the corresponding unique local solution extends globally in time to a unique solution $\mathbf{u} = \mathbf{u}(t)$ that is defined for all $t \in \mathbb{R}$ and with values $\mathbf{u}(t)$ in X. For such invariant sets the CFT theory applies and, following Constantin et al. [1985a,b], we can prove that the fractal and Hausdorff dimensions of any invariant set bounded in V have bounds that are exactly of the same order as the number of degrees of freedom predicted in the conventional theory of turbulence. As explained earlier, the number of degrees of freedom of a 3-dimensional turbulent flow is

$$N \sim \left(\frac{\ell_0}{\ell_d}\right)^3, \tag{3.36}$$

where ℓ_0 is a characteristic macroscopic length for the flow and ℓ_d is the Kolmogorov dissipation length. The dissipation length is defined heuristically through dimensional analysis as $\ell_d = (\nu^3/\epsilon)^{1/4}$, where ϵ is the energy dissipation rate.

Consider, then, a bounded set X in V that is invariant in the sense just described. Within the set X we have global existence and uniqueness of solutions, so the CFT theory can be applied in the same way as in the 2-dimensional case. We need to estimate the dimension of volume elements whose volume decays in time. We define the quantities

$$q_m = \limsup_{t\to\infty} \sup_{\mathbf{u}_0 \in X} \sup_{\substack{\boldsymbol{\xi}_0^i \in H \\ |\boldsymbol{\xi}_0^i| \le 1 \\ i=1,\ldots,m}} \left\{\frac{1}{t}\int_0^t \operatorname{Tr}(D\mathbf{F}(\mathbf{u}(s)) \circ Q_m(s; \mathbf{u}_0, \boldsymbol{\xi}_0^1, \ldots, \boldsymbol{\xi}_0^m))\, ds\right\}, \tag{3.37}$$

and we look for m such that

$$q_m < 0. \tag{3.38}$$

In the expression for q_m, $\mathbf{u} = \mathbf{u}(t)$ is the unique global solution with the initial condition $\mathbf{u}(0) = \mathbf{u}_0 \in X$; the $\boldsymbol{\xi}_0^j$ $(j = 1, \ldots, m)$ are initial conditions for the linearized Navier–Stokes system

$$\frac{d\boldsymbol{\xi}^i}{dt} = D\mathbf{F}(\mathbf{u})\boldsymbol{\xi}^i, \qquad \boldsymbol{\xi}^i(0) = \boldsymbol{\xi}_0^i, \tag{3.39}$$

which is globally well-posed; and $Q_m = Q_m(t; \mathbf{u}_0, \boldsymbol{\xi}_0^1, \ldots, \boldsymbol{\xi}_0^m)$ is the orthogonal projector of H onto the space spanned by the corresponding solutions $\boldsymbol{\xi}^1(t), \ldots, \boldsymbol{\xi}^m(t)$, at time t, of the linearized equation. Similarly to the 2-dimensional case, we obtain the estimate

$$q_m \le -\kappa_1 m^{5/3} + \kappa_2, \tag{3.40}$$

where

$$\kappa_1 = \frac{\nu}{2c_7|\Omega|^{2/3}}, \qquad \kappa_2 = \frac{c_8|\Omega|}{\nu^{11/4}}\epsilon^{5/4}, \tag{3.41}$$

the constants c_7 and c_8 depend only on the shape of the domain Ω, and the rate of energy dissipation ϵ is rigorously interpreted as

$$\epsilon = \nu \limsup_{t\to\infty} \sup_{\mathbf{u}_0 \in X} \frac{1}{|\Omega| t}\int_0^t \int_\Omega |\nabla \mathbf{u}(\mathbf{x}, s)|^2\, d\mathbf{x}\, ds. \tag{3.42}$$

For m given by

$$m-1 < c_9|\Omega|\left(\frac{\epsilon}{\nu^3}\right)^{3/4} \le m \tag{3.43}$$

with $c_9 = (4c_7c_8)^{3/5}$, we have

$$q_m < 0, \qquad \frac{(q_j)_+}{|q_m|} \le 1, \quad j = 1, \ldots, m-1. \tag{3.44}$$

Therefore, by taking $\ell_0 = |\Omega|^{1/3}$ to be the characteristic macroscopic length and with $\ell_d = (\nu^3/\epsilon)^{1/4}$ as the dissipation length, the Hausdorff and fractal dimensions of the invariant set X can be estimated by

$$\dim_H(X) \le c_{10}\left(\frac{\ell_0}{\ell_d}\right)^3, \qquad \dim_f(X) \le 2c_{10}\left(\frac{\ell_0}{\ell_d}\right)^3, \tag{3.45}$$

where c_{10} is some suitable constant that depends only on the shape of Ω.

The Weak Global Attractor in Dimension 3

In the 3-dimensional case, following Foias and Temam [1985], we may define a notion of global attractor with respect to the weak topology. This attractor is denoted the weak global attractor, $\mathcal{A}_w$, and is defined by

$$\begin{aligned}\mathcal{A}_w = \big\{\mathbf{u}_0 \in H;\ &\text{there exists a weak solution } \mathbf{u} = \mathbf{u}(t), \text{ defined on } \mathbb{R}\\ &\text{and uniformly bounded in } H, \text{ such that } \mathbf{u}(\mathbf{x},0) = \mathbf{u}_0(\mathbf{x})\big\}.\end{aligned} \tag{3.46}$$

The role of the weak global attractor is to encompass all of the asymptotic behavior of the system. This is achieved by the properties of attraction (in the weak topology) and invariance. These two properties will be made precise shortly. First, let us notice that $\mathcal{A}$ is a bounded set in H. Indeed, one has

$$\mathcal{A}_w \subset \left\{\mathbf{u} \in H;\ |\mathbf{u}| \le \frac{1}{\nu\lambda_1}|\mathbf{f}|\right\}. \tag{3.47}$$

This follows from the a priori estimate (II.A.41). The invariance is a trivial consequence of the following definition.

> Let $\mathbf{u}_0$ belong to $\mathcal{A}_w$ and let $\mathbf{u} = \mathbf{u}(t)$ be a global weak solution with $\mathbf{u}(0) = \mathbf{u}_0$ and uniformly bounded in H, as required in the definition of $\mathcal{A}$; then, at any other time $t \in \mathbb{R}$, the velocity field $\mathbf{u}(t)$ also belongs to $\mathcal{A}_w$. (3.48)

As for the attraction property, $\mathcal{A}_w$ is weakly attracting in the following sense:

> For every weak solution $\mathbf{u}$ on $[0,\infty)$, it follows that $\mathbf{u}(t) \to \mathcal{A}_w$ in H_w as $t \to \infty$. (3.49)

This means that every weak neighborhood of $\mathcal{A}_w$ will eventually contain $\mathbf{u}(t)$. The proof of this property is slightly more technical and is given in the Appendix to this chapter.

Finally, the weak global attractor is also weakly compact:

$$\mathcal{A}_w \text{ is compact for the weak topology of } H. \tag{3.50}$$

The proof of this weak compactness property is similar to that of the attraction property and is also given in the Appendix.

An important open question concerning the weak global attractor is whether it is included and bounded in V. This would imply, in particular, that all the trajectories inside the weak global attractor are strong solutions and hence are unique.

Alternate approaches to the notion of an attractor for the 3-dimensional Navier–Stokes equations are due to Ball [1997] and Sell [1996].

4 Approximate Inertial Manifolds

The notion that turbulent flows have a finite number of degrees of freedom – as indicated by the Landau–Lifshitz description, or by the existence of a finite number of determining modes or nodes and a finite-dimensional attractor – strongly suggests that, for practical purposes, it might be possible to describe the evolution of a turbulent flow by a finite, reasonably sized set of parameters. Approximate inertial manifolds give (in an approximate sense) a practical answer to this question.

Consider the decomposition of the velocity field of the flow, say $\mathbf{u} = \mathbf{u}(\mathbf{x}, t)$, into two parts,

$$\mathbf{u} = \mathbf{y} + \mathbf{z}, \tag{4.1}$$

where $\mathbf{y} = \mathbf{y}(\mathbf{x}, t)$ is a variable with relatively few dimensions (in the sense that $t \mapsto \mathbf{y}(\cdot, t)$ has its values in a vector space with relatively few dimensions), while $\mathbf{z} = \mathbf{z}(\mathbf{x}, t)$, the remaining part, is somehow enslaved by $\mathbf{y}$. The relation between these two parts of the flow could be represented by a functional relation of the form

$$\mathbf{z} = \Phi(\mathbf{y}). \tag{4.2}$$

Generally, such a decomposition can be carried out exactly when there is an adequate separation of scales. That is not the case for turbulent flows, but an approximate decomposition is possible. A natural candidate to start with stems from the Fourier representation of a flow:

$$\mathbf{u}(\mathbf{x}, t) = \sum_{k=1}^{\infty} \hat{u}_k(t)\mathbf{w}_k(\mathbf{x}),$$

where the $\mathbf{w}_k$ are eigenfunctions of the Stokes operator (remember that, in the periodic case, this expansion coincides with the usual Fourier series expansion). The finite, low-dimensional part can consist of the low eigenmodes,

$$\mathbf{y}(\mathbf{x}, t) = P_m\mathbf{u}(\mathbf{x}, t) = \sum_{k=1}^{m} \hat{u}_k(t)\mathbf{w}_k(\mathbf{x}), \tag{4.3}$$

while $\mathbf{z}$ contains the remaining modes,

$$\mathbf{z}(\mathbf{x},t) = \mathbf{u}(\mathbf{x},t) - \mathbf{y}(\mathbf{x},t) = \sum_{k=m+1}^{\infty} \hat{u}_k(t)\mathbf{w}_k(\mathbf{x}). \tag{4.4}$$

Each eigenfunction $\mathbf{w}_k$ is associated with an eigenvalue λ_k of the Stokes operator. Those eigenvalues are positive and nondecreasing, with $\lambda_k \to \infty$ as $k \to \infty$ (see Section II.6). The quantity $\ell_k = 1/\sqrt{\lambda_k}$, which has the dimension of length, measures in some manner the characteristic wavelength associated with $\mathbf{w}_k$, while $\kappa_k = 1/\ell_k$ represents the spatial frequency (or wavenumber) of the corresponding eigenfunction. In the periodic case, these interpretations are exact. The low modes, represented by $\mathbf{y}$, are thus related to the large-scale structures of the flow; the high modes, represented by $\mathbf{z}$, are related to the small scales.

An exact relation of the form (4.2) leads to a system of ordinary differential equations for the evolution of the flow. Indeed, recall the functional equation form of the NSE given in Section II.3:

$$\frac{d\mathbf{u}}{dt} + \nu A\mathbf{u} + B(\mathbf{u}) = \mathbf{f}. \tag{4.5}$$

With the decomposition (4.1) in mind – and with $\mathbf{y}$ and $\mathbf{z}$ given by (4.3) and (4.4), respectively – we can formally obtain the equations corresponding to the evolution of the low and high modes by applying to (4.5) the Galerkin projector P_m and its complement $Q_m = I - P_m$. Hence, we find

$$\frac{d\mathbf{y}}{dt} + \nu A\mathbf{y} + P_m B(\mathbf{y}+\mathbf{z}) = P_m\mathbf{f}, \tag{4.6}$$

$$\frac{d\mathbf{z}}{dt} + \nu A\mathbf{z} + Q_m B(\mathbf{y}+\mathbf{z}) = Q_m\mathbf{f}. \tag{4.7}$$

If an exact relation of the form (4.2) holds, then the high modes are given in terms of the low modes and hence we need only consider the evolution equation for the low modes; this takes the form

$$\frac{d\mathbf{y}}{dt} + \nu A\mathbf{y} + P_m B(\mathbf{y}+\Phi(\mathbf{y})) = P_m\mathbf{f}. \tag{4.8}$$

This is a system of ordinary differential equations for the unknowns $\hat{u}_1, \hat{u}_2, \ldots, \hat{u}_m$. The portion or "surface" of the phase space (the infinite-dimensional function space for the velocity field $\mathbf{u}$) defined by the relation (4.2) is known as an *inertial manifold.* The corresponding finite-dimensional system (4.8) is known as the *inertial form* of the system (equation (4.5)). This picture is consistent with the Kolmogorov description of turbulence in 3-dimensional flows (and, similarly, with the Kraichnan description of turbulence in 2-dimensional flows). In that picture we have, in particular, an inertial range, say $[\kappa_L, \kappa_d]$, and the dissipation range $[\kappa_d, \infty)$. In the dissipation range, the viscous effects dominate and so the energy is thought to be dissipated by viscosity without interacting further with other modes. The bulk of the energy relevant for the motion of the flow is then mostly contained in the modes with wavenumbers lower than the Kolmogorov cutoff wavenumber κ_d. In accordance with this theory, one could expect the relation (4.2) to occur for κ_m (the highest mode among the low modes in the separation of scales) of the order of κ_d.

However, the exact relation (4.2) is not known to exist. The existence of such a relation has been proved (in a mathematically rigorous way) for a number of partial differential equations modeling turbulent phenomena in mechanics, chemistry, and other fields. It is still an open question whether the 2-dimensional or 3-dimensional Navier–Stokes equations possesses such an inertial manifold.

More plausible is the existence of an approximate relation of the form (4.2), that is,

$$\mathbf{z} \approx \tilde{\Phi}(\mathbf{y}). \tag{4.9}$$

More precisely, one could have that the flow $\mathbf{u}(t) = \mathbf{y}(t) + \mathbf{z}(t)$ is close to $\mathbf{y}(t) + \tilde{\Phi}(\mathbf{y}(t))$ in the sense that, for all times (or, at least, for large t),

$$|\mathbf{z}(t) - \tilde{\Phi}(\mathbf{y}(t))| \leq \zeta \tag{4.10}$$

in some suitable norm $|\cdot|$, where ζ is a small number. When (4.10) holds, the manifold $\tilde{\mathbf{z}} = \tilde{\Phi}(\tilde{\mathbf{y}})$ is called an *approximate inertial manifold.* It provides us with an approximate law relating the high modes to the low modes. In constrast with inertial manifolds, a number of approximate inertial manifolds are known to exist and their explicit expressions have been derived (see the references cited in Remark 4.1). They have been used also in the development of several multilevel numerical methods. In fact, the Galerkin approximation of the Navier–Stokes equations corresponds to the flat manifold $\tilde{\mathbf{z}} \equiv 0$. For a given approximate inertial manifold $\tilde{\mathbf{z}} = \tilde{\Phi}(\tilde{\mathbf{y}})$, one can consider an approximation to the NSE by the finite-dimensional system

$$\frac{d\tilde{\mathbf{y}}}{dt} + \nu A\tilde{\mathbf{y}} + P_m B(\tilde{\mathbf{y}} + \tilde{\Phi}(\tilde{\mathbf{y}})) = P_m \mathbf{f}. \tag{4.11}$$

Note that the solution $\tilde{\mathbf{y}} = \tilde{\mathbf{y}}(t)$ of this system does not coincide with the low-mode part of the exact flow, $\mathbf{y}(t) = P_m\mathbf{u}(t)$, because the enslaving is not exact. Nevertheless, one can expect that, the smaller the error ζ in (4.10), the better the approximation (4.11) in the sense that $\tilde{\mathbf{y}}(t) + \tilde{\Phi}(\tilde{\mathbf{y}}(t))$ would be closer to the exact solution $\mathbf{u}(t)$. Starting from this simple idea, the time multilevel approximation of the Navier–Stokes equations has developed in directions that are beyond the scope of this work (see Remark 4.1). The decomposition (4.1) is at the heart of such methods, although approximations like (4.11) are not directly used.

Another practical aspect of the concept of an approximate inertial manifold is that, even when an exact inertial manifold is known to exist, it might be more useful (for computational purposes) to have on hand an explicit form for an approximate inertial manifold, since inertial manifolds are usually obtained implicitly. Furthermore, although the proof of existence of the approximate relation (4.9) requires that $\mathbf{z}$ be well inside the dissipation range, practical experience suggests that, by choosing the cutoff wavenumber κ_m well inside the inertial range, one can still derive a reasonably good approximation (4.10) with a relatively low-dimensional system (4.11).

As just mentioned, there are a number of explicit approximate inertial manifolds for the Navier–Stokes equations in the 2- and 3-dimensional cases. We now describe the first nontrivial example of such a manifold (called the FMT manifold after Foias, Manley, and Temam [1987c, 1988b]).

We consider the 2-dimensional NSE with either no-slip boundary conditions or periodic boundary conditions with vanishing space average. The forcing term is assumed to be time-independent and to belong to the space H. We proceed in a somewhat heuristic manner; the details of the proof are given in Appendix A.6. The first observation (proved result) is that – in the decomposition (4.1), (4.3), and (4.4) – with m sufficiently large, the component $\mathbf{z}(t)$ is small after a finite time $t \geq T$:

$$|\mathbf{z}(t)| \leq M_0 \frac{\lambda_1}{\lambda_{m+1}} \left(1 + \log \frac{\lambda_m}{\lambda_1}\right)^{1/2} \tag{4.12}$$

for $t \geq T$ and for some constant[10] M_0 depending on the data ν, $|\mathbf{f}|$, and Ω; here, the time T depends on these data and on $|\mathbf{u}_0|$. With this in mind, we consider the bilinear term

$$B(\mathbf{u}) = B(\mathbf{y}+\mathbf{z}) = B(\mathbf{y},\mathbf{y}) + B(\mathbf{y},\mathbf{z}) + B(\mathbf{z},\mathbf{y}) + B(\mathbf{z},\mathbf{z}).$$

We see that, for large times, the term involving only $\mathbf{z}$ should be much smaller than those involving both $\mathbf{y}$ and $\mathbf{z}$, which, in turn, should be much smaller than the term involving only $\mathbf{y}$. In other words,

$$|B(\mathbf{z},\mathbf{z})| \ll |B(\mathbf{y},\mathbf{z})|, \qquad |B(\mathbf{z},\mathbf{y})| \ll |B(\mathbf{y},\mathbf{y})|. \tag{4.13}$$

Hence, from (4.7) we expect that

$$\frac{d\mathbf{z}}{dt} + \nu A\mathbf{z} + Q_m B(\mathbf{y}) \approx Q_m \mathbf{f}. \tag{4.14}$$

Moreover, the relaxation time for $\mathbf{z}$ is of the order of $\nu\lambda_{m+1}$, which is much larger than the relaxation time $\nu\lambda_1$ of the large eddies; hence the time derivative $d\mathbf{z}/dt$ is very small. Indeed, one can show that, for $t \geq T$,

$$\left|\frac{d\mathbf{z}(t)}{dt}\right| \leq M_0' \frac{\lambda_1}{\lambda_{m+1}} \left(1 + \log \frac{\lambda_m}{\lambda_1}\right)^{1/2} \tag{4.15}$$

for another constant M_0' (see the comments in Remark 4.2). Hence, we neglect this term in (4.14) and obtain the FMT manifold:

$$\nu A\mathbf{z} + Q_m B(\mathbf{y}) \approx Q_m \mathbf{f}. \tag{4.16}$$

Since $A\mathbf{z} = AQ_m\mathbf{z}$ and AQ_m is invertible, we can rewrite (4.16) in the form

$$\mathbf{z} \approx \tilde{\Phi}_m(\mathbf{y}) \equiv (\nu AQ_m)^{-1}(Q_m\mathbf{f} - Q_m B(\mathbf{y})). \tag{4.17}$$

In fact, as shown in Foias et al. [1988b] and as we prove in the Appendix to this chapter, the following estimate holds for large times (say, $t \geq T$):

$$|\mathbf{z}(t) - \tilde{\Phi}_m(\mathbf{y}(t))| \leq N_0 \left(\frac{\lambda_1}{\lambda_m}\right)^{3/2} \left(1 + \log \frac{\lambda_m}{\lambda_1}\right), \tag{4.18}$$

[10] See the comments in Remark 4.2.

where N_0 is a constant depending on the data ν, Ω, and $|\mathbf{f}|$; T depends on the same quantities and also on the norm $|\mathbf{u}_0|$ of the initial condition. The estimate (4.18) is to be compared with the estimate (4.12) corresponding to the flat manifold $\tilde{\Phi} \equiv 0$ associated with the Galerkin method.

As mentioned at the beginning of this derivation, the estimate (4.18) holds for m large enough. In the periodic case we can give an explicit upper bound estimate, in terms of the Grashof number, of this minimum number m of modes to be contained in the low-mode component of $\mathbf{u}$ (i.e., an estimate for λ_{m+1}). This estimate in the periodic case reads

$$\frac{\lambda_{m+1}}{\lambda_1} \geq c_0 \mathrm{G}^2, \tag{4.19}$$

where G is the Grashof number defined by

$$\mathrm{G} = \frac{|\mathbf{f}|}{\nu^2 \lambda_1} \tag{4.20}$$

and c_0 is a constant that depends only on the shape of the domain Ω (i.e., the ratio between the periods).

The manifold $\tilde{\mathbf{z}} = \tilde{\Phi}_m(\tilde{\mathbf{y}})$ can be regarded as part of a family $\{\tilde{\Phi}_m\}_m$, which provides better and better approximations by increasing m. Several other families of approximate inertial manifolds have been constructed after this first one (see e.g. Titi [1988], Foias, Jolly, Kevrekidis, and Titi [1988a], Temam [1989], Debussche and Temam [1994], Debussche and Dubois [1994], and Rosa [1995]). In the last three works, the error for the families decreases exponentially with λ_{m+1} and decreases as a power of λ_{m+1}^{-1} for all other manifolds.

Remark 4.1 Starting from the simple ideas just described, the time multilevel approximation of the Navier–Stokes equations has developed in directions that are beyond the scope of this work. In particular, multilevel decompositions of the form

$$\mathbf{u} = \mathbf{y} + \mathbf{z}_1 + \cdots + \mathbf{z}_r,$$

with

$$\mathbf{y} = \sum_{k=1}^{m_0} \hat{u}_k \mathbf{w}_k, \; \mathbf{z}_1 = \sum_{k=m_0+1}^{m_1} \hat{u}_k \mathbf{w}_k, \; \ldots, \; \mathbf{z}_r = \sum_{k=m_{r-1}+1}^{\infty} \hat{u}_k \mathbf{w}_k$$

have been considered with $r = 2$ or 3 or larger. Also, in the case where $r = 1$, the decomposition (4.1)–(4.3) has been used for values of m for which the approximations (4.13) and (4.15) (relying on the idea that $d\mathbf{z}/dt$ is small) are no longer valid. Without entering too much into the details, let us point out some references on the numerical applications. The nonlinear Galerkin method was proposed by Marion and Temam [1989, 1990] and further studied by a number of authors; see the description and references in Marion and Temam [1998]. The dynamical multilevel method was introduced and implemented by Dubois, Jauberteau, and Temam [1998, 1999]. These methods can be seen as a *dynamical version* of the multigrid and multilevel wavelet methods, which are static methods (i.e., conceived for *stationary* problems). The extension to spectral and pseudo-spectral methods is developed by Costa et al. [2001].

A totally different route (the Galerkin postprocessing method) has been followed by Titi and his collaborators; see for example García-Archilla et al. [1999].

The next remark is also relevant to the numerical aspects.

Remark 4.2 The previous analysis is described from a mathematical perspective. Concerning a physical perspective, two observations are important. First, the physical value of estimates like (4.12), (4.15), and (4.18) depend on the value of the "constants" M_0, M_0', and N_0, which involve the Grashof (or Reynolds) number of the flow and which may be quite high. In particular, it is believed in turbulence theory that the small eddies evolve rapidly so that, for an estimate like (4.15) to imply that $d\mathbf{z}/dt$ is small, it will probably be necessary for λ_m to be very large, with $\lambda_m^{1/2}$ much larger than κ_d. However, we compensate for this discouraging physical remark with a more encouraging one: $\mathbf{z}$ and $d\mathbf{z}/dt$ are highly oscillating functions (in time and space), and their time average values are believed to be much smaller than their maximum value. Hence, beyond the limitations of current mathematical techniques, one can hope to derive (e.g., for $|\mathbf{z}(t) - \tilde{\Phi}_m(\mathbf{y}(t))|$) an estimate much better than (4.18), or perhaps like (4.18) but with a "small" constant N_0. In numerical applications, it is found that λ_m is much smaller than required by the analysis here; $\lambda_m^{1/2}$, smaller than κ_d, is usually somewhere inside the inertial range.

A practical result of obtaining an approximate inertial manifold, both in the 2- and 3-dimensional cases, is that one can obtain an effective viscosity arising from the severance of the high-wavenumber components $\mathbf{z}$. For the sake of simplicity, we limit ourselves to the cases where the driving force $\mathbf{f}$ has no components in common with $\mathbf{z}$ (i.e., where $Q_m\mathbf{f} = 0$).

Consider first the 2-dimensional case. We have recognized (4.11) as an approximate Navier–Stokes equation. In that spirit, we want this approximation to retain as many of the properties of the exact equation as possible. In particular, for periodic boundary conditions, we want the nonlinear term in (4.11) to be orthogonal to $A\mathbf{y}$. This can be accomplished by subtracting a suitable quantity from the nonlinear term. After some elementary manipulations (see Foias, Manley, and Temam [1991]), that quantity takes the form

$$\nu\frac{|A\tilde{\mathbf{z}}|^2}{|A\tilde{\mathbf{y}}|^2}A\tilde{\mathbf{y}},$$

where $\tilde{\mathbf{z}} = \tilde{\Phi}(\tilde{\mathbf{y}})$. This quantity can be combined with $\nu A\tilde{\mathbf{y}}$ to yield an *effective viscosity*

$$\nu_{\text{eff-2D}} = \nu\left(1 + \frac{|A\tilde{\mathbf{z}}|^2}{|A\tilde{\mathbf{y}}|^2}\right). \tag{4.21}$$

On carrying out a similar exercise in three dimensions, where we would wish to retain the orthogonality of the nonlinear term with respect to $\tilde{\mathbf{y}}$ itself, we find that in this case the effective viscosity takes the form

$$\nu_{\text{eff-3D}} = \nu\left(1 + \frac{\|\tilde{\mathbf{z}}\|^2}{\|\tilde{\mathbf{y}}\|^2}\right). \tag{4.22}$$

Appendix A Proofs of Technical Results in Chapter III

We now present the more technical (mathematical) proofs omitted from the preceding sections of this chapter.

A.1 Proof of the Generalized Gronwall Lemma 1.1

In view of assumptions (1.9) and (1.10), we let $t_0 > 0$ be such that

$$\frac{1}{T}\int_t^{t+T} \alpha(\tau)\,d\tau \geq \gamma \tag{A.1}$$

and

$$\frac{1}{T}\int_t^{t+T} \alpha^-(\tau)\,d\tau \leq \Gamma \tag{A.2}$$

for all $t \geq t_0$ and for some $\gamma > 0$ and $\Gamma < \infty$.

By the classical Gronwall lemma, we deduce from (1.12) that

$$0 \leq \xi(t) \leq \xi(t_0)\exp\biggl(-\int_{t_0}^t \alpha(\sigma)\,d\sigma\biggr) + \int_{t_0}^t \exp\biggl(-\int_\tau^t \alpha(\sigma)\,d\sigma\biggr)\beta^+(\tau)\,d\tau. \tag{A.3}$$

For $t \geq \tau \geq t_0$, we let k be an integer such that $\tau + kT \leq t \leq \tau + (k+1)T$. Then

$$\begin{aligned}\exp\biggl(-\int_\tau^t \alpha(\sigma)\,d\sigma\biggr) &\leq \exp\biggl(-\int_\tau^{t+kT} \alpha(\sigma)\,d\sigma\biggr)\exp\biggl(-\int_{t+kT}^t \alpha(\sigma)\,d\sigma\biggr)\\ &\leq e^{-\gamma kT}e^{\Gamma T} \leq e^{-\gamma(k+1)T}e^{(\Gamma+\gamma)T} \leq \Gamma' e^{-\gamma(t-\tau)},\end{aligned}$$

where we have set $\Gamma' = \exp((\Gamma+\gamma)T)$. Thus, from (A.3) we obtain

$$0 \leq \xi(t) \leq \xi(t_0)\Gamma' e^{-\gamma(t-t_0)} + \Gamma'\int_{t_0}^t e^{-\gamma(t-\tau)}\beta^+(\tau)\,d\tau. \tag{A.4}$$

Now, we take k_0 to be an integer such that $t_0 + (k_0-1)T \leq t \leq t_0 + k_0 T$. Then we can bound the second term in the RHS of (A.4) as follows:

$$\begin{aligned}\int_{t_0}^t e^{-\gamma(t-\tau)}\beta^+(\tau)\,d\tau &\leq \sum_{k=1}^{k_0}\int_{t_0+(k-1)T}^{t_0+kT} e^{-\gamma(t-\tau)}\beta^+(\tau)\,d\tau\\ &\leq \sum_{k=1}^{k_0}\int_{t_0+(k-1)T}^{t_0+kT} e^{-\gamma(k_0-k-1)T}\,\beta^+(\tau)\,d\tau\\ &\leq e^{\gamma T}\max_{k=1,\ldots,k_0}\int_{t_0+(k-1)T}^{t_0+kT} \beta^+(\tau)\,d\tau\\ &\leq Te^{\gamma T}\sup_{s\geq t_0}\frac{1}{T}\int_s^{s+T}\beta^+(\tau)\,d\tau.\end{aligned}$$

Thus, from (A.4) we find

$$0 \leq \xi(t) \leq \xi(t_0)\Gamma' e^{-\gamma(t-t_0)} + \Gamma' Te^{\gamma T}\sup_{s\geq t_0}\frac{1}{T}\int_s^{s+T}\beta^+(\tau)\,d\tau. \tag{A.5}$$

Let t go to infinity in (A.5) to obtain

$$0 \leq \limsup_{t\to\infty} \xi(t) \leq \Gamma' T e^{\gamma T} \sup_{t\geq t_0} \frac{1}{T}\int_t^{t+T} \beta^+(\tau)\,d\tau. \tag{A.6}$$

Finally, let t go to infinity and deduce, using the assumption (1.11) on β^+, that

$$\lim_{t\to\infty} \xi(t) = 0.$$

This completes the proof of Lemma 1.1.

A.2 Proof of Lemma 2.1

Consider first a square $Q = (0,\ell)\times(0,\ell)$, with $\ell > 0$. Fix $\mathbf{x}^0 = (x_0, y_0)$ in Q, and let $\mathbf{u} = \mathbf{u}(\mathbf{x})$ be a smooth function defined on Q. For any $\mathbf{x} = (x, y)$ in Q, we have

$$\mathbf{u}(x,y) - \mathbf{u}(x_0,y_0) = \int_{x_0}^{x} \mathbf{u}_x(\xi,y)\,d\xi + \int_{y_0}^{y} \mathbf{u}_y(x_0,\eta)\,d\eta.$$

Hence, using the Cauchy–Schwarz inequality,

$$|\mathbf{u}(x,y) - \mathbf{u}(x_0,y_0)|^2 \leq 2\ell \int_0^\ell |\mathbf{u}_x(\xi,y)|^2\,d\xi + 2\ell\int_0^\ell |\mathbf{u}_y(x_0,\eta)|^2\,d\eta.$$

After integration on Q on the dummy variables (x, y), we find

$$\begin{aligned}\int_0^\ell\int_0^\ell |\mathbf{u}(x,y) - \mathbf{u}(x_0,y_0)|^2\,dx\,dy &\leq 2\ell^2\int_0^\ell\int_0^\ell |\mathbf{u}_x(\xi,y)|^2\,d\xi\,dy \\ &\quad + 2\ell^3\int_0^\ell |\mathbf{u}_y(x_0,\eta)|^2\,d\eta. \end{aligned}\tag{A.7}$$

We need to estimate the second term on the RHS of (A.7). It will be an estimate of the type of trace theorem (traces in the sense of Sobolev spaces). First, we write

$$\begin{aligned}\mathbf{u}_y(x_0,\eta)^2 &= \mathbf{u}_y(x,\eta)^2 + 2\int_x^{x_0} \mathbf{u}_y(\xi,\eta)\mathbf{u}_{yx}(\xi,\eta)\,d\xi \\ &\leq \mathbf{u}_y(x,\eta)^2 + 2\int_0^\ell |\mathbf{u}_y(\xi,\eta)||\mathbf{u}_{yx}(\xi,\eta)|\,d\xi.\end{aligned}$$

Then, upon integration in x and η over Q and by using the Cauchy–Schwarz and Young inequalities,

$$\begin{aligned}&\ell\int_0^\ell \mathbf{u}_y(x_0,\eta)^2\,d\eta \\ &\quad\leq \int_0^\ell\int_0^\ell \mathbf{u}_y(x,\eta)^2\,dx\,d\eta + 2\ell\int_0^\ell\int_0^\ell |\mathbf{u}_y(\xi,\eta)||\mathbf{u}_{yx}(\xi,\eta)|\,d\xi\,d\eta \\ &\quad\leq 2\|\mathbf{u}_y\|^2_{L^2(Q)^2} + \ell^2\|\mathbf{u}_{xy}\|^2_{L^2(Q)^2}.\end{aligned}\tag{A.8}$$

Inserting (A.8) into (A.7), we obtain

$$\|\mathbf{u}\|^2_{L^2(Q)^2} - \ell^2|\mathbf{u}(x_0, y_0)|^2 \le 2\ell^2\|\mathbf{u}_x\|^2_{L^2(Q)^2} + 4\ell^2\|\mathbf{u}_y\|^2_{L^2(Q)^2} + 2\ell^4\|\mathbf{u}_{xy}\|^2_{L^2(Q)^2}. \tag{A.9}$$

Now, consider the domain Ω and let it be covered by N identical squares with sides of length ℓ. Let also $\mathcal{E} = \{\mathbf{x}^1, \mathbf{x}^2, \ldots, \mathbf{x}^N\}$ be a set of points in Ω, distributed one in each of the squares. By applying (A.9) to each of the squares – with (x_0, y_0) replaced by the corresponding point $\mathbf{x}^j$ within the square – and summing over all the squares, we find

$$\|\mathbf{u}\|^2_{L^2(\Omega)^2} - \ell^2\sum_{j=1}^N|\mathbf{u}(\mathbf{x}^j)|^2 \le 2\ell^2\|\mathbf{u}_x\|^2_{L^2(\Omega)^2} + 4\ell^2\|\mathbf{u}_y\|^2_{L^2(\Omega)^2} + 2\ell^4\|\mathbf{u}_{xy}\|^2_{L^2(\Omega)^2}.$$

Recalling that the norm of $D(A)$ is equivalent to that of $H^2(\Omega)^2$ (see (6.2) and (6.16) in Chapter II), we obtain

$$|\mathbf{u}|^2 \le \ell^2\sum_{j=1}^N|\mathbf{u}(\mathbf{x}^j)|^2 + 4\ell^2\|\mathbf{u}\|^2 + c\ell^4|A\mathbf{u}|^2. \tag{A.10}$$

Using the interpolation inequality

$$\|\mathbf{u}\| \le c|\mathbf{u}|^{1/2}|A\mathbf{u}|^{1/2} \tag{A.11}$$

(see (A.49) in Chapter II) and Young's inequality, we infer from (A.10) that

$$|\mathbf{u}|^2 \le 2\ell^2\sum_{j=1}^N|\mathbf{u}(\mathbf{x}^j)|^2 + c\ell^4|A\mathbf{u}|^2. \tag{A.12}$$

By a straightforward application of the interpolation inequality (A.11) and Agmon's inequality (II.A.48), we deduce from (A.12) the inequalities

$$\|\mathbf{u}\|^2 \le \sum_{j=1}^N|\mathbf{u}(\mathbf{x}^j)|^2 + c\ell^2|A\mathbf{u}|^2 \tag{A.13}$$

and

$$\|\mathbf{u}\|^2_{L^\infty(\Omega)} \le \sum_{j=1}^N|\mathbf{u}(\mathbf{x}^j)|^2 + c\ell^2|A\mathbf{u}|^2 \tag{A.14}$$

Here the constant c depends only on the shape of the domain Ω.

The number N and the length ℓ of the sides of the squares that appear in equations (A.12)–(A.14) are obviously related. Indeed, the area $|\Omega|$ of the domain is of the order of $N\ell^2$. The Grashof number is defined in terms of λ_1, the first eigenvalue of the Stokes operator, so we would like to relate ℓ and N to λ_1. We may do so using the relation

$$\lambda_1 = \frac{c}{|\Omega|},$$

where c is a constant that depends only on the shape of Ω (i.e., the constant does not change if we rescale the domain). Hence, we have

$$\ell^2 = \frac{c}{\lambda_1 N}$$

for a possibly different constant c, which also depends only on the shape of the domain. By substituting for ℓ^2 in the aforementioned inequalities and then using the simple estimate

$$\sum_{j=1}^{N} |\mathbf{u}(\mathbf{x}^j)|^2 \le N\eta(\varphi),$$

we obtain the estimates claimed in Lemma 2.1.

A.3 Estimates for the Dimension of the Global Attractor

We now sketch the estimate of the trace (3.17) and of q_m that we used for estimating the dimension of the global attractor in the 2-dimensional case with either no-slip or space-periodic boundary conditions.

We compute the trace in the definition of q_m by observing that, if $\{\varphi_1(t), \dots, \varphi_m(t)\}$ is an orthonormal basis of $Q_m H = \operatorname{span}\{\boldsymbol{\xi}^1(t), \dots, \boldsymbol{\xi}^m(t)\}$, then we have

$$\operatorname{Tr}(D\mathbf{F}(\mathbf{u}(t)) \circ Q_m(t)) = \sum_{j=1}^{m} (D\mathbf{F}(\mathbf{u}(t))\varphi_j(t), \varphi_j(t)). \tag{A.15}$$

For each $j = 1, \dots, m$ we can write, using the orthogonality property (II.A.33),

$$\begin{aligned}(D\mathbf{F}(\mathbf{u})\varphi_j, \varphi_j) &= -\nu(A\varphi_j, \varphi_j) - (B(\mathbf{u}, \varphi_j) + B(\varphi_j, \mathbf{u}), \varphi_j) \\ &= -\nu\|\varphi_j\|^2 - b(\varphi_j, \mathbf{u}, \varphi_j).\end{aligned}$$

The trilinear terms add up to

$$\sum_{j=1}^{m} b(\varphi_j, \mathbf{u}, \varphi_j) = \int_\Omega \sum_{j=1}^{m} \sum_{i,k=1}^{2} \varphi_{ji}(\mathbf{x}) \partial_{x_i} u_k(\mathbf{x}) \varphi_{jk}(\mathbf{x})\, d\mathbf{x},$$

where $\varphi_{ji}(\mathbf{x})$ $(i = 1, 2)$ are the space components of $\varphi_j(\mathbf{x})$ and the $u_k(\mathbf{x})$ are those of $\mathbf{u}(\mathbf{x})$. For each $\mathbf{x}$ in Ω, we can estimate

$$\left| \sum_{j=1}^{m} \sum_{i,k=1}^{2} \varphi_{ji}(\mathbf{x}) \frac{\partial u_k(\mathbf{x})}{\partial_{x_i}} \varphi_{jk}(\mathbf{x}) \right| \le \left\{ \sum_{i,k=1}^{2} |\partial_{x_i} u_k(\mathbf{x})|^2 \right\}^{1/2} \rho(\mathbf{x}), \tag{A.16}$$

where

$$\rho(\mathbf{x}) = \sum_{i,k=1}^{2} \varphi_{ji}(\mathbf{x})^2. \tag{A.17}$$

Thus,

$$\left| \sum_{j=1}^{m} b(\varphi_j, \mathbf{u}, \varphi_j) \right| \le \|\mathbf{u}\| |\rho|_{L^2}, \tag{A.18}$$

where $|\rho|_{L^2}$ is the $L^2(\Omega)$-norm of ρ. Therefore,

$$\mathrm{Tr}(D\mathbf{F}(\mathbf{u}(t)) \circ Q_m(t)) \leq -\nu \sum_{j=1}^{m} \|\boldsymbol{\varphi}_j(t)\|^2 + \|\mathbf{u}(t)\| |\rho(t)|_{L^2}. \tag{A.19}$$

At this point, we need the following collective Sobolev inequality (of Lieb–Thirring type) to obtain a sharp estimate of the $L^2(\Omega)$-norm of ρ:

$$|\rho(t)|_{L^2}^2 \leq c_3 \sum_{j=1}^{m} \|\boldsymbol{\varphi}_j(t)\|^2, \tag{A.20}$$

where c_3 is a constant depending only on the shape of Ω. The original inequality was obtained by Lieb and Thirring [1976] in the context of quantum mechanics, with the domain being the whole space; the proof of (A.20) is reported and simplified in Temam [1997].

Using the Cauchy–Schwarz inequality, we thus obtain

$$\mathrm{Tr}(D\mathbf{F}(\mathbf{u}(t)) \circ Q_m(t)) \leq -\frac{\nu}{2} \sum_{j=1}^{m} \|\boldsymbol{\varphi}_j(t)\|^2 + \frac{c_3}{2\nu} \|\mathbf{u}(t)\|^2. \tag{A.21}$$

Because $\{\boldsymbol{\varphi}_1, \ldots, \boldsymbol{\varphi}_m\}$ is an orthonormal family in H, one can show that the following estimate is valid:

$$\sum_{j=1}^{m} \|\boldsymbol{\varphi}_j(t)\|^2 \geq \lambda_1 + \cdots + \lambda_m. \tag{A.22}$$

From Section II.6, the eigenvalues of the Stokes operator satisfy

$$\lambda_j \sim c_4 \lambda_1 j \ \text{ as } j \to \infty \tag{A.23}$$

(see (II.6.25)), where c_4 is a constant depending only on the shape of Ω. One can check that

$$\lambda_1 + \cdots + \lambda_m \geq c_5 \lambda_1 m^2 \tag{A.24}$$

for another constant c_5 depending only on the shape of Ω. Therefore,

$$\mathrm{Tr}(D\mathbf{F}(\mathbf{u}(t)) \circ Q_m(t)) \leq -c_5 \frac{\nu \lambda_1}{2} m^2 + \frac{c_3}{2\nu} \|\mathbf{u}(t)\|^2. \tag{A.25}$$

From (A.25), we can derive the following estimate for q_m:

$$q_m \leq -\kappa_1 m^2 + \kappa_2, \tag{A.26}$$

where

$$\kappa_1 = \frac{c_5}{2} \nu \lambda_1, \qquad \kappa_2 = \frac{c_3}{2} \frac{\epsilon}{\nu^2 \lambda_1}, \tag{A.27}$$

and ϵ is given by (3.10). For m given by

$$m - 1 < \left(\frac{2\kappa_2}{\kappa_1} \right)^{1/2} \leq m, \tag{A.28}$$

one can check that

$$q_m < 0, \qquad \frac{(q_j)_+}{|q_m|} \leq 1, \quad j = 1, \ldots, m-1. \tag{A.29}$$

Therefore, the Hausdorff and fractal dimensions of the global attractor can be estimated by

$$\dim_H(\mathcal{A}) \le m, \qquad \dim_f(\mathcal{A}) \le 2m. \tag{A.30}$$

Since

$$\left(\frac{2\kappa_2}{\kappa_1}\right)^{1/2} = \left(2\frac{c_3}{c_5}\frac{\epsilon}{\nu^3\lambda_1^2}\right)^{1/2} = \left(\frac{2c_3}{c_5}\right)^{1/2}\left(\frac{\ell_0}{\ell_d}\right)^2,$$

we obtain

$$\dim_H(\mathcal{A}) \le c_1\left(\frac{\ell_0}{\ell_d}\right)^2, \qquad \dim_f(\mathcal{A}) \le 2c_1\left(\frac{\ell_0}{\ell_d}\right)^2, \tag{A.31}$$

where c_1 is some constant dependent only on the shape of Ω.

A.4 *Proof of the Triviality of the Attractor with Force in the First Mode*

We now consider the 2-dimensional space-periodic Navier–Stokes equations with the particular forcing term of the form (3.31). We want to prove that the global attractor is reduced to the fixed point (3.34), which we do by establishing the asymptotic decay (3.35).

Let $\mathbf{u} = \mathbf{u}(t)$ be an arbitrary solution of the Navier–Stokes equations with initial condition $\mathbf{u}(0) = \mathbf{u}_0$. The energy and enstrophy equations for $\mathbf{u} = \mathbf{u}(t)$ (see (4.8) and (A.65) in Chapter II) read

$$\frac{1}{2}\frac{d}{dt}|\mathbf{u}(t)|^2 + \nu\|\mathbf{u}(t)\|^2 = (\mathbf{f}, \mathbf{u}(t))$$

and

$$\frac{1}{2}\frac{d}{dt}\|\mathbf{u}(t)\|^2 + \nu|A\mathbf{u}(t)|^2 = (\mathbf{f}, A\mathbf{u}(t)) = (A\mathbf{f}, \mathbf{u}(t)) = \lambda_1(\mathbf{f}, \mathbf{u}(t)).$$

Multiplying the energy equation by λ_1 and subtracting it from the enstrophy equation, we obtain

$$\frac{1}{2}\frac{d}{dt}(\|\mathbf{u}\|^2 - \lambda_1|\mathbf{u}|^2) + \nu(|A\mathbf{u}|^2 - \lambda_1\|\mathbf{u}\|^2) \le 0, \tag{A.32}$$

where we have omitted the time dependence in $\mathbf{u} = \mathbf{u}(t)$.

Using the expansion of a vector field in terms of the basis $\{\mathbf{w}_k,\ k = 1, 2, ...\}$ of eigenfunctions of the Stokes operator (see (6.4) in Chapter II), that is,

$$\mathbf{u} = \sum_{k=1}^{\infty} \hat{u}_k \mathbf{w}_k,$$

we see that

$$\|\mathbf{u}\|^2 - \lambda_1|\mathbf{u}|^2 = \sum_{k=5}^{\infty}(\lambda_k - \lambda_1)|\hat{\mathbf{u}}_k|^2.$$

Similarly,

$$|A\mathbf{u}|^2 - \lambda_1\|\mathbf{u}\|^2 = \sum_{k=5}^{\infty}\lambda_k(\lambda_k - \lambda_1)|\hat{\mathbf{u}}_k|^2$$
$$\ge \lambda_5\sum_{k=5}^{\infty}(\lambda_k - \lambda_1)|\hat{\mathbf{u}}_k|^2 = \lambda_5(\|\mathbf{u}\|^2 - \lambda_1|\mathbf{u}|^2).$$

Thus, (A.32) yields

$$\frac{1}{2}\frac{d}{dt}(\|\mathbf{u}\|^2 - \lambda_1|\mathbf{u}|^2) + \nu\lambda_5(\|\mathbf{u}\|^2 - \lambda_1|\mathbf{u}|^2) \le 0.$$

Therefore,

$$\lim_{t\to\infty}\left(\|\mathbf{u}(t)\|^2 - \lambda_1|\mathbf{u}(t)|^2\right) = 0. \tag{A.33}$$

Consider the Galerkin projector P_4 associated with the first four eigenvalues (which coincide), and let $Q_4 = I - P_4$ be its complement. We have

$$|Q_4\mathbf{u}|^2 = \sum_{k=5}^{\infty}|\hat{\mathbf{u}}_k|^2 \le \frac{1}{\lambda_5 - \lambda_1}\sum_{k=5}^{\infty}(\lambda_k - \lambda_1)|\hat{\mathbf{u}}_k|^2 = \frac{1}{\lambda_5 - \lambda_1}(\|\mathbf{u}\|^2 - \lambda_1|\mathbf{u}|^2).$$

Hence, by (A.33),

$$\lim_{t\to\infty}|Q_4\mathbf{u}(t)| = 0. \tag{A.34}$$

Define $\mathbf{v}(t) = \mathbf{u}(t) - \bar{\mathbf{u}}$. From (3.32), we have that $P_4\bar{\mathbf{u}} = \bar{\mathbf{u}}$ and hence $Q_4\mathbf{v}(t) = Q_4\mathbf{u}(t)$. Thus, $|Q_4\mathbf{v}(t)|$ decays to zero asymptotically in time. It remains to show that $|P_4\mathbf{v}(t)|$ decays to zero as well. Subtracting the stationary Navier–Stokes equations for $\bar{\mathbf{u}}$ from those for $\mathbf{u}$, we obtain

$$\frac{d}{dt}\mathbf{v} + \nu A\mathbf{v} + B(\mathbf{v},\mathbf{v}) + B(\mathbf{v},\bar{\mathbf{u}}) + B(\bar{\mathbf{u}},\mathbf{v}) = 0.$$

Taking the inner product in H of this equation with $P_4\mathbf{v}$ yields

$$\frac{1}{2}\frac{d}{dt}|P_4\mathbf{v}|^2 + \lambda_1|P_4\mathbf{v}|^2 + b(\mathbf{v},\mathbf{v},P_4\mathbf{v}) + b(\mathbf{v},\bar{\mathbf{u}},P_4\mathbf{v}) + b(\bar{\mathbf{u}},\mathbf{v},P_4\mathbf{v}) = 0.$$

We now decompose the trilinear terms according to $\mathbf{v} = P_4\mathbf{v} + Q_4\mathbf{v}$. Recall that the solutions of the 2-dimensional NSE are uniformly bounded in time in the norms associated with both H and V. Therefore, since $|Q_4\mathbf{v}(t)|$ decays asymptotically to zero in time (see (A.34)), we see that all the trilinear terms with $Q_4\mathbf{v}$ in one of the entries decay asymptotically to zero as well. Hence, we can write

$$\begin{aligned}\frac{1}{2}\frac{d}{dt}|P_4\mathbf{v}|^2 + \nu\lambda_1|P_4\mathbf{v}|^2 &+ b(P_4\mathbf{v},P_4\mathbf{v},P_4\mathbf{v})\\ &+ b(P_4\mathbf{v},\bar{\mathbf{u}},P_4\mathbf{v}) + b(\bar{\mathbf{u}},P_4\mathbf{v},P_4\mathbf{v}) = \beta(t),\end{aligned}$$

where $\beta(t)$ vanishes asymptotically in time. From the orthogonality property of the trilinear term, we have

$$b(P_4\mathbf{v},P_4\mathbf{v},P_4\mathbf{v}) = 0.$$

From differentiation of this relation with respect to $P_4\mathbf{v}$, we also find that

$$b(P_4\mathbf{v},\bar{\mathbf{u}},P_4\mathbf{v}) + b(\bar{\mathbf{u}},P_4\mathbf{v},P_4\mathbf{v}) = -b(P_4\mathbf{v},P_4\mathbf{v},\bar{\mathbf{u}}).$$

Moreover, from the Fourier representation of the space-periodic vector fields in terms of the eigenfunctions (6.17) in Chapter II, one can check that

$$b(P_4\mathbf{v},P_4\mathbf{v},\bar{\mathbf{u}}) = 0.$$

This holds simply because all three entries are eigenfunctions associated with the smallest eigenvector λ_1. We are then left with

$$\frac{1}{2}\frac{d}{dt}|P_4\mathbf{v}|^2 + \nu\lambda_1|P_4\mathbf{v}|^2 = \beta(t),$$

whence we deduce that

$$\lim_{t\to\infty}|P_4\mathbf{u}(t)| = 0.$$

This, together with (A.34), implies that

$$\lim_{t\to\infty}|\mathbf{u}(t) - \bar{\mathbf{u}}| = 0,$$

which proves (3.35). Since this holds for the solution $\mathbf{u} = \mathbf{u}(t)$ with an arbitrary initial condition $\mathbf{u}_0$ in H, we deduce that the global attractor consists of exactly $\{\bar{\mathbf{u}}\}$.

A.5 Attraction and Compactness of the 3-Dimensional Weak Global Attractor

We start with the proof of the weak attraction, where we present a stronger result than the one announced in Section 3. Then we give the proof of weak compactness.

Weak Attraction of the Weak Global Attractor

The weak attraction holds in a stronger sense, namely, $\mathcal{A}_w$ attracts all the solutions in the weak topology *uniformly* for initial conditions bounded in H. More precisely, we will show that if $\{\mathbf{u}_{0n}\}_n$ is a sequence of initial conditions bounded in H, say

$$|\mathbf{u}_{0n}| \leq C_1 \quad \text{for all } n \in \mathbb{N}, \tag{A.35}$$

and if $\mathbf{u}_n = \mathbf{u}_n(t)$ are weak solutions on $[0,\infty)$ with $\mathbf{u}_n(0) = \mathbf{u}_{0n}$, then for every neighborhood $\mathcal{U}$ of $\mathcal{A}_w$ in the weak topology there exists a time $T > 0$ such that $\mathbf{u}_n(t) \in \mathcal{U}$ for all $t \geq T$ and all $n \in \mathbb{N}$.

The proof of this result is by contradiction. Assume it is not true. Then we can find a neighborhood $\mathcal{U}$ of $\mathcal{A}_w$ in the weak topology as well as two subsequences, $\{n_j\}_j$ ($n_j \in \mathbb{N}$) and $\{t_j\}_j$ ($t_j \to \infty$), such that $\mathbf{u}_{n_j}(t_j)$ does not belong to $\mathcal{U}$. From (A.35) and the a priori estimates (II.A.42) and (II.A.40), we find that

$$\mathbf{u}_n \text{ is bounded in } L^\infty(0,\infty; H) \cap L^2(0,T; V) \text{ for all } T > 0. \tag{A.36}$$

In particular, $\mathbf{u}_{n_j}(t_j)$ is bounded in H. Hence, extracting a subsequence from $\{n_j\}_j$ and $\{t_j\}_j$ (and still denoting it $\{n_j\}_j, \{t_j\}_j$), there exists $\mathbf{v}_0$ in H such that

$$\mathbf{u}_{n_j}(t_j) \rightharpoonup \mathbf{v}_0 \text{ weakly in } H. \tag{A.37}$$

Since $\mathbf{u}_{n_j}(t_j)$ does not belong to $\mathcal{U}$ and since $\mathcal{U}$ is a neighborhood of $\mathcal{A}_w$ in the weak topology, it follows that

$$\mathbf{v}_0 \notin \mathcal{A}_w. \tag{A.38}$$

We will now show that $\mathbf{v}_0$ does belong to $\mathcal{A}_w$, establishing a contradiction. Define $\mathbf{v}_j(t) = \mathbf{u}_{n_j}(t + t_j)$ for $t \geq -t_j$, with $\mathbf{v}_j(t) = 0$ for $t < -t_j$. From (A.36), it follows that

$$\mathbf{v}_j \text{ is bounded in } L^\infty(\mathbb{R}; H) \cap L^2(-T, T; V) \text{ for all } T > 0. \tag{A.39}$$

Moreover, each $\mathbf{v}_j$ is a solution of the Navier–Stokes equations. Exactly as in the proof of the existence of weak solutions, one can pass to the limit in the corresponding equations for a suitable subsequence of $\mathbf{v}_j$ to deduce that there exists a weak solution $\mathbf{v} = \mathbf{v}(t)$, defined on all $\mathbb{R}$, that is the limit of that subsequence. Finally, one can check that $\mathbf{v}(0) = \mathbf{v}_0$ and that $\mathbf{v}(t)$ is bounded in H, uniformly for t in $\mathbb{R}$. Therefore, by the definition of the weak global attractor, $\mathbf{v}_0 \in \mathcal{A}_w$, which contradicts (A.38). Hence, the weak global attractor does attract all the orbits in the weak topology, uniformly for initial conditions bounded in H.

Weak Compactness of the Weak Global Attractor

The proof of the weak compactness of $\mathcal{A}_w$ is similar to that for weak attraction. Consider an arbitrary sequence $\{\mathbf{u}_{0n}\}_n$ in $\mathcal{A}_w$. By (3.47), $\mathcal{A}_w$ is bounded; hence, there exists a subsequence $\{\mathbf{u}_{0n_j}\}_j$ that converges weakly in H to some element $\mathbf{v}_0$. We need to show that $\mathbf{v}_0$ belongs to $\mathcal{A}_w$. For that purpose, we proceed as in the proof of the weak attraction property. We consider the global solutions $\mathbf{u}_n = \mathbf{u}_n(t)$ on $\mathbb{R}$ that correspond to each element $\mathbf{u}_{n0}$, as provided in the definition of $\mathcal{A}_w$. Since $\mathcal{A}_w$ is bounded in H, one can obtain the appropriate a priori estimates to show that those solutions have a subsequence that converges to another solution $\mathbf{v}$, which is also defined on $\mathbb{R}$, with the property that $\mathbf{v}(0) = \mathbf{v}_0$. Thus, $\mathbf{v}_0 \in \mathcal{A}_w$. Therefore, $\mathcal{A}_w$ is weakly compact in H.

A.6 Error Bounds for the FMT Approximate Inertial Manifold

We now prove in a rigorous way the estimate (4.18) for the FMT approximate inertial manifold. For the periodic case we also prove the explicit upper bound estimate (4.19), in terms of the Grashof number, of the minimum number of modes to be contained in the low-mode component of $\mathbf{u}$ in order for relation (4.18) to hold.

Assume for the moment that (4.12) and (4.15) hold. Likewise, assume that the following two estimates also hold for $t \geq T$:

$$\|\mathbf{z}(t)\| \leq M_1 \left(\frac{\lambda_1}{\lambda_{m+1}} \right)^{1/2} \left(1 + \log \frac{\lambda_m}{\lambda_1} \right)^{1/2}, \tag{A.40}$$

$$|A\mathbf{z}(t)| \leq M_2 \left(1 + \log \frac{\lambda_m}{\lambda_1} \right)^{1/2}. \tag{A.41}$$

We want to estimate the difference between the solution $\mathbf{u} = \mathbf{y} + \mathbf{z}$ and the approximation $\mathbf{y} + \tilde{\mathbf{z}}$, where $\tilde{\mathbf{z}} = \tilde{\Phi}(\mathbf{y})$. This amounts to estimating $\mathbf{z} - \tilde{\mathbf{z}}$. From the definition of $\tilde{\Phi}$, we see that

$$\nu A\tilde{\mathbf{z}} + Q_m B(\mathbf{y}) = Q_m \mathbf{f}.$$

Subtracting this equation from the equation (4.6) for $\mathbf{z}$, we find

$$\nu A(\mathbf{z} - \tilde{\mathbf{z}}) = -\mathbf{z}' - Q_m B(\mathbf{y}, \mathbf{z}) - Q_m B(\mathbf{z}, \mathbf{y}) - Q_m B(\mathbf{z}, \mathbf{z}), \tag{A.42}$$

where (for simplicity) we have set $\mathbf{z}' = d\mathbf{z}/dt$. Using estimates (4.12), (4.15), (A.40), and (A.41) for $\mathbf{z}$ together with estimates (A.46a), (A.46b), and (A.51a) from Chapter II for the nonlinear term, we bound the RHS of (A.42) as follows:

$$\begin{aligned}
|\nu A(\mathbf{z} - \tilde{\mathbf{z}})| &\le |\mathbf{z}'| + c_1\|\mathbf{y}\|\,\|\mathbf{z}\|\left(1 + \log\frac{|A\mathbf{y}|^2}{\lambda_1\|\mathbf{y}\|^2}\right)^{1/2} + c_1|\mathbf{z}|^{1/2}\|\mathbf{z}\|^{1/2}\|\mathbf{y}\|^{1/2}|A\mathbf{y}|^{1/2} \\
&\quad + c_1|\mathbf{z}|^{1/2}\|\mathbf{z}\||A\mathbf{z}|^{1/2} \\
&\le M_1\left(\frac{\lambda_1}{\lambda_{m+1}}\right)^{1/2}\left(1 + \log\frac{\lambda_m}{\lambda_1}\right)^{1/2} \\
&\quad + c_1 M_1\|\mathbf{y}\|\left(\frac{\lambda_1}{\lambda_{m+1}}\right)^{1/2}\left(1 + \log\frac{\lambda_m}{\lambda_1}\right)^{1/2}\left(1 + \log\frac{|A\mathbf{y}|^2}{\lambda_1\|\mathbf{y}\|^2}\right)^{1/2} \\
&\quad + c_1 M_0^{1/2} M_1^{1/2}\left(\frac{\lambda_1}{\lambda_{m+1}}\right)^{3/4}\left(1 + \log\frac{\lambda_m}{\lambda_1}\right)^{1/2}\|\mathbf{y}\|^{1/2}|A\mathbf{y}|^{1/2}.
\end{aligned}$$

By (A.56) in Chapter II we know that, for large t, $\mathbf{u}(t)$ is uniformly bounded in V by a constant that is independent of the initial data – say,

$$\|\mathbf{u}(t)\| \le C_1 \tag{A.43}$$

for $t \ge T$. In particular,

$$\|\mathbf{y}(t)\| \le C_1$$

for $t \ge T$; since $\mathbf{y}$ contains only the lower modes, we also have the estimate

$$|A\mathbf{y}(t)| \le \lambda_m^{1/2}\|\mathbf{y}(t)\| \le C_1\lambda_m^{1/2}.$$

Thus, we find that

$$\begin{aligned}
|\nu A(\mathbf{z} - \tilde{\mathbf{z}})| &\le M_1\left(\frac{\lambda_1}{\lambda_{m+1}}\right)^{1/2}\left(1 + \log\frac{\lambda_m}{\lambda_1}\right)^{1/2} + c_1 C_1 M_1\left(\frac{\lambda_1}{\lambda_{m+1}}\right)^{1/2}\left(1 + \log\frac{\lambda_m}{\lambda_1}\right) \\
&\quad + c_1 C_1 M_0^{1/2} M_1^{1/2}\lambda_1^{1/4}\left(\frac{\lambda_1}{\lambda_{m+1}}\right)^{1/2}\left(1 + \log\frac{\lambda_m}{\lambda_1}\right)^{1/2} \\
&\le (M_1 + c_1 C_1(M_1^{1/2} + \lambda_1 M_0^{1/2})M_1^{1/2})\left(\frac{\lambda_1}{\lambda_{m+1}}\right)^{1/2}\left(1 + \log\frac{\lambda_m}{\lambda_1}\right).
\end{aligned}$$

This implies that

$$\left|A\big(\mathbf{z}(t) - \tilde{\Phi}_m(\mathbf{y}(t))\big)\right| \le N_2\left(\frac{\lambda_1}{\lambda_{m+1}}\right)^{1/2}\left(1 + \log\frac{\lambda_m}{\lambda_1}\right) \tag{A.44}$$

for $t \ge T$, where

$$N_2 = \frac{M_1 + c_1 C_1(M_1^{1/2} + \lambda_1 M_0^{1/2})M_1^{1/2}}{\nu}. \tag{A.45}$$

The inverse of the Stokes operator is a bounded operator; when restricted to the space $Q_m H$, its norm in H is given by λ_{m+1}^{-1} and the norm of its square root, $A^{-1/2}Q_m$, is $\lambda_{m+1}^{-1/2}$. Thus, we obtain via (A.44) the estimates

$$|\mathbf{z}(t) - \tilde{\Phi}_m(\mathbf{y}(t))| \leq N_0\left(\frac{\lambda_1}{\lambda_{m+1}}\right)^{1/2}\left(1 + \log\frac{\lambda_m}{\lambda_1}\right) \tag{A.46}$$

and

$$\|\mathbf{z}(t) - \tilde{\Phi}_m(\mathbf{y}(t))\| \leq N_1\left(\frac{\lambda_1}{\lambda_{m+1}}\right)^{1/2}\left(1 + \log\frac{\lambda_m}{\lambda_1}\right) \tag{A.47}$$

for $t \geq T$, where

$$N_0 = \frac{N_2}{\lambda_1} \quad \text{and} \quad N_1 = \frac{N_2}{\lambda_1^{1/2}}. \tag{A.48}$$

This proves, in particular, (4.18).

We now prove the estimates (4.12), (4.15), (A.40), and (A.41). We start with (4.12). Taking the inner product in H of the evolution equation (4.7) with $\mathbf{z}$ and using the orthogonality property of the trilinear operator, $b(\mathbf{y}+\mathbf{z}, \mathbf{z}, \mathbf{z}) = 0$, we find

$$\frac{1}{2}\frac{d}{dt}|\mathbf{z}|^2 + \nu\|\mathbf{z}\|^2 = (Q_m\mathbf{f}, \mathbf{z}) - b(\mathbf{y}, \mathbf{y}, \mathbf{z}) - b(\mathbf{z}, \mathbf{y}, \mathbf{z}). \tag{A.49}$$

We estimate the terms in the right-hand side by using the estimates for the trilinear term as in (A.44):

$$\begin{aligned}
&(Q_m\mathbf{f}, \mathbf{z}) - b(\mathbf{y}, \mathbf{y}, \mathbf{z}) - b(\mathbf{z}, \mathbf{y}, \mathbf{z}) \\
&\quad \leq |Q_m\mathbf{f}||\mathbf{z}| + c_1\|\mathbf{y}\|^2\left(1 + \log\frac{|A\mathbf{y}|^2}{\lambda_1\|\mathbf{y}\|^2}\right)^{1/2}|\mathbf{z}| + c_1|\mathbf{z}|\|\mathbf{z}\|\|\mathbf{y}\| \\
&\quad \leq \frac{1}{\lambda_{m+1}^{1/2}}|\mathbf{f}|\|\mathbf{z}\| + \frac{c_1}{\lambda_{m+1}^{1/2}}\|\mathbf{u}\|^2\|\mathbf{z}\|\left(1 + \log\frac{\lambda_m}{\lambda_1}\right)^{1/2} + \frac{c_1}{\lambda_{m+1}^{1/2}}\|\mathbf{z}\|^2\|\mathbf{u}\| \\
&\quad \leq \frac{\nu}{4}\|\mathbf{z}\|^2 + \frac{2}{\nu\lambda_{m+1}}|Q_m\mathbf{f}|^2 + \frac{2c_1^2}{\nu\lambda_{m+1}}\|\mathbf{u}\|^4\left(1 + \log\frac{\lambda_m}{\lambda_1}\right) + \frac{c_1}{\lambda_{m+1}^{1/2}}\|\mathbf{z}\|^2\|\mathbf{u}\|.
\end{aligned}$$

Thus, using also (A.43), we deduce from (A.49) that

$$\frac{d}{dt}|\mathbf{z}|^2 + \left(\frac{3\nu}{2} - \frac{c_1C_1}{\lambda_{m+1}^{1/2}}\right)\|\mathbf{z}\|^2 \leq \frac{4}{\nu\lambda_{m+1}}|Q_m\mathbf{f}|^2 + \frac{4c_1^2C_1^4}{\nu\lambda_{m+1}}\left(1 + \log\frac{\lambda_m}{\lambda_1}\right). \tag{A.50}$$

Therefore, by assuming

$$\lambda_{m+1} \geq \left(\frac{2c_1C_1}{\nu}\right)^2, \tag{A.51}$$

we obtain

$$\frac{d}{dt}|\mathbf{z}|^2 + \nu\|\mathbf{z}\|^2 \leq \frac{4}{\nu\lambda_{m+1}}\left(|Q_m\mathbf{f}|^2 + c_1^2C_1^4\left(1 + \log\frac{\lambda_m}{\lambda_1}\right)\right). \tag{A.52}$$

Hence,

$$\frac{d}{dt}|\mathbf{z}|^2 + \nu\lambda_{m+1}|\mathbf{z}|^2 \leq \frac{4}{\nu\lambda_{m+1}}\left(|Q_m\mathbf{f}|^2 + c_1^2C_1^4\left(1 + \log\frac{\lambda_m}{\lambda_1}\right)\right). \tag{A.53}$$

We infer from (A.53) that, for $t \geq t_0 \geq T$,

$$|\mathbf{z}(t)|^2 \leq |\mathbf{z}(t_0)|^2 e^{-\nu\lambda_{m+1}(t-t_0)} + \frac{4}{\nu^2\lambda_{m+1}^2}\left(|Q_m\mathbf{f}|^2 + c_1^2 C_1^4\left(1+\log\frac{\lambda_m}{\lambda_1}\right)\right). \tag{A.54}$$

By increasing T if necessary we find that, for $t \geq T$,

$$|\mathbf{z}(t)|^2 \leq \frac{5}{\nu^2\lambda_{m+1}^2}\left(|Q_m\mathbf{f}|^2 + c_1^2 C_1^4\left(1+\log\frac{\lambda_m}{\lambda_1}\right)\right). \tag{A.55}$$

This implies (4.12) with

$$M_0 = \frac{(|Q\mathbf{f}|^2 + c_1^2 C_1^4)^{1/2}}{\nu\lambda_{m+1}}. \tag{A.56}$$

The proof of (A.40) makes use of the enstrophy equation and is similar to the previous proof. We take the inner product in H of the evolution equation (4.7) with $A\mathbf{z}$ to find

$$\frac{1}{2}\frac{d}{dt}\|\mathbf{z}\|^2 + \nu|A\mathbf{z}|^2 = (Q_m\mathbf{f}, A\mathbf{z}) - b(\mathbf{y}+\mathbf{z}, \mathbf{y}+\mathbf{z}, A\mathbf{z}). \tag{A.57}$$

Recall from (A.53) in Chapter II that $\mathbf{u}(t)$ is uniformly bounded in H, for large t, by a constant that is independent of the initial condition:

$$|\mathbf{u}(t)| \leq C_0 = \frac{2}{\nu\lambda_1}|\mathbf{f}|. \tag{A.58}$$

Then, we estimate the RHS of (A.57) in a similar way:

$$\begin{aligned}
&(Q_m\mathbf{f}, A\mathbf{z}) - b(\mathbf{y}+\mathbf{z}, \mathbf{y}+\mathbf{z}, A\mathbf{z})\\
&\quad\leq |Q_m\mathbf{f}||A\mathbf{z}| + |b(\mathbf{y}, \mathbf{y}+\mathbf{z}, A\mathbf{z})| + |b(\mathbf{z}, \mathbf{y}+\mathbf{z}, A\mathbf{z})|\\
&\quad\leq |Q_m\mathbf{f}||A\mathbf{z}| + c_1\|\mathbf{y}\|\|\mathbf{y}+\mathbf{z}\|\left(1+\log\frac{|A\mathbf{y}|^2}{\lambda_1\|\mathbf{y}\|^2}\right)^{1/2}|A\mathbf{z}| + c_1|\mathbf{z}|^{1/2}|A\mathbf{z}|^{3/2}\|\mathbf{y}+\mathbf{z}\|\\
&\quad\leq |Q_m\mathbf{f}||A\mathbf{z}| + 2c_1C_1^2\left(1+\log\frac{\lambda_{m+1}}{\lambda_1}\right)^{1/2}|A\mathbf{z}| + 2c_1C_0^{1/2}C_1|A\mathbf{z}|^{3/2}\\
&\quad\leq \frac{\nu}{2}|A\mathbf{z}|^2 + \frac{3}{2\nu}|Q_m\mathbf{f}|^2 + \frac{3c_1^2C_1^4}{\nu}\left(1+\log\frac{\lambda_{m+1}}{\lambda_1}\right) + \frac{9^3c_1^4C_0^2C_1^4}{2\nu}.
\end{aligned}$$

We thus have from (A.57) that

$$\frac{d}{dt}\|\mathbf{z}\|^2 + \nu|A\mathbf{z}|^2 \leq \frac{3}{\nu}|Q_m\mathbf{f}|^2 + \frac{3c_1^2C_1^4}{\nu}\left(1+\log\frac{\lambda_{m+1}}{\lambda_1}\right) + \frac{9^3c_1^4C_0^2C_1^4}{\nu}. \tag{A.59}$$

Hence,

$$\frac{d}{dt}\|\mathbf{z}\|^2 + \nu\lambda_{m+1}\|\mathbf{z}\|^2 \leq \frac{3}{\nu}|Q_m\mathbf{f}|^2 + \frac{3c_1^2C_1^4}{\nu}\left(1+\log\frac{\lambda_{m+1}}{\lambda_1}\right) + \frac{9^3c_1^4C_0^2C_1^4}{\nu}. \tag{A.60}$$

From (A.60) we conclude that, for $t \geq t_0 \geq T$,

$$\begin{aligned}
\|\mathbf{z}(t)\|^2 &\leq \|\mathbf{z}(t_0)\|^2 e^{-\nu\lambda_{m+1}(t-t_0)}\\
&\quad + \frac{3}{\nu^2\lambda_{m+1}}\left(|Q_m\mathbf{f}|^2 + c_1^2C_1^4\left(1+\log\frac{\lambda_{m+1}}{\lambda_1}\right) + 243c_1^4C_0^2C_1^4\right).
\end{aligned} \tag{A.61}$$

By increasing T if necessary we deduce that, for $t \geq T$,

$$\|\mathbf{z}(t)\|^2 \leq \frac{4}{\nu^2 \lambda_{m+1}} \left(|Q_m \mathbf{f}|^2 + c_1^2 C_1^4 \left(1 + \log \frac{\lambda_{m+1}}{\lambda_1}\right) + 243 c_1^4 C_0^2 C_1^4 \right). \tag{A.62}$$

This proves (A.40) with

$$M_1 = \frac{2}{\nu \lambda_1^{1/2}} (|Q_m \mathbf{f}|^2 + c_1^2 C_1^4 + 243 c_1^4 C_0^2 C_1^4)^{1/2}. \tag{A.63}$$

For (4.15) we use that the solutions of the 2-dimensional Navier–Stokes equations are analytic in time and use the Cauchy formula for the time derivative, as established in Section II.8:

$$\mathbf{z}'(t) = \frac{1}{2\pi i} \int_\Gamma \frac{\mathbf{z}(\zeta)}{(t-\zeta)^2} \, d\zeta, \tag{A.64}$$

where Γ is a circle in the complex plane centered at t (for the complexified time). Thus, using (4.12),

$$|\mathbf{z}'(t)| \leq \frac{1}{2\pi} \int_\Gamma \frac{|\mathbf{z}(\zeta)|}{|t-\zeta|^2} |d\zeta| \leq c_2 M_0 \frac{\lambda_1}{\lambda_{m+1}} \left(1 + \log \frac{\lambda_m}{\lambda_1}\right)^{1/2} \tag{A.65}$$

for some appropriate constant c_2. This proves (4.15).

Finally, for (A.41) we use the evolution equation (4.7) and write

$$\nu A \mathbf{z} = Q_m \mathbf{f} - Q_m B(\mathbf{y} + \mathbf{z}) - \frac{d\mathbf{z}}{dt}. \tag{A.66}$$

Thus,

$$|A\mathbf{z}| \leq \frac{1}{\nu} |Q_m \mathbf{f}| + \frac{1}{\nu} |Q_m B(\mathbf{y} + \mathbf{z})| + \frac{1}{\nu} \left| \frac{d\mathbf{z}}{dt} \right|. \tag{A.67}$$

Upon using the appropriate inequalities for the nonlinear term and using the estimates obtained so far, one can deduce (A.41). We omit the details.

The upper bound estimate (4.19) on λ_{m+1} in the periodic case follows from the estimate (A.51) and an explicit estimate for C_1 in (A.43). This explicit estimate is given in Chapter II by equation (A.66), which implies, for large times (say, $t \geq T$),

$$\|\mathbf{u}(t)\|^2 \leq \frac{2}{\nu^2 \lambda_1} |\mathbf{f}|^2. \tag{A.68}$$

Thus, $C_1 = \sqrt{2/\nu^2 \lambda_1} |\mathbf{f}|$ and, from (A.51), it suffices that

$$\frac{\lambda_{m+1}}{\lambda_1} \geq 8 c_1^2 \mathrm{G}^2. \tag{A.69}$$

IV

Stationary Statistical Solutions of the Navier–Stokes Equations, Time Averages, and Attractors

Introduction

As mentioned earlier in this text, we take for granted that the Navier–Stokes equations (NSE), together with the associated boundary and initial conditions, embody all the macroscopic physics of fluid flows. In particular, the evolution of any measured property of a turbulent flow must be relatable to the solutions of those equations. In turbulent flow regimes, the physical properties are universally recognized as randomly varying and characterized by some suitable probability distribution functions. In this and the following chapter, we discuss how those probability distribution functions (also called probability distributions or measures, or Borel measures, in the mathematical terminology; see Appendix A.1) are determined by the underlying Navier–Stokes equations. Although in many cases such distributions may not be known explicitly, their existence and many useful properties may be readily established. For many practical purposes, such partial knowledge may be all that is needed. Thus we note that the issue of an explicit form of the distribution function – in particular, whether this measure is unique or depends on the initial data – is still an incompletely solved mathematical problem. But there are enough firm results available assuring that many of the widely accepted experimental results are meaningful and in consonance with the theory of the Navier–Stokes equations.

For instance, measurements of various aspects of turbulent flows (e.g., the turbulent boundary layer) are actually measurements of time-averaged quantities. On the other hand, theoretical considerations are often couched in terms of ensemble averages – that is, averages with respect to some (presumably existing, but unavailable) probability distribution function. Thus it is assumed more or less explicitly that: (a) suitable probability distribution functions exist, assuring the existence of meaningful ensemble averages; and (b) there is an established connection between time-averaged and ensemble-averaged quantities.

Suitable probability distributions are known to exist, subject to some (still unresolved) technical mathematical questions; as revealed in this and the following chapter, they have many properties that are assumed in practice by physicists on intuitive grounds.

Conventionally, the question of the relationship between the two types of averages has been dealt with, in analogy with classical statistical mechanics, by an ad hoc "ergodic hypothesis." According to that hypothesis, the solutions of the NSE characterizing a given turbulent flow will, over time, sample almost all of the attractor in the phase space and so time averages are precisely equal to ensemble averages. An alternative statement of the ergodic hypothesis in current use is that time and ensemble averages (moments) are equal to one another. Note that this alternative statement of the ergodic property does not necessarily imply that, in the phase space, the trajectories of the turbulent solutions come arbitrarily close to all the points in the attractor. However, as discussed in this chapter, the equality of the two kinds of averages can be proved in a suitable sense – without invoking any ad hoc "ergodic hypothesis" – provided that one is careful in defining the time averages (using Banach limits). This clearly strengthens reliance on conventional measurement techniques, because the nature of interactions leading to the statistical description of gas kinetics (classical statistical mechanics) is quite different from the mechanisms underlying the statistics of turbulence, and the usual argument by analogy may not be appropriate (cf. Rose and Sulem [1978]).

Before delving into the mathematical intricacies of statistical solutions of the Navier–Stokes equations, we present a simplified, heuristic overview of the results obtained in this chapter. We focus our attention here on stationary statistical solutions of the NSE – that is, the statistics of turbulent flows at times long after the flow was initiated and (in the terminology of our earlier discussion) when the flow reaches the attractor. Under these conditions it is natural to expect that, to within the expected experimental errors, measurements of any aspect of the flow made at different time intervals should yield identical values.

For the sake of completeness, this chapter starts by recalling, from Chapter II, the meaning of a solution $\mathbf{u} = \mathbf{u}(t)$ with a given initial condition $\mathbf{u}(0) = \mathbf{u}_0$. If the initial conditions are given according to a probability distribution $\mu_0 = \mu_0(\mathbf{u})$ on the phase space, then the solutions at some later time t will be distributed according to another probability distribution $\mu_t = \mu_t(\mathbf{u})$. For an abstract form of the autonomous NSE

$$\mathbf{u}' = \mathbf{F}(\mathbf{u}) \tag{0.1}$$

in some space H, the evolution of the probability distributions μ_t $(t \geq 0)$ is given in terms of a Liouville-type equation:

$$\frac{d}{dt}\int \Phi(\mathbf{u})\, d\mu_t(\mathbf{u}) = \int (\mathbf{F}(\mathbf{u}), \Phi'(\mathbf{u}))\, d\mu_t(\mathbf{u}) = 0 \tag{0.2}$$

for all suitably defined moments Φ. When statistical equilibrium is considered, the probability distribution μ_t is time-independent. This leads us to introduce the concept of a stationary statistical solution, which is a time-independent solution of (0.2). In other words, the stationary statistical solutions are probability measures $\mu = \mu(\mathbf{u})$ satisfying the stationary Liouville-type equation

$$\int (\mathbf{F}(\mathbf{u}), \Phi'(\mathbf{u}))\, d\mu(\mathbf{u}) = 0. \tag{0.3}$$

As shown in this chapter, if solutions of the NSE generate a semigroup (i.e., a well-defined time-dependent family of maps from the initial condition $\mathbf{u}_0$ to the corresponding solution $\mathbf{u}(t)$ at a later time t), as they do in the 2-dimensional case, then the stationary statistical solutions coincide with the usual notion of an invariant measure for a dynamical system. For the 3-dimensional NSE, however, the measures associated to stationary statistical behaviors are as shown in (0.3).

Then, preparatory to showing the relationship between time and ensemble averages, the limiting operation defining a time average is specified very carefully. With the time average thus identified, a probability distribution function $\mu = \mu(\mathbf{u})$ is defined as a so-called time-average measure:

$$\mathop{\mathrm{LIM}}_{T\to\infty} \frac{1}{T}\int_0^T \Phi(\mathbf{u}(t))\,dt = \int \Phi(\mathbf{u})\,d\mu(\mathbf{u}) \tag{0.4}$$

for any suitable Φ. Here $\mathrm{LIM}_{T\to\infty}$ represents a generalized concept of limit that we define later on in the chapter. The text then proceeds to demonstrate that any time-average measure is, in fact, a stationary statistical solution of the NSE. In this way, a simple relationship (viz., equality) between time and ensemble averages of the physical properties of fluid flows is established. In the 2-dimensional case, even more can be said: time-average measures (defined by (0.4)) and ensemble averages (defined by (0.3)) are, in fact, equivalent.

Because of the known difficulties with the solutions of NSE in three dimensions, one would expect difficulties with establishing the existence of time-average measures. Somewhat surprisingly, this is not the case. Thus, the possible lack of regularity of 3-dimensional solutions for long times does not preclude the existence of time averages.

The chapter ends with a discussion of the relationship between the asymptotic behavior of the solutions in both three and two dimensions, their corresponding global attractors, and the stationary statistical solutions.

As a final note to this heuristic overview, it is important to add the warning that in both the 2- and 3-dimensional cases, there are no uniqueness theorems for the corresponding stationary statistical solutions of the Navier–Stokes equations. The resulting ambiguities are resolved by appealing to the existence of the so-called Sinai–Ruelle–Bowen (SRB) measure. This measure is unique and, in many systems underlying physical phenomena, is appropriate for characterizing those phenomena. However, it is still an open question whether the SRB measure exists for the NSE. There are several references on this subject; see, for instance, the reviews by Viana [1997], Young [1997], and Ruelle [1998] as well as the original works by Sinai [1972], Bowen and Ruelle [1975], and Ruelle [1976].

There is an extensive literature related to statistical hydrodynamics in fluid mechanics and physics. We cannot review this literature here, but we would like to mention a few classical references: the original articles of Kolmogorov [1941a,b, 1962] and Kraichnan [1967, 1972]; the review article of Orszag [1970]; the books of Batchelor [1959, 1988], Landau and Lifshitz [1974], and Monin and Yaglom [1975]; the more recent books of Frisch [1995] and Lesieur [1997]; and the references therein. At the interface of mathematics and physics, the book of Dubois, Jauberteau, and Temam [1999] contains some review chapters on the statistical theory of turbulence. On the mathematical side – and beside the articles, cited elsewhere, written by the authors of this book – we should mention the significant contribution of Hopf [1952], the pioneering work of Prodi [1961], and the book by Vishik and Fursikov [1988], which is fully devoted to the mathematical aspects of statistical hydrodynamics and contains many important results on statistical and stochastic solutions of the NSE.

1 Mathematical Framework, Definition of Stationary Statistical Solutions, and Banach Generalized Limits

This section is divided into three parts. In the first part we recall the mathematical framework for the 2- and 3-dimensional Navier–Stokes equations and recall the main existence and uniqueness results to be used in the following sections. Although this part (Section 1.1) reproduces material already given in previous chapters, we believe this summary might be useful for those who wish to read this chapter but not the previous ones, which are much different in their content. The second part of this section motivates the definition of a stationary statistical solution and gives its rigorous definition. In the last part we recall the concept of generalized limits.

1.1 Weak and Strong Solutions of the Navier–Stokes Equations

We consider the Navier–Stokes equations

$$\mathbf{u}_t - \nu\Delta\mathbf{u} + (\mathbf{u}\cdot\nabla)\mathbf{u} + \nabla p = \mathbf{f}, \tag{1.1a}$$

$$\nabla\cdot\mathbf{u} = 0, \tag{1.1b}$$

in $\Omega \subset \mathbb{R}^d$ ($d = 2$ or 3) with $\nu > 0$ and $\mathbf{f} = \mathbf{f}(\mathbf{x})$. We endow the system with either the no-slip boundary conditions or the space-periodic boundary conditions. In the first case we assume more precisely that

$$\Omega \subset \mathbb{R}^d \text{ is open, bounded, and connected, with a } \mathcal{C}^2 \text{ boundary } \partial\Omega, \tag{1.2}$$

and we require that

$$\mathbf{u} = 0 \ \text{ on } \partial\Omega. \tag{1.3}$$

In the space-periodic case, we assume that

$$\Omega = \prod_{i=1}^{d}\left(\frac{L_i}{2}, \frac{L_i}{2}\right), \quad L_i > 0,\ i = 1, \dots, d, \tag{1.4}$$

and require that

$$\mathbf{u} = \mathbf{u}(\mathbf{x}, t) \text{ and } p = p(\mathbf{x}, t),\ \mathbf{x} = (x_1, \dots, x_d), \text{ are } L_i\text{-periodic in each variable } x_1, \dots, x_d \tag{1.5}$$

and, for simplicity,

$$\int_\Omega \mathbf{f}(\mathbf{x})\,d\mathbf{x} = \int_\Omega \mathbf{u}(\mathbf{x})\,d\mathbf{x} = 0. \tag{1.6}$$

We now recall the mathematical setting of the problem. Let

$$\mathcal{V} = \{\mathbf{u} \in \mathcal{C}_c^\infty(\Omega)^d;\ \nabla\cdot\mathbf{u} = 0\}$$

in the no-slip case and

$$\mathcal{V} = \left\{ \mathbf{u} \in \mathcal{C}_c^\infty(\bar{\Omega})^d;\ u_i|_{\Gamma_i} = u_i|_{\Gamma_{i+d}}\ (i = 1, \dots, d),\ \int_\Gamma \mathbf{u}(\mathbf{x})\, d\mathbf{x} = 0,\ \nabla \cdot \mathbf{u} = 0 \right\}$$

in the space-periodic case, where Γ_i and Γ_{i+d} $(i = 1, \dots, d)$ are the faces $x_i = 0$ and $x_i = L_i$ (respectively) of $\partial\Omega$. In either case we let H be the closure of $\mathcal{V}$ in $L^2(\Omega)^d$ and let V be the closure of $\mathcal{V}$ in $H^1(\Omega)^d$. In H and V we consider (respectively) the scalar products

$$(\mathbf{u}, \mathbf{v}) = \int_\Omega \mathbf{u}(x) \cdot \mathbf{v}(x)\, d\mathbf{x} \quad \text{and} \quad ((\mathbf{u}, \mathbf{v})) = \int_\Omega \sum_{i=1}^d \frac{\partial \mathbf{u}}{\partial x_i} \cdot \frac{\partial \mathbf{v}}{\partial x_i}\, d\mathbf{x}$$

as well as the associated norms, denoted by

$$|\mathbf{u}| = (\mathbf{u}, \mathbf{u})^{1/2} \quad \text{for } \mathbf{u} \in H, \qquad \|\mathbf{v}\| = ((\mathbf{v}, \mathbf{v}))^{1/2} \quad \text{for } \mathbf{v} \in V.$$

We recall that the spaces H and V can be characterized as follows. In the no-slip case,

$$H = \{\mathbf{u} \in L^2(\Omega)^d;\ \nabla \cdot \mathbf{u} = 0,\ \mathbf{u} \cdot \mathbf{n}|_{\partial\Omega} = 0\}$$

and

$$V = \{\mathbf{u} \in H_0^1(\Omega)^d;\ \nabla \cdot \mathbf{u} = 0\},$$

where $\mathbf{n}$ denotes the outward unit normal to $\partial\Omega$. We remark that $\mathbf{u} \cdot \mathbf{n}|_{\partial\Omega}$ makes sense when $\mathbf{u} \in L^2(\Omega)^d$ and $\nabla \cdot \mathbf{u} = 0$ in the distribution sense. In the space-periodic case,

$$H = \left\{ \mathbf{u} \in L^2_{\text{per}}(\Omega)^d;\ \nabla \cdot \mathbf{u} = 0,\ \int_\Omega \mathbf{u}(\mathbf{x})\, d\mathbf{x} = 0 \right\}$$

and

$$V = \left\{ \mathbf{v} \in H^1_{\text{per}}(\Omega)^d;\ \nabla \cdot \mathbf{u} = 0,\ \int_\Omega \mathbf{u}(\mathbf{x})\, d\mathbf{x} = 0 \right\},$$

where $H^1_{\text{per}}(\Omega)^d$ is the space of $\mathbb{R}^d$-valued functions u defined on $\mathbb{R}^d$ that are L_i-periodic in each variable x_i $(i = 1, \dots, d)$ and such that $\mathbf{u}|_{\mathcal{O}} \in H^1(\mathcal{O})^d$ for every bounded open set $\mathcal{O}$. The functions in $H^1_{\text{per}}(\Omega)^d$ are easily characterized by their Fourier series expansion:

$$H^1_{\text{per}}(\Omega)^d = \left\{ \mathbf{u} = \sum_{\mathbf{k} \in \mathbb{Z}^d} \hat{\mathbf{u}}_\mathbf{k} e^{2\pi i \frac{\mathbf{k}}{\mathbf{L}} \cdot \mathbf{x}};\ \hat{\mathbf{u}}_{-\mathbf{k}} = \bar{\hat{\mathbf{u}}}_\mathbf{k},\ \sum_{\mathbf{k} \in \mathbb{Z}^d} \left(\frac{1}{L^2} + 2\pi \left| \frac{\mathbf{k}}{\mathbf{L}} \right|^2 \right) |\hat{\mathbf{u}}_\mathbf{k}|^2 < \infty \right\},$$

where $L = \min\{L_1, \dots, L_d\}$ and

$$\frac{\mathbf{k}}{\mathbf{L}} = \left(\frac{k_1}{L_1}, \dots, \frac{k_d}{L_d} \right).$$

Hence, we also have

$$V = \left\{ \mathbf{u} = \sum_{\mathbf{k} \in \mathbb{Z}^d \setminus \{0\}} \hat{\mathbf{u}}_\mathbf{k} e^{2\pi i \frac{\mathbf{k}}{\mathbf{L}} \cdot \mathbf{x}};\ \hat{\mathbf{u}}_{-\mathbf{k}} = \bar{\hat{\mathbf{u}}}_\mathbf{k},\ \frac{\mathbf{k}}{\mathbf{L}} \cdot \hat{\mathbf{u}}_\mathbf{k} = 0,\ \sum_{\mathbf{k} \in \mathbb{Z}^d \setminus \{0\}} \left| \frac{\mathbf{k}}{\mathbf{L}} \right|^2 |\hat{\mathbf{u}}_\mathbf{k}|^2 < \infty \right\}$$

and

$$H = \left\{ \mathbf{u} = \sum_{\mathbf{k} \in \mathbb{Z}^d \setminus \{0\}} \hat{\mathbf{u}}_{\mathbf{k}} e^{2\pi i \frac{\mathbf{k}}{\mathbf{L}} \cdot \mathbf{x}};\ \hat{\mathbf{u}}_{-\mathbf{k}} = \overline{\hat{\mathbf{u}}}_{\mathbf{k}},\ \frac{\mathbf{k}}{\mathbf{L}} \cdot \hat{\mathbf{u}}_{\mathbf{k}} = 0,\ \sum_{\mathbf{k} \in \mathbb{Z}^d \setminus \{0\}} |\hat{\mathbf{u}}_{\mathbf{k}}|^2 < \infty \right\}.$$

The problem posed in the weak formulation of the NSE (with either no-slip or space-periodic boundary conditions) is the following: Given $T > 0$, $\mathbf{u}_0$ in H, and $\mathbf{f}$ in $L^2(0, T; H)$, find a function $\mathbf{u}$ in $L^\infty(0, T; H) \cap L^2(0, T; V)$ such that

$$\frac{d}{dt}(\mathbf{u}, \mathbf{v}) + \nu((\mathbf{u}, \mathbf{v})) + b(\mathbf{u}, \mathbf{u}, \mathbf{v}) = (\mathbf{f}, \mathbf{v}) \quad \forall \mathbf{v} \in V, \tag{1.7}$$

in the distribution sense on $(0, T)$, and with $\mathbf{u}(0) = \mathbf{u}_0$ in some suitable sense. Here

$$b(\mathbf{u}, \mathbf{v}, \mathbf{w}) = \sum_{i,j=1}^{d} \int_\Omega u_i \frac{\partial v_j}{\partial x_i} w_j \, d\mathbf{x};$$

if $\mathbf{f}$ is square integrable but not in H then we can replace it by its Leray projection on H, so that $\mathbf{f}$ is always assumed to be in H.

Note that the pressure term disappears in the weak formulation of the problem. This is because the gradient of a function is orthogonal to the space of solenoidal functions. More precisely, with respect to the inner product in $L^2(\Omega)^d$, we have

$$H^\perp = \{\mathbf{u} \in L^2(\Omega)^d;\ \mathbf{u} = \nabla p,\ p \in H^1(\Omega)^d\}$$

in the no-slip case and

$$H^\perp = \{\mathbf{u} \in L^2(\Omega)^d;\ \mathbf{u} = \nabla p,\ p \in H^1_{\text{per}}(\Omega)^d\}$$

in the space-periodic case. In any case, we have the so-called Leray projector

$$P_L \colon L^2(\Omega)^d \to H,$$

which is the orthogonal projector onto H in $L^2(\Omega)^d$.

The trilinear operator $b = b(\mathbf{u}, \mathbf{v}, \mathbf{w})$ defined previously can be extended to a continuous trilinear operator defined on V. Moreover,

$$b(\mathbf{u}, \mathbf{v}, \mathbf{v}) = 0 \quad \text{for all } \mathbf{u}, \mathbf{v} \in V, \tag{1.8}$$

which implies that

$$b(\mathbf{u}, \mathbf{v}, \mathbf{w}) = -b(\mathbf{u}, \mathbf{w}, \mathbf{v}) \quad \text{for all } \mathbf{u}, \mathbf{v}, \mathbf{w} \in V. \tag{1.9}$$

An equivalent formulation for the NSE is achieved with the definition of the Stokes operator

$$A\mathbf{u} = -P_L \Delta \mathbf{u} \quad \text{for all } \mathbf{u} \in D(A) = V \cap H^2(\Omega)^d. \tag{1.10}$$

The Stokes operator is a positive self-adjoint operator, so we can work with fractional powers of A. We will consider, as explained in Chapter II, the positive and negative powers of A: A^α for $\alpha \in \mathbb{R}$ with $\alpha \geq 0$ or $\alpha < 0$; we also explained in Chapter II how $D(A^\alpha)$ and $D(A^{-\alpha})$ are paired through the "duality" product (a bilinear form from $D(A^\alpha) \times D(A^{-\alpha})$ into $\mathbb{R}$). In particular, we have $D(A^{1/2}) = V$, and it follows that $A \colon D(A^{1/2}) \to D(A^{-1/2})$ with

$$(A\mathbf{u}, \mathbf{v}) = ((\mathbf{u}, \mathbf{v})) \quad \text{for all } \mathbf{u}, \mathbf{v} \in D(A^{1/2}),$$

where $(\cdot, \cdot)$ is the duality product between $D(A^{-1/2})$ and $D(A^{1/2})$. Moreover, the trilinear operator $b(\mathbf{u}, \mathbf{v}, \mathbf{w})$ leads to the bilinear operator $B\colon D(A^{1/2}) \times D(A^{1/2}) \mapsto D(A^{-1/2})$ defined by

$$(B(\mathbf{u}, \mathbf{v}), \mathbf{w}) = b(\mathbf{u}, \mathbf{v}, \mathbf{w}) \quad \text{for all } \mathbf{u}, \mathbf{v}, \mathbf{w} \in D(A^{1/2}).$$

We then have the following functional formulation of the Navier–Stokes equations: Given $T > 0$, $\mathbf{u}_0$ in H, and $\mathbf{f}$ in $L^2(0, T; H)$, find $\mathbf{u}$ in $L^\infty(0, T; H) \cap L^2(0, T; V)$ with $\mathbf{u}' \in L^1(0, T; D(A^{-1/2}))$ such that

$$\mathbf{u}' + \nu A\mathbf{u} + B(\mathbf{u}) = \mathbf{f} \tag{1.11}$$

and $\mathbf{u}(0) = \mathbf{u}_0$ in some suitable sense, where $B(\mathbf{u}) = B(\mathbf{u}, \mathbf{u})$.

Concerning the initial condition $\mathbf{u}(0) = \mathbf{u}_0$, if $\mathbf{u} \in L^2(0, T; V)$ and satisfies (1.7) then $\mathbf{u} \in \mathcal{C}([0, T]; D(A^{-1/2}))$, so that $\mathbf{u}(0) = \mathbf{u}_0$ makes sense. In the other formulation, if $\mathbf{u} \in L^2(0, T; V)$ and $\mathbf{u}' \in L^1(0, T; D(A^{-1/2}))$, then $\mathbf{u}$ is weakly continuous from $[0, T]$ into H (i.e., $t \mapsto (\mathbf{u}(t), \mathbf{v})$ is continuous for each $\mathbf{v} \in H$), so that $\mathbf{u}(0) = \mathbf{u}_0$ again makes sense. For simplicity, we denote by H_w the space H endowed with its weak topology; thus the weak continuity of $\mathbf{u} = \mathbf{u}(t)$ can be expressed as $\mathbf{u} \in \mathcal{C}([0, T], H_w)$.

We will have the opportunity to use the following inequality for the bilinear operator:

$$\|B(\mathbf{u})\|_{D(A^{-1/2})} \leq c|\mathbf{u}|^{1/2}\|\mathbf{u}\|^{3/2} \quad \text{for all } \mathbf{u} \in V, \tag{1.12}$$

where c is a constant depending on Ω.

Since the domain Ω is assumed to be bounded, it follows from the Rellich lemma that V is compactly embedded in H, so that A^{-1} is compact as a closed operator in H. Hence, there exists an orthonormal basis $\{\mathbf{w}_m\}_{m=1}^\infty$ in H such that

$$A\mathbf{w}_m = \lambda_m \mathbf{w}_m, \quad m = 1, 2, \ldots.$$

Moreover, we have that

$$0 < \lambda_1 \leq \lambda_2 \leq \cdots \leq \lambda_m \leq \cdots, \quad \lambda_m \to +\infty \text{ as } m \to +\infty.$$

We also have the Poincaré inequality

$$|\mathbf{u}|^2 \leq \frac{1}{\lambda_1}\|\mathbf{u}\|^2 \quad \text{for all } u \in V. \tag{1.13}$$

In what follows we denote by P_m the orthogonal projector of H onto the space spanned by $\mathbf{w}_1, \mathbf{w}_2, \ldots, \mathbf{w}_m$. The following inequalities will be useful:

$$|\mathbf{u} - P_m\mathbf{u}|^2 \leq \frac{1}{\lambda_m}\|\mathbf{u} - P_m\mathbf{u}\|^2 \quad \forall \mathbf{u} \in V,\ \forall m \in \mathbb{N}; \tag{1.14}$$

$$\|P_m\mathbf{u}\|^2 \leq \lambda_m|\mathbf{u}|^2 \quad \forall \mathbf{u} \in H,\ \forall m \in \mathbb{N}. \tag{1.15}$$

In fact, as shown in Chapter II, we need to define solutions of (1.11) that possess further properties; in this way we introduced in Section II.7 the concepts of *weak* and *strong* solutions to the Navier–Stokes equations (1.11). We have seen also that the results of existence and uniqueness of solutions are different in space dimensions 2 and 3 (see Theorems 7.1–7.4 in Chapter II) and that, for technical reasons, even the notion of weak solution differs from one space dimension to the other. In space dimension 2, the weak solutions of (1.11) are also assumed to satisfy

$$\mathbf{u}' \in L^2(0, T; D(A^{-1/2})) \quad \text{for all } T > 0, \tag{1.16}$$

and this suffices for all technical purposes. In space dimension 3, we can only obtain weak solutions of (1.11) that satisfy, instead of (1.16),

$$\mathbf{u}' \in L^{4/3}(0, T; D(A^{-1/2})) \quad \text{for all } T > 0. \tag{1.17}$$

Since this information is not sufficient, we also require the 3-dimensional weak solutions to satisfy further properties, which we proved in Chapter II (see Theorem 7.1 and the discussion in the section entitled "Further properties of the solutions in dimension 3"). Particularly important for many purposes are the energy inequalities (II.7.5) and (II.7.7), together with their consequences (II.A.38) and (II.A.43).

On a more technical note, we will also need from Chapter II (Sections 7–9 and Appendix A) several estimates for the inertial term, either through $B(\mathbf{u}, \mathbf{v})$ or through $b(\mathbf{u}, \mathbf{v}, \mathbf{w})$.

1.2 Definition of Stationary Statistical Solution

In order to motivate the definition of a stationary statistical solution, we first consider the 2-dimensional Navier–Stokes equation and its associated semigroup $\{S(t)\}_{t\geq 0}$ in H. For this and for later purposes, we first recall the definition of the image of a measure by a mapping. If μ is a probability distribution on H and g is a continuous (measurable) mapping from H into itself, then $g(\mu)$ is the probability distribution defined by

$$g(\mu)(E) = \mu(g^{-1}(E)) \tag{1.18}$$

for every set $E \subset H$ that is μ-measurable. Alternatively, if Φ is a continuous and bounded function from H into $\mathbb{R}$, then

$$\int_H \Phi(\mathbf{u})\, dg(\mu)(\mathbf{u}) = \int_H \Phi(g(\mathbf{u}))\, d\mu(\mathbf{u}). \tag{1.19}$$

We next use this concept to define, in the 2-dimensional case, the evolution of a probability distribution governed by the Navier–Stokes equations.

Suppose we are given a probability distribution μ_0 as initial data. Then, since the solutions to the NSE evolve according to $\mathbf{u}' = \mathbf{F}(\mathbf{u})$ (see (1.11)), it follows that (i) $\mathbf{v}_0 = S(t)\mathbf{u}_0$ if and only if $\mathbf{u}_0 \in S(t)^{-1}\mathbf{v}_0$ and (ii) the probability distribution $\mu_t(E)$ for $\mathbf{v}_0$ to be in the set $E \subset H$ is the same as the probability $\mu_0(S(t)^{-1}E)$ for $\mathbf{u}_0$ to be in $S(t)^{-1}E$. Therefore,

$$\mu_t(E) = \mu_0(S(t)^{-1}E),$$

which means, according to (1.18), that μ_t is equal to $S(t)\mu_0$. Remember that we can extract information from the system through the averages

$$\int \Phi(\mathbf{u})\, d\mu_t(\mathbf{u}),$$

where $\Phi \in L^1(\mu_t)$, for all $t \geq 0$. Hence, information on the evolution of such moments is particularly useful.

Remark 1.1 For the sake of simplicity, we assume throughout this chapter and the next that, whenever the domain of integration is not specified, the integral is over H.

It follows promptly from (1.19) that

$$\int \Phi(\mathbf{u})\, d\mu_t(\mathbf{u}) = \int \Phi(S(t)\mathbf{u})\, d\mu_0(\mathbf{u}) \quad \forall \Phi \in L^1(\mu_t),\ \forall t \geq 0. \tag{1.20}$$

Using that $\mathbf{u}' = \mathbf{F}(\mathbf{u})$, we can formally differentiate (with respect to time) the term on the RHS of (1.20). Thus we find that those quantities vary in time according to

$$\begin{aligned}\frac{d}{dt}\int \Phi(\mathbf{u})\, d\mu_t(\mathbf{u}) &= \frac{d}{dt}\int \Phi(S(t)\mathbf{u})\, d\mu_0(\mathbf{u}) \\ &= \int \left(\frac{d}{dt}S(t)\mathbf{u},\, \Phi'(S(t)\mathbf{u})\right) d\mu_0(\mathbf{u}) \\ &= \int (\mathbf{F}(S(t)\mathbf{u}),\, \Phi'(S(t)\mathbf{u}))\, d\mu_0(\mathbf{u}) \\ &= \int (\mathbf{F}(\mathbf{u}),\, \Phi'(\mathbf{u}))\, d\mu_t(\mathbf{u})\end{aligned}$$

for suitable differentiable functions Φ. We have applied a generalized chain differentiation rule to differentiate the term $t \mapsto \Phi(S(t)\mathbf{u})$, for which the appropriate notation is

$$\frac{d}{dt}\Phi(S(t)\mathbf{u}) = \left(\frac{d}{dt}S(t)\mathbf{u},\, \Phi'(S(t)\mathbf{u})\right).$$

The preceding sequence of equalities leads to the Liouville-type equation

$$\frac{d}{dt}\int \Phi(\mathbf{u})\, d\mu_t(\mathbf{u}) = \int (\mathbf{F}(\mathbf{u}),\, \Phi'(\mathbf{u}))\, d\mu_t(\mathbf{u}). \tag{1.21}$$

If statistical equilibrium has been reached by the system, then the moments of Φ (i.e., the statistical information) do not change with time, so the time derivative in (1.21) must be zero:

$$\frac{d}{dt}\int \Phi(\mathbf{u})\, d\mu_t(\mathbf{u}) = 0; \tag{1.22}$$

in other words,

$$\int \Phi(\mathbf{u})\, d\mu_t(\mathbf{u}) = \int \Phi(\mathbf{u})\, d\mu_0(\mathbf{u}) \quad \forall \Phi \in L^1(\mu_t),\ \forall t \geq 0. \tag{1.23}$$

From (1.23), we easily deduce that $\mu_t = \mu_0$ for every positive time t. Hence, after renaming μ_0 to μ, it follows from (1.20) that

$$\int \Phi(\mathbf{u})\, d\mu(\mathbf{u}) = \int \Phi(S(t)\mathbf{u})\, d\mu(\mathbf{u}) \quad \forall \Phi \in L^1(\mu). \tag{1.24}$$

The relation (1.24) is equivalent to $\mu(E) = \mu(S(t)^{-1}E)$ for every measurable set E in H. This leads us to the definition of an invariant measure.

Definition 1.1 *A probability density (measure) μ on H is called an* invariant measure *for the semigroup $\{S(t)\}_{t\geq 0}$ if*

$$\mu(E) = \mu(S(t)^{-1}E) \quad \forall t \geq 0 \tag{1.25}$$

for every measurable set E in H.

Remark 1.2 In view of the definition of probability distribution as a Borel measure (see Appendix A.1), it suffices to require in Definition 1.1 that (1.25) hold for all closed or all open subsets E of H.

Remark 1.3 The definition of invariant measure involves $S(t)^{-1}$ instead of $S(t)$ because the important relation is (1.24), which is equivalent to (1.25). The condition that $\mu(E) = \mu(S(t)E)$ for all $t \geq 0$ and all measurable sets E does not, in general, imply (1.24). One can nonetheless show that, if a semigroup $\{S(t)\}_{t\geq 0}$ is one-to-one, then (1.25) implies that $\mu(E) = \mu(S(t)E)$ for all $t \geq 0$ and all measurable sets E; if the semigroup is onto, then the converse implication holds. If the semigroup is not one-to-one then from (1.25) we can deduce only that $\mu(S(t)E) = \mu(S(t)^{-1}S(t)E) \geq \mu(E)$. A measure satisfying this inequality is called *accretive.* The concept of accretivity can be extended to the 3-dimensional NSE, where a semigroup is not well-defined (see Appendix C for more details).

From (1.21) and (1.22) we deduce that, in statistical equilibrium,

$$\int (\mathbf{F}(\mathbf{u}), \Phi'(\mathbf{u}))\, d\mu(\mathbf{u}) = 0. \tag{1.26}$$

Now, in contrast to equation (1.25), equation (1.26) makes sense even if a semigroup is not well-defined, as in the case of the 3-dimensional Navier–Stokes equations. We can thus generalize the concept of invariant measure to the 3-dimensional case by defining the stationary statistical solutions as the solutions to (1.26). Similarly, (1.21) also makes sense when a semigroup is not defined, and a solution $\{\mu_t\}_{t\geq 0}$ of (1.21) is called a (time-dependent) statistical solution of the equation $\mathbf{u}' = \mathbf{F}(\mathbf{u})$, a concept that we study in the next chapter. At this point, and in order to make these definitions rigorous, we must first define a class of functionals Φ for which (1.26) and (1.21) make sense. In this chapter we consider only stationary statistical solutions; time-dependent statistical solutions are treated in the next chapter.

Definition 1.2 *We define the class $\mathcal{T}$ of* test functions *to be the set of real-valued functionals $\Phi = \Phi(\mathbf{u})$ on H that are bounded on bounded subsets of H and such that the following conditions hold.*

(i) *For any* $\mathbf{u} \in V$, *the Fréchet derivative* $\Phi'(\mathbf{u})$ *taken in* H *along* V *exists. More precisely, for each* $\mathbf{u} \in V$, *there exists an element in* H *denoted* $\Phi'(\mathbf{u})$ *such that*

$$\frac{|\Phi(\mathbf{u}+\mathbf{v}) - \Phi(\mathbf{u}) - (\Phi'(\mathbf{u}), \mathbf{v})|}{|\mathbf{v}|} \to 0 \quad \text{as } |\mathbf{v}| \to 0, \ \mathbf{v} \in V. \tag{1.27}$$

(ii) $\Phi'(\mathbf{u}) \in V$ *for all* $\mathbf{u} \in V$, *and* $\mathbf{u} \mapsto \Phi'(\mathbf{u})$ *is continuous and bounded as a function from* V *into* V.

For example, we can take

$$\Phi(\mathbf{u}) = \psi((\mathbf{u}, \mathbf{g}_1), \ldots, (\mathbf{u}, \mathbf{g}_m)),$$

where $\psi \in \mathcal{C}_c^1(\mathbb{R}^m)$, $m \in \mathbb{N}$, and $\mathbf{g}_1, \ldots, \mathbf{g}_m \in V$. In this case,

$$\Phi'(\mathbf{u}) = \sum_{j=1}^{m} \partial_j \psi((\mathbf{u}, \mathbf{g}_1), \ldots, (\mathbf{u}, \mathbf{g}_m))\mathbf{g}_j,$$

where ∂_j indicates the derivative of ψ with respect to its jth argument. This example includes the case $\Phi(\mathbf{u}) = \rho(|P_m\mathbf{u}|^2)$ with $\rho \in \mathcal{C}_c^1(\mathbb{R})$, and then $\Phi'(\mathbf{u}) = 2\rho'(|P_m\mathbf{u}|^2)P_m\mathbf{u}$.

Let now μ denote a probability measure on H such that

$$\int \|\mathbf{u}\|^2 \, d\mu(\mathbf{u}) < \infty, \tag{1.28}$$

where $\|\mathbf{u}\| = \infty$ for $\mathbf{u} \in H \setminus V$. Note that this implies that $\mu(H \setminus V) = 0$. Then, for any test function Φ,

$$\mathbf{u} \mapsto (\nu A\mathbf{u} + B(\mathbf{u}) - \mathbf{f}, \Phi'(\mathbf{u}))$$

is continuous in V and

$$\begin{aligned}
|(\nu A\mathbf{u} + B(\mathbf{u}) - \mathbf{f}, \Phi'(\mathbf{u}))| &= |\nu((\mathbf{u}, \Phi'(\mathbf{u}))) + b(\mathbf{u}, \mathbf{u}, \Phi'(\mathbf{u})) - (\mathbf{f}, \Phi'(\mathbf{u}))| \\
&\le \nu\|\mathbf{u}\|\|\Phi'(\mathbf{u})\| + c_1|\mathbf{u}|^{1/2}\|\mathbf{u}\|^{3/2}\|\Phi'(\mathbf{u})\| + |\mathbf{f}||\Phi'(\mathbf{u})| \\
&\le (\nu\|\mathbf{u}\| + c_1\lambda_1^{1/4}\|\mathbf{u}\|^2 + \lambda_1^{1/2}|\mathbf{f}|) \sup_{\mathbf{u}\in V}\|\Phi'(\mathbf{u})\|,
\end{aligned}$$

where we used the inequalities (1.12) and (1.13). Thus, since $\mu(H \setminus V) = 0$, it follows from (1.28) and the estimate just given that

$$\int (\nu A\mathbf{u} + B(\mathbf{u}) - \mathbf{f}, \Phi'(\mathbf{u})) \, d\mu(\mathbf{u})$$

makes sense and is finite. Thus, we propose the following definition.

Definition 1.3 *A* stationary statistical solution *of the Navier–Stokes equation is a probability measure* μ *on* H *such that*

(i) $\int \|\mathbf{u}\|^2 \, d\mu(\mathbf{u}) < \infty;$ (1.29)

(ii) $\int (\mathbf{F}(\mathbf{u}), \Phi'(\mathbf{u})) \, d\mu(\mathbf{u}) = 0$ (1.30)

for any test functional $\Phi \in \mathcal{T}$, *where* $\mathbf{F}(\mathbf{u})$ *is as in* (1.11); *and*

(iii) $\int_{E_1 \leq |\mathbf{u}|^2 < E_2} \{\nu \|\mathbf{u}\|^2 - (\mathbf{f}, \mathbf{u})\} \, d\mu(\mathbf{u}) \leq 0$

$\forall E_1, E_2, \; 0 \leq E_1 < E_2 \leq +\infty.$ (1.31)

The condition (1.31) is a weak form of an energy-type inequality. It leads to a bound on the integral in (1.29) and to an estimate of the support of the measure μ on H. In fact, let E denote the set $E = \{E_1 \leq |\mathbf{u}|^2 < E_2\}$ for given $0 \leq E_1 < E_2 \leq +\infty$. We then have

$$\begin{aligned}\int_E (\mathbf{f}, \mathbf{u}) \, d\mu(\mathbf{u}) &\leq |\mathbf{f}| \int_E |\mathbf{u}| \, d\mu(\mathbf{u}) \\ &\leq |\mathbf{f}| \left(\int_E |\mathbf{u}|^2 \, d\mu(\mathbf{u}) \right)^{1/2} \\ &\leq \frac{|\mathbf{f}|}{\lambda_1^{1/2}} \left(\int_E \|\mathbf{u}\|^2 \, d\mu(\mathbf{u}) \right)^{1/2} \quad \text{(by (1.13))},\end{aligned}$$

which implies, taking (1.29) into account, that the μ-integral of $(\mathbf{f}, \mathbf{u})$ on E is finite. Hence, we can deduce from (1.31) that

$$\nu \int_E \|\mathbf{u}\|^2 \, d\mu(\mathbf{u}) \leq \int_E (\mathbf{f}, \mathbf{u}) \, d\mu(\mathbf{u}),$$

so that, from the previous estimate,

$$\nu \int_E \|\mathbf{u}\|^2 \, d\mu(\mathbf{u}) \leq \frac{|\mathbf{f}|}{\lambda_1^{1/2}} \left(\int_E \|\mathbf{u}\|^2 \, d\mu(\mathbf{u}) \right)^{1/2}.$$

Hence,

$$\int_E \|\mathbf{u}\|^2 \, d\mu(\mathbf{u}) \leq \frac{|\mathbf{f}|^2}{\nu^2 \lambda_1}. \tag{1.32}$$

In particular, for $E_1 = 0$ and $E_2 = +\infty$ we find, since $\mu(\{\mathbf{u} \in H; \; \|\mathbf{u}\| = +\infty\}) = \mu(H \setminus V) = 0$, that

$$\int \|\mathbf{u}\|^2 \, d\mu(\mathbf{u}) \leq \frac{|\mathbf{f}|^2}{\nu^2 \lambda_1}. \tag{1.33}$$

Now, from (1.32) and (1.13), we deduce that

$$\int_E |\mathbf{u}|^2 \, d\mu(\mathbf{u}) \leq \frac{|\mathbf{f}|^2}{\nu^2 \lambda_1^2}$$

and hence

$$\int_E \left\{ |\mathbf{u}|^2 - \frac{|\mathbf{f}|^2}{\nu^2 \lambda_1^2} \right\} d\mu(\mathbf{u}) \leq 0.$$

On the other hand, if we choose $E_1 = |\mathbf{f}|^2/\nu^2\lambda_1^2$ and $E_2 = +\infty$, we see that

$$|\mathbf{u}|^2 - \frac{|\mathbf{f}|^2}{\nu^2\lambda_1^2} \geq 0$$

for all $\mathbf{u}$ in E, so that necessarily $\mu(E) = 0$. Therefore,

$$\mu(\{\mathbf{u} \in H;\ |\mathbf{u}| \leq |\mathbf{f}|/\nu\lambda_1\}) = 1.$$

This shows that the measure μ is carried by a bounded set in H:

$$\operatorname{supp} \mu \subset \left\{\mathbf{u} \in H;\ |\mathbf{u}| \leq \frac{|\mathbf{f}|}{\nu\lambda_1}\right\}. \tag{1.34}$$

We emphasize that our definition of stationary statistical solutions, as well as properties (1.32) and (1.34), are valid for both the 2- and 3-dimensional Navier–Stokes equations.

Note that the existence of stationary statistical solutions follows readily from the well-known facts that (i) in both two and three dimensions, the set of stationary (weak) solutions is not empty (see e.g. Constantin and Foias [1988] and Temam [1997]) and (ii) a delta function supported by $\{\mathbf{u}_*\}$, where $\mathbf{u}_*$ is a stationary solution of the NSE, is a stationary statistical solution. Indeed, if $\mathbf{u}_*$ is a stationary solution of the NSE then $\mathbf{F}(\mathbf{u}_*) = 0$, where $\mathbf{F}(\cdot)$ is given by (1.11). Then, if $\mu(\mathbf{u}) = \delta(\mathbf{u} - \mathbf{u}_*)$ (i.e., $d\mu(\mathbf{u}) = \delta(\mathbf{u} - \mathbf{u}_*)\, d\mathbf{u}$), we have

$$\begin{aligned}\int (\mathbf{F}(\mathbf{u}), \Phi'(\mathbf{u}))\, d\mu(\mathbf{u}) &= \int (\mathbf{F}(\mathbf{u}), \Phi'(\mathbf{u}))\delta(\mathbf{u} - \mathbf{u}_*)\, d\mathbf{u} \\ &= (\mathbf{F}(\mathbf{u}_*), \Phi'(\mathbf{u}_*)) = 0.\end{aligned}$$

However, the problem that we face in the next two sections is not the existence of stationary statistical solutions but rather their determination via time averages, which allows us to relate time averages with ensemble (phase-space) averages.

Remark 1.4 Note that the suppport of an invariant measure consists only of points that are nonwandering. More precisely, let μ be an invariant measure for the semigroup $\{S(t)\}_{t\geq 0}$. Let $F = \operatorname{supp} \mu$, which is the smallest closed set of full μ-measure, $\mu(F) = 1$. We claim that, for any $\mathbf{u} \in F$ and any set A containing $\mathbf{u}$ and relatively open in F, there exists a sequence $t_n \to \infty$ of positive times such that $S(t_n)A \cap A \neq \emptyset$ for all $n \in \mathbb{N}$. Indeed, first note that $\mu(A) > 0$; otherwise, $F \setminus A$ would be a closed set of full measure, properly included in F, so that F would not be the support of μ. Consider then the positive orbit of A, $\gamma^+(A) = \bigcup_{s\geq 0} S(s)A$. Since μ is invariant and $S(t)^{-1}S(t)B \supset B$ for every set B in E, we find that

$$\mu(S(t)\gamma^+(A)) = \mu(S(t)^{-1}S(t)\gamma^+(A)) \geq \mu(\gamma^+(A)) \quad \forall t \geq 0.$$

Since

$$S(t)\gamma^+(A) \subset \gamma^+(A) \quad \text{for } t \geq 0,$$

we obtain

$$0 \leq \mu(\gamma^+(A) \setminus S(t)\gamma^+(A)) = \mu(\gamma^+(A)) - \mu(S(t)\gamma^+(A)) \leq 0.$$

Hence,

$$\mu(\gamma^+(A) \setminus S(t)\gamma^+(A)) = 0.$$

On the other hand, if $S(t)A \cap A = \emptyset$ for all t large enough (say, $t \geq T$), then $A \cap S(T)\gamma^+(A) = \emptyset$ and so $A \subset \gamma^+(A) \setminus S(T)\gamma^+(A)$. Hence, from the relation just displayed, $\mu(A) \leq \mu(\gamma^+(A) \setminus S(T)\gamma^+(A)) = 0$, which would contradict the fact that $\mu(A) > 0$. Therefore, we must have $S(t)A \cap A \neq \emptyset$ for arbitrarily large t.

1.3 Definition and Properties of Generalized Limits

We conclude this section with the concept of generalized limit. For the physics-oriented reader, let us point out that the introduction of this generalized limit serves to mitigate concerns over some mathematical technicalities. It has no effect on the practice of actual laboratory measurements of time averages.

In short, a generalized limit is any positive linear functional that extends the usual limit. The use of such limits enables us to avoid any assumption of ergodicity of orbits in phase space and still relate time averages with ensemble averages in the phase space. The use of generalized limits will become clear in the next section; for the moment, let us make its definition more precise in the setting relevant to us and recall a few of its properties.

Definition 1.4 A generalized limit *is any linear functional, denoted* $\textsc{Lim}_{T\to\infty}$, *defined on the space* $\mathcal{B}([0,\infty))$ *of all bounded real-valued functions on* $[0,\infty)$ *and satisfying*

$$\text{(i)}\quad \underset{T\to\infty}{\textsc{Lim}}\, g(T) \geq 0 \quad \forall g \in \mathcal{B}([0,\infty)) \textit{ with } g(s) \geq 0\ \forall s \geq 0; \tag{1.35}$$

$$\text{(ii)}\quad \underset{T\to\infty}{\textsc{Lim}}\, g(T) = \lim_{T\to\infty} g(T) \quad \forall g \in \mathcal{B}([0,\infty)) \tag{1.36}$$

such that the usual limit, denoted $\lim_{T\to\infty} g(T)$, *exists.*

We have mentioned that the generalized limits are, in the case of interest to us, extensions of the classical limit to all bounded real-valued functions on $[0,\infty)$. Hence, any bounded real-valued function on $[0,\infty)$ admits a generalized limit, which coincides with the classical limit whenever the classical limit exists. Note that this generalized limit may not be unique for a given function because there may be different such extensions, but it will be uniquely determined once we fix a particular extension (see Appendix A.2 for more details).

Any such generalized limit can be shown to possess also the following properties:

$$\text{(iii)}\quad \liminf_{T\to\infty} g(T) \leq \underset{T\to\infty}{\textsc{Lim}}\, g(T) \leq \limsup_{T\to\infty} g(T) \quad \forall g \in \mathcal{B}([0,\infty)); \tag{1.37}$$

$$\text{(iv)}\quad \left|\underset{T\to\infty}{\textsc{Lim}}\, g(T)\right| \leq \limsup_{T\to\infty} |g(T)| \leq \sup_{T\geq 0} |g(T)| \quad \forall g \in \mathcal{B}([0,\infty)). \tag{1.38}$$

Property (1.38) follows easily from (1.37). For the proofs of the existence of generalized limits and of property (1.37), see Appendix A.2.

In what follows, we are interested only in generalized limits of time averages – that is, when $g(T)$ is of the form

$$g(T) = \frac{1}{T}\int_0^T f(t)\,dt \tag{1.39}$$

for bounded, measurable functions f. In this case, we have also the following invariance property:

(v) For $f \in L^\infty(0,\infty)$ and $g(T) = (1/T)\int_0^T f(t)\,dt$, we have

$$\operatorname*{LIM}_{T\to\infty} g(T+\tau) = \operatorname*{LIM}_{T\to\infty} g(T) \quad \forall \tau \geq 0. \tag{1.40}$$

A proof that the invariance property (1.40) follows from the other properties can be found in Appendix A.2. One can prove the existence of generalized limits that satisfy (1.40) for any $g \in \mathcal{B}([0,\infty))$; however, we will consider only g in the form of a time average, so we can consider all generalized limits satisfying only the properties (i)–(v) just listed.

Remark 1.5 The generalized limit described in Definition 1.4 allows some freedom in its construction. For instance, given a particular $g_0 \in \mathcal{B}([0,\infty))$ and a sequence $t_j \to \infty$ for which $g_0(t_j)$ converges to a number ℓ, we can construct a generalized limit $\mathrm{LIM}_{T\to\infty}$ for which $\mathrm{LIM}_{T\to\infty}\, g_0 = \ell$. On the other hand, if we require that (1.40) hold for any $g \in \mathcal{B}([0,\infty))$, then $\mathrm{LIM}_{T\to\infty}$ is unique for a class of functions larger than those that admit a classical limit.

2 Invariant Measures and Stationary Statistical Solutions in Dimension 2

In this section we consider only the 2-dimensional case. Our aims in this case are to show that invariant measures can be obtained from time averages and that invariant measures and stationary statistical solutions are actually equivalent concepts. This equivalence in dimension 2 justifies to some extent the use of stationary statistical solutions in the 3-dimensional case considered in the next section, where we cannot work with invariant measures because the semigroup is not well-defined.

Generating invariant measures as limits, in a suitable sense, of time averages of solutions of the 2-dimensional Navier–Stokes equations is accomplished in a way akin to the Birkhoff ergodic theorem and the Krylov–Bogoliubov theory: relating time averages with ensemble (phase-space) averages, but without assuming any ergodicity or chaoticity of the flow. This is made possible by the use of generalized limits (introduced in the Section 1.3), a tool from functional analysis that allows us to bypass the usual ad hoc conjectures of ergodicity.

Note that we do not solve the question of ergodicity. A particular invariant measure may or may not be ergodic, which means that the limit of the time averages may be taken in the classical sense or in the generalized sense. In the generalized sense, this limit might be some weighted average of the accumulation points. For example, for an attracting homoclinic cycle – heteroclinic orbits and corresponding limit points forming a closed curve, with nearby orbits winding around and approaching it – the invariant measure obtained as a generalized limit of the time averages of any solution converging to this attractor is a convex combination of the delta functions supported on each fixed point. It is important to observe that, even in a generalized sense, an invariant measure exists and we can still relate time averages with ensemble averages. In what follows, we use $\text{LIM}_{T\to\infty}$ to denote any such generalized limit.

2.1 Invariant Measures and Stationary Statistical Solutions Generated by Time Averages

The ergodicity problem addresses the question of whether we can relate time averages with ensemble averages via the classical notion of limit. Our aim is to relax the use of the classical limit to allow for generalized limits, obtaining in this way certain measures that we call time-average measures. Indeed, we will show that, for a solution $t \mapsto S(t)\mathbf{u}_0$, the time averages

$$\frac{1}{T}\int_0^T \varphi(S(t)\mathbf{u}_0)\,dt$$

are bounded uniformly, for $T > 0$, for any continuous function $\varphi\colon H \to \mathbb{R}$. This implies that their generalized limit

$$\operatorname*{LIM}_{T\to\infty} \frac{1}{T}\int_0^T \varphi(S(t)\mathbf{u}_0)\,dt$$

exists. We do not know, however, whether this limit holds in the classical sense. Nonetheless, note that a map exists which associates to each function φ the generalized limit just displayed; this map is a linear functional on the space of continuous functions $\varphi\colon H \to \mathbb{R}$. Using the compactness of the semigroup $\{S(t)\}_{t\geq 0}$, we then show that this map can be represented by a measure μ on H in the sense that

$$\operatorname*{LIM}_{T\to\infty} \frac{1}{T}\int_0^T \varphi(S(t)\mathbf{u}_0)\,dt = \int_H \varphi(\mathbf{u})\,d\mu(\mathbf{u}) \tag{2.1}$$

for any φ. This is how time averages can be related to ensemble averages. Note that the existence of a probability distribution μ satisfying (2.1) is based, in the 2- and 3-dimensional cases, on a characterization of measures known in the mathematical literature as the Kakutani–Riesz theorem (see Bourbaki [1969] and Rudin [1987]); this theorem is recalled in Apppendix A, which also includes another technical (mathematical) tool repeatedly used hereafter, the Tietze extension theorem.

We now set the following definition in the 2-dimensional case.

Definition 2.1 *A probability measure μ on H such that*

$$\operatorname*{LIM}_{T\to\infty} \frac{1}{T}\int_0^T \varphi(S(t)\mathbf{u}_0)\,dt = \int_H \varphi(\mathbf{u})\,d\mu(\mathbf{u}) \quad \forall\varphi\in\mathcal{C}(H), \tag{2.2}$$

for some $\mathbf{u}_0 \in H$ *and some generalized limit* $\text{LIM}_{T\to\infty}$, *is called a* time-average measure *of the solution through* $\mathbf{u}_0$.

Note that (2.2) implies that such a time-average measure must be a probability measure – namely, a measure that is positive and of unit mass. These properties are true for the left-hand side of (2.2) and hence must be true for the right-hand side, as well. Here $\mathcal{C}(H)$ denotes the space of continous real-valued (not necessarily bounded) functions on H.

We first show that, for any $\mathbf{u}_0 \in H$, there exists a time-average measure of the solution through $\mathbf{u}_0$. In fact, we show that, for any $\mathbf{u}_0$ and any generalized limit $\text{LIM}_{T\to\infty}$, there exists a time-average measure of the solution through u_0 for which (2.2) holds with this choice of generalized limit. Note that, as mentioned in the Introduction, the time-average measures may not be unique for a given $\mathbf{u}_0 \in H$, since they depend on the choice of a generalized limit. We later prove that any such time-average measure is, in fact, an invariant measure for the semigroup $\{S(t)\}_{t\geq 0}$.

Proposition 2.1 *For any initial condition* $\mathbf{u}_0 \in H$ *and any choice of* $\text{LIM}_{T\to\infty}$, *a time-average measure* μ *of the solution through* $\mathbf{u}_0$ *exists for which* (2.2) *holds with this choice of generalized limit* $\text{LIM}_{T\to\infty}$; *that is, the time averages of the orbit starting at* $\mathbf{u}_0$ *converge to a probability distribution in the sense of this limit.*

Proof. Let $\mathbf{u}_0 \in H$ be fixed. In the 2-dimensional case, we know that the semigroup $\{S(t)\}_{t\geq 0}$ possesses an absorbing set $\mathcal{B}_a \subset H$ that is bounded in $D(A)$ (see Foias and Temam [1979] and the estimates in Section 8 of Chapter II). Therefore, the closure $X = \overline{\mathcal{B}_a}$ of $\mathcal{B}_a$ in H is compact in H. Let $T_a \geq 0$ be such that $S(t)\mathbf{u}_0 \in \mathcal{B}_a \subset X$ for all $t \geq T_a$.

Let now $\varphi \in \mathcal{C}(H)$, the space of continuous real-valued functions on H. Since $t \mapsto \mathbf{u}(t)$ is continuous in H and $S(t)\mathbf{u}_0$ belongs to a compact set in H for $t \geq T_a$, it follows that the whole orbit $\{S(t)\mathbf{u}_0\}_{t\geq 0}$ is compact in H. Hence, $t \to \varphi(S(t)\mathbf{u}_0)$ is continuous and bounded. Thus, the time average

$$\frac{1}{T}\int_0^T \varphi(S(t)\mathbf{u}_0)\, dt$$

makes sense and is bounded for $T > 0$. Its generalized limit

$$L(\varphi) = \underset{T\to\infty}{\text{LIM}} \frac{1}{T}\int_0^T \varphi(S(t)\mathbf{u}_0)\, dt$$

is therefore well-defined for a given $\text{LIM}_{T\to\infty}$. Note that $L = L(\varphi)$ is linear in φ, since $\text{LIM}_{T\to\infty}$ is a linear functional and the integral operation is also linear.

We also observe that, for any given φ, $L(\varphi)$ depends only on the restriction of φ to X. Indeed, if $\tilde{\varphi} \in \mathcal{C}(H)$ is another function such that $\tilde{\varphi}|_X = \varphi|_X$, then $\varphi - \tilde{\varphi} = 0$ on X and

$$\varphi(S(t)\mathbf{u}_0) - \tilde{\varphi}(S(t)\mathbf{u}_0) = 0 \quad \forall t \geq T_a,$$

where T_a, as defined before, is the "absorbing time" after which $S(t)\mathbf{u}_0$ has entered X. Hence,

$$\int_0^T \{\varphi(S(t)\mathbf{u}_0) - \tilde{\varphi}(S(t)\mathbf{u}_0)\}\,dt = \int_0^{T_a} \{\varphi(S(t)\mathbf{u}_0) - \tilde{\varphi}(S(t)\mathbf{u}_0)\}\,dt$$

for all $T \geq T_a$. Therefore,

$$\begin{aligned} L(\varphi - \tilde{\varphi}) &= \operatorname*{LIM}_{T\to\infty} \frac{1}{T}\int_0^T \{\varphi(S(t)\mathbf{u}_0) - \tilde{\varphi}(S(t)\mathbf{u}_0)\}\,dt \\ &= \operatorname*{LIM}_{T\to\infty} \frac{1}{T}\int_0^{T_a} \{\varphi(S(t)\mathbf{u}_0) - \tilde{\varphi}(S(t)\mathbf{u}_0)\}\,dt \\ &= 0. \end{aligned}$$

Thus, $L(\tilde{\varphi}) = L(\varphi)$, which shows that $L(\varphi)$ depends only on the values of φ on X.

A useful consequence of this remark is the following. Let ψ be a continuous real-valued function on X, that is, $\psi \in \mathcal{C}(X)$. Since X is closed in H, we can extend ψ to a continuous function φ in $\mathcal{C}(H)$ such that φ equals ψ on X.[1] Therefore, we can define a functional $G = G(\psi)$ on $\mathcal{C}(X)$ through $G(\psi) = L(\varphi)$, where φ is any extension of ψ to H.

Like L, this functional $G = G(\psi)$ is obviously linear and positive; namely, G is a positive linear functional on $\mathcal{C}(X)$. Because X is compact, it follows from the Kakutani–Riesz representation theorem (see Theorem A.1) that there exists a measure μ on X such that

$$G(\psi) = \int_X \psi(\mathbf{u})\,d\mu(\mathbf{u})$$

for all ψ in $\mathcal{C}(X)$.

The measure μ can be naturally extended to a measure on H by simply setting $\mu(E) = \mu(E \cap X)$ for any measurable subset E of H. Clearly, $\mu(H \setminus X) = 0$. Moreover, for any $\varphi \in \mathcal{C}(H)$ we have

$$\begin{aligned} \operatorname*{LIM}_{T\to\infty} \frac{1}{T}\int_0^T \varphi(S(t)\mathbf{u}_0)\,dt &= L(\varphi) = G(\varphi|_X) \\ &= \int_X \varphi|_X(\mathbf{u})\,d\mu(\mathbf{u}) \\ &= \int_H \varphi(\mathbf{u})\,d\mu(\mathbf{u}), \end{aligned}$$

where the last step follows from the fact that μ is carried by X (i.e., $\mu(H \setminus X) = 0$).

Finally, note that for $\varphi \equiv 1$ we obtain $\mu(H) = \operatorname{LIM}_{T\to\infty} 1 = 1$, so that μ is a positive measure on H of total mass 1; it is a probability measure on H. ■

We have obtained the existence of measures by considering the limits of time averages, and we now proceed to show that these measures are invariant for $\{S(t)\}_{t\geq 0}$. Toward that end, let μ be a time-average measure of the solution through some $\mathbf{u}_0 \in H$. Let φ be a continuous real-valued function on H and let $\tau > 0$. Note that the function $\varphi \circ S(\tau)\colon \mathbf{u} \mapsto \varphi(S(\tau)\mathbf{u})$ is also a continuous real-valued function on H, and it is bounded because the orbits $\{S(t)\mathbf{u}_0\}_{t\geq 0}$ are all bounded in H. Hence, from (2.2) with φ replaced by $\varphi \circ S(\tau)$, we find

[1] We use here the Tietze extension theorem, recalled as Theorem A.7 in Appendix A.

$$\begin{aligned}
\int \varphi(S(\tau)u)\, d\mu(\mathbf{u})
&= \operatorname{LIM}_{T\to\infty} \frac{1}{T}\int_0^T \varphi(S(\tau+t)\mathbf{u}_0)\, dt \\
&= \operatorname{LIM}_{T\to\infty} \frac{1}{T}\int_\tau^{\tau+T} \varphi(S(t)\mathbf{u}_0) dt \\
&= \operatorname{LIM}_{T\to\infty} \left\{ \frac{1}{T}\int_0^T \varphi(S(t)\mathbf{u}_0)\, dt + \frac{1}{T}\int_T^{\tau+T} \varphi(S(t)\mathbf{u}_0)\, dt - \frac{1}{T}\int_0^\tau \varphi(S(t)\mathbf{u}_0)\, dt \right\} \\
&= \operatorname{LIM}_{T\to\infty} \frac{1}{T}\int_0^T \varphi(S(t)\mathbf{u}_0)\, dt \\
&\quad + \operatorname{LIM}_{T\to\infty} \left\{ \frac{1}{T}\left[\int_T^{\tau+T} \varphi(S(t)\mathbf{u}_0)\, dt - \int_0^\tau \varphi(S(t)\mathbf{u}_0)\, dt \right]\right\} \\
&= \int \varphi(\mathbf{u})\, d\mu(\mathbf{u}).
\end{aligned}$$

Thus,

$$\int \varphi(S(\tau)\mathbf{u})\, d\mu(\mathbf{u}) = \int \varphi(\mathbf{u})\, d\mu(\mathbf{u})$$

for any $\tau > 0$ and any $\varphi \in \mathcal{C}(H)$. By a suitable approximation argument and using the density of $\mathcal{C}(H)$ in $L^1(\mu)$, we obtain this same relation for any $\varphi \in L^1(\mu)$. Then, considering $\varphi = \varphi_E$ as the characteristic function of E (i.e., the function equal to 1 on E and to 0 on the complement $H \setminus E$), we deduce that

$$\mu(E) = \mu(S(t)^{-1}E) \quad \forall \tau \geq 0$$

for any measurable set E in H. This proves that $S(t)\mu = \mu$ (see (1.18)), and so μ is an invariant measure for the semigroup $\{S(t)\}_{t\geq 0}$. We have thus obtained the following result.

Theorem 2.1 *Any time-average measure is invariant for the Navier–Stokes semigroup* $\{S(t)\}_{t\geq 0}$.

It remains to show (in the 2-dimensional case) that any invariant measure is a stationary statistical solution – as well as the converse of this claim. The proof, which is quite technical, is left to Appendix B.1. Here, we simply state the result as follows.

Theorem 2.2 *Let μ be a probability measure on H in the 2-dimensional case. Then μ is invariant for $\{S(t)\}_{t\geq 0}$ if and only if it is a stationary statistical solution.*

2.2 Regularity of the Support of an Invariant Measure

We end this section with a result that extends the relation (2.2) between time averages and ensemble averages to functions φ that are defined only on V or on the domain of the Stokes operator A, $D(A) = \{\mathbf{v} \in V;\ A\mathbf{v} \in H\}$. This will enable us to apply formula (2.2) with, for example, the functions $\eta_{j,\mathbf{x}_0}$ defined on $D(A)$ by $\eta_{j,\mathbf{x}_0}(\mathbf{v}) = v_j(\mathbf{x}_0)$ for some component v_j $(j = 1, 2)$ and some point $\mathbf{x}_0$ in Ω. These functions are important in the study of correlation functions and power spectra in the statistical study of turbulence.

First, the following lemma states that an invariant probability measure is carried by an absorbing set of the semigroup $\{S(t)\}_{t\geq 0}$ (see Section III.3 for the concept of an absorbing set).

Lemma 2.1 *Let μ be an invariant probability measure for $\{S(t)\}_{t\geq 0}$, and let $\mathcal{B}_a$ be a bounded set in $D(A)$ that is absorbing for $\{S(t)\}_{t\geq 0}$. Then $\mu(H \setminus \mathcal{B}_a) = 0$.*

Proof. For each $r > 0$, let $B_r = \{\mathbf{u} \in H;\ |\mathbf{u}| \leq r\}$. Since $\mathcal{B}_a$ is absorbing, there exists a time $t_r \geq 0$ such that $S(t)B_r \subset \mathcal{B}_a$ for all $t \geq t_r$. Hence, $B_r \subset S(t_r)^{-1}\mathcal{B}_a$, so that $\mu(S(t_r)^{-1}\mathcal{B}_a) \geq \mu(B_r)$. Since μ is an invariant measure – that is, we have (1.25) for all $t \geq 0$ and all measurable (e.g., closed or open) sets E – we also have $\mu(\mathcal{B}_a) = \mu(S(t_r)^{-1}\mathcal{B}_a)$. Hence,

$$1 \geq \mu(\mathcal{B}_a) = \mu(S(t_r)^{-1}\mathcal{B}_a) \geq \mu(B_r).$$

On the other hand, since $H = \bigcup_{r>0} B_r$ with $B_{r_1} \subset B_{r_2}$ if $r_1 \leq r_2$, then $\mu(B_r) \to \mu(H) = 1$ as $r \to \infty$. Hence $\mu(\mathcal{B}_a) = 1$, which implies that $\mu(H \setminus \mathcal{B}_a) = 0$. ∎

Proposition 2.2 *Let $\mathbf{u}_0 \in D(A) \subset H$ and let μ be a time-average measure of the solution through $\mathbf{u}_0$. Then*

$$\operatorname*{L{\scriptstyle IM}}_{T\to\infty} \frac{1}{T}\int_0^T \varphi(S(t)\mathbf{u}_0)\,dt = \int_H \varphi(\mathbf{u})\,d\mu(\mathbf{u}) \tag{2.3}$$

for any $\varphi \in \mathcal{C}(D(A))$.

Note that the integral on the RHS of (2.3) makes sense for φ in $\mathcal{C}(D(A))$, the space of continuous real-valued (not necessarily bounded) functions defined on $D(A)$, because μ is an invariant measure and we infer from Lemma 2.1 that $\mu(H \setminus D(A)) = 0$.

Proof of Proposition 2.2. Since $\mathbf{u}_0 \in D(A)$, it follows that the whole orbit through $\mathbf{u}_0$ belongs to a closed ball X in $D(A)$, which can be made large enough to contain an absorbing set $\mathcal{B}_a$ bounded in $D(A)$. Note that, since X is a closed ball in $D(A)$, it is also closed in H.[2]

Let X be endowed with the H-norm and let $\varphi \in \mathcal{C}(D(A))$. Since $D(A) \subset H$, it is obvious that $\varphi|_X \in \mathcal{C}(X)$. Thus, since X is closed in H, it can be extended to a function continuous on H with $\tilde{\varphi}|_X = \varphi|_X$.[3] Then, since $S(t)\mathbf{u}_0 \in X$ for all $t \geq 0$, it follows that

[2] Indeed, suppose without loss of generality that $X = \{\mathbf{u} \in D(A);\ |A\mathbf{u}| \leq r\}$. Let $\mathbf{u}_n$ be a sequence in X converging to an element $\mathbf{u}$ in H. Since $A\mathbf{u}_n$ is bounded in H, it has a weakly convergent subsequence in H, say $A\mathbf{u}_{n_j} \rightharpoonup \mathbf{v}$. Since A^{-1} is compact, we have that $\mathbf{u}_n \to A^{-1}\mathbf{v}$ strongly in H. Thus $\mathbf{u} = A^{-1}\mathbf{v}$ belongs to $D(A)$ with $|A\mathbf{u}| = \lim_{n\to\infty}|A\mathbf{u}_n| \leq r$; this shows that $\mathbf{u} \in X$ and hence that X is closed in H.

[3] See footnote 1.

$$\int_0^T \tilde{\varphi}(S(t)\mathbf{u}_0)\,dt = \int_0^T \varphi(S(t)\mathbf{u}_0)\,dt$$

for all $T > 0$. Hence,

$$\underset{T\to\infty}{\text{LIM}} \frac{1}{T}\int_0^T \tilde{\varphi}(S(t)\mathbf{u}_0)\,dt = \underset{T\to\infty}{\text{LIM}} \frac{1}{T}\int_0^T \varphi(S(t)\mathbf{u}_0)\,dt. \tag{2.4}$$

On the other hand, by Lemma 2.1, $\mu(H \setminus \mathcal{B}_a) = 0$; since we have chosen X to contain $\mathcal{B}_a$, we thus find that $\mu(H \setminus X) = 0$. Therefore, the integral

$$\int_H \varphi(\mathbf{u})\,d\mu(\mathbf{u})$$

makes sense and, moreover,

$$\begin{aligned}\int_H \tilde{\varphi}(\mathbf{u})\,d\mu(\mathbf{u}) &= \int_X \tilde{\varphi}(\mathbf{u})\,d\mu(\mathbf{u})\\ &= \int_X \varphi(\mathbf{u})\,d\mu(\mathbf{u})\\ &= \int_H \varphi(\mathbf{u})\,d\mu(\mathbf{u}).\end{aligned} \tag{2.5}$$

The identity (2.3) now follows from (2.2), (2.4), and (2.5). ■

One can similarly prove the following result. It is not a straightforward consequence of Proposition 2.2 because the functions φ are only continuous in the V-topology, which is weaker than the $D(A)$-topology.

Proposition 2.3 *Let $\mathbf{u}_0 \in V$ and let μ be a time-average measure of the solution through $\mathbf{u}_0$. Then*

$$\underset{T\to\infty}{\text{LIM}} \frac{1}{T}\int_0^T \varphi(S(t)\mathbf{u}_0)\,dt = \int_H \varphi(\mathbf{u})\,d\mu(\mathbf{u}) \tag{2.6}$$

for any $\varphi \in \mathcal{C}(V)$.

Remark 2.1 Note that, owing to the regularization property of solutions of the Navier–Stokes equations (they belong to $D(A)$ for $t > 0$ even if u_0 belongs only to H; see equation (A.61) in Chapter II), the requirements that $u_0 \in V$ in Proposition 2.3 and that $u_0 \in D(A)$ in Proposition 2.2 are not very restrictive. By this we mean that, for any u_0 in H, we can shift the solution by a positive time t_0 and consider as the initial condition the vector field $u(t_0) = S(t_0)u_0$, which belongs to $D(A)$. The generalized limit in Proposition 2.2 remains the same, and we can therefore consider functions φ in $\mathcal{C}(D(A))$ or $\mathcal{C}(V)$ for any time-average measure.

3 Stationary Statistical Solutions in Dimension 3

We follow the approach of Section 2 to obtain stationary statistical solutions for the 3-dimensional Navier–Stokes equations as generalized limits of time averages. The proof is essentially the same; the main difference is that we do not have a priori estimates guaranteeing the regularity of the solutions for all time $t > 0$, and for this reason we must work with the weak topology of H to obtain a compact absorbing set. A compact absorbing set is needed to allow our use of the Kakutani–Riesz representation theorem[4] to deduce the existence of a measure associated with a generalized limit of time averages. The other major difference is the lack of uniqueness of the weak solutions, which prevents us from working with invariant measures. That, in turn, is one of the reasons for introducing the stationary statistical solutions.

3.1 Stationary Statistical Solutions Generated by Time Averages

We recall that H_w denotes the space H endowed with its weak topology. In the 3-dimensional case, the appropriate definition of a time-average measure is as follows.

Definition 3.1 *A probability measure μ on H such that*

$$\operatorname{LIM}_{T\to\infty} \frac{1}{T}\int_0^T \varphi(\mathbf{u}(t))\,dt = \int_H \varphi(\mathbf{u})\,d\mu(\mathbf{u}) \tag{3.1}$$

for any $\varphi \in \mathcal{C}(H_w)$, for a given weak solution $\mathbf{u} = \mathbf{u}(t)$ of the NSE on $[0, \infty)$, is called a time-average measure *of the solution* $\mathbf{u}$.

The existence of time-average measures is assured by the following result.

Proposition 3.1 *For any weak solution $\mathbf{u}$ on $[0, \infty)$ and any given generalized limit* $\operatorname{LIM}_{T\to\infty}$, *there exists a time-average measure μ of $\mathbf{u}$ satisfying* (3.1) *with this* $\operatorname{LIM}_{T\to\infty}$.

Proof. Let $\mathbf{u}_0 = \mathbf{u}(0)$. Consider

$$K_w = \{\mathbf{v} \in H;\ |\mathbf{v}|^2 \le |\mathbf{u}_0|^2 + |\mathbf{f}|^2/\nu^2\lambda_1^2\}$$

endowed with the weak topology of H. Since K_w is bounded, closed, and convex in H, it follows that K_w is compact. Also, from the a priori estimate (II.A.41), we see that $\mathbf{u}(t) \in K_w$ for all $t \ge 0$.

Let now $\psi \in \mathcal{C}(K_w)$. Since $t \mapsto \mathbf{u}(t)$ is weakly continuous and $\mathbf{u}(t)$ belongs to the compact set K_w in H_w, the function $t \to \psi(\mathbf{u}(t))$ is well-defined, continuous, and bounded. Thus,

$$\frac{1}{T}\int_0^T \psi(\mathbf{u}(t))\,dt$$

makes sense and is bounded for $T \ge 0$. Therefore, its generalized limit

$$G(\psi) = \operatorname{LIM}_{T\to\infty} \frac{1}{T}\int_0^T \psi(\mathbf{u}(t))\,dt$$

is well-defined. Note that $G = G(\psi)$ is linear in ψ, since $\operatorname{LIM}_{T\to\infty}$ is a linear functional. Note also that $G = G(\psi)$ is positive. Hence, G is a positive linear functional

[4] See Appendix A and the comment before Definition 2.1.

on $\mathcal{C}(K_w)$. Therefore, by the the Kakutani–Riesz representation theorem (explained in Appendix A), there exists a measure μ on K_w such that

$$G(\psi) = \int_{K_w} \psi(\mathbf{u})\, d\mu(\mathbf{u}) \quad \forall \psi \in \mathcal{C}(K_w).$$

The measure μ can be easily extended to a measure on H by simply setting $\mu(E) = \mu(E \cap K_w)$ for any measurable set E in H. Clearly, $\mu(H \setminus K_w) = 0$.

Now, if $\varphi \in \mathcal{C}(H_w)$, then obviously $\varphi|_{K_w} \in \mathcal{C}(K_w)$. Therefore,

$$\begin{aligned}
\underset{T\to\infty}{\text{LIM}} \frac{1}{T}\int_0^T \varphi(\mathbf{u}(t))\, dt &= \underset{T\to\infty}{\text{LIM}} \frac{1}{T}\int_0^T \varphi|_{K_w}(\mathbf{u}(t))\, dt \\
&= G(\varphi|_{K_w}) = \int_{K_w} \varphi|_{K_w}(\mathbf{u})\, d\mu(\mathbf{u}) \\
&= \int_H \varphi(\mathbf{u})\, d\mu(\mathbf{u}),
\end{aligned}$$

where the last equality follows from the fact that $\mu(H \setminus K_w) = 0$.

Finally, if we take $\varphi \equiv 1$ then $\mu(H) = \text{LIM}_{T\to\infty} 1 = 1$, so that μ is a probability measure on H. ■

The proof of Proposition 3.1 enables a straightforward deduction of the following corollary, which is used in the proof of Theorem 4.2.

Corollary 3.1 *Let* $\mathbf{u}$ *be a weak solution of the NSE defined for all* $t \geq 0$*, and let* μ *be a time-average measure of* $\mathbf{u}$*. Set* $\mathbf{u}_0 = \mathbf{u}(0)$ *and consider the set*

$$K_w = \{\mathbf{v} \in H;\ |\mathbf{v}|^2 \leq |\mathbf{u}_0|^2 + |\mathbf{f}|^2/\nu^2\lambda_1^2\} \tag{3.2}$$

endowed with the weak topology of H*. Then* (3.1) *holds for all* φ*, which restricted to* K_w *belongs to* $\mathcal{C}(K_w)$*. In particular,* (3.1) *holds for all* $\varphi\colon H \to \mathbb{R}$ *that are weakly continuous on bounded subsets of* H *and all* $\varphi \in \mathcal{C}(D(A^{-1/2}))$.

Once we have obtained the existence of time-average measures, we may state the following important result.

Theorem 3.1 *Any time-average measure is a stationary statistical solution.*

The proof that the time-average measures are stationary statistical solutions is much more involved than in the 2-dimensional case and is given in Appendix B.2. Here we shall simply illustrate a few ideas involved in the proof. We need to prove properties (1.29), (1.30), and (1.31). The idea of the proof of (1.30) is simply to show that

$$\begin{aligned}
\int (\mathbf{F}(\mathbf{u}), \Phi'(\mathbf{u}))\, d\mu(\mathbf{u}) &= \underset{T\to\infty}{\text{LIM}} \frac{1}{T}\int_0^T (\mathbf{F}(\mathbf{u}(t)), \Phi'(\mathbf{u}(t)))\, dt \\
&= \underset{T\to\infty}{\text{LIM}} \frac{1}{T}\int_0^T \left(\frac{d}{dt}\Phi(\mathbf{u}(t))\right) dt \\
&= \underset{T\to\infty}{\text{LIM}} \frac{1}{T}\big(\Phi(\mathbf{u}(T)) - \Phi(\mathbf{u}(0))\big) \\
&= 0 \quad (\text{since } t \to \Phi(\mathbf{u}(t)) \text{ is bounded});
\end{aligned}$$

however, some of these steps – although intuitively valid – are not actually legitimate mathematically. For instance, depending on Φ, the function $\mathbf{u} \mapsto (\mathbf{F}(\mathbf{u}), \Phi'(\mathbf{u}))$ may not be continuous on H, so that the first step does not follow directly from μ being a time-average measure (see (3.1)). For the next two steps, the difficulty is that the map $t \mapsto \Phi(\mathbf{u}(t))$ might not be differentiable. What we do then is approximate $\Phi = \Phi(\mathbf{u})$ by $\Phi(P_m \mathbf{u})$ and $\mathbf{F} = \mathbf{F}(\mathbf{u})$ by some suitable $\mathbf{F}_k(\mathbf{u})$ using also a Galerkin projector P_k. Then, letting $m, k \to \infty$ in the appropriate order, we prove (1.30).

Properties (1.29) and (1.31) are treated similarly. In order to illustrate the technique of using the Galerkin projectors we prove (1.29), which is simpler. We leave (1.30) and (1.31) to Appendix B.

Let μ be a time-average measure of a weak solution $\mathbf{u} = \mathbf{u}(t)$ with $\mathbf{u}(0) = \bar{\mathbf{u}}_0$. Let $\Phi(\mathbf{u}) = \|P_m \mathbf{u}\|^2$ for $m \in \mathbb{N}$. Thus, $\Phi \in \mathcal{C}(H_w)$ and, from (3.1),

$$\int \|P_m \mathbf{u}\|^2 \, d\mu(\mathbf{u}) = \operatorname*{LIM}_{T\to\infty} \frac{1}{T} \int_0^T \|P_m \mathbf{u}(t)\|^2 \, dt. \tag{3.3}$$

From the estimate (II.A.40), we have

$$\frac{1}{T} \int_0^T \|\mathbf{u}(t)\|^2 \, dt \le \frac{1}{\nu T} |\mathbf{u}_0|^2 + \frac{1}{\nu^2 \lambda_1} |\mathbf{f}|^2. \tag{3.4}$$

Since $\|P_m \mathbf{u}\| \le \|\mathbf{u}\|$, it follows that

$$\frac{1}{T} \int_0^T \|P_m \mathbf{u}(t)\|^2 \, dt \le \frac{1}{T} \int_0^T \|\mathbf{u}(t)\|^2 \, dt \le \frac{1}{\nu T} |\mathbf{u}_0|^2 + \frac{1}{\nu^2 \lambda_1} |\mathbf{f}|^2.$$

Then, by the positivity of the generalized limit $\operatorname{LIM}_{T\to\infty}$, we obtain

$$\begin{aligned} \operatorname*{LIM}_{T\to\infty} \frac{1}{T} \int_0^T \|P_m \mathbf{u}(t)\|^2 \, dt &\le \operatorname*{LIM}_{T\to\infty} \left\{ \frac{1}{\nu T} |\mathbf{u}_0|^2 + \frac{1}{\nu^2 \lambda_1} |\mathbf{f}|^2 \right\} \\ &= \frac{1}{\nu^2 \lambda_1} |\mathbf{f}|^2; \end{aligned}$$

together with (3.3), this implies

$$\int \|P_m \mathbf{u}\|^2 \, d\mu(\mathbf{u}) \le \frac{1}{\nu^2 \lambda_1} |\mathbf{f}|^2.$$

Because $\|P_m \mathbf{u}\|^2 \nearrow \|\mathbf{u}\|^2$ as $m \to \infty$ pointwise in H (with $\|\mathbf{u}\| = \infty$ on $H \setminus V$), it follows by the monotone convergence theorem (see Rudin [1987]) that

$$\int \|\mathbf{u}\|^2 \, d\mu(\mathbf{u}) = \lim_{m\to\infty} \int \|P_m \mathbf{u}\|^2 \, d\mu(\mathbf{u}) \le \frac{1}{\nu^2 \lambda_1} |\mathbf{f}|^2 < \infty.$$

This proves (1.29). As already mentioned, the remaining part of the proof of Theorem 3.1 is left to Appendix B.

The estimate (3.4) has another consequence. It allows us to extend the relation (3.1) between time averages and ensemble averages to more general functions φ. Indeed, we have the following result.

Proposition 3.2 *Let* $\mathbf{u}$ *be a weak solution of the NSE defined for all* $t \ge 0$, *and let* μ *be a time-average measure of* $\mathbf{u}$. *Let*

$$K_0 = \{\mathbf{v} \in H;\ |\mathbf{v}|^2 \le |\mathbf{u}_0|^2 + |\mathbf{f}|^2/\nu^2\lambda_1^2\}. \tag{3.5}$$

Then (3.1) *holds for all* $\varphi\colon H \to \mathbb{R}$ *that are Lipschitz on* K_0 *– that is, such that there exists a constant* $L > 0$ *with*

$$|\varphi(\mathbf{u}) - \varphi(\mathbf{v})| \le L|\mathbf{u} - \mathbf{v}|$$

for all $\mathbf{u}, \mathbf{v}$ *in* K_0.

An important example of a function φ allowed by Proposition 3.2 is $\varphi(\mathbf{u}) = |\mathbf{u}|^2$, which is associated with the kinetic energy of a flow. The proof of this proposition is given in Appendix B.3.

3.2 Regularity of the Support of Time-Average Measures

We show in Section 4 that any time-average measure is carried by the weak global attractor in the 3-dimensional case. However, unlike the global attractor in the 2-dimensional case, it is not known whether this weak global attractor is regular – that is, if it is included in $D(A)$ or even in V. For this reason, the result we present next is a bit surprising; it states that any time-average measure μ is nevertheless carried by $D(A)$ (i.e., $\mu(H \setminus D(A)) = 0$). This is an important regularity result, since it allows us to relate time averages to ensemble averages via (3.1) with functions φ that are only defined and weakly continuous in $D(A)$, such as the functions $\eta_{j,\mathbf{x}_0}(\mathbf{u}) = u_j(\mathbf{x}_0)$ used in the pointwise correlation functions in the statistical study of turbulence.

Proposition 3.3 *Let* μ *be a time-average measure of a weak solution of the NSE defined for all* $t \ge 0$. *Then* μ *is carried by* $D(A)$; *that is,* $\mu(H \setminus D(A)) = 0$.

Proof. Let $\mathbf{u} = \mathbf{u}(t)$ be a weak solution on $[0, \infty)$, and let μ be a time-average measure of $\mathbf{u}$. Our aim is to show that

$$\int_H |A\mathbf{u}|^{2/3}\, d\mu(\mathbf{u}) < \infty, \tag{3.6}$$

from which we can deduce that $\mu(H \setminus D(A)) = 0$. The proof of (3.6) is based on the formal relation between the ensemble average

$$\int_H |A\mathbf{u}|^{2/3}\, d\mu(\mathbf{u}) \tag{3.7}$$

and the time average

$$\operatorname*{LIM}_{T\to\infty} \frac{1}{T}\int_0^T |A\mathbf{u}(t)|^{2/3}\, dt, \tag{3.8}$$

and on a priori estimates for the NSE showing that the generalized limit (3.8) is finite. However, we do not prove that (3.7) equals (3.8). That equality does not follow from the fact that μ is a time-average measure, because the function $\mathbf{u} \to |A\mathbf{u}|^{2/3}$ is not weakly continuous on H (see Proposition 3.1 and Corollary 3.1). What we do instead is approximate this function by $|AP_m\mathbf{u}|^{2/3}$ for $m \in \mathbb{N}$, which is weakly continuous on H, so that

$$\int_H |AP_m\mathbf{u}|^{2/3}\, d\mu(\mathbf{u}) = \operatorname*{LIM}_{T\to\infty} \frac{1}{T}\int_0^T |AP_m\mathbf{u}(t)|^{2/3}\, dt. \tag{3.9}$$

Then, since $|AP_m\mathbf{u}| \nearrow |A\mathbf{u}|$ pointwise on H (with $|A\mathbf{u}| = \infty$ on $H \setminus D(A)$), it follows by the monotone convergence theorem (see Rudin [1987]) that

$$\int_H |A\mathbf{u}|^{2/3}\, d\mu(\mathbf{u}) = \lim_{m\to\infty} \int_H |AP_m\mathbf{u}|^{2/3}\, d\mu(\mathbf{u}). \tag{3.10}$$

Therefore, we need only show that

$$\operatorname*{LIM}_{T\to\infty} \frac{1}{T}\int_0^T |AP_m\mathbf{u}(t)|^{2/3}\, dt$$

remains bounded when $m \to \infty$. For that purpose, let $T > 0$. This boundedness follows from the a priori estimate (see (A.43))

$$\int_0^T |A\mathbf{u}(t)|^{2/3}\, dt \le c_3(1+T) \quad \forall T \ge 0 \tag{3.11}$$

for some constant $c_3 \ge 0$ independent of T. Indeed, from (3.11) and using that $|P_m\mathbf{u}| \le |\mathbf{u}|$ for all $\mathbf{u}$ in H and all $m \in \mathbb{N}$, we find that

$$\frac{1}{T}\int_0^T |P_m A\mathbf{u}(t)|^{2/3}\, dt \le c_3\frac{1+T}{T} \le 2c_3 \quad \forall T \ge 1,$$

where c_3 obviously does not depend on m either. Therefore, we deduce that

$$\limsup_{T\to\infty} \frac{1}{T}\int_0^T |P_m A\mathbf{u}(t)|^{2/3}\, dt \le 2c_3.$$

By property (1.37) of the generalized limit, this gives us the bound

$$\operatorname*{LIM}_{T\to\infty} \frac{1}{T}\int_0^T |P_m A\mathbf{u}(t)|^{2/3}\, dt \le \limsup_{T\to\infty} \frac{1}{T}\int_0^T |P_m A\mathbf{u}(t)|^{2/3}\, dt \le 2c_3. \tag{3.12}$$

From (3.12), (3.9), and (3.10), we conclude that

$$\int_H |A\mathbf{u}|^{2/3}\, d\mu(\mathbf{u}) \le 2c_3 < \infty.$$

Hence $\mu(H \setminus D(A)) = 0$, and the proof is complete. ■

4 Attractors and Stationary Statistical Solutions

Stationary statistical solutions, invariant measures, and global attractors are all related to the asymptotic behavior of a system. Stationary statistical solutions and invariant measures are measures on the phase space, whereas global attractors are sets in the phase space. Our aim in this section is to relate these objects by showing that, in dimension 2, any stationary statistical solution (or, equivalently, any invariant measure) has its support included in the global attractor; in dimension 3, all the stationary statistical solutions generated by time averages have their support included in the (weak) global attractor.

This result is expected, of course – at least in the 2-dimensional case, where we have a well-defined semigroup. The support of an invariant measure is made only of points that are, in particular, nonwandering (see Remark 1.4), and there are only wandering points outside the global attractor. In principle, however, the global attractor may also contain some wandering points (e.g., some heteroclinic orbits); hence, in the absence of ergodicity, we may have several different stationary statistical solutions supported on different parts of the global attractor and with their support properly included in the global attractor. As we said in the Introduction to this chapter, the existence of multiple attractors – each with its own basin of attraction – or the existence of multiple stationary statistical solutions is a theoretical (mathematical) possibility that we cannot disprove at our present level of understanding. However, as discussed in an extensive literature that is beyond the scope of this monograph, many dynamical systems are endowed with a unique invariant measure – that is, the Sinai–Ruelle–Bowen (SRB) measure – suited for computing physically significant moments. For the sake of completeness, we mention here that such SRB measures may be computed by means of a variational principle, maximizing the following functional over all invariant measures ρ:

$$h(\rho) = \int \sum_{\lambda_i > 0} \lambda_i(\mathbf{u})\rho(\mathbf{u}),$$

where $h(\rho)$ is the measure-theoretical entropy of a map with an invariant measure ρ and the $\lambda_i(\mathbf{u})$ are the Lyapunov exponents for trajectories starting at $\mathbf{u}$. Of course, it is not known whether the dynamical system underlying the Navier–Stokes equations (in two *or* three dimensions) satisfies the conditions sufficient for the existence of the corresponding SRB measure.

4.1 The 2-Dimensional Case: The Support of an Invariant Measure Is Included in the Global Attractor

We start with the 2-dimensional case, which is simpler. Let $\mathcal{B}_a$ be a bounded subset of $D(A)$ that is absorbing for the semigroup $\{S(t)\}_{t\geq 0}$ associated to the 2-dimensional NSE. In other words, $\mathcal{B}_a$ is such that, for any given bounded set B in H, there exists a time T such that $S(t)B \subset \mathcal{B}_a$ for all $t \geq T$. Recall that in Chapter III we introduced the global attractor for $\{S(t)\}_{t\geq 0}$, defined as the ω-limit set of $\mathcal{B}_a$:

$$\mathcal{A} = \bigcap_{t\geq 0} \overline{\bigcup_{s\geq t} S(s)\mathcal{B}_a}, \tag{4.1}$$

where the closure (indicated by the overline) is taken with respect to the H-topology.

Before proceeding, note the following alternate expression of the global attractor.

Lemma 4.1 *Let $\{t_k\}_{k\in\mathbb{N}}$ be a sequence of positive numbers such that $t_k \to \infty$ as $k \to \infty$. Then the global attractor $\mathcal{A}$ defined by* (4.1) *is also equal to the set*

$$\bigcap_{k\in\mathbb{N}} S(t_k)\mathcal{B}_a.$$

Proof. Let us temporarily denote the latter set by $\mathcal{A}'$; we want to show that $\mathcal{A}' = \mathcal{A}$. We have $\mathcal{A} \subset \mathcal{B}_a$ and so, using the invariance of $\mathcal{A}$, we see that $\mathcal{A} = S(t_k)\mathcal{A} \subset S(t_k)\mathcal{B}_a$ for all $k \in \mathbb{N}$. Thus, $\mathcal{A} \subset \mathcal{A}'$. On the other hand,

$$S(t_k)\mathcal{B}_a \subset \overline{\bigcup_{s\geq t} S(s)\mathcal{B}_a} \quad \forall t \ (0 \leq t \leq t_k),\ \forall k \in \mathbb{N}.$$

Hence, since $t_k \to \infty$,

$$\mathcal{A}' = \bigcap_{k\in\mathbb{N}} S(t_k)\mathcal{B}_a \subset \bigcap_{k\in\mathbb{N}} \bigcap_{0\leq t\leq t_k} \overline{\bigcup_{s\geq t} S(s)\mathcal{B}_a}$$
$$= \bigcap_{t\geq 0} \overline{\bigcup_{s\geq t} S(s)\mathcal{B}_a} = \mathcal{A}.$$

Therefore, $\mathcal{A} = \mathcal{A}'$. ■

Theorem 4.1 *In dimension 2, any probability measure μ that is invariant for the Navier–Stokes equations (i.e., for the semigroup $\{S(t)\}_{t\geq 0}$) has its support included in the global attractor $\mathcal{A}$.*

Proof. Let $T_a > 0$ be such that $S(t)\mathcal{B}_a \subset \mathcal{B}_a$ for all $t \geq T_a$. Hence, the sequence of sets $\{S(kT_a)\mathcal{B}_a\}_{k\in\mathbb{N}}$ is decreasing. Therefore, from Lemma 4.1 with $t_k = kT_a$, we see that

$$\mu(\mathcal{A}) = \lim_{k\to\infty} \mu(S(kT_a)\mathcal{B}_a). \tag{4.2}$$

Now, since $\mathcal{B}_a$ is absorbing, the set $S(kT_a)\mathcal{B}_a$ is also absorbing for each $k \in \mathbb{N}$. Then, by Lemma 2.1, it follows that $\mu(S(kT_a)\mathcal{B}_a) = 1$ for all $k \in \mathbb{N}$. Hence, using (4.2), we find that $\mu(\mathcal{A}) = 1$. Since $\mathcal{A}$ is closed in H and since $\operatorname{supp}\mu$ is the smallest closed set of full μ-measure, we conclude that $\operatorname{supp}\mu \subset \mathcal{A}$. ■

4.2 The 3-Dimensional Case: The Support of a Time-Average Measure Is Included in the Weak Global Attractor

Consider now the 3-dimensional case. Recall that in Section III.3 we introduced the weak global attractor, defined by

$$\mathcal{A}_w = \big\{\mathbf{u}_0 \in H;\ \text{there exists a weak solution } \mathbf{u} = \mathbf{u}(t) \text{ of the NSE, defined on all of } \mathbb{R}, \text{ uniformly bounded in } H, \text{ and such that } \mathbf{u}(0) = \mathbf{u}_0\big\}. \tag{4.3}$$

Recall also that $\mathcal{A}_w$ has the properties that

$$\mathcal{A}_w \subset \left\{\mathbf{u} \in H;\ |\mathbf{u}| \leq \frac{1}{\nu\lambda_1}|\mathbf{f}|\right\}, \tag{4.4}$$

so that $\mathcal{A}_w$ is bounded in H; it is invariant in the following sense:

> If $\mathbf{u}_0$ belongs to $\mathcal{A}_w$ and $\mathbf{u} = \mathbf{u}(t)$ is a global weak solution with $\mathbf{u}(0) = \mathbf{u}_0$ and uniformly bounded in H, as required in the definition of $\mathcal{A}$, then the velocity field $\mathbf{u}(t)$ also belongs to $\mathcal{A}_w$ at any other time $t \in \mathbb{R}$. (4.5)

The attractor $\mathcal{A}_w$ is weakly attracting in the sense that, for every weak solution $\mathbf{u}$ on $[0, \infty)$, we have

$$\mathbf{u}(t) \to \mathcal{A}_w \text{ in } H_w \text{ as } t \to \infty \tag{4.6}$$

and

$$\mathcal{A}_w \text{ is compact for the weak topology of } H. \tag{4.7}$$

We next prove that any time-average measure has its support included in the weak global attractor $\mathcal{A}_w$.

Theorem 4.2 *In dimension* 3, *any time-average measure* μ *of the NSE has its support included in the weak global attractor* $\mathcal{A}_w$.

Proof. Let u be a weak solution on $[0, \infty)$ associated with the time-average measure μ, and let $\mathbf{u}_0 = \mathbf{u}(0)$. Consider the set

$$K_w = \{\mathbf{v} \in H;\ |\mathbf{v}|^2 \le |\mathbf{u}_0|^2 + |\mathbf{f}|^2/\nu^2\lambda_1^2\}$$

endowed with the weak topology of H. Since K_w is bounded, closed, and convex in H, it follows that K_w is compact. Also, it follows from the a priori estimate (II.A.41) that $\mathbf{u}(t) \in K_w$ for all $t \ge 0$ and that, from (4.4), $\mathcal{A}_w \subset K_w$.

Let $\mathcal{O} \supset \mathcal{A}_w$ be weakly open in H. Then $\mathcal{O} \cap K_w$ is an open neighborhood of $\mathcal{A}_w$ relative to K_w. Because $\mathcal{A}_w$ is closed in K_w, we can separate $\mathcal{A}_w$ and the relatively closed set $K_w \setminus \mathcal{O}$ by open sets in K_w.[5] Hence, there exists another weakly open set $\mathcal{O}' \supset \mathcal{A}_w$ in H such that its weak closure $\overline{\mathcal{O}'}^w \cap K_w$ in K_w is contained in $\mathcal{O} \cap K_w$.

Now $\overline{\mathcal{O}'}^w \cap K_w$ and $K_w \setminus \mathcal{O}$ are disjoint closed subsets in K_w. Hence there exists a continuous function[6] $\varphi \in \mathcal{C}(K_w)$ with $0 \le \varphi \le 1$ and such that $\varphi(\mathbf{v}) = 1$ for $v \in \overline{\mathcal{O}'}^w \cap K_w$ and $\varphi(\mathbf{v}) = 0$ for $v \in K_w \setminus \mathcal{O}$. Thus, since $\mu(H \setminus K_w) = 0$ (see (1.34)), $\varphi \le 1$, and φ vanishes outside $\mathcal{O} \cap K_w$, we obtain

$$\mu(\mathcal{O}) = \mu(\mathcal{O} \cap K_w) = \int_{\mathcal{O} \cap K_w} d\mu(\mathbf{u}) \ge \int_{\mathcal{O} \cap K_w} \varphi(\mathbf{u})\, d\mu(\mathbf{u}) = \int \varphi(\mathbf{u})\, d\mu(\mathbf{u}).$$

Since $\varphi \in \mathcal{C}(K_w)$ we find, thanks to Corollary 3.1,

$$\mu(\mathcal{O}) \ge \int \varphi(\mathbf{u})\, d\mu(\mathbf{u}) = \operatorname*{LIM}_{T \to \infty} \frac{1}{T} \int_0^T \varphi(\mathbf{u}(t))\, dt.$$

By the weak attraction property (4.6) of $\mathcal{A}_w$, we now have $\mathbf{u}(t) \in \mathcal{O}' \cap K_w$ for all $t \ge T_0$ for some T_0 sufficiently large. Then, since $\varphi = 1$ on $\mathcal{O}' \cap K_w$, we find

$$\begin{aligned}
\mu(\mathcal{O}) &\ge \operatorname*{LIM}_{T \to \infty} \frac{1}{T} \int_0^T \varphi(\mathbf{u}(t))\, dt \\
&= \operatorname*{LIM}_{T \to \infty} \frac{1}{T} \left[\int_0^{T_0} \varphi(\mathbf{u}(t))\, dt + \int_{T_0}^T \varphi(\mathbf{u}(t))\, dt \right] \\
&= \operatorname*{LIM}_{T \to \infty} \left[\frac{1}{T} \int_0^{T_0} \varphi(\mathbf{u}(t))\, dt + \frac{T - T_0}{T} \right] = 1.
\end{aligned}$$

[5] We use here that K_w is a compact Hausdorff space and hence a normal topological space, so that disjoint closed sets can be separated by disjoint open sets; see e.g. Dunford and Schwartz [1958].

[6] We use here Urysohn's lemma; see Theorem A.6.

Hence, $\mu(\mathcal{O}) = 1$. Thus, we have shown that

$$\mu(\mathcal{O}) = 1 \quad \forall \mathcal{O} \supset \mathcal{A}_w \text{ weakly open in } H.$$

Since the measure μ is regular in the sense given in (A.4), we obtain $\mu(\mathcal{A}_w) = 1$. Since $\mathcal{A}_w$ is closed (in both the weak and the strong topologies in H), this implies that $\operatorname{supp}\mu \subset \mathcal{A}_w$, which completes the proof. ■

5 Average Transfer of Energy and the Cascades in Turbulent Flows

One of the aspects of the Kolmogorov theory of turbulence is that in 3-dimensional, fully developed turbulence, energy within the inertial range is transferred from large to small eddies. Another aspect is the energy cascade, where energy is transferred to smaller and smaller eddies – nearly without viscous dissipation – until eddies become sufficiently small and reach the dissipation range. In this section, our aim is to put these notions on rigorous mathematical grounds as consequences of the properties of the Navier–Stokes equations.

In turbulent flows, it is understood that the energy introduced at the low end of the spectrum is transferred in average to the high-end (high-wavenumber) portion of the spectrum, where it is eventually dissipated by molecular viscosity. In a statistical steady state, conservation of energy demands that the dissipation rate match (on average) the rate at which energy is supplied to the flow. When there is extensive separation between the portion of the spectrum where molecular dissipation dominates and the portion where energy is injected into the flow, we have a region of the spectrum in which the effects of small finite viscosity are negligible. In that spectral region, the energy dissipation rate is governed by the inertial term acting to break up flow velocity structures of any length into smaller ones, evoking the image of an energy cascade (Onsager [1945]). Although this picture originally evolved in the context of 3-dimensional turbulence, it has also prevailed in studies of 2-dimensional turbulence, where of necessity (owing to enstrophy conservation) the cascading entity is predominantly enstrophy rather than energy. In both cases, those parts of the spectra where cascades occur are referred to as inertial ranges. These concepts of turbulence have been based primarily on empirical and heuristic grounds. As we said before, we now aim to put these notions on rigorous mathematical grounds and to relate them to the study of statistical solutions of the Navier–Stokes equations.

We consider the time-average measures constructed earlier in this chapter and show that, on average with respect to those measures, the energy is indeed transferred from large to small eddies in the range beyond that where energy is injected into the system. This is obtained for the NSE with either periodic boundary conditions that have vanishing space average or with no-slip boundary conditions. We also show that there is an inverse transfer of energy to lower modes, in the portion of the spectrum below the injection of energy.

These results hold independently of the existence of the inertial range. In case the inertial range exists, the average net transfer of energy to higher modes extends indefinitely, beyond the inertial range and into the dissipative range. What characterizes the inertial range (when it exists) is that, within it, the average net transfer of energy occurs with a certain specific and nearly constant rate, leading to what is called the energy cascade, an issue that we also address in this section. These results were announced in Foias, Manley, Rosa, and Temam [2001b,c]. It is particularly interesting to observe that, in the 3-dimensional case, we can prove these results rigorously, using only the current state of the mathematical theory of the Navier–Stokes equations. In particular, as discussed in Section II.7, the energy equation is not available and we are content with the energy inequality (7.7) of Chapter II (see (5.9) in this chapter).

We also address 2-dimensional turbulence. We show, in particular, the direct and inverse transfers of energy and enstrophy within appropriate ranges of length scales. The direct enstrophy cascade and the inverse energy cascade in this case are also addressed. We should mention here a result by Constantin, Foias, and Manley [1994] asserting that, in order for the Kraichnan inertial range spectrum to be sustained within a statistically stationary turbulent flow, the forcing term must have at least two eigenmodes: one acting as a power source and the other as a power sink. A single mode is not sufficient to sustain such a regime. Most of the results presented in this section have been arrived at previously in the general literature addressing turbulent flows on heuristic and/or intuitive grounds.

5.1 Energy Transfer and the Cascade in 3-Dimensional Turbulence

We consider both the no-slip and the periodic boundary conditions. In the case of periodic boundary conditions, for the sake of simplicity we assume that the space average of the velocity field vanishes.

As we have seen in Section II.6, we can expand each velocity field $\mathbf{u}$ in H in the form

$$\mathbf{u} = \sum_{k=1}^{\infty} \hat{u}_k \mathbf{w}_k,$$

where the $\mathbf{w}_k$ are the eigenfunctions of the Stokes operator. Each eigenfunction $\mathbf{w}_k$ is associated with an eigenvalue λ_k, which has the dimension of $(\text{length})^{-2}$. The quantity $\ell_k = 1/\sqrt{\lambda_k}$ reflects (in some sense) the characteristic wavelength or size associated with $\mathbf{w}_k$, while $\kappa_k = 1/\ell_k$ represents the spatial frequency, or wavenumber, of the corresponding eigenfunction.

Because the family $\{\mathbf{w}_k\}_k$ of eigenfunctions of the Stokes operator is an orthonormal family in the space H, the total kinetic energy of the flow is given by

$$e(\mathbf{u}) = \frac{1}{2}|\mathbf{u}|^2 = \frac{1}{2}\sum_{k=1}^{\infty} |\hat{u}_k|^2.$$

Similarly, the total enstrophy of the flow is given by

$$E(\mathbf{u}) = \|\mathbf{u}\|^2 = \sum_{k=1}^{\infty} \lambda_k |\hat{u}_k|^2 = \sum_{k=1}^{\infty} \kappa_k^2 |\hat{u}_k|^2.$$

In general, for $1 \leq m' \leq m'' \leq \infty$, we denote the constituent of $\mathbf{u}$ with wavenumbers between $\kappa_{m'} = \sqrt{\lambda_{m'}}$ and $\kappa_{m''} = \sqrt{\lambda_{m''}}$ by

$$\mathbf{u}_{\kappa_{m'},\kappa_{m''}} = \sum_{k=m'}^{m''} \hat{u}_k \mathbf{w}_k. \tag{5.1}$$

The corresponding energy and enstrophy are respectively

$$e(\mathbf{u}_{\kappa_{m'},\kappa_{m''}}) = \frac{1}{2} |\mathbf{u}_{\kappa_{m'},\kappa_{m''}}|^2 = \frac{1}{2} \sum_{k=m'}^{m''} |\hat{u}_k|^2$$

and

$$E(\mathbf{u}_{\kappa_{m'},\kappa_{m''}}) = \|\mathbf{u}_{\kappa_{m'},\kappa_{m''}}\|^2 = \sum_{k=m'}^{m''} \lambda_k |\hat{u}_k|^2 = \sum_{k=m'}^{m''} \kappa_k^2 |\hat{u}_k|^2.$$

Note that $\mathbf{u}_{\kappa_1,\infty} = \mathbf{u}$.

As for the forcing term $\mathbf{f}$, we assume it to be time-independent and to contain only a finite number of modes – say, between $\kappa_{\underline{m}}$ and $\kappa_{\bar{m}}$ for some $1 \leq \underline{m} \leq \bar{m} < \infty$. More precisely, using the notation just introduced, we assume that

$$\mathbf{f} = \mathbf{f}_{\kappa_{\underline{m}},\kappa_{\bar{m}}}. \tag{5.2}$$

As usual, we denote by P_m the spectral projector onto the first m eigenfunctions of the Stokes operator, and we set $Q_m = I - P_m$.

We start with the transfer of kinetic energy on average with respect to time-average measures. We show that in the range $[\kappa_{\bar{m}+1}, \infty)$ – that is, beyond the injection of energy – the average net transfer of energy is from lower modes to higher modes (from large to small eddies), while in the range $[\kappa_1, \kappa_{\underline{m}-1})$, provided $\underline{m} > 1$, the transfer is from higher modes to lower modes (large to small eddies). Next, we address the direct energy cascade between neighboring length scales within the inertial range.

Average Net Transfer of Kinetic Energy

We split the velocity field $\mathbf{u}$ into two components: a large-scale component $\mathbf{y}$ containing the large structures (eddies) and a small-scale component $\mathbf{z}$ containing the small structures (eddies). Thus,

$$\mathbf{u} = \mathbf{y} + \mathbf{z}, \tag{5.3}$$

where

$$\mathbf{y} = \sum_{k=1}^{m} \hat{u}_k \mathbf{w}_k, \qquad \mathbf{z} = \sum_{k=m+1}^{\infty} \hat{u}_k \mathbf{w}_k. \tag{5.4}$$

The kinetic energy contained in the large and in the small eddies is given by

$$e(\mathbf{y}) = \frac{1}{2}|\mathbf{y}|^2 = \frac{1}{2}\sum_{k=1}^{m}|\hat{u}_k|^2, \qquad e(\mathbf{z}) = \frac{1}{2}|\mathbf{z}|^2 = \frac{1}{2}\sum_{k=m+1}^{\infty}|\hat{u}_k|^2.$$

We are interested in how energy is transferred between $\mathbf{y}$ and $\mathbf{z}$. First, we consider the case where $m \geq \bar{m}$, that is, beyond the injection of energy.

Recall the Navier–Stokes equations in their functional form:

$$\frac{d\mathbf{u}}{dt} + \nu A\mathbf{u} + B(\mathbf{u}, \mathbf{u}) = \mathbf{f}. \tag{5.5}$$

On taking into account (5.3), (5.4), and (5.2), the Navier–Stokes equations (5.5) lead to the following coupled equation for the evolution of the large and small eddies:

$$\frac{d\mathbf{y}}{dt} + \nu A\mathbf{y} + P_m B(\mathbf{y}+\mathbf{z}, \mathbf{y}+\mathbf{z}) = P_m\mathbf{f}, \tag{5.6}$$

$$\frac{d\mathbf{z}}{dt} + \nu A\mathbf{z} + Q_m B(\mathbf{y}+\mathbf{z}, \mathbf{y}+\mathbf{z}) = 0. \tag{5.7}$$

The associated energy equations read

$$\frac{1}{2}\frac{d}{dt}|\mathbf{y}|^2 + \nu\|\mathbf{y}\|^2 + b(\mathbf{y}+\mathbf{z}, \mathbf{y}+\mathbf{z}, \mathbf{y}) = (\mathbf{f}, \mathbf{y}), \tag{5.8}$$

$$\frac{1}{2}\frac{d}{dt}|\mathbf{z}|^2 + \nu\|\mathbf{z}\|^2 + b(\mathbf{y}+\mathbf{z}, \mathbf{y}+\mathbf{z}, \mathbf{z}) \leq 0. \tag{5.9}$$

Note that in the 3-dimensional case we can rigorously guarantee only the inequality in (5.9) (see (7.7) in Chapter II).[7] However, for the large eddies (represented by $\mathbf{y}$), we do have an equality in (5.8) because $\mathbf{y}$ is finite-dimensional.

Owing to the orthogonality property (II.A.33) of the nonlinear term, we find for almost every t that

$$b(\mathbf{y}+\mathbf{z}, \mathbf{y}, \mathbf{y}) = 0, \qquad b(\mathbf{y}+\mathbf{z}, \mathbf{z}, \mathbf{z}) = 0. \tag{5.10}$$

Another consequence of the orthogonality property is the skew symmetry of the trilinear term with respect to the last two variables (see (II.A.34)), which implies, for almost every t,

$$b(\mathbf{y}, \mathbf{z}, \mathbf{y}) = -b(\mathbf{y}, \mathbf{y}, \mathbf{z}), \qquad b(\mathbf{z}, \mathbf{y}, \mathbf{z}) = -b(\mathbf{z}, \mathbf{z}, \mathbf{y}). \tag{5.11}$$

Define, then,

$$\begin{aligned} \Phi_{\mathbf{z}}(\mathbf{y}) &= -b(\mathbf{z}, \mathbf{z}, \mathbf{y}) + b(\mathbf{y}, \mathbf{y}, \mathbf{z}), \\ \Phi_{\mathbf{y}}(\mathbf{z}) &= -b(\mathbf{y}, \mathbf{y}, \mathbf{z}) + b(\mathbf{z}, \mathbf{z}, \mathbf{y}). \end{aligned} \tag{5.12}$$

Note that, for almost every t,

[7] An equality holds in (5.9) if we replace $\frac{1}{2}(d/dt)|\mathbf{z}(t)|^2$ by $(\mathbf{z}', \mathbf{z})$. However, as explained in Appendix A.2 of Chapter II (see, in particular, after (A.37)), the lack of regularity in dimension 3 means we know only that $\frac{1}{2}(d/dt)|\mathbf{z}(t)|^2 \leq (\mathbf{z}', \mathbf{z})$.

$$\Phi_{\mathbf{z}}(\mathbf{y}) + \Phi_{\mathbf{y}}(\mathbf{z}) = 0.$$

Using (5.12) and (5.11), we rewrite the energy equations (5.8) and (5.9) in the following form:

$$\frac{d}{dt}e(\mathbf{y}) + \nu E(\mathbf{y}) = \Phi_{\mathbf{z}}(\mathbf{y}) + (\mathbf{f}, \mathbf{y}), \tag{5.13}$$

$$\frac{d}{dt}e(\mathbf{z}) + \nu E(\mathbf{z}) \leq \Phi_{\mathbf{y}}(\mathbf{z}), \tag{5.14}$$

where $E(\mathbf{y})$ and $E(\mathbf{z})$ denote the enstrophy associated with $\mathbf{y}$ and $\mathbf{z}$, respectively.

In (5.13), the term $\nu E(\mathbf{y})$ represents the rate of energy dissipation by viscous effects per unit time for the large length scales, the term $(\mathbf{f}, \mathbf{y})$ represents the energy flow injected into the large length scales by the forcing term, and the term $\Phi_{\mathbf{z}}(\mathbf{y})$ clearly represents the net amount of kinetic energy per unit time that is transferred from the small length scales to the large length scales. Similarly, in (5.14), $\nu E(\mathbf{z})$ represents the dissipation of kinetic energy per unit time by viscous effects for the small scales, while $\Phi_{\mathbf{y}}(\mathbf{z})$ represents the net amount of kinetic energy per unit time that is transferred from the large scales to the small scales. Moreover, note that

$$-b(\mathbf{y}, \mathbf{y}, \mathbf{z}) = (-B(\mathbf{y}, \mathbf{y}), \mathbf{z}), \qquad -b(\mathbf{z}, \mathbf{z}, \mathbf{y}) = (-B(\mathbf{z}, \mathbf{z}), \mathbf{y});$$

hence the term in the first expression is the energy flow induced in the higher modes by inertial forces associated with the lower modes, while the term in the second expression is the energy flow induced in the lower modes by inertial forces associated with the higher modes. We can thus see, for instance, that the net transfer of energy $\Phi_{\mathbf{y}}(\mathbf{z})$ is the energy flow in the higher modes induced by the inertial forces associated with the lower modes *minus* the energy flow in the lower modes induced by the inertial forces associated with the higher modes.

In a more precise notation, we can write the rate of net transfer of energy as

$$\mathfrak{e}_{\kappa_m}(\mathbf{u}) = \mathfrak{e}^{\rightarrow}_{\kappa_m}(\mathbf{u}) - \mathfrak{e}^{\leftarrow}_{\kappa_m}(\mathbf{u}), \tag{5.15}$$

where

$$\mathfrak{e}^{\rightarrow}_{\kappa_m}(\mathbf{u}) = -b(\mathbf{u}_{\kappa_1,\kappa_m}, \mathbf{u}_{\kappa_1,\kappa_m}, \mathbf{u}_{\kappa_{m+1},\infty}), \tag{5.16}$$

$$\mathfrak{e}^{\leftarrow}_{\kappa_m}(\mathbf{u}) = b(\mathbf{u}_{\kappa_{m+1},\infty}, \mathbf{u}_{\kappa_{m+1},\infty}, \mathbf{u}_{\kappa_1,\kappa_m}). \tag{5.17}$$

The term $\mathfrak{e}^{\rightarrow}_{\kappa_m}(\mathbf{u})$ represents the rate of transfer of energy from the lower modes to the higher modes; $\mathfrak{e}^{\leftarrow}_{\kappa_m}(\mathbf{u})$ represents the rate of transfer of energy from the higher modes to the lower modes. Then we can rewrite (5.13) and (5.14) as

$$\frac{d}{dt}e(\mathbf{u}_{\kappa_1,\kappa_m}) + \nu E(\mathbf{u}_{\kappa_1,\kappa_m}) = -\mathfrak{e}_{\kappa_m}(\mathbf{u}) + (\mathbf{f}, \mathbf{u}_{\kappa_1,\kappa_m}), \tag{5.18}$$

$$\frac{d}{dt}e(\mathbf{u}_{\kappa_{m+1},\infty}) + \nu E(\mathbf{u}_{\kappa_{m+1},\infty}) \leq \mathfrak{e}_{\kappa_m}(\mathbf{u}). \tag{5.19}$$

The claim is that, on average, this net transfer of energy occurs only into the small scales. Hence, if $\langle\cdot\rangle$ denotes a suitable ensemble average, we should have

$$\langle \mathfrak{e}_{\kappa_m}(\mathbf{u}) \rangle = \langle \Phi_{\mathbf{y}}(\mathbf{z}) \rangle = -\langle \Phi_{\mathbf{z}}(\mathbf{y}) \rangle \geq 0. \tag{5.20}$$

We want to show rigorously that, with respect to an arbitrary time-average measure, beyond the range of injection of energy the average net transfer of energy does indeed occur in this sense – a result obtained heuristically by Kolmogorov [1941a,b].

Let μ be a time-average measure of the Navier–Stokes equations. Then (see Theorems 5.1 and 2.1), there exists a weak solution $\mathbf{u} = \mathbf{u}(t)$ defined on $[0, \infty)$ such that

$$\mathop{\mathrm{LIM}}_{T\to\infty} \frac{1}{T} \int_0^T \varphi(\mathbf{u}(t))\, dt = \int_H \varphi(\mathbf{u})\, d\mu(\mathbf{u}) \tag{5.21}$$

for all test functions φ in $\mathcal{C}(H_w)$.[8] The functions $\varphi(\mathbf{z}) = E(\mathbf{z}) = \|\mathbf{z}\|^2$ and $\Phi_{\mathbf{y}}(\mathbf{z})$ are not allowed in this relation, but we proceed formally and assume, in particular, that those choices for φ in (5.21) are permissible. The rigourous justification of the results to be obtained here is given in Appendix B.4.

We integrate (5.14) in time, from 0 to $T > 0$, for this particular solution associated with the time-average measure. We find

$$e(\mathbf{z}(T)) + \nu \int_0^T E(\mathbf{z}(t))\, dt \leq e(\mathbf{z}(0)) + \int_0^T \Phi_{\mathbf{y}(t)}(\mathbf{z}(t))\, dt.$$

Dividing this inequality by T and taking the generalized limit as T goes to infinity, we obtain, using that the energy $e(\mathbf{z}(t)) \leq e(\mathbf{u}(t))$ is uniformly bounded in time,

$$\nu \mathop{\mathrm{LIM}}_{T\to\infty} \frac{1}{T} \int_0^T E(\mathbf{z}(t))\, dt \leq \mathop{\mathrm{LIM}}_{T\to\infty} \frac{1}{T} \int_0^T \Phi_{\mathbf{y}(t)}(\mathbf{z}(t))\, dt.$$

Hence, from (5.21) we see that

$$\langle \Phi_{\mathbf{y}}(\mathbf{z}) \rangle \geq \nu \langle E(\mathbf{z}) \rangle \geq 0. \tag{5.22}$$

This proves (5.20).

We now consider the modes that are below the injection of energy. For that purpose, we assume $\underline{m} > 1$ and consider m such that $1 \leq m < \underline{m}$. In this case, $P_m \mathbf{f} = 0$ and the Navier–Stokes equations split as follows:

$$\frac{d\mathbf{y}}{dt} + \nu A\mathbf{y} + P_m B(\mathbf{y}+\mathbf{z}, \mathbf{y}+\mathbf{z}) = 0, \tag{5.23}$$

$$\frac{d\mathbf{z}}{dt} + \nu A\mathbf{z} + Q_m B(\mathbf{y}+\mathbf{z}, \mathbf{y}+\mathbf{z}) = Q_m \mathbf{f}. \tag{5.24}$$

By taking the inner product in H of equation (5.23) with $\mathbf{y}$, we obtain the following energy equation for the lower modes:

$$\frac{d}{dt} e(\mathbf{y}) + \nu E(\mathbf{y}) = \Phi_{\mathbf{z}}(\mathbf{y}), \tag{5.25}$$

where $\Phi_{\mathbf{z}}(\mathbf{y})$ is defined, as before, according to (5.12).

[8] This is the space of continuous real-valued (not necessarily bounded) functions on H_w; see Section 2 after Definition 2.1.

Then, taking the time average of the energy equation (5.25), we obtain

$$\langle \Phi_{\mathbf{z}}(\mathbf{y})\rangle = \nu\langle E(\mathbf{y})\rangle \geq 0. \tag{5.26}$$

In other words, the transfer of energy is from higher modes to lower modes, which is called an *inverse* transfer of energy; this is in contrast to the previous, *direct* energy transfer. This inverse transfer of energy into the large scales may be responsible for the observed persistent low-wavenumber coherent structures in turbulent flows.

We summarize this result as follows.

Theorem 5.1 *Consider the 3-dimensional Navier–Stokes equations with either periodic boundary conditions having vanishing space average or with no-slip boundary conditions. Assume that the forcing term* $\mathbf{f}$ *contains only lower-order modes in the sense that* $\mathbf{f} = \mathbf{f}_{\kappa_{\underline{m}},\kappa_{\bar{m}}}$ *for some* $1 \leq \underline{m} \leq \bar{m} < \infty$. *Let* μ *be a time-average measure corresponding to this* $\mathbf{f}$, *and let* $\langle\cdot\rangle$ *denote the ensemble average associated with this probability measure. Then, on average with respect to* μ *and for any* $m \geq \bar{m}$, *energy is transferred from the lower modes (large eddies)* $\mathbf{u}_{\kappa_1,\kappa_m} = P_m\mathbf{u}$ *to the higher modes (small eddies)* $\mathbf{u}_{\kappa_{m+1},\infty} = Q_m\mathbf{u}$ *in the sense that*

$$\langle \mathfrak{e}_{\kappa_m}(\mathbf{u})\rangle \geq \nu\langle E(\mathbf{u}_{\kappa_{m+1},\infty})\rangle \geq 0. \tag{5.27}$$

Here $\mathfrak{e}_{\kappa_m}(\mathbf{u})$, *defined by* (5.15), *represents the net amount of kinetic energy per unit time that is transferred from the large scales into the small scales (wavenumber higher than* κ_m*); the term* $\nu\langle E(\mathbf{u}_{\kappa_{m+1},\infty})\rangle$ *represents the average dissipation of kinetic energy per unit time by viscous effects among the higher modes.*

Furthermore, if $\underline{m} > 1$ *then, for any* $1 \leq m < \underline{m}$, *energy is transferred into the lower modes (large eddies)* $\mathbf{u}_{\kappa_1,\kappa_m} = P_m\mathbf{u}$ *in the sense that*

$$\langle \mathfrak{e}_{\kappa_m}(\mathbf{u})\rangle = -\nu\langle E(\mathbf{u}_{\kappa_1,\kappa_m})\rangle \leq 0. \tag{5.28}$$

Direct Energy Cascade

When the part of the spectrum dominated by molecular dissipation is distant from the wavenumber range where energy is injected, the intervening spectrum (the inertial range) is governed by the nonlinear inertial term in the Navier–Stokes equations. There, the spectrum is determined by an energy cascade wherein the eddies break up into successively smaller ones. For such a cascade picture, the velocity field is separated into several distinct parts, each containing eddies within a relatively small range of length scales. Our aim is to study the transfer of energy between consecutive ranges. According to the Kolmogorov theory, there is no dissipation of energy by viscous effects within the inertial range, and the energy is simply transferred to smaller and smaller length scales. In that theory, there is no constraint due to enstrophy conservation as in the 2-dimensional case that we will consider shortly.

For our purposes, it is sufficient to split the flow into three parts, according to three length-scale ranges. Recall that we assume $Q_{\bar{m}}\mathbf{f} = 0$ for some $\bar{m} \in \mathbb{N}$. Then, we consider $m', m'' \in \mathbb{N}$ such that $m'' \geq m' \geq \bar{m}$. Write

$$\mathbf{u} = \mathbf{y} + \mathbf{w} + \mathbf{z}, \tag{5.29}$$

where

$$\mathbf{y} = \mathbf{u}_{\kappa_1,\kappa_{m'-1}}, \quad \mathbf{w} = \mathbf{u}_{\kappa_{m'},\kappa_{m''}}, \quad \mathbf{z} = \mathbf{u}_{\kappa_{m''+1},\infty}. \tag{5.30}$$

Upon decomposing the NSE according to (5.29), and taking (5.2) into consideration, we see that the velocity field associated with the wavenumbers between $\kappa_{m'}$ and $\kappa_{m''}$ (i.e., $\mathbf{w}$) evolves according to

$$\frac{d\mathbf{w}}{dt} + \nu A\mathbf{w} + B(\mathbf{y}+\mathbf{w}+\mathbf{z}, \mathbf{y}+\mathbf{w}+\mathbf{z}) = 0. \tag{5.31}$$

By taking the inner product in H of (5.31) with $\mathbf{w}$, we obtain an evolution equation for the energy associated with the modes between $\kappa_{m'}$ and $\kappa_{m''}$:

$$\frac{1}{2}\frac{d}{dt}e(\mathbf{w}) + \nu E(\mathbf{w}) = -b(\mathbf{y}+\mathbf{w}+\mathbf{z}, \mathbf{y}+\mathbf{w}+\mathbf{z}, \mathbf{w}). \tag{5.32}$$

Here, as in the equation for $|\mathbf{y}|^2$ in (5.8), we obtain an equality because $\mathbf{w}$ contains only a finite number of modes.

The RHS of (5.32) can be written as

$$-b(\mathbf{y}+\mathbf{w}+\mathbf{z}, \mathbf{y}+\mathbf{w}+\mathbf{z}, \mathbf{w}) = \Phi_{\mathbf{y}}(\mathbf{w}+\mathbf{z}) - \Phi_{\mathbf{y}+\mathbf{w}}(\mathbf{z}), \tag{5.33}$$

where

$$\begin{aligned} \Phi_{\mathbf{y}}(\mathbf{w}+\mathbf{z}) &= -b(\mathbf{y}, \mathbf{y}, \mathbf{w}+\mathbf{z}) + b(\mathbf{w}+\mathbf{z}, \mathbf{w}+\mathbf{z}, \mathbf{y}), \\ -\Phi_{\mathbf{y}+\mathbf{w}}(\mathbf{z}) &= \Phi_{\mathbf{z}}(\mathbf{y}+\mathbf{w}) = -b(\mathbf{z}, \mathbf{z}, \mathbf{y}+\mathbf{w}) + b(\mathbf{y}+\mathbf{w}, \mathbf{y}+\mathbf{w}, \mathbf{z}). \end{aligned} \tag{5.34}$$

As we have seen earlier in this section, $\Phi_{\mathbf{y}}(\mathbf{w}+\mathbf{z})$ represents the net amount of kinetic energy per unit time that is transferred into the wavenumbers larger than or equal to $\kappa_{m'}$, while $\Phi_{\mathbf{y}+\mathbf{w}}(\mathbf{z}) = -\Phi_{\mathbf{z}}(\mathbf{y}+\mathbf{w})$ represents the net amount of kinetic energy per unit time that is transferred into the wavenumbers larger than $\kappa_{m''}$. Hence, the term $\Phi_{\mathbf{y}}(\mathbf{w}+\mathbf{z}) - \Phi_{\mathbf{y}+\mathbf{w}}(\mathbf{z})$ represents the net amount of kinetic energy per unit time that is transferred into the wavenumbers between $\kappa_{m'}$ and $\kappa_{m''}$.

By expanding the trilinear terms in the RHS of (5.33) and using the skew-symmetry property of the trilinear term with respect to the last two variables, one can readily check that (5.33) indeed holds. Actually, it holds for arbitrary functions $\mathbf{y}, \mathbf{w}, \mathbf{z}$, which are not necessarily associated with a spectral decomposition of $\mathbf{u}$.

We know from the preceeding results (see Theorem 5.1) that

$$\langle \Phi_{\mathbf{y}}(\mathbf{w}+\mathbf{z}) \rangle \geq \nu \langle E(\mathbf{w}+\mathbf{z}) \rangle = \nu \langle E(\mathbf{w}) \rangle + \nu \langle E(\mathbf{z}) \rangle \geq 0$$

and

$$-\langle \Phi_{\mathbf{y}+\mathbf{w}}(\mathbf{z}) \rangle \leq -\nu \langle E(\mathbf{z}) \rangle \leq 0.$$

Since $\mathbf{w}$ contains only a finite number of modes, one can obtain a more precise relation (see Appendix B.4 for details), namely,

$$\langle \Phi_{\mathbf{y}}(\mathbf{w}+\mathbf{z}) - \Phi_{\mathbf{y}+\mathbf{w}}(\mathbf{z}) \rangle = \nu \langle E(\mathbf{w}) \rangle \geq 0. \tag{5.35}$$

We can rewrite this relation in the form

$$\langle \Phi_{\mathbf{y}+\mathbf{w}}(\mathbf{z}) \rangle = \langle \Phi_{\mathbf{y}}(\mathbf{w}+\mathbf{z}) \rangle - \nu \langle E(\mathbf{w}) \rangle. \tag{5.36}$$

In a more explicit form using definition (5.15), we have

$$\langle \mathfrak{e}_{\kappa_{m''}}(\mathbf{u}) \rangle = \langle \mathfrak{e}_{\kappa_{m'-1}}(\mathbf{u}) \rangle - \nu \langle E(\mathbf{u}_{\kappa_{m'},\kappa_{m''}}) \rangle. \tag{5.37}$$

In other words, the average net transfer of energy into the modes higher than $\kappa_{m''}$ is equal to the average net transfer of energy to the modes higher than $\kappa_{m'}$ minus the energy lost to viscous dissipation within the range $[\kappa_{m'}, \kappa_{m''}]$.

In the energy cascade, for $[\kappa_{m'}, \kappa_{m''}]$ inside the inertial range, the viscous dissipation is negligible and the energy is simply transferred to smaller and smaller length scales. This cascade occurs whenever

$$\langle \mathfrak{e}_{\kappa_{m''}}(\mathbf{u}) \rangle \gtrapprox \langle \mathfrak{e}_{\kappa_{m'-1}}(\mathbf{u}) \rangle \gg \nu \langle E(\mathbf{u}_{\kappa_{m'},\kappa_{m''}}) \rangle. \tag{5.38}$$

The range of wavenumbers $\kappa > \kappa_{\bar{m}}$ up to where (5.38) holds is called the *inertial range* in 3-dimensional turbulence.

5.2 Enstrophy Transfer and the Cascade in 2-Dimensional Turbulence

The energy transfer in the 2-dimensional case is exactly as in the 3-dimensional case: above the range where energy is injected into the system, energy goes from lower modes to higher modes; in the range below the injection of energy, energy goes from higher modes to lower modes.

As for the cascade, however, there is an important modification. Because of vorticity constraints in the 2-dimensional case, there is a similar transfer of enstrophy between neighboring modes: beyond the modes where energy is injected into the system, enstrophy goes from lower modes to higher modes, whereas for the modes below the injection of energy, enstrophy goes from higher modes to lower modes. Then, within a certain range above the injection of energy, there is a much stronger direct transfer of enstrophy, leading to a direct enstrophy cascade instead of a direct energy cascade. On the other hand, below the injection of energy, an inverse energy cascade takes place. The region containing both an inverse energy cascade and a direct enstrophy cascade is what defines the inertial range in the 2-dimensional case. Beyond the inertial range, the cascade ceases and viscous effects dissipate the enstrophy.

Next we consider the details of the transfer of enstrophy and of the enstrophy cascade. Then we discuss conditions on the forcing term that enable the existence of the Kraichnan inertial range spectrum. For further results on the Kraichnan spectrum, see Foias, Manley, Jolly, and Rosa [2001a].

The direct and inverse transfers of energy hold in the 2-dimensional case with either no-slip or periodic boundary conditions (with vanishing space average), exactly as in the 3-dimensional case (we omit the details). However, for the transfer of enstrophy, we consider only the 2-dimensional periodic case. Indeed, only in this case is the necessary orthogonality property needed for our proof available (see Appendix II.A, equations (A.62)–(A.64)):

$$b(\mathbf{u}, \mathbf{u}, A\mathbf{u}) = 0 \tag{5.39}$$

for $\mathbf{u}$ in $D(A)$. This relation is an important one that leads to the conservation of enstrophy by the nonlinear term. This, in turn, seems to be responsible for the appearance of the enstrophy cascade characteristic of 2-dimensional turbulence. As seen in Appendix II.A, the identity (5.39) also leads to the relations

$$b(\mathbf{v}, \mathbf{u}, A\mathbf{u}) + b(\mathbf{u}, \mathbf{v}, A\mathbf{u}) + b(\mathbf{u}, \mathbf{u}, A\mathbf{v}) = 0 \tag{5.40}$$

and

$$\begin{aligned} b(\mathbf{v}, \mathbf{w}, A\mathbf{u}) &+ b(\mathbf{v}, \mathbf{u}, A\mathbf{w}) + b(\mathbf{w}, \mathbf{v}, A\mathbf{u}) \\ &+ b(\mathbf{u}, \mathbf{v}, A\mathbf{w}) + b(\mathbf{w}, \mathbf{u}, A\mathbf{v}) + b(\mathbf{u}, \mathbf{w}, A\mathbf{v}) = 0 \end{aligned} \tag{5.41}$$

for $\mathbf{u}, \mathbf{v}, \mathbf{w}$ in $D(A)$. These identities will be used in the definition of the enstrophy transfer functions.

As in the 3-dimensional case, we assume that $\mathbf{f}$ is time-independent and that it contains only a finite number of modes, between $\kappa_{\underline{m}}$ and $\kappa_{\bar{m}}$, for some $1 \le \underline{m} \le \bar{m} < \infty$; that is,

$$\mathbf{f} = \mathbf{f}_{\kappa_{\underline{m}}, \kappa_{\bar{m}}}. \tag{5.42}$$

Before discussing the transfer of enstrophy, we state the results on the transfer of energy in the 2-dimensional case.

Theorem 5.2 *The results in Theorem 5.1 hold also for the 2-dimensional Navier–Stokes equations with either periodic boundary conditions having vanishing space average or with no-slip boundary conditions: If $\underline{m} > 1$, we obtain the following relation for all $1 \le m < \underline{m}$:*

$$\langle \mathfrak{e}_{\kappa_m}(\mathbf{u}) \rangle = -\nu \langle E(\mathbf{u}_{\kappa_1, \kappa_m}) \rangle \le 0. \tag{5.43}$$

Furthermore, instead of simply (5.27), we obtain the following more precise relation for all $m \ge \bar{m}$:

$$\langle \mathfrak{e}_{\kappa_m}(\mathbf{u}) \rangle = \nu \langle E(\mathbf{u}_{\kappa_{m+1}, \infty}) \rangle \ge 0. \tag{5.44}$$

Average Net Transfer of Enstrophy

We split the velocity field $\mathbf{u}$ into two components, a large-scale component $\mathbf{y}$ (containing the large eddies) and a small-scale component $\mathbf{z}$ (containing the small eddies):

$$\mathbf{u} = \mathbf{y} + \mathbf{z}, \tag{5.45}$$

where

$$\mathbf{y} = \sum_{k=1}^{m} \hat{u}_k \mathbf{w}_k, \qquad \mathbf{z} = \sum_{k=m+1}^{\infty} \hat{u}_k \mathbf{w}_k. \tag{5.46}$$

The enstrophy contained in the large and small eddies are given (respectively) by

$$E(\mathbf{y}) = \|\mathbf{y}\|^2 = \sum_{k=1}^{m} \lambda_k |\hat{u}_k|^2 = \sum_{k=1}^{m} \kappa_k^2 |\hat{u}_k|^2$$

and

$$E(\mathbf{z}) = \|\mathbf{z}\|^2 = \sum_{k=m+1}^{\infty} \lambda_k |\hat{u}_k|^2 = \sum_{k=m+1}^{\infty} \kappa_k^2 |\hat{u}_k|^2.$$

We are interested in how the enstrophy is transferred between **y** and **z**. First, we consider the case where $m \geq \bar{m}$ – that is, beyond the injection of energy.

In order to find the evolution of the enstrophy associated with each component of the velocity field, we take the inner product in H of the NSE with each component. We thus obtain (see (5.6) and (5.7))

$$\frac{1}{2}\frac{d}{dt}\|\mathbf{y}\|^2 + \nu|A\mathbf{y}|^2 = (\mathbf{f}, A\mathbf{y}) - b(\mathbf{y}+\mathbf{z}, \mathbf{y}+\mathbf{z}, A\mathbf{y}), \tag{5.47}$$

$$\frac{1}{2}\frac{d}{dt}\|\mathbf{z}\|^2 + \nu|A\mathbf{z}|^2 = -b(\mathbf{y}+\mathbf{z}, \mathbf{y}+\mathbf{z}, A\mathbf{z}). \tag{5.48}$$

In contrast to the 3-dimensional case, we have equality in both enstrophy expressions (5.47) and (5.48) (as well as in the energy equations).

Similarly to the analysis for the transfer of energy, we write, using first (5.39) and then (5.40),

$$\begin{aligned}-b(\mathbf{y}+\mathbf{z}, \mathbf{y}+\mathbf{z}, A\mathbf{z}) &= -b(\mathbf{y}, \mathbf{y}, A\mathbf{z}) - b(\mathbf{z}, \mathbf{y}, A\mathbf{z}) - b(\mathbf{y}, \mathbf{z}, A\mathbf{z}) \\ &= -b(\mathbf{y}, \mathbf{y}, A\mathbf{z}) + b(\mathbf{z}, \mathbf{z}, A\mathbf{y}).\end{aligned} \tag{5.49}$$

We can also write

$$\begin{aligned}-b(\mathbf{y}+\mathbf{z}, \mathbf{y}+\mathbf{z}, A\mathbf{y}) &= -b(\mathbf{z}, \mathbf{z}, A\mathbf{y}) - b(\mathbf{y}, \mathbf{z}, A\mathbf{y}) - b(\mathbf{z}, \mathbf{y}, A\mathbf{y}) \\ &= -b(\mathbf{z}, \mathbf{z}, A\mathbf{y}) + b(\mathbf{y}, \mathbf{y}, A\mathbf{z}).\end{aligned} \tag{5.50}$$

Define, then,

$$\begin{aligned}\Psi_{\mathbf{z}}(\mathbf{y}) &= -b(\mathbf{z}, \mathbf{z}, A\mathbf{y}) + b(\mathbf{y}, \mathbf{y}, A\mathbf{z}), \\ \Psi_{\mathbf{y}}(\mathbf{z}) &= -b(\mathbf{y}, \mathbf{y}, A\mathbf{z}) + b(\mathbf{z}, \mathbf{z}, A\mathbf{y}).\end{aligned} \tag{5.51}$$

Note that, for almost every t,

$$\Psi_{\mathbf{z}}(\mathbf{y}) + \Psi_{\mathbf{y}}(\mathbf{z}) = 0.$$

Next, we rewrite the enstrophy equations (5.47) and (5.48) in the following form:

$$\frac{1}{2}\frac{d}{dt}\|\mathbf{y}\|^2 + \nu|A\mathbf{y}|^2 = \Psi_{\mathbf{z}}(\mathbf{y}) + (\mathbf{f}, A\mathbf{y}), \tag{5.52}$$

$$\frac{1}{2}\frac{d}{dt}\|\mathbf{z}\|^2 + \nu|A\mathbf{z}|^2 = \Psi_{\mathbf{y}}(\mathbf{z}). \tag{5.53}$$

In (5.52), the term $\nu|A\mathbf{y}|^2$ represents the dissipation of enstrophy per unit time by viscous effects for the large scales, and the term $\Psi_{\mathbf{z}}(\mathbf{y})$ represents the net amount of enstrophy per unit time that is transferred into the lower modes. In (5.53), $\nu|A\mathbf{z}|^2$ represents the dissipation of enstrophy by viscous effects, and $\Psi_{\mathbf{y}}(\mathbf{z})$ represents the net amount of enstrophy per unit time that is transferred into the higher modes.

As in the transfer of energy, we can write (5.52) and (5.53) in a more explicit form. We define

$$\mathfrak{E}_{\kappa_m}(\mathbf{u}) = \mathfrak{E}^{\rightarrow}_{\kappa_m}(\mathbf{u}) - \mathfrak{E}^{\leftarrow}_{\kappa_m}(\mathbf{u}), \tag{5.54}$$

where

$$\mathfrak{E}^{\rightarrow}_{\kappa_m}(\mathbf{u}) = -b(\mathbf{u}_{\kappa_1,\kappa_m}, \mathbf{u}_{\kappa_1,\kappa_m}, A\mathbf{u}_{\kappa_{m+1},\infty}), \tag{5.55}$$

$$\mathfrak{E}^{\leftarrow}_{\kappa_m}(\mathbf{u}) = b(\mathbf{u}_{\kappa_{m+1},\infty}, \mathbf{u}_{\kappa_{m+1},\infty}, A\mathbf{u}_{\kappa_1,\kappa_m}). \tag{5.56}$$

The term $\mathfrak{E}^{\rightarrow}_{\kappa_m}(\mathbf{u})$ represents the rate of transfer of enstrophy from the lower modes to the higher modes; $\mathfrak{E}^{\leftarrow}_{\kappa_m}(\mathbf{u})$ represents the rate of transfer of enstrophy from the higher modes to the lower modes.

Now we can rewrite (5.52) and (5.53) as

$$\frac{1}{2}\frac{d}{dt}\|\mathbf{u}_{\kappa_1,\kappa_m}\| + \nu|A\mathbf{u}_{\kappa_1,\kappa_m}|^2 = -\mathfrak{E}_{\kappa_m}(\mathbf{u}) + (\mathbf{f}, A\mathbf{u}_{\kappa_1,\kappa_m}), \tag{5.57}$$

$$\frac{1}{2}\frac{d}{dt}\|\mathbf{u}_{\kappa_{m+1},\infty}\|^2 + \nu|A\mathbf{u}_{\kappa_{m+1},\infty}|^2 = \mathfrak{E}_{\kappa_m}(\mathbf{u}). \tag{5.58}$$

We want to show rigorously that, with respect to any time-average measure corresponding to $\mathbf{f}$, the transfer of enstrophy in the range beyond that of the injection of energy occurs only from the lower modes to the higher modes.

Let μ be a time-average measure of the Navier–Stokes equations. Then, in the 2-dimensional case (see Definition 2.1 and Remark 2.1), there exists a weak solution $\mathbf{u} = \mathbf{u}(t) = S(t)\mathbf{u}_0$ defined on $[0, \infty)$ with $\mathbf{u}_0 \in D(A)$ such that

$$\operatorname*{LIM}_{T\to\infty} \frac{1}{T}\int_0^T \varphi(S(t)\mathbf{u}_0)\,dt = \int_H \varphi(\mathbf{u})\,d\mu(\mathbf{u}) \tag{5.59}$$

for all φ in $\mathcal{C}(D(A))$.[9]

We integrate (5.14) in time (from 0 to $T > 0$) for this particular solution, which is associated with the time-average measure. We find

$$\frac{1}{2}\|\mathbf{z}(T)\|^2 + \nu\int_0^T |A\mathbf{z}(t)|^2\,dt = \frac{1}{2}\|\mathbf{z}(0)\|^2 + \int_0^T \Psi_{\mathbf{y}(t)}(\mathbf{z}(t))\,dt.$$

Dividing this inequality by T and taking the generalized limit as T goes to infinity, we obtain, using that the enstrophy $\|\mathbf{z}(t)\|^2 \le \|\mathbf{u}(t)\|^2$ is bounded uniformly in time,

$$\nu \operatorname*{LIM}_{T\to\infty}\frac{1}{T}\int_0^T |A\mathbf{z}(t)|^2\,dt = \operatorname*{LIM}_{T\to\infty}\frac{1}{T}\int_0^T \Psi_{\mathbf{y}(t)}(\mathbf{z}(t))\,dt.$$

Hence, from (5.59) we see that

$$\langle\Psi_{\mathbf{y}}(\mathbf{z})\rangle = \nu\langle|A\mathbf{z}|^2\rangle \ge 0. \tag{5.60}$$

This proves the *direct* transfer of enstrophy in the range beyond that of injection of energy.

We now consider the modes below the injection of energy. For this purpose, we assume that $\underline{m} > 1$ and consider m such that $1 \le m < \underline{m}$. In this case, $P_m\mathbf{f} = 0$ and the Navier–Stokes equations split as follows:

[9] This is the space of continuous real-valued (not necessarily bounded) functions on $D(A)$; see Definition 2.1.

$$\frac{d\mathbf{y}}{dt} + \nu A\mathbf{y} + P_m B(\mathbf{y}+\mathbf{z}, \mathbf{y}+\mathbf{z}) = 0, \tag{5.61}$$

$$\frac{d\mathbf{z}}{dt} + \nu A\mathbf{z} + Q_m B(\mathbf{y}+\mathbf{z}, \mathbf{y}+\mathbf{z}) = Q_m\mathbf{f}. \tag{5.62}$$

By taking the inner product in H of equation (5.61) with $A\mathbf{y}$, we obtain the following enstrophy equation for the lower modes:

$$\frac{1}{2}\frac{d}{dt}|\mathbf{y}|^2 + \nu|A\mathbf{y}|^2 = \Psi_{\mathbf{z}}(\mathbf{y}). \tag{5.63}$$

Then, taking the time average of the enstrophy equation (5.63), we have

$$\langle\Psi_{\mathbf{z}}(\mathbf{y})\rangle = -\langle\Psi_{\mathbf{y}}(\mathbf{z})\rangle = \nu\langle|A\mathbf{y}|^2\rangle \geq 0. \tag{5.64}$$

In other words, the average net transfer of enstrophy is from higher modes to lower modes, which is called an *inverse* transfer of enstrophy.

We summarize this rigorous result as follows.

Theorem 5.3 *Consider the 2-dimensional Navier–Stokes equations with periodic boundary conditions and vanishing space average. Assume that the forcing term* $\mathbf{f}$ *contains only lower-order modes in the sense that* $\mathbf{f} = \mathbf{f}_{\kappa_{\underline{m}},\kappa_{\bar{m}}}$ *for some* $1 \leq \underline{m} \leq \bar{m} < \infty$*. Let* μ *be a time-average measure corresponding to this* $\mathbf{f}$*, and let* $\langle\cdot\rangle$ *denote the ensemble average associated with this probability measure. Then, on average with respect to* μ *and for any* $m \geq \bar{m}$*, enstrophy is transferred from the lower modes (large eddies)* $\mathbf{u}_{\kappa_1,\kappa_m} = P_m\mathbf{u}$ *to the higher modes (small eddies)* $\mathbf{u}_{\kappa_{m+1},\infty} = Q_m\mathbf{u}$ *in the sense that*

$$\langle\mathfrak{E}_{\kappa_m}(\mathbf{u})\rangle = \nu\langle|A\mathbf{u}_{\kappa_{m+1},\infty}|^2\rangle \geq 0. \tag{5.65}$$

Here $\mathfrak{E}_{\kappa_m}(\mathbf{u})$*, defined by (5.54), represents the net amount of enstrophy per unit time that is transferred into the small scales; the term* $\nu\langle|A\mathbf{u}_{\kappa_{m+1},\infty}|^2\rangle$ *represents the average dissipation of enstrophy per unit time by viscous effects among the higher modes.*

Furthermore, if $\underline{m} > 1$ *then, for any* $1 \leq m < \underline{m}$*, enstrophy is transferred into lower modes (large eddies)* $\mathbf{u}_{\kappa_1,\kappa_m} = P_m\mathbf{u}$ *in the sense that*

$$\langle\mathfrak{E}_{\kappa_m}(\mathbf{u})\rangle = -\nu\langle|A\mathbf{u}_{\kappa_1,\kappa_m}|^2\rangle \leq 0. \tag{5.66}$$

Direct Enstrophy Cascade

The enstrophy cascade in two dimensions is similar to the energy cascade in three dimensions. The velocity field is separated into several distinct parts, with each part containing eddies within a relatively small range of length scales. Then, within a certain range above the injection of energy (the so-called inertial range), there is hardly any dissipation of energy by viscous effects; the energy is simply transferred to smaller and smaller length scales.

We split the flow into three parts, according to three length-scale ranges. Recall that we are assuming $Q_{\bar{m}}\mathbf{f} = 0$ for some $\bar{m} \in \mathbb{N}$. Then, consider $m', m'' \in \mathbb{N}$ such that $m'' \geq m' \geq \bar{m}$. We write

$$\mathbf{u} = \mathbf{y} + \mathbf{w} + \mathbf{z}, \tag{5.67}$$

where

$$\mathbf{y} = \mathbf{u}_{\kappa_1,\kappa_{m'-1}}, \quad \mathbf{w} = \mathbf{u}_{\kappa_{m'},\kappa_{m''}}, \quad \mathbf{z} = \mathbf{u}_{\kappa_{m''+1},\infty}. \tag{5.68}$$

We now decompose the NSE in accordance with (5.67), taking (5.42) into consideration. The velocity field associated with the wavenumbers between $\kappa_{m'}$ and $\kappa_{m''}$ (i.e., $\mathbf{w}$) evolves according to

$$\frac{d\mathbf{w}}{dt} + \nu A\mathbf{w} + B(\mathbf{y}+\mathbf{w}+\mathbf{z}, \mathbf{y}+\mathbf{w}+\mathbf{z}) = 0. \tag{5.69}$$

Taking the inner product in H of (5.69) with $A\mathbf{w}$, we obtain an evolution equation for the enstrophy associated with the modes between $\kappa_{m'}$ and $\kappa_{m''}$:

$$\frac{1}{2}\frac{d}{dt}\|\mathbf{w}\|^2 + \nu|A\mathbf{w}|^2 = -b(\mathbf{y}+\mathbf{w}+\mathbf{z}, \mathbf{y}+\mathbf{w}+\mathbf{z}, A\mathbf{w}). \tag{5.70}$$

We claim that the RHS of (5.70) can be written as

$$\begin{aligned} -b(\mathbf{y}+\mathbf{w}+\mathbf{z}, \mathbf{y}+\mathbf{w}+\mathbf{z}, A\mathbf{w}) &= \mathfrak{E}_{\kappa_{m'-1}}(\mathbf{u}) - \mathfrak{E}_{\kappa_{m''+1}}(\mathbf{u}) \\ &= \Psi_{\mathbf{y}}(\mathbf{w}+\mathbf{z}) - \Psi_{\mathbf{y}+\mathbf{w}}(\mathbf{z}), \end{aligned} \tag{5.71}$$

where, according to definition (5.54),

$$\begin{aligned} \mathfrak{E}_{\kappa_{m'-1}}(\mathbf{u}) &= \Psi_{\mathbf{y}}(\mathbf{w}+\mathbf{z}) = -b(\mathbf{y}, \mathbf{y}, A(\mathbf{w}+\mathbf{z})) + b(\mathbf{w}+\mathbf{z}, \mathbf{w}+\mathbf{z}, A\mathbf{y}), \\ -\mathfrak{E}_{\kappa_{m''+1}}(\mathbf{u}) &= -\Psi_{\mathbf{y}+\mathbf{w}}(\mathbf{z}) = -b(\mathbf{z}, \mathbf{z}, A(\mathbf{y}+\mathbf{w})) + b(\mathbf{y}+\mathbf{w}, \mathbf{y}+\mathbf{w}, \mathbf{z}). \end{aligned} \tag{5.72}$$

The term $\mathfrak{E}_{\kappa_{m'-1}}(\mathbf{u})$ represents the net amount of enstrophy transferred per unit time into the modes higher than or equal to $\kappa_{m'}$, and similarly for $\mathfrak{E}_{\kappa_{m''+1}}(\mathbf{u})$. The term $\mathfrak{E}_{\kappa_{m'-1}}(\mathbf{u}) - \mathfrak{E}_{\kappa_{m''+1}}(\mathbf{u})$ represents, then, the net amount of enstrophy per unit time that is transferred into the wavenumbers between $\kappa_{m'}$ and $\kappa_{m''}$.

The proof of (5.71) follows from properties (5.39), (5.40), and (5.41) of the trilinear term. Indeed, we can expand the RHS of (5.71) as follows:

$$\begin{aligned} \Psi_{\mathbf{y}}(\mathbf{w}+\mathbf{z}) - \Psi_{\mathbf{y}+\mathbf{w}}(\mathbf{z}) &= -b(\mathbf{y}, \mathbf{y}, A(\mathbf{w}+\mathbf{z})) + b(\mathbf{w}+\mathbf{z}, \mathbf{w}+\mathbf{z}, A\mathbf{y}) \\ &\quad + b(\mathbf{y}+\mathbf{w}, \mathbf{y}+\mathbf{w}, A\mathbf{z}) - b(\mathbf{z}, \mathbf{z}, A(\mathbf{y}+\mathbf{w})) \\ &= -b(\mathbf{y}, \mathbf{y}, A\mathbf{w}) - b(\mathbf{y}, \mathbf{y}, A\mathbf{z}) + b(\mathbf{w}, \mathbf{w}, A\mathbf{y}) \\ &\quad + b(\mathbf{w}, \mathbf{z}, A\mathbf{y}) + b(\mathbf{z}, \mathbf{w}, A\mathbf{y}) \\ &\quad + b(\mathbf{z}, \mathbf{z}, A\mathbf{y}) + b(\mathbf{y}, \mathbf{y}, A\mathbf{z}) + b(\mathbf{y}, \mathbf{w}, A\mathbf{z}) \\ &\quad + b(\mathbf{w}, \mathbf{y}, A\mathbf{z}) + b(\mathbf{w}, \mathbf{w}, A\mathbf{z}) - b(\mathbf{z}, \mathbf{z}, A\mathbf{y}) \\ &\quad - b(\mathbf{z}, \mathbf{z}, A\mathbf{w}) \\ &= -b(\mathbf{y}, \mathbf{y}, A\mathbf{w}) + b(\mathbf{w}, \mathbf{w}, A\mathbf{y}) + b(\mathbf{w}, \mathbf{z}, A\mathbf{y}) \\ &\quad + b(\mathbf{z}, \mathbf{w}, A\mathbf{y}) + b(\mathbf{y}, \mathbf{w}, A\mathbf{z}) + b(\mathbf{w}, \mathbf{y}, A\mathbf{z}) \\ &\quad + b(\mathbf{w}, \mathbf{w}, A\mathbf{z}) - b(\mathbf{z}, \mathbf{z}, A\mathbf{w}). \end{aligned}$$

We leave the first and last terms unchanged, apply (5.40) to the second and seventh terms, and apply (5.41) to the remaining terms. This yields

$$\begin{aligned}\Psi_{\mathbf{y}}(\mathbf{w}+\mathbf{z})-\Psi_{\mathbf{y}+\mathbf{w}}(\mathbf{z}) &= -b(\mathbf{y},\mathbf{y},A\mathbf{w})-b(\mathbf{y},\mathbf{w},A\mathbf{w})-b(\mathbf{w},\mathbf{y},A\mathbf{w})\\ &\quad -b(\mathbf{z},\mathbf{w},A\mathbf{w})-b(\mathbf{w},\mathbf{z},A\mathbf{w})-b(\mathbf{z},\mathbf{y},A\mathbf{w})\\ &\quad -b(\mathbf{y},\mathbf{z},A\mathbf{w})-b(\mathbf{z},\mathbf{z},A\mathbf{w})\\ &= -b(\mathbf{y}+\mathbf{z},\mathbf{w},A\mathbf{w})-b(\mathbf{y}+\mathbf{z}+\mathbf{w},\mathbf{y}+\mathbf{z},A\mathbf{w}).\end{aligned}$$

Finally, owing to (5.39), we have

$$b(\mathbf{w},\mathbf{w},A\mathbf{w})=0;$$

we subtract this vanishing term from the previous expression and find

$$\Psi_{\mathbf{y}}(\mathbf{w}+\mathbf{z})-\Psi_{\mathbf{y}+\mathbf{w}}(\mathbf{z})=-b(\mathbf{y}+\mathbf{w}+\mathbf{z},\mathbf{y}+\mathbf{w}+\mathbf{z},A\mathbf{w}),$$

which proves (5.71). Note that this equality actually holds for arbitrary $\mathbf{v}, \mathbf{w}, \mathbf{z}$, not only those that are associated with a spectral decomposition of an arbitrary vector field $\mathbf{u}$; such a decomposition was not used in the foregoing computations.

We thus have the equation

$$\frac{1}{2}\frac{d}{dt}\|\mathbf{w}\|^2+\nu|A\mathbf{w}|^2=\Psi_{\mathbf{y}}(\mathbf{w}+\mathbf{z})-\Psi_{\mathbf{y}+\mathbf{w}}(\mathbf{z}) \tag{5.73}$$

or, in a more explicit form,

$$\frac{1}{2}\frac{d}{dt}\|\mathbf{u}_{\kappa_{m'},\kappa_{m''}}\|^2+\nu|A\mathbf{u}_{\kappa_{m'},\kappa_{m''}}|^2=\mathfrak{E}_{\kappa_{m'-1}}(\mathbf{u})-\mathfrak{E}_{\kappa_{m''+1}}(\mathbf{u}). \tag{5.74}$$

Integrate (5.73) in time from 0 to T and divide by T to find

$$\begin{aligned}&\frac{1}{2T}\|\mathbf{w}(T)\|^2+\frac{\nu}{T}\int_0^T|A\mathbf{w}(s)|^2\,ds\\ &\qquad=\frac{1}{2T}\|\mathbf{w}(0)\|^2+\frac{1}{T}\int_0^T\big(\Psi_{\mathbf{y}}(s)(\mathbf{w}(s)+\mathbf{z}(s))-\Psi_{\mathbf{y}(s)+\mathbf{w}(s)}(\mathbf{z}(s))\big)\,ds.\end{aligned}$$

Take the generalized limit of this relation as T goes to infinity; then, because the enstrophy $\|\mathbf{w}(t)\|\leq\|\mathbf{u}(t)\|$ is uniformly bounded in time in the 2-dimensional case, we find

$$\langle\Psi_{\mathbf{y}}(\mathbf{w}+\mathbf{z})-\Psi_{\mathbf{y}+\mathbf{w}}(\mathbf{z})\rangle=\nu\langle|A\mathbf{w}|^2\rangle\geq 0. \tag{5.75}$$

In a more explicit form, we have

$$\langle\mathfrak{E}_{\kappa_{m''}}(\mathbf{u})\rangle=\langle\mathfrak{E}_{\kappa_{m'-1}}(\mathbf{u})\rangle-\nu\langle|A\mathbf{u}_{\kappa_{m'},\kappa_{m''}}|^2\rangle. \tag{5.76}$$

In other words, the average net transfer of enstrophy into the modes higher than $\kappa_{m''}$ is equal to the average net transfer of enstrophy into the modes higher than or equal to $\kappa_{m'}$ minus the enstrophy lost to viscous dissipation within the range $[\kappa_{m'},\kappa_{m''}]$.

In the enstrophy cascade, for $[\kappa_{m'},\kappa_{m''}]$ inside the inertial range, the viscous dissipation is negligible and the enstrophy is simply transferred to smaller and smaller length scales. This occurs whenever

$$\langle \mathfrak{E}_{\kappa_{m''}}(\mathbf{u})\rangle \gtrapprox \langle \mathfrak{E}_{\kappa_{m'-1}}(\mathbf{u})\rangle \gg \nu\langle |A\mathbf{u}_{\kappa_{m'},\kappa_{m''}}|^2\rangle. \tag{5.77}$$

The range of wavenumbers $\kappa > \kappa_{\bar{m}}$ up to where (5.77) holds is called the *inertial range* in 2-dimensional turbulence, as in 3-dimensional turbulence.

As proved in Theorem 5.2 (see (5.44)), within the enstrophy cascade range there is still an average net transfer of energy to higher modes:

$$\langle \mathfrak{e}_{\kappa_m}(\mathbf{u})\rangle = \nu\langle E(\mathbf{u}_{\kappa_{m+1},\infty})\rangle \geq 0 \tag{5.78}$$

for $\kappa_m > \kappa_{\bar{m}}$. Because

$$E(\mathbf{u}_{\kappa_{m+1},\infty}) = \|\mathbf{u}_{\kappa_{m+1},\infty}\|^2 \leq \frac{1}{\kappa_{m+1}^2}|A\mathbf{u}_{\kappa_{m+1},\infty}|^2,$$

we find from (5.78) and (5.65) that

$$\langle \mathfrak{e}_{\kappa_m}(\mathbf{u})\rangle \leq \frac{1}{\kappa_{m+1}^2}\langle \mathfrak{E}_{\kappa_m}(\mathbf{u})\rangle. \tag{5.79}$$

Hence, for large κ_{m+1}, the average net transfer of energy to higher modes is significantly smaller than the corresponding transfer of enstrophy. In other words, in the range beyond that of the injection of energy, the transfer of enstrophy to higher modes is more significant and yields the characteristic direct enstrophy cascade of 2-dimensional turbulence.

Inverse Energy Cascade

The inverse energy cascade takes place in the range below that of the injection of energy. Let us consider, then, $1 \leq m' \leq m'' < \underline{m}$ with $\underline{m} > 1$. Recall that $\underline{m}$ is a lower bound on the modes that are active in the forcing term. Typically, for this cascade to occur, we need $\underline{m} \gg 1$. We decompose the flow as in the direct energy cascade; that is,

$$\mathbf{u} = \mathbf{u}_{\kappa_1,\kappa_{m'-1}} + \mathbf{u}_{\kappa_{m'},\kappa_{m''}} + \mathbf{u}_{\kappa_{m''+1},\infty}. \tag{5.80}$$

Proceeding as for (5.37), we obtain

$$\langle \mathfrak{e}_{\kappa_{m''}}(\mathbf{u})\rangle = \langle \mathfrak{e}_{\kappa_{m'-1}}(\mathbf{u})\rangle - \nu\langle E(\mathbf{u}_{\kappa_{m'},\kappa_{m''}})\rangle. \tag{5.81}$$

Now, as we have seen in Theorem 5.1 (see also Theorem 5.2), we have in this case an inverse transfer of energy:

$$\langle \mathfrak{e}_{\kappa_{m''}}(\mathbf{u})\rangle \leq 0, \qquad \langle \mathfrak{e}_{\kappa_{m'-1}}(\mathbf{u})\rangle \leq 0.$$

Then, as long as

$$\langle \mathfrak{e}_{\kappa_{m''}}(\mathbf{u})\rangle \lessapprox \langle \mathfrak{e}_{\kappa_{m'-1}}(\mathbf{u})\rangle \ll -\nu\langle E(\mathbf{u}_{\kappa_{m'},\kappa_{m''}})\rangle, \tag{5.82}$$

there is an *inverse* energy cascade – from higher modes to lower modes. This is consistent with the heuristic conclusions in Kraichnan [1967].

As we have shown previously, within the range corresponding to the energy cascade (hence, below that of the injection of energy), both energy and enstrophy are transferred to the lower modes. Indeed, from (5.43) and (5.66) we have the following relations for $1 \le m < \underline{m}$:

$$\langle -\mathfrak{e}_{\kappa_m}(\mathbf{u}) \rangle = \nu \langle E(\mathbf{u}_{\kappa_1,\kappa_m}) \rangle \ge 0, \tag{5.83}$$

$$\langle -\mathfrak{E}_{\kappa_m}(\mathbf{u}) \rangle = \nu \langle |A\mathbf{u}_{\kappa_1,\kappa_m}|^2 \rangle \ge 0. \tag{5.84}$$

Here, $-\mathfrak{e}_{\kappa_m}(\mathbf{u})$ (resp., $-\mathfrak{E}_{\kappa_m}(\mathbf{u})$) denotes the transfer of energy (resp., enstrophy) to the lower modes. Since $\mathbf{u}_{\kappa_1,\kappa_m}$ contains only modes larger than or equal to κ_m, we see that

$$|A\mathbf{u}_{\kappa_1,\kappa_m}|^2 \le \kappa_m^2 \|\mathbf{u}_{\kappa_1,\kappa_m}\|^2 = \kappa_m^2 E(\mathbf{u}_{\kappa_1,\kappa_m}).$$

Thus, using (5.83) and (5.84), we obtain

$$\langle -\mathfrak{E}_{\kappa_m}(\mathbf{u}) \rangle \le \kappa_m \langle -\mathfrak{e}_{\kappa_m}(\mathbf{u}) \rangle. \tag{5.85}$$

Hence, for κ_m very small, we see that the (inverse) average net transfer of energy to lower modes is much stronger than the corresponding transfer of enstrophy, so that – below the injection of energy – the inverse energy cascade prevails.

5.3 Kraichnan's Cascade Mechanism

We now turn to a more physical (heuristic) discussion and consider a remarkable mechanism for explaining the cascades in 2- and 3-dimensional turbulence. This mechanism was introduced by Kraichnan [1972], and it was later exploited by Frisch, Nelkin, and Sulem [1978]. Here we follow the presentation in Foias [1997]. We start with the 2-dimensional periodic case with vanishing space average, and then we treat the 3-dimensional case. We denote the upper bound on the wavenumber components present on the forcing term as $\bar{\kappa} = \kappa_{\bar{m}}$.

In the 2-dimensional case, the mechanism that we describe is associated with the conservation of enstrophy by the inertial term, which imposes the direct enstrophy cascade in the inertial range beyond the injection of energy. For κ within the inertial range, the flow of enstrophy per unit mass[10] to higher modes is nearly constant:

$$\langle \kappa_1^2 \mathfrak{E}_\kappa(\mathbf{u}) \rangle \approx \eta,$$

where η is the total rate of enstrophy dissipation per unit mass,

$$\eta = \nu \langle \kappa_1^2 |A\mathbf{u}|^2 \rangle = \langle \kappa_1^2 (\mathbf{f}, A\mathbf{u}) \rangle.$$

Conventionally, the dissipation of enstrophy is thought to occur only at very small scales – below a certain length scale ℓ_η. The associated wavenumber is κ_η, and the range $[\kappa_\eta, \infty)$ is the so-called *dissipation range*. This κ_η is the Kraichnan dissipation wavenumber (see Kraichnan [1967]; see also the previous discussions in Sections III.3 and I.3).

[10] Note that $\kappa_1^2 = \lambda_1$ has the dimension of (length)$^{-2}$. Since the density has been normalized to 1, multiplication by κ_1^2 has the effect of normalizing the mass to 1.

Consider the components of the velocity field of the form $\mathbf{u}_{\kappa,2\kappa}$, that is, with wavenumbers in the range $[\kappa, 2\kappa)$. The wavenumber κ has the physical dimension of $(\text{length})^{-1}$, and the component of the velocity field with wavenumber κ is considered to represent eddies of linear size about κ^{-1}. Thus, the component $\mathbf{u}_{\kappa,2\kappa}$ of the velocity field is thought to represent the system of eddies of linear size in the range $(1/2\kappa, 1/\kappa]$. The energy in this component can be written as

$$e(\mathbf{u}_{\kappa,2\kappa}) = \frac{1}{2}|\mathbf{u}_{\kappa,2\kappa}|^2 = \frac{1}{2}\sum_{\kappa \le \chi < 2\kappa} |\hat{u}_k|^2, \tag{5.86}$$

where $|\hat{u}_k|^2/2$ is the kinetic energy corresponding to the wavenumber χ, and $\chi = \lambda_k^{1/2}$ (recall that λ_k is the kth eigenvalue of the Stokes operator; see e.g. (5.4)). For notational simplicity, we write the sum (5.86) in the expression of the energy *per unit mass* as an integral in the wavenumbers:

$$\kappa_1^2 e(\mathbf{u}_{\kappa,2\kappa}) = \int_\kappa^{2\kappa} \mathcal{S}(\chi)\,d\chi.$$

The function $\mathcal{S}(\kappa)$ is precisely the energy spectrum function. Similarly, we can write

$$\kappa_1^2 E(\mathbf{u}_{\kappa,2\kappa}) = \|\mathbf{u}_{\kappa,2\kappa}\|^2 = \int_\kappa^{2\kappa} \chi^2 \mathcal{S}(\chi)\,d\chi$$

and

$$\kappa_1^2 |A\mathbf{u}_{\kappa,2\kappa}|^2 = \int_\kappa^{2\kappa} \chi^4 \mathcal{S}(\chi)\,d\chi.$$

Notice that, from these identities,

$$\kappa_1^2 e(\mathbf{u}_{\kappa,2\kappa}) \approx \kappa\mathcal{S}(\kappa), \quad \kappa_1^2 E(\mathbf{u}_{\kappa,2\kappa}) \approx 2\kappa^3\mathcal{S}(\kappa), \quad \kappa_1^2 |A\mathbf{u}_{\kappa,2\kappa}|^2 \approx \kappa^5\mathcal{S}(\kappa).$$

The condition (5.77) for $[\kappa, 2\kappa)$ to be within the region of direct enstrophy cascade then reads

$$\nu\kappa^5\mathcal{S}(\kappa) \ll \kappa_1^2 \langle \mathfrak{E}_\kappa(\mathbf{u})\rangle.$$

Moreover, as we have discussed, the transfer of enstrophy per unit mass within the inertial range is nearly constant and equal to η, so we see from (5.77) that $[\kappa, 2\kappa)$ is within the inertial range as long as

$$\nu\kappa^5\mathcal{S}(\kappa) \ll \eta. \tag{5.87}$$

Now we describe the mechanism explaining the cascade in the 2-dimensional case. The transfer of enstrophy is considered to be produced by the breakup of eddies with wavenumbers in $[\kappa, 2\kappa)$ into eddies of linear size smaller than or equal to $1/2\kappa$. This breakup is assumed to occur after the eddy travels a distance comparable to its linear size. Since the energy of the eddies with linear size in the range $(1/2\kappa, 1/\kappa]$ is on average about

$$\int_\kappa^{2\kappa} \mathcal{S}(\chi)\,d\chi \approx \kappa\mathcal{S}(\kappa),$$

we see that the average velocity of those eddies is

$$V_\kappa \approx (\kappa \mathcal{S}(\kappa))^{1/2}.$$

Hence, the time necessary for those eddies to travel a distance comparable to their linear size is

$$t_\kappa \approx \frac{1}{\kappa V_\kappa} \approx \frac{1}{\kappa^{2/3} \mathcal{S}(\kappa)^{1/2}}.$$

On the other hand, the enstrophy of eddies with linear size in the range $(1/2\kappa, 1/\kappa]$ is

$$\int_\kappa^{2\kappa} \chi^2 \mathcal{S}(\chi)\, d\chi \approx 2\kappa^3 \mathcal{S}(\kappa).$$

Therefore, according to the breakup mechanism, the average dissipation of enstrophy per unit time – within the inertial range where the direct enstrophy cascade occurs – should be

$$\eta \approx \frac{\kappa^3 \mathcal{S}(\kappa)}{t_\kappa} \approx \kappa^3 (\kappa \mathcal{S}(\kappa))^{3/2}.$$

From this it follows that

$$\mathcal{S}(\kappa) \approx \frac{\eta^{2/3}}{\kappa^3}. \tag{5.88}$$

According to the previous argument (see (5.87)), one expects (5.88) as long as

$$\kappa > \bar{\kappa} \quad \text{and} \quad \nu \kappa^5 \mathcal{S}(\kappa) \ll \eta. \tag{5.89}$$

Therefore, from (5.88) and (5.89), we see that

$$\mathcal{S}(\kappa) \approx \frac{\eta^{2/3}}{\kappa^3} \quad \text{as long as } \underline{\kappa} < \kappa \ll \kappa_\eta \equiv \left(\frac{\eta}{\nu^3}\right)^{1/6}, \tag{5.90}$$

which also gives an explicit expression for the cutoff wavenumber κ_η.

In the 3-dimensional case, the role played by the enstrophy in the preceeding argument is taken over by the energy. Condition (5.38) for the direct energy cascade reads

$$\nu \kappa^3 \mathcal{S}(\kappa) \ll \epsilon, \tag{5.91}$$

where ϵ is the total average rate of energy dissipation per unit mass:

$$\epsilon = \kappa_1^3 \nu \langle E(\mathbf{u}) \rangle = \kappa_1^3 \nu \langle \|\mathbf{u}\|^2 \rangle.$$

Within the inertial range, energy is simply transferred to higher modes at a nearly constant rate that is approximately equal to ϵ.

The breakup mechanism is such that a transfer of energy is produced by the breakup of the eddies with size between $(1/2\kappa, 1/\kappa]$ into eddies of linear size smaller than or equal to $1/2\kappa$. Proceeding as in the 2-dimensional case, this mechanism leads to the following expression for the average dissipation of energy per unit time:

$$\epsilon \approx \frac{\kappa \mathcal{S}(\kappa)}{t_\kappa} \approx (\kappa \mathcal{S}(\kappa))^{3/2} \kappa.$$

Thus,

$$\mathcal{S}(\kappa) \approx \frac{\epsilon^{2/3}}{\kappa^{5/3}}$$

as long as $\kappa > \bar{\kappa}$, with

$$\nu\kappa^{4/3}\epsilon^{2/3} \ll \epsilon.$$

Therefore,

$$\mathcal{S}(\kappa) \approx \frac{\epsilon^{2/3}}{\kappa^{5/3}} \quad \text{as long as } \underline{\kappa} < \kappa \ll \kappa_d \equiv \left(\frac{\epsilon}{\nu^3}\right)^{1/4}. \tag{5.92}$$

The cutoff wavenumber κ_d is Kolmogorov's dissipation wavenumber (see Section 3 in Chapter I). The spectra given by (5.92) and (5.90) are called, respectively, the *Kolmogorov spectrum for 3-dimensional turbulence* (Kolmogorov [1941a]) and the *Kraichnan spectrum for 2-dimensional turbulence* (Kraichnan [1967]); see also Monin and Yaglom [1975] and Rose and Sulem [1978].

The Effects of Walls on Kraichnan's Dissipation Wavenumber

As discussed in Section III.3, the technique for calculating the number of degrees of freedom (in two dimensions) in the presence of a no-slip boundary condition is hampered by the difficulty in estimating the Kraichnan cutoff wavenumber κ_η. The mathematical method used in Section III.3 for estimating the dimension of the attractor involved the wavenumber κ_d, whose definition was strikingly similar to that of the Kolmogorov wavenumber for 3-dimensional fluid flows. This certainly seems to be at odds with the dominance of the Kraichnan spectrum in 2-dimensional turbulence. Fortunately, the results developed in this section offer an approach that provides an explanation for why – although, in the spectral portion of interest, the energy cascade is negligible when compared with the enstrophy cascade – the Kolmogorov-like wavenumber (i.e., one that is dependent on the energy dissipation rate ϵ) is a "close" approximation of the total Kraichnan dissipation number.

As seen in the earlier portions of this section, the net average flow of energy from some point in the spectrum (say, $\kappa > \bar{\kappa}$, where $\bar{\kappa}$ is the highest wavenumber present in the driving force $\mathbf{f}$) to the end of the inertial range at κ_η is given by

$$\langle \nu\kappa_1^2 \|\mathbf{u}_{\kappa,\kappa_\eta}\|^2 \rangle = \kappa_1^2[\langle \mathfrak{e}_\kappa(\mathbf{u})\rangle - \langle \mathfrak{e}_{\kappa_\eta}(\mathbf{u})\rangle], \tag{5.93}$$

where $\kappa_1^2\mathfrak{e}_\kappa(\mathbf{u})$ is the rate of net energy transfer per unit mass at the wavenumber κ within the inertial range. Henceforth, we limit ourselves to the portion of the Kraichnan inertial range in which the spectrum is given by $\eta^{2/3}\kappa^{-3}$ and where, as follows from our earlier discussion, the rate of net transfer of enstrophy is $\langle \kappa_1^2 \mathfrak{E}_\kappa(\mathbf{u})\rangle \approx \eta$. Here we neglect the departure from this spectrum in the boundary layer abutting the solid boundary. In a sense, we assume that the characteristic length (say, L) of the flow domain is by far much larger than the thickness of the boundary layer. Using that spectrum allows us to obtain useful results involving some norms (which are integrals over the flow domain) by carrying out the required calculations in the corresponding Fourier domain.

To proceed further, recall that we have $\langle \mathfrak{e}_\kappa(\mathbf{u})\rangle \le \langle \mathfrak{E}_\kappa(\mathbf{u})\rangle/\kappa^2$ for all $\kappa > \bar{\kappa}$. Now we rewrite the LHS of (5.93) in the Fourier domain as

$$\langle \nu\kappa_1^2 \|\mathbf{u}_{\kappa,\kappa_\eta}\|^2\rangle = \nu\eta^{2/3} \int_\kappa^{\kappa_\eta} \frac{\chi^2}{\chi^3}\, d\chi = \nu\eta^{2/3} \ln\left(\frac{\kappa_\eta}{\kappa}\right).$$

Remember that the integral in this expression is actually a sum (Fourier series). However, since κ_η is large, we can consider this sum as the Riemann sum of an integral; whence the abuse of notation by using this integral sign. We then rearrange the terms to obtain

$$\begin{aligned}\kappa_1^2\langle \mathfrak{e}_\kappa(\mathbf{u})\rangle &= \nu\eta^{2/3} \ln\left(\frac{\kappa_\eta}{\kappa}\right) + \kappa_1^2 \langle \mathfrak{e}_{\kappa_\eta}(\mathbf{u})\rangle \\ &\leq \nu\eta^{2/3} \ln\left(\frac{\kappa_\eta}{\kappa}\right) + \langle \mathfrak{E}_{\kappa_\eta}\rangle \frac{\kappa_1^2}{\kappa_\eta^2}. \end{aligned} \tag{5.94}$$

Now, since $\kappa_1^2 \langle \mathfrak{E}_{\kappa_\eta}(\mathbf{u})\rangle \approx \eta$,

$$\begin{aligned}\bar{\epsilon} \equiv \kappa_1^2 \langle \mathfrak{e}_{\kappa}\rangle &= \nu\eta^{2/3} \ln\left(\frac{\kappa_\eta}{\kappa}\right) + \frac{\eta}{\kappa_\eta^2} \\ &\leq \nu^3 \left(\frac{\eta}{\nu^3}\right)^{2/3} \left[\ln\left(\frac{\kappa_\eta}{\kappa}\right) + 1\right], \end{aligned} \tag{5.95}$$

where we have used the definition of $\kappa_\eta \equiv (\eta/\nu^3)^{1/6}$. Hence, we have

$$\kappa_{\bar{\epsilon}}^4 \equiv \frac{\bar{\epsilon}}{\nu^3} = \kappa_\eta^4 \left(\ln\left(\frac{\kappa_\eta}{\kappa}\right) + 1\right). \tag{5.96}$$

We have thus arrived at a characteristic wavenumber $\kappa_{\bar{\epsilon}}$ associated with the small but finite energy transfer in parallel with the enstrophy cascade. We find that this characteristic wavenumber is expressed in terms of the Kraichnan cutoff wavenumber κ_η. This estimate is not affected by the exclusion of the boundary layer from the preceding discussion. Indeed, the integral of

$$\sum_{i,j} \left(v_i \frac{\partial v_j}{\partial x_i} w_j - w_i \frac{\partial w_j}{\partial x_i} v_j \right), \quad \mathbf{v} = P_\kappa \mathbf{u}, \quad \mathbf{w} = Q_\kappa \mathbf{u}$$

over the boundary layer (which should have been excluded from the computation of $\langle \mathfrak{e}_\kappa \rangle$ in the definition (5.95)) of $\bar{\epsilon}$ is small when compared to $\mathfrak{e}_\kappa$. (This fact involves a sophisticated mathematical argument that is beyond the scope of the present heuristic discussion.)

Finally, using the same argument as in the case of periodic flows, we estimate the dimension of the attractor for the no-slip case as $(\kappa_\eta/\kappa_1)^2$, that is, $(\kappa_{\bar{\epsilon}}/\kappa_1)^2$.

Appendix A New Concepts and Results Used in Chapter IV

The purpose of this appendix is first, in Section A.1, to recall a few fundamental facts about measure theory that are used throughout this chapter and the next. In Section A.2 we discuss the concept of generalized limits, which we use in defining the limits of time averages.

A.1 Background on Measure Theory

In the first part of this chapter we used the expression "probability distribution," which is commonly seen in the physical literature; the usual mathematical expression is that of a probability *measure* – in particular, the Borel measures (Rudin [1987]) or the Radon measures (Bourbaki [1969]).

For the convenience of the reader, we recall here a few basic facts in measure theory. Borel measures are defined as measures on the Borel family of a topological set X (the σ-algebra generated by the open and closed subsets of X); Radon measures are defined as linear functionals on the set of continuous functions on X. We adopt the Borel point of view for this presentation, but the equivalence between Borel and Radon measures follows from the Kakutani–Riesz representation theorem. Indeed, we have used this equivalence (and the theorem) throughout this chapter, when time-average measures were defined as linear functionals on $\mathcal{C}(X)$; see Sections 2 and 3.

A *measurable space* is a couple $(X, \mathcal{M})$ where X is a set and $\mathcal{M}$ is a collection of subsets of X (the "events") with the properties that (i) the empty set is in $\mathcal{M}$, (ii) complements of sets in $\mathcal{M}$ belong to $\mathcal{M}$, and (iii) countable unions of sets in $\mathcal{M}$ belong to $\mathcal{M}$. From the first two properties, the whole set X must belong to $\mathcal{M}$; from this and the previous properties it follows that countable intersections of sets in $\mathcal{M}$ also belong to $\mathcal{M}$. A collection $\mathcal{M}$ with these properties is called a σ-algebra.

A *measure* on a measurable space $(X, \mathcal{M})$ is a nonnegative set function $\mu : \mathcal{M} \to [0, \infty)$ with the properties that $\mu(\emptyset) = 0$ and $\mu\big(\bigcup_{E\in\mathcal{E}} E\big) = \sum_{E\in\mathcal{E}} \mu(E)$ for all countable, pairwise disjoint collections $\mathcal{E} \subset \mathcal{M}$. A *probability measure* on a measurable space $(X, \mathcal{M})$ is a measure μ on $(X, \mathcal{M})$ with the property that $\mu(X) = 1$.

For our purposes, it suffices to consider measures on the space H or on subsets of H. Because we wish to take into account the topology of H, it is natural to require that the open and closed sets belong to the corresponding σ-algebra. Therefore, it is natural to consider the so-called Borel σ-algebra of H. By definition, the *Borel σ-algebra* of a given topological space X is the smallest σ-algebra that contains the open subsets (and hence also the closed subsets) of X. The sets in the Borel σ-algebra of X are called the Borel sets of X or the Borel measurable sets of X. A measure defined on a Borel σ-algebra is called a *Borel measure,* and a probability measure defined on Borel σ-algebra is called a *Borel probability measure.*

We also consider the space H endowed with the weak topology – that is, the space H_w. This could, in principle, lead to a different Borel σ-algebra. However, that does not happen. The σ-algebra generated by the weakly open sets in H coincides with that generated by the strongly open sets in H. Indeed, since weakly open sets are open for the strong topology, it is obvious that Borel sets in H_w are Borel sets in H. On the other hand, since H is separable, any open set in H is a countable union of open balls. Moreover, from the Parseval identity $|\mathbf{u}|^2 = \sum_{m\in\mathbb{N}} |(\mathbf{u}, \mathbf{w}_m)|^2$, any open ball in H, say $\{\mathbf{u} \in H;\ |\mathbf{u} - \mathbf{u}_0| < r\}$, can be written as a countable intersection of weakly open sets:

$$\{\mathbf{u} \in H;\ |\mathbf{u} - \mathbf{u}_0| < r\} = \bigcap_{m \in \mathbb{N}} \left\{ \mathbf{u} \in H;\ \sum_{j=1}^{m} |(\mathbf{u} - \mathbf{u}_0, \mathbf{w}_j)|^2 < r^2 \right\}.$$

Therefore, any open set in H belongs to the Borel σ-algebra of H_w.

We also want to take into account the topologies of V, $D(A)$ and $D(A^{-1/2})$; hence we wish to compare the corresponding Borel sets. For instance, any open set in V can be written as a countable union of closed balls in V, and any closed ball in V, say $\{\mathbf{u} \in V;\ \|\mathbf{u} - \mathbf{u}_0\| \leq r\}$, can be written as a countable intersection of closed sets in H:

$$\{\mathbf{u} \in V;\ \|\mathbf{u} - \mathbf{u}_0\| \leq r\} = \bigcap_{m \in \mathbb{N}} \{\mathbf{u} \in H;\ \|P_m \mathbf{u} - P_m \mathbf{u}_0\| \leq r\}.$$

Therefore, any Borel set in V is a Borel set in H. On the other hand, since any set in V that is open with respect to the topology of H is also open in V, it follows that the intersection of V with any Borel set of H is a Borel set of V. One consequence of this fact is that any Borel measure μ on V can be extended to a Borel measure μ' on H through the relation $\mu'(E) = \mu(E \cap V)$ for every Borel set E in H. Similar statements hold for $D(A)$, $D(A^{-1/2})$, and for subsets of $D(A)$, V, H, H_w, and $D(A^{-1/2})$.

A concept we will use often is that of a support of a measure. For a Borel probability measure μ on H, the *support* of μ (denoted $\operatorname{supp} \mu$) is the smallest closed set of full measure. A nontopological concept is that of a *carrier* of a measure: we say that a measurable set $E \subset H$ *carries* μ if E is a set of full measure – that is, if $\mu(E) = 1$ or $\mu(H \setminus E) = 0$. Notice that carriers are not unique, in general. Another important concept is that of *regularity* of the measure. Since H is separable, any Borel probability measure μ on H is automatically regular in the sense that, for every Borel set E in H, we have

$$\mu(E) = \sup\{\mu(K);\ K \subset E,\ K \text{ compact in } H\}, \tag{A.1}$$

$$\mu(E) = \inf\{\mu(\mathcal{O});\ E \subset \mathcal{O},\ \mathcal{O} \text{ open in } H\}. \tag{A.2}$$

This regularity result holds not only for H but for any metrizable, separable, complete space – also called Polish spaces (see Bourbaki [1969]). Therefore, since sets in H can be written as countable collections of bounded sets in H and since bounded sets in H endowed with the weak topology are metrizable as well as separable, we can obtain this regularity also with respect to the weak topology. More precisely, we have

$$\mu(E) = \sup\{\mu(K);\ K \subset E,\ K \text{ weakly compact in } H\}, \tag{A.3}$$

$$\mu(E) = \inf\{\mu(\mathcal{O});\ E \subset \mathcal{O},\ \mathcal{O} \text{ weakly open in } H\}. \tag{A.4}$$

Actually, the regularity (A.3) follows directly from (A.1) because compact sets in H are weakly compact. Similar regularity results follow with respect to the topologies of $D(A)$, V, and $D(A^{-1/2})$.

An important consequence of the regularity of a Borel probability measure is the density of the continuous functions (or just weakly continuous functions) in the space of integrable functions. In the case of interest to us, if μ is a Borel probability measure on H and if φ is a μ-integrable function on H then, given any $\varepsilon > 0$, there exists a function ψ defined on H and continuous (or just weakly continuous) such that

$$\int |\varphi(\mathbf{u}) - \psi(\mathbf{u})| \, d\mu(\mathbf{u}) < \varepsilon.$$

With these remarks in mind and for the sake of simplicity: throughout this work, *a measure, probability measure, density, or probability distribution always means a Borel probability measure on H, whereas a measurable set always means a Borel measurable set,* unless otherwise stated.

We now quote (without proof) the classical theorems that are used in Chapters IV and V. Each theorem is followed by a (nonexhaustive) list of places where it has been used.

Theorems in Measure Theory

Theorem A.1 (Kakutani–Riesz Representation Theorem) *Let X be a locally compact Hausdorff space. Let Λ be a positive linear functional on $\mathcal{C}_c(X)$, the space of compactly supported, continuous real-valued functions on X. Then there exists a σ-algebra $\mathcal{M}$ in X that contains all Borel sets in X, and there also exists a unique positive measure μ on $\mathcal{M}$ which represents Λ in the sense that*

(a) $\Lambda f = \int_X f \, d\mu(\mathbf{u})$ *for every* $f \in \mathcal{C}_c(X)$

and which has the following additional properties:

(b) $\mu(K) < \infty$ *for every compact set* $K \subset X$;
(c) μ *is outer regular in the sense that, for every* $E \in \mathcal{M}$, *we have*

$$\mu(E) = \inf\{\mu(\mathcal{O});\ E \subset \mathcal{O},\ \mathcal{O} \text{ open}\};$$

(d) μ *is inner regular in the sense that, for every open set E and for every* $E \subset \mathcal{M}$ *with* $\mu(E) < \infty$,

$$\mu(E) = \sup\{\mu(K);\ K \subset E,\ K \text{ compact}\};$$

(e) μ *is complete – that is, if* $E \in \mathcal{M}$, $A \subset E$, *and* $\mu(E) = 0$, *then* $A \in \mathcal{M}$.

We recall here that a Hausdorff topological space is one in which points can be separated by disjoint open neighborhoods. The space H is Hausdorff, as is any Hilbert space. For a proof of Theorem A.1, see Rudin [1987].

The Kakutani–Riesz representation theorem was used, for example, in the proof of the existence of time-average measures (Propositions 2.1–2.3 and 3.1).

Theorem A.2 (Monotone Convergence Theorem) *Let $(X, \mathcal{M}, \mu)$ be a measure space. Let $\{g_n\}$ be a monotone increasing sequence of nonnegative, real-valued, μ-measurable functions (not necessarily integrable) that converge μ-almost everywhere to a function g. Then*

$$\lim_{n\to\infty} \int_X g_n(x)\, d\mu(x) = \int_X g(x)\, d\mu(x).$$

For a proof of the monotone convergence theorem, see Dunford and Schwartz [1958] or Rudin [1987]. This theorem was used on several occasions; see Section 3.

Theorem A.3 (Lebesgue Dominated Convergence Theorem) *Let $1 \le p < \infty$, let $(X, \mathcal{M}, \mu)$ be a measure space, and let Y be a Banach space. Let $\{g_n\}$ be a sequence of functions in $L^p(X, \mathcal{M}, \mu; Y)$ (i.e., functions from X into Y that are in L^p with respect to the measure μ). Suppose $\{g_n\}$ converges μ-almost everywhere to a function g, and suppose there exists a function $\bar{g}$ in $L^p(X, \mathcal{M}, \mu; Y)$ such that $|g_n(s)| \le |\bar{g}(s)|$ μ-almost everywhere. Then g belongs to $L^p(X, \mathcal{M}, \mu; Y)$ and $\{g_n\}$ converges to g in $L^p(X, \mathcal{M}, \mu; Y)$.*

This theorem is used on several occasions, in particular for the proof of Theorem 3.1 (see Appendix B.2). For a proof of Theorem A.3, see Dunford and Schwartz [1958].

Theorem A.4 *Let $(X, \Sigma_X, \rho) = (X_1, \Sigma_{X_1}, \mu) \times (X_2, \Sigma_{X_2}, \lambda)$ be the product of two positive σ-finite measure spaces. Then we have the following results.*

(i) (Tonelli Theorem) *If f is a nonnegative ρ-measurable funcion on X then, for μ-almost all x_1 in X_1: the function $f(x_1, \cdot)$ is λ-measurable; the (extended real-valued) function $\int_{X_2} f(\cdot, x_2)\, d\lambda(x_2)$ is μ-measurable; and, with $x = (x_1, x_2)$,*

$$\int_{X_1} \left\{ \int_{X_2} f(x_1, x_2)\, d\lambda(x_2) \right\} d\mu(x_1) = \int_X f(x)\, d\rho(x),$$

irrespective of whether the integrals have finite or infinite values.

(ii) (Fubini Theorem) *If f is a ρ-integrable function on X then, for μ-almost all x_1 in X_1: the function $f(x_1, \cdot)$ is λ-integrable on X_2; the (extended real-valued) function $\int_{X_2} f(\cdot, x_2)\, d\lambda(x_2)$ is μ-integrable on X_1; and, with $x = (x_1, x_2)$,*

$$\int_{X_1} \left\{ \int_{X_2} f(x_1, x_2)\, d\lambda(x_2) \right\} d\mu(x_1) = \int_X f(x)\, d\rho(x).$$

For a proof of the Tonelli and Fubini theorems, see Dunford and Schwartz [1958]. Tonelli's theorem is used in the proof of Proposition C.3, Fubini's in the proof of Theorem V.1.1.

Theorem A.5 (Lebesgue Differentiation Theorem) *If $f \in L^1_{\text{loc}}(0, \infty)$ then, for almost every $t \in [0, \infty)$, it follows that*

$$\lim_{h\to 0} \frac{1}{h} \int_t^{t+h} f(s)\, ds = f(t);$$

the points t where the convergence holds are called the Lebesgue points *of f.*

This is a simplified version of the Lebesgue differentiation theorem that will be sufficient for our purposes (we use it, e.g., in Theorem 3.1). For its proof, see Hewitt and Stromberg [1975].

Theorems in Topology

Theorem A.6 (Urysohn Lemma) *Let A and B be disjoint closed sets in a normal topological space X. Then there exists a continuous real-valued function f, defined on X, such that* $0 \leq f(x) \leq 1$ *with* $f(A) = 0$ *and* $f(B) = 1$.

We recall that a normal topological space is one with the properties that (i) singletons (sets consisting of a single point) are closed and (ii) for every pair of disjoint closed sets A and B, there exist disjoint neighborhoods of A and B.

For a proof of the Urysohn lemma, see Dunford and Schwartz [1958]. This lemma is used in the proofs that invariant measures have their support included in the global attractor for the 2-dimensional case (Theorem 4.1) and that time-average measures have their support included in the weak global attractor for the 3-dimensional case (Theorem 4.2).

Theorem A.7 (Tietze Extension Theorem) *Let X be a normal topological space, let A be a closed subset of X, and let f be a bounded real-valued continuous function defined on A. Then there exists a continuous real-valued function F defined on X such that* (i) $F(x) = f(x)$ *for all x in A and* (ii) $\sup_{x\in X}|F(x)| = \sup_{x\in A}|f(x)|$.

For a proof of the Tietze extension theorem, see Dunford and Schwartz [1958, Cor. I.5.4]. This theorem is used in proofs of the existence of time-average measures (Propositions 2.1–2.3 and 3.1).

Theorems in Functional Analysis

Theorem A.8 (Hahn–Banach Theorem) *Let a real-valued function p defined on a real linear space X satisfy*

$$p(x + y) \leq p(x) + p(y) \quad \textit{and} \quad p(\alpha x) = \alpha p(x)$$

for $\alpha \geq 0$ *and* $x, y \in X$. *Let f be a real-valued linear functional defined on a subspace Y of X with*

$$f(x) \leq p(x), \quad x \in Y.$$

Then there exists a real-valued linear functional F on X for which

$$F(x) = f(x), \quad x \in Y; \qquad F(x) \leq p(x), \quad x \in X.$$

For a proof of the Hahn–Banach theorem, see Dunford and Schwartz [1958]. This theorem is used in the definition of the Banach limit $\textsc{Lim}_{T\to\infty}$; see Section A.2.

Theorem A.9 (Krein–Milman Theorem) *If K is a compact subset of a locally convex topological vector space X and if E is the set of extremal points of K, then the closed convex hull of E is equal to the closed convex hull of K. If K is convex, then K itself is equal to the closed convex hull of E.*

An extremal point of K is a point x such that $x = ax_1 + (1-a)x_2$ with $0 < a < 1$ and x_1, x_2 in X if and only if $x_1 = x_2 = x$. A topological vector space is a linear vector space endowed with a topology that renders continuous the operations of vector addition and scalar multiplication. A locally convex topological vector space is a topological vector space with the property that any neighborhood of the origin contains a convex neighborhood.

For a proof of Krein–Milman theorem, see Dunford and Schwartz [1958]. This theorem is used in the proof of existence of time-dependent statistical solutions of the Navier–Stokes equations (Theorems 1.1 and 1.3).

Theorem A.10 (Arzelà–Ascoli Theorem) *Let (X, d_1) be a compact metric space and let (Y, d_2) be a complete metric space. Let $\mathcal{C}(X, Y)$ denote the space of continuous functions from X into Y, endowed with the metric*

$$\rho(f_1, f_2) = \sup_{x\in X} d_2(f_1(x), f_2(x)).$$

Let $\mathcal{S}$ be a subset of $\mathcal{C}(X, Y)$. Suppose that $\mathcal{S}$ is pointwise compact (i.e., for each $x \in X$ the set $\{f(x);\ f \in \mathcal{S}\}$ has compact closure in Y) and equicontinuous (i.e., for each $\varepsilon > 0$ there exists $\delta > 0$ such that if $d_1(x_1, x_2) < \delta$ then $d_2(f(x_1), f(x_2)) < \varepsilon$ for every f in $\mathcal{S}$). Then $\mathcal{S}$ has compact closure in $\mathcal{C}(X, Y)$.

For a proof of the Arzelà–Ascoli theorem, see for example Naylor and Sell [1982]. This theorem is used in the proof of Theorem V.1.1 on the existence of time-dependent statistical solutions.

Theorem A.11 (Aubin Compactness Theorem) *Let X_0, X, X_1 be three Banach spaces and suppose that X_0 and X_1 are reflexive, $X_0 \subset X$ with compact injection and $X \subset X_1$ with continuous injection. Let $T > 0$ and $p_0, p_1 > 1$. Consider the space*

$$\mathcal{Y} = \{\mathbf{v} \in L^{p_0}(0, T; X_0);\ \mathbf{v}' = d\mathbf{v}/dt \in L^{p_1}(0, T; X_1)\}$$

endowed with the norm

$$\|\mathbf{v}\|_{\mathcal{Y}} = \|\mathbf{v}\|_{L^{p_0}(0,T;X_0)} + \|\mathbf{v}'\|_{L^{p_1}(0,T;X_1)}.$$

Then the injection of $\mathcal{Y}$ into $L^{p_0}(0, T; X)$ is compact.

For a proof, see Aubin [1963] or Temam [2001]. This theorem is used in the proof of Theorem V.1.1, and it is also used in the proof of existence of weak solutions of the Navier–Stokes equations. Another very useful compactness theorem of this type is Theorem 13.2 in Temam [1995].

Theorem A.12 (Prokhorov Theorem) *Let $\{\rho_n\}_n$ be a sequence of Borel probability measures on a complete and separable metric space S. Then $\{\rho_n\}_n$ has a weakly convergent subsequence if and only if, for each $\varepsilon > 0$, there exists a compact set K_ε in S such that $\rho_n(K_\varepsilon) \geq 1 - \varepsilon$ for all n in $\mathbb{N}$.*

We recall that a sequence of Borel probability measures $\{\rho_n\}_n$ on a metric space S is said to converge weakly to a Borel probability measure ρ on S if, for every continuous bounded real-valued function ψ on S,

$$\int_S \psi(s)\, d\rho_n(s) \to \int_S \psi(s)\, d\rho(s)$$

as n goes to infinity. For a proof of the Prokhorov theorem, see Prohorov [1956, Thm. 1.12] or Gihman and Skorohod [1974, Thm. VI.1.1]. This theorem is used in the proof of Theorem V.2.3 on the existence of homogeneous time-dependent statistical solutions on the whole space.

A.2 Banach Generalized Limits

We prove here the existence and a few properties of generalized limits as defined in Section 1.3; these are also called Banach generalized limits. For notational simplicity, let $G = \mathcal{B}([0, \infty))$ be the set of bounded real-valued functions on $[0, \infty)$. Define on G the functional

$$p(g) = \limsup_{t\to 0} g(t) \quad \forall g \in G. \tag{A.5}$$

It is easy to see that the functional $p = p(g)$ is a gauge function on G; that is, $p(f + g) \leq p(f) + p(g)$ and $p(\alpha g) = \alpha p(g)$ for all $f, g \in G$ and all $\alpha \geq 0$.

Consider now the subspace

$$G_0 = \left\{ g \in G;\ \lim_{t\to\infty} g(t) \text{exists} \right\},$$

and define on G_0 the linear functional

$$\Lambda_0(g) = \lim_{t\to\infty} g(t) \quad \forall g \in G_0.$$

Note that $\Lambda_0(g) = p(g)$ for all g in G_0. Hence, by the Hahn–Banach extension theorem, there exists a linear functional Λ defined on the whole space G such that

$$\Lambda(g) = \Lambda_0(g) = \lim_{t\to\infty} g(t) \quad \forall g \in G_0 \tag{A.6}$$

and

$$\Lambda(g) \leq p(g) \quad \forall g \in G.$$

We show now that Λ is a generalized limit in the sense of our Definition 1.4. In view of (A.6), we need only prove property (1.35). For that purpose, note that if $g(t) \geq 0$ for all $t \geq 0$ then $-g \leq 0$ and, clearly, $p(-g) \leq 0$; hence $\Lambda(-g) \leq p(-g) \leq 0$ and thus $\Lambda(g) \geq 0$, which proves the positivity of Λ. This completes the proof that Λ is a generalized limit.

It remains to prove properties (1.37), (1.38), and (1.40) for an arbitrary generalized limit $\text{LIM}_{T\to\infty}$. Let $g \in \mathcal{B}([0, \infty))$. Define

$$\bar{g}(T) = \sup_{t\geq T} g(t) \quad \forall T \geq 0.$$

Note that $\bar{g}(T) - g(T) \geq 0$ for all $T \geq 0$. Therefore, by (1.35), it follows that

$$\operatorname*{LIM}_{T\to\infty} g(T) \leq \operatorname*{LIM}_{T\to\infty} \bar{g}(T).$$

But $\mathrm{LIM}_{T\to\infty} \bar{g}(T) = \lim_{T\to\infty} \bar{g}(T) = \limsup_{T\to\infty} g(T)$. Thus,

$$\operatorname*{LIM}_{T\to\infty} g(T) \leq \limsup_{T\to\infty} g(T).$$

Similarly, using $\underline{g}(T) = \inf_{t\geq T} g(t)$, one can show that

$$\operatorname*{LIM}_{T\to\infty} g(T) \geq \liminf_{T\to\infty} g(T),$$

proving (1.37). Property (1.38) follows easily from (1.37) by noting that

$$\limsup_{T\to\infty} g(T) \leq \sup_{T\geq 0}|g(T)| \quad \text{and} \quad \liminf_{T\to\infty} g(T) \geq -\sup_{T\geq 0}|g(T)|.$$

Finally, let us prove (1.40). Let $f \in L^\infty(0,\infty)$, and set

$$g(T) = \frac{1}{T}\int_0^T f(t)\,dt.$$

Clearly $g \in \mathcal{B}([0,\infty))$, so $\mathrm{LIM}_{T\to\infty} g(T)$ makes sense. Now take $\tau > 0$ and write

$$\begin{aligned}
&\operatorname*{LIM}_{T\to\infty} g(T+\tau)\\
&\quad = \operatorname*{LIM}_{T\to\infty} \frac{1}{T+\tau}\int_0^{T+\tau} f(t)\,dt\\
&\quad = \operatorname*{LIM}_{T\to\infty} \frac{1}{T}\int_0^T f(t)\,dt + \operatorname*{LIM}_{T\to\infty}\left[\frac{1}{T+\tau}\int_0^{T+\tau} f(t)\,dt - \frac{1}{T}\int_0^T f(t)\,dt\right]\\
&\quad = \operatorname*{LIM}_{T\to\infty} g(T) + \operatorname*{LIM}_{T\to\infty}\left[\left(\frac{1}{T+\tau} - \frac{1}{T}\right)\int_0^{T+\tau} f(t)\,dt + \frac{1}{T}\int_T^{T+\tau} f(t)\,dt\right]\\
&\quad = \operatorname*{LIM}_{T\to\infty} g(T) + \operatorname*{LIM}_{T\to\infty}\left[-\frac{\tau}{T(T+\tau)}\int_0^{T+\tau} f(t)\,dt + \frac{1}{T}\int_T^{T+\tau} f(t)\,dt\right].
\end{aligned}$$

Observe that, by (1.37) and the fact that $f \in L^\infty(0,\infty)$,

$$\begin{aligned}
&\left|\operatorname*{LIM}_{T\to\infty}\left[-\frac{\tau}{T(T+\tau)}\int_0^{T+\tau} f(t)\,dt + \frac{1}{T}\int_T^{T+\tau} f(t)\,dt\right]\right|\\
&\qquad\qquad \leq \lim_{T\to\infty}\left[\frac{2\tau}{T}\|f\|_{L^\infty(0,\infty)}\right] = 0.
\end{aligned}$$

Thus,

$$\operatorname*{LIM}_{T\to\infty} g(T+\tau) = \operatorname*{LIM}_{T\to\infty} g(T),$$

which proves (1.40).

In order to prove the existence of a generalized limit with the property described in Remark 1.5, one need only replace G_0 by $G_0 + \operatorname{span} g_0$ and set $\Lambda_0(g_0) = \ell$.

Appendix B Proofs of Technical Results in Chapter IV

We now present the more technical proofs omitted from the previous sections of this chapter.

B.1 Equivalence between Invariant Measures and Stationary Statistical Solutions in Dimension 2 (Proof of Theorem 2.2)

In what follows, we restrict ourselves to the 2-dimensional case. Let μ be a probability measure on H. Assume first that μ is invariant for the semigroup $\{S(t)\}_{t\geq 0}$ on H generated by the 2-dimensional Navier–Stokes equations. From Lemma 2.1, it follows that $\mu(H \setminus \mathcal{B}_a) = 0$ for an absorbing set $\mathcal{B}_a$ bounded in $D(A)$. On $\mathcal{B}_a$, the function $\mathbf{u} \mapsto \|\mathbf{u}\|^2$ is bounded and so

$$\int_H \|\mathbf{u}\|^2 \, d\mu(\mathbf{u}) < \infty,$$

which is condition (1.29) for μ to be a stationary statistical solution. For (1.31), we proceed as follows. Let $0 \leq E_1 < E_2 \leq \infty$, and let $E = \{\mathbf{u} \in H;\, E_1 \leq |\mathbf{u}| < E_2\}$. Since μ is an invariant measure and since, from (1.29), $\mathbf{u} \mapsto \nu\|\mathbf{u}\|^2 - (\mathbf{f}, \mathbf{u})$ belongs to $L^1(\mu)$, we have that

$$\int_E \{\nu\|\mathbf{u}\|^2 - (\mathbf{f}, \mathbf{u})\}\, d\mu(\mathbf{u}) = \int_E \{\nu\|S(t)\mathbf{u}\|^2 - (\mathbf{f}, S(t)\mathbf{u})\}\, d\mu(\mathbf{u}) \quad \forall t \geq 0.$$

This means that the right-hand side of the equality just displayed is constant in t, so it does not change if we integrate in t from 0 to T and divide it by T. Thus, we find

$$\begin{aligned}\int_E \{\nu\|\mathbf{u}\|^2 - (\mathbf{f}, \mathbf{u})\}\, d\mu(\mathbf{u}) &= \frac{1}{T}\int_0^T \int_E \{\nu\|S(t)\mathbf{u}\|^2 - (\mathbf{f}, S(t)\mathbf{u})\}\, d\mu(\mathbf{u})\, dt \\ &= \int_E \frac{1}{T}\int_0^T \{\nu\|S(t)\mathbf{u}\|^2 - (\mathbf{f}, S(t)\mathbf{u})\}\, dt\, d\mu(\mathbf{u}),\end{aligned}$$

where in the last step we are allowed to use the Fubini theorem (see Theorem A.4), since the map $(t, \mathbf{u}) \mapsto S(t)\mathbf{u}$ is (jointly) continuous and

$$(t, \mathbf{u}) \mapsto \nu\|S(t)\mathbf{u}\|^2 - (\mathbf{f}, S(t)\mathbf{u})$$

belongs to L^1 on $[0, T] \times H$ for all $T > 0$. Then, using the energy equation (II.A.52), we obtain

$$\int_E \{\nu\|\mathbf{u}\|^2 - (\mathbf{f}, \mathbf{u})\}\, d\mu(\mathbf{u}) = \frac{1}{2}\int_E \frac{1}{T}\{|\mathbf{u}|^2 - |S(t)\mathbf{u}|^2\}\, d\mu(\mathbf{u}) \quad \forall T > 0.$$

Since $t \mapsto S(t)\mathbf{u}$ is bounded in H, we let $T \to \infty$ and find, using the Lebesgue dominated convergence theorem, that actually

$$\int_E \{\nu\|\mathbf{u}\|^2 - (\mathbf{f}, \mathbf{u})\}\, d\mu(\mathbf{u}) = 0,$$

which implies (1.31).

Let us now prove (1.30). Let Φ be a test functional, $\Phi \in \mathcal{T}$. For $m \in \mathbb{N}$, set $\Phi_m(\mathbf{u}) = \Phi(P_m\mathbf{u})$. Note that $\Phi'_m(\mathbf{u}) = \Phi'(P_m\mathbf{u})P_m$ and that, in fact, Φ_m is a $\mathcal{C}^1$ functional on H. On the other hand, for any $\mathbf{u} \in D(A)$, the solution $t \mapsto S(t)\mathbf{u}$ is $\mathcal{C}^1$ from $[0, \infty)$ into H. Thus, for $\mathbf{u} \in D(A)$ we see that $t \mapsto \Phi_m(S(t)\mathbf{u})$ is of class $\mathcal{C}^1$ with

$$\frac{d}{dt}\Phi_m(S(t)\mathbf{u}) = (\mathbf{F}(S(t)\mathbf{u}), \Phi'_m(S(t)\mathbf{u})) \quad \forall t \geq 0,\ \forall \mathbf{u} \in D(A). \tag{B.1}$$

Since μ is invariant, we have

$$\int (\mathbf{F}(S(t)\mathbf{u}), \Phi'_m(S(t)\mathbf{u}))\, d\mu(\mathbf{u}) = \int (\mathbf{F}(\mathbf{u}), \Phi'_m(\mathbf{u}))\, d\mu(\mathbf{u}).$$

Thus,

$$\begin{aligned}\int (\mathbf{F}(\mathbf{u}), \Phi'_m(\mathbf{u}))\, d\mu(\mathbf{u}) &= \frac{1}{T}\int_0^T \int (\mathbf{F}(S(t)\mathbf{u}), \Phi_m(S(t)\mathbf{u}))\, d\mu(\mathbf{u})\, dt \\ &= \iint \frac{1}{T}(\mathbf{F}(S(t)\mathbf{u}), \Phi'_m(S(t)\mathbf{u}))\, dt\, d\mu(\mathbf{u}),\end{aligned}$$

where we have again used Fubini's theorem in the last step. Then, from (B.1) and the fact that $\mu(H \setminus D(A)) = 0$ (from Lemma 2.1),

$$\int (\mathbf{F}(\mathbf{u}), \Phi'_m(\mathbf{u}))\, d\mu(\mathbf{u}) = \int \frac{1}{T}\{\Phi_m(S(T)\mathbf{u}) - \Phi_m(\mathbf{u})\}\, d\mu(\mathbf{u}).$$

Because Φ_m is bounded, letting $T \to \infty$ in this equation yields

$$\int (\mathbf{F}(\mathbf{u}), \Phi'_m(\mathbf{u}))\, d\mu(\mathbf{u}) = 0.$$

Then, since $\Phi'_m(\mathbf{u}) = \Phi'(P_m\mathbf{u})P_m$, we see that

$$\int (P_m F(\mathbf{u}), \Phi'(P_m\mathbf{u}))\, d\mu(\mathbf{u}) = \int (F(\mathbf{u}), P_m\Phi'(P_m\mathbf{u}))\, d\mu(\mathbf{u}) = 0.$$

But $P_m\mathbf{u} \to \mathbf{u}$ in V as $m \to \infty$ for all $\mathbf{v} \in V$, so that

$$(\mathbf{F}(\mathbf{u}), P_m\Phi'(P_m\mathbf{u})) \to (\mathbf{F}(\mathbf{u}), \Phi'(\mathbf{u})) \ \text{ as } m \to \infty,\ \mu\text{-a.e.};$$

here the convergence μ-almost everywhere follows because $\mu(H \setminus V) = 0$. Moreover, since Φ' is bounded, the following estimate holds with some positive constant c:

$$(\mathbf{F}(\mathbf{u})P_m\Phi'(P_m\mathbf{u})) \leq c(\|\mathbf{u}\| + \|\mathbf{u}\|^2 + |\mathbf{f}|),$$

where the right-hand side is μ-integrable.[11] Therefore, we can apply (the Lebesgue dominated convergence) Theorem A.3 to find

$$\int (\mathbf{F}(\mathbf{u}), \Phi'(\mathbf{u}))\, d\mu(\mathbf{u}) = \lim_{m\to\infty} \int (\mathbf{F}(\mathbf{u}), P_m\Phi'(P_m\mathbf{u}))\, d\mu(\mathbf{u}) = 0,$$

[11] (For the physics-oriented reader.) It is assumed here that we work with the nondimensional Navier–Stokes equations, in which case all quantities (including $\mathbf{u}$ and $\mathbf{f}$) are nondimensional and ν is the inverse of the Reynolds number (see (1.8) in Chapter I). Otherwise, some dimensional constants should appear in front of $\|\mathbf{u}\|^2$ and $|\mathbf{f}|$ in the inequality just displayed and in other places that follow (e.g., (B.13) and thereafter).

an equality that holds for every test functional Φ. This proves (1.30), which completes the proof that μ is a stationary statistical solution.

For the converse, let μ be a probability measure on H and assume that μ is a stationary statistical solution. In order to show that μ is an invariant measure for $\{S(t)\}_{t\geq 0}$, we show first that, for any test functional $\Phi \in \mathcal{T}$,

$$\int \Phi(S(t)\mathbf{u})\, d\mu(\mathbf{u}) = \int \Phi(\mathbf{u})\, d\mu(\mathbf{u}) \quad \forall t \geq 0. \tag{B.2}$$

Toward this end, consider for each $m \in \mathbb{N}$ the Galerkin approximation (see Section 6) of the Navier–Stokes equation:

$$\frac{d\mathbf{u}_m}{dt} = P_m\mathbf{F}(\mathbf{u}_m) = P_m\mathbf{f} - P_mB(\mathbf{u}_m) - \nu A\mathbf{u}_m,$$
$$\mathbf{u}_m(0) = P_m\mathbf{u} \in P_mH,$$

and denote by $S_m(t)\mathbf{u} = \mathbf{u}_m(t)$ its solution operator. Note that $S_m(t)\mathbf{u}$ is defined for all $\mathbf{u}$ in H, with $S_m(0) = P_m$. By taking the inner product in H of the equation for $\mathbf{u}_m$ with $\mathbf{u}_m$ itself, one finds an energy equation which reveals that the solution $\mathbf{u}_m(t)$ exists for all $t \in \mathbb{R}$, so that $\{S_m(t)\}_{t\in\mathbb{R}}$ is a group in P_mH. This group property and the regularity of the solution operator are the main reasons for the adequacy of this approximation.

The operator $u \mapsto S_m(t)\mathbf{u}$, for each $t \in \mathbb{R}$, is also Fréchet differentiable as an operator in H. Denote its differential at $\mathbf{u} \in H$ by $D_\mathbf{u}S_m(t)\mathbf{u}$; this is a bounded linear operator on H. We can also consider the adjoint in H of $D_\mathbf{u}S_m(t)\mathbf{u}$, denoted $(D_\mathbf{u}S_m(t)\mathbf{u})^*$, which is also a bounded linear operator on H. We will show that $D_\mathbf{u}S_m(t)\mathbf{u}$ is a bounded operator from $D(A^{-1/2})$ into itself, so that its adjoint, $(D_\mathbf{u}S_m(t)\mathbf{u})^*$, is a bounded operator from V into itself.

Let Φ be a given test functional, $\Phi \in \mathcal{T}$. Since $S_m(t)\mathbf{u} \in V$ for all $t \in \mathbb{R}$ and all $\mathbf{u} \in H$, and since $t \mapsto S_m(t)\mathbf{u}$ is continuously differentiable from $\mathbb{R}$ into H for any $\mathbf{u} \in H$, we see that

$$\frac{d}{dt}\Phi(S_m(t)\mathbf{u}) = (P_m\mathbf{F}(S_m(t)\mathbf{u}), \Phi'(S_m(t)\mathbf{u})) \quad \forall (t, \mathbf{u}) \in \mathbb{R} \times H.$$

Therefore,

$$\begin{aligned}\int \Phi(S_m(T)\mathbf{u})\, d\mu(\mathbf{u}) &\\ = \int \Phi(P_m\mathbf{u})\, d\mu(\mathbf{u}) &\\ + \int_0^T \int (P_m\mathbf{F}(S_m(t)\mathbf{u}), \Phi'(S_m(t)\mathbf{u}))\, d\mu(\mathbf{u})\, dt &\quad \forall T > 0.\end{aligned} \tag{B.3}$$

Note that Φ, as a test functional in $\mathcal{T}$, is continuous on V with respect to the norm of H; thus, since $\mu(H \setminus V) = 0$, it follows that Φ is μ-integrable. It is therefore legitimate to integrate with respect to μ, as we have done.

Define

$$\Phi_m(t, \mathbf{u}) = \Phi(S_m(t)\mathbf{u}) \quad \forall (t, \mathbf{u}) \in \mathbb{R} \times H.$$

One can show that, for each $t \in \mathbb{R}$, the map $\mathbf{u} \mapsto \Phi_m(t, \mathbf{u})$, $\mathbf{u} \in H$, is a test functional with Fréchet derivative

$$(\Phi_m)'_{\mathbf{u}}(t, \mathbf{u}) = (D_{\mathbf{u}} S_m(t)\mathbf{u})^* \Phi'(S_m(t)\mathbf{u}) \quad \forall (t, \mathbf{u}) \in \mathbb{R} \times H. \tag{B.4}$$

We want to rewrite (B.3) by showing that

$$(P_m \mathbf{F}(S_m(t)\mathbf{u}), \Phi'(S_m(t)\mathbf{u})) = (P_m \mathbf{F}(P_m \mathbf{u}), (\Phi_m)'_{\mathbf{u}}(t, \mathbf{u})) \quad \forall (t, \mathbf{u}) \in \mathbb{R} \times H. \tag{B.5}$$

In order to prove this identity we first note, using the group property of $\{S_m(t)\}_{t \in \mathbb{R}}$ in $P_m H$, that

$$\Phi_m(t, S_m(-t)\mathbf{v}) = \Phi(S_m(t) S_m(-t)\mathbf{v}) = \Phi(P_m \mathbf{v}) \quad \forall (t, \mathbf{v}) \in \mathbb{R} \times H.$$

Thus, $\Phi_m(t, S_m(-t)\mathbf{v})$ is time-independent. On the other hand,

$$\begin{aligned} \frac{d}{dt} \Phi_m(t, S_m(-t)\mathbf{v}) &= (\Phi_m)'_t(t, S_m(-t)\mathbf{v}) \\ &\quad - (P_m \mathbf{F}(S_m(-t)\mathbf{v}), (\Phi_m)'_{\mathbf{u}}(t, S_m(-t)\mathbf{v})) \quad \forall (t, \mathbf{v}) \in \mathbb{R} \times H, \end{aligned}$$

where $(\Phi_m)'_t(\cdot, \cdot)$ denotes the derivative of Φ_m with respect to the first variable. Since the function $\Phi_m(t, S_m(-t)\mathbf{v})$ is time-independent (as we have just shown), its time derivative vanishes and we find

$$(\Phi_m)'_t(t, S_m(-t)\mathbf{v}) - (P_m \mathbf{F}(S_m(-t)\mathbf{v}), (\Phi_m)'_{\mathbf{u}}(t, S_m(-t)\mathbf{v})) = 0 \quad \forall (t, \mathbf{v}) \in \mathbb{R} \times H.$$

For any $\mathbf{u} \in H$, we can take $\mathbf{v} = S_m(t)\mathbf{u} \in P_m H$ and so, since $S_m(-t)\mathbf{v} = P_m \mathbf{u}$, we obtain

$$(\Phi_m)'_t(t, P_m \mathbf{u}) - (P_m \mathbf{F}(P_m \mathbf{u}), (\Phi_m)'_{\mathbf{u}}(t, P_m \mathbf{u})) = 0 \quad \forall (t, \mathbf{u}) \in \mathbb{R} \times H. \tag{B.6}$$

On the other hand,

$$(\Phi_m)'_t(t, P_m \mathbf{u}) = \frac{d}{dt} \Phi(S_m(t)\mathbf{u}) = (P_m \mathbf{F}(S_m(t)\mathbf{u}), \Phi'(S_m(t)\mathbf{u})),$$

so that (B.6) leads to (B.5).

Taking (B.5) into account, it follows from (B.3) that

$$\begin{aligned} &\int \Phi(S_m(T)\mathbf{u})\, d\mu(\mathbf{u}) \\ &\quad = \int \Phi(P_m \mathbf{u})\, d\mu(\mathbf{u}) \\ &\qquad + \int_0^T \int (P_m \mathbf{F}(P_m \mathbf{u}), (\Phi_m)'_{\mathbf{u}}(t, \mathbf{u}))\, d\mu(\mathbf{u})\, dt \quad \forall T > 0. \end{aligned} \tag{B.7}$$

Because $\mathbf{u} \mapsto \Phi_m(t, \mathbf{u})$ is a test functional in $\mathcal{T}$ for each $t \in \mathbb{R}$, it follows from our assumption that μ is a stationary statistical solution that

$$\int (\mathbf{F}(\mathbf{u}), (\Phi_m)'_{\mathbf{u}}(t, \mathbf{u}))\, d\mu(\mathbf{u}) = 0 \quad \forall t \in \mathbb{R}.$$

Hence, we can rewrite (B.7) as

$$\int \Phi(S_m(T)\mathbf{u})\,d\mu(\mathbf{u})$$
$$= \int \Phi(P_m\mathbf{u})\,d\mu(\mathbf{u})$$
$$+ \int_0^T \int \left[(P_m\mathbf{F}(P_m\mathbf{u}), (\Phi_m)'_{\mathbf{u}}(t, P_m\mathbf{u})) - (\mathbf{F}(\mathbf{u}), (\Phi_m)'_{\mathbf{u}}(t, \mathbf{u}))\right] d\mu(\mathbf{u})\,dt.$$

Since $\Phi_m(t, \mathbf{u}) = \Phi_m(t, P_m\mathbf{u})$, it is easy to deduce that

$$(\Phi_m)'_{\mathbf{u}}(t, \mathbf{u}) = (\Phi_m)'_{\mathbf{u}}(t, P_m\mathbf{u}) = P_m(\Phi_m)'_{\mathbf{u}}(t, P_m\mathbf{u}) = P_m(\Phi_m)'_{\mathbf{u}}(t, \mathbf{u}). \tag{B.8}$$

Then, using also that $(\cdot, P_m\cdot) = (P_m\cdot, \cdot)$, we have

$$\int \Phi(S_m(T)\mathbf{u})\,d\mu(\mathbf{u})$$
$$= \int \Phi(P_m\mathbf{u})\,d\mu(\mathbf{u})$$
$$+ \int_0^T \int (P_m\mathbf{F}(P_m\mathbf{u}) - P_m\mathbf{F}(\mathbf{u}), (\Phi_m)'_{\mathbf{u}}(t, \mathbf{u}))\,d\mu(\mathbf{u})\,dt \quad \forall T > 0.$$

Now $P_m\mathbf{F}(P_m\mathbf{u}) - P_m\mathbf{F}(\mathbf{u}) = P_mB(\mathbf{u}) - P_mB(P_m\mathbf{u})$; using again (B.8) and the fact that $(\cdot, P_m\cdot) = (P_m\cdot, \cdot)$, we obtain

$$\int \Phi(S_m(T)\mathbf{u})\,d\mu(\mathbf{u})$$
$$= \int \Phi(P_m\mathbf{u})\,d\mu(\mathbf{u})$$
$$+ \int_0^T \int (P_mB(\mathbf{u}) - P_mB(P_m\mathbf{u}), (\Phi_m)'_{\mathbf{u}}(t, \mathbf{u}))\,d\mu(\mathbf{u})\,dt$$
$$= \int \Phi(P_m\mathbf{u})\,d\mu(\mathbf{u})$$
$$+ \int_0^T \int (B(\mathbf{u}) - B(P_m\mathbf{u}), (\Phi_m)'_{\mathbf{u}}(t, \mathbf{u}))\,d\mu(\mathbf{u})\,dt \quad \forall T > 0. \tag{B.9}$$

The next step is to show that, as m goes to infinity, the time integral term in (B.9) vanishes and we are left with (B.2).

In the 2-dimensional case, the following inequality holds for the trilinear operator (see (A.46d) in Chapter II):

$$|b(\mathbf{u}, \mathbf{v}, \mathbf{w})| \leq c_1|\mathbf{u}|^{1/2}\|\mathbf{u}\|^{1/2}\|\mathbf{v}\||\mathbf{w}|^{1/2}\|\mathbf{w}\|^{1/2} \quad \forall \mathbf{u}, \mathbf{v}, \mathbf{w} \in V \tag{B.10}$$

for a suitable positive constant c_1. Hence, using (B.10) and (1.9), we find that

$$\begin{aligned}
&|(B(\mathbf{u}) - B(P_m\mathbf{u}), (\Phi_m)'_{\mathbf{u}}(t, \mathbf{u}))| \\
&\quad = |b(\mathbf{u}, \mathbf{u}, (\Phi_m)'_{\mathbf{u}}(t, \mathbf{u})) - b(P_m\mathbf{u}, P_m\mathbf{u}, (\Phi_m)'_{\mathbf{u}}(t, \mathbf{u}))| \\
&\quad = |b(\mathbf{u} - P_m\mathbf{u}, \mathbf{u}, (\Phi_m)'_{\mathbf{u}}(t, \mathbf{u})) + b(P_m\mathbf{u}, \mathbf{u} - P_m\mathbf{u}, (\Phi_m)'_{\mathbf{u}}(t, \mathbf{u}))| \\
&\quad \leq 2c_1|\mathbf{u} - P_m\mathbf{u}|^{1/2}\|\mathbf{u} - P_m\mathbf{u}\|^{1/2}|\mathbf{u}|^{1/2}\|\mathbf{u}\|^{1/2}\|(\Phi_m)'_{\mathbf{u}}(t, \mathbf{u}))\| \\
&\qquad + c_1|P_m\mathbf{u}|^{1/2}\|P_m\mathbf{u}\|^{1/2}|\mathbf{u} - P_m\mathbf{u}|^{1/2}\|\mathbf{u} - P_m\mathbf{u}\|^{1/2}\|(\Phi_m)'_{\mathbf{u}}(t, \mathbf{u}))\| \\
&\quad \leq 4c_1|\mathbf{u} - P_m\mathbf{u}|^{1/2}\|\mathbf{u} - P_m\mathbf{u}\|^{1/2}|\mathbf{u}|^{1/2}\|\mathbf{u}\|^{1/2}\|(\Phi_m)'_{\mathbf{u}}(t, \mathbf{u}))\|.
\end{aligned}$$

Then, using (1.14),

$$\begin{aligned}|(B(\mathbf{u}) - B(P_m\mathbf{u}), (\Phi_m)'_{\mathbf{u}}(t, \mathbf{u}))| \\ &\le \frac{4c_1}{\lambda_m^{1/4}} \|\mathbf{u} - P_m\mathbf{u}\| |\mathbf{u}|^{1/2} \|\mathbf{u}\|^{1/2} \|(\Phi_m)'_{\mathbf{u}}(t, \mathbf{u}))\| \\ &\le \frac{4c_1}{\lambda_m^{1/4}} \|\mathbf{u}\|^2 \|(\Phi_m)'_{\mathbf{u}}(t, \mathbf{u}))\| \quad \forall (t, \mathbf{u}) \in \mathbb{R} \times H. \qquad \text{(B.11)}\end{aligned}$$

We need now an estimate for $\|(\Phi_m)'_{\mathbf{u}}(t, \mathbf{u}))\|$. We have

$$\begin{aligned}\|(\Phi_m)'_{\mathbf{u}}(t, \mathbf{u}))\| &= |A^{1/2}(\Phi_m)'_{\mathbf{u}}(t, \mathbf{u})| \\ &= \sup_{\mathbf{v}\in H, |\mathbf{v}|\le 1} (A^{1/2}(\Phi_m)'_{\mathbf{u}}(t, \mathbf{u}), \mathbf{v}) \\ &= \sup_{\mathbf{v}\in V, |\mathbf{v}|\le 1} (\Phi_m)'_{\mathbf{u}}(t, \mathbf{u}), A^{1/2}\mathbf{v}) \quad \text{(by the density of } V \text{ in } H) \\ &= \sup_{\mathbf{v}\in V, |\mathbf{v}|\le 1} ((D_{\mathbf{u}}S_m(t)\mathbf{u})^* \Phi'(S_m(t)\mathbf{u}), A^{1/2}\mathbf{v}) \quad \text{(from (B.4))} \\ &= \sup_{\mathbf{v}\in V, |\mathbf{v}|\le 1} (\Phi'(S_m(t)\mathbf{u}), (D_{\mathbf{u}}S_m(t)\mathbf{u})A^{1/2}\mathbf{v}) \\ &= \sup_{\mathbf{v}\in V, |\mathbf{v}|\le 1} (A^{1/2}\Phi'(S_m(t)\mathbf{u}), A^{-1/2}(D_{\mathbf{u}}S_m(t)\mathbf{u})A^{1/2}\mathbf{v}) \\ &= \sup_{\mathbf{v}\in V, |\mathbf{v}|\le 1} \|\Phi'(S_m(t)\mathbf{u})\| \|(D_{\mathbf{u}}S_m(t)\mathbf{u})A^{1/2}\mathbf{v}\|_{D(A^{-1/2})}.\end{aligned}$$

Thus, since $\Phi'(\cdot)$ is bounded on V (because Φ is a test functional in $\mathcal{T}$), we find that

$$\|(\Phi_m)'_{\mathbf{u}}(t, \mathbf{u}))\| \le c_2 \sup_{\mathbf{v}\in V, |\mathbf{v}|\le 1} \|(D_{\mathbf{u}}S_m(t)\mathbf{u})A^{1/2}\mathbf{v}\|_{D(A^{-1/2})} \qquad \text{(B.12)}$$

for some positive constant c_2. It remains to estimate

$$\sup_{\mathbf{v}\in V, |\mathbf{v}|\le 1} \|(D_{\mathbf{u}}S_m(t)\mathbf{u})A^{1/2}\mathbf{v}\|_{D(A^{-1/2})} = \sup_{\mathbf{v}\in H, |A^{-1/2}\mathbf{v}|\le 1} \|(D_{\mathbf{u}}S_m(t)\mathbf{u})\mathbf{v}\|_{D(A^{-1/2})},$$

which is essentially the norm of $D_{\mathbf{u}}S_m(t)\mathbf{u}$ regarded as an operator from $D(A^{-1/2})$ into itself. Let then $\mathbf{v} \in H$ and set

$$\mathbf{v}_m(t) = (D_{\mathbf{u}}S_m(t)\mathbf{u})\mathbf{v},$$

which solves the equation

$$\frac{d}{dt}\mathbf{v}_m + \nu A\mathbf{v}_m + P_m B(\mathbf{v}_m, \mathbf{u}_m) + P_m B(\mathbf{u}_m, \mathbf{v}_m) = 0 \quad \forall t \in \mathbb{R}, \ \mathbf{v}_m(0) = \mathbf{v},$$

where $\mathbf{u}_m = \mathbf{u}_m(t) = S_m(t)\mathbf{u}$. Take the inner product in H of the expression just displayed with $A^{-1}\mathbf{v}_m$ to find

$$\begin{aligned}\frac{1}{2}\frac{d}{dt}|A^{-1/2}\mathbf{v}_m|^2 + \nu|\mathbf{v}_m|^2 &= -b(\mathbf{v}_m, \mathbf{u}_m, A^{-1}\mathbf{v}_m) - b(\mathbf{u}_m, \mathbf{v}_m, A^{-1}\mathbf{v}_m) \\ &= b(\mathbf{v}_m, A^{-1}\mathbf{v}_m, \mathbf{u}_m) + b(\mathbf{u}_m, A^{-1}\mathbf{v}_m, \mathbf{v}_m).\end{aligned}$$

Since we treat here the 2-dimensional case, the last term can be estimated as

$$\begin{aligned}
&b(\mathbf{v}_m, A^{-1}\mathbf{v}_m, \mathbf{u}_m) + b(\mathbf{u}_m, A^{-1}\mathbf{v}_m, \mathbf{v}_m) \\
&\quad \le c_3 |\mathbf{v}_m| |A^{1/2}A^{-1}\mathbf{v}_m|^{1/2} |AA^{-1}\mathbf{v}_m|^{1/2} |\mathbf{u}_m|^{1/2} \|\mathbf{u}_m\|^{1/2} \\
&\quad \le c_3 |\mathbf{v}_m|^{3/2} |A^{-1/2}\mathbf{v}_m|^{1/2} |\mathbf{u}_m|^{1/2} \|\mathbf{u}_m\|^{1/2} \\
&\quad \le \frac{\nu}{2}|\mathbf{v}_m|^2 + \left(\frac{c_4}{\nu}\right)^3 |A^{-1/2}\mathbf{v}_m|^2 |\mathbf{u}_m|^2 \|\mathbf{u}_m\|^2,
\end{aligned}$$

with suitable constants c_3 and c_4. Hence,

$$\frac{d}{dt}|A^{-1/2}\mathbf{v}_m|^2 \le c_5 |A^{-1/2}\mathbf{v}_m|^2 |\mathbf{u}_m|^2 \|\mathbf{u}_m\|^2$$

for a constant c_5 depending on ν. Integrating this inequality, we find

$$|A^{-1/2}\mathbf{v}_m(t)|^2 \le |A^{-1/2}\mathbf{v}_m(0)|^2 \exp\left(c_5 \int_0^t |\mathbf{u}_m(s)|^2 \|\mathbf{u}_m(s)\|^2 \, ds\right) \quad \forall t \ge 0.$$

Similarly to (II.A.41) and (II.A.40), using that $|\mathbf{u}_m(0)| = |P_m\mathbf{u}| \le |\mathbf{u}|$ yields[12]

$$|\mathbf{u}_m(s)|^2 \le |\mathbf{u}|^2 + \frac{|\mathbf{f}|^2}{\nu^2\lambda_1^2} \tag{B.13}$$

and

$$\nu \int_0^t \|\mathbf{u}_m(s)\|^2 \, ds \le |\mathbf{u}|^2 + \frac{|\mathbf{f}|^2}{\nu\lambda_1^2} t.$$

Thus,

$$|A^{-1/2}\mathbf{v}_m(t)|^2 \le |A^{-1/2}\mathbf{v}_m(0)|^2 \exp(c_7(1 + |\mathbf{u}|^2)(t + |\mathbf{u}|^2)) \quad \forall t \ge 0$$

for a constant c_7 depending on ν, $|\mathbf{f}|$, and λ_1. Since the support of μ is bounded in H (see Theorem 4.1), we find

$$|A^{-1/2}(D_{\mathbf{u}}S_m(t)\mathbf{u})\mathbf{v}|^2 \le |A^{-1/2}\mathbf{v}|^2 \exp(c_8(1 + t)) \quad \forall t \ge 0, \ \forall \mathbf{u} \in \operatorname{supp} \mu,$$

for another positive constant c_8. Then,

$$\begin{aligned}
\sup_{\mathbf{v}\in V, |\mathbf{v}|\le 1} \|(D_{\mathbf{u}}S_m(t)\mathbf{u})A^{1/2}\mathbf{v}\|_{D(A^{-1/2})} &= \sup_{\mathbf{v}\in H, |A^{-1/2}\mathbf{v}|\le 1} |A^{-1/2}(D_{\mathbf{u}}S_m(t)\mathbf{u})\mathbf{v}| \\
&\le e^{c_8(1+t)} \quad \forall t \ge 0, \ \forall \mathbf{u} \in \operatorname{supp} \mu.
\end{aligned}$$

Hence, from (B.12),

$$\|(\Phi_m)'_{\mathbf{u}}(t, \mathbf{u}))\| \le c_2 e^{c_8(1+t)} \quad \forall t \ge 0, \ \forall \mathbf{u} \in \operatorname{supp} \mu.$$

From (B.11) we now have

$$|(B(\mathbf{u}) - B(P_m\mathbf{u}), (\Phi_m)'_{\mathbf{u}}(t, \mathbf{u}))| \le \frac{4c_1}{\lambda_m^{1/4}} \|\mathbf{u}\|^2 c_2 e^{c_8(1+t)} \quad \forall t \in \mathbb{R}, \ \forall \mathbf{u} \in \operatorname{supp} \mu.$$

This inequality holds for all $\mathbf{u}$ in supp μ, so it holds for $\mathbf{u}$ μ-almost everywhere in H. Therefore,

[12] See footnote 11.

$$\int_0^T \int |(B(\mathbf{u}) - B(P_m\mathbf{u}), (\Phi_m)'_{\mathbf{u}}(t, \mathbf{u}))| \, d\mu(\mathbf{u}) \, dt$$
$$\leq \frac{4c_1c_2}{\lambda_m^{1/4}} \int \|\mathbf{u}\|^2 \, d\mu(\mathbf{u}) \int_0^T e^{c_8(1+t)} \, dt \leq \frac{c_T}{\lambda_m^{1/4}}$$

for some constant (with respect to m) c_T depending also on T, where at the last step we used (1.29). We may let m go to infinity to deduce from (B.9) and the above that

$$\int \Phi(S_m(T)\mathbf{u}) \, d\mu(\mathbf{u}) - \int \Phi(P_m\mathbf{u}) \, d\mu(\mathbf{u}) \to 0 \text{ as } m \to \infty \tag{B.14}$$

for every $T > 0$. On the other hand, the support of μ is bounded and so, from (B.13), $S_m(T)\mathbf{u}$ is uniformly bounded in H for $\mathbf{u} \in \operatorname{supp} \mu$ and $m \in \mathbb{N}$. Moreover, Φ is bounded on bounded sets; thus $\Phi(S_m(T)\mathbf{u})$ is uniformly bounded for $m \in H$ and for $\mathbf{u}$, μ-almost everywhere on H. Hence, we can use that $S_m(T)\mathbf{u} \to S(T)\mathbf{u}$ for all $\mathbf{u}$ in H as m goes to infinity; using then the Lebesgue dominated convergence theorem, we obtain

$$\int \Phi(S_m(T)\mathbf{u}) \, d\mu(\mathbf{u}) \to \int \Phi(S(T)\mathbf{u}) \, d\mu(\mathbf{u}) \text{ as } m \to \infty \tag{B.15}$$

for all $T > 0$. Similarly,

$$\int \Phi(P_m\mathbf{u}) \, d\mu(\mathbf{u}) \to \int \Phi(\mathbf{u}) \, d\mu(\mathbf{u}) \text{ as } m \to \infty \tag{B.16}$$

for all $T > 0$. From (B.14), (B.15), and (B.16), we find

$$\int \Phi(S(T)\mathbf{u}) \, d\mu(\mathbf{u}) = \int \Phi(\mathbf{u}) \, d\mu(\mathbf{u}) \quad \forall T > 0,$$

thus proving (B.2).

Finally, one can show that the test functionals of $\mathcal{T}$ are dense in the space of continuous functionals on H, which in turn is dense in $L^1(\mu)$. Hence, we can take Φ in (B.2) to be the characteristic function of any given measurable set E in H. Doing so, we find that $\mu(S(T)^{-1}E) = \mu(E)$ for any $T > 0$. Therefore, the stationary statistical solution μ is an invariant measure for the 2-dimensional Navier–Stokes equations.

B.2 Time-Average Measures Are Stationary Statistical Solutions in Dimension 3 (Proof of Theorem 3.1)

Let us now consider the 3-dimensional case. Let μ be a time-average measure of a weak solution $\mathbf{u} = \mathbf{u}(t)$ on $[0, \infty)$, and let $\mathbf{u}_0 = \mathbf{u}(0)$. Thus μ is a probability measure; in order to show that μ is a stationary statistical solution, we need to prove properties (1.29), (1.30), and (1.31). Property (1.29) was proved in Section 3, so we start by proving (1.30).

Let Φ be a test functional in $\mathcal{T}$. One would like to show that

$$\begin{aligned}\int (\mathbf{F}(\mathbf{u}), \Phi'(\mathbf{u}))\, d\mu(\mathbf{u}) &= \operatorname*{LIM}_{T\to\infty} \frac{1}{T} \int_0^T \big(\mathbf{F}(\mathbf{u}(t)), \Phi'(\mathbf{u}(t))\big)\, dt \\ &= \operatorname*{LIM}_{T\to\infty} \frac{1}{T} \int_0^T \frac{d}{dt}\Phi(\mathbf{u}(t))\, dt \\ &= \operatorname*{LIM}_{T\to\infty} \frac{1}{T}\{\Phi(\mathbf{u}(t)) - \Phi(\mathbf{u}(0))\} \\ &= 0,\end{aligned}$$

but the problem is that $u \mapsto \Phi'(\mathbf{u})$ and $\mathbf{u} \mapsto \mathbf{F}(\mathbf{u})$ are not sufficiently regular, so that the function $u \mapsto (\mathbf{F}(\mathbf{u}), \Phi'(\mathbf{u}))$ is not necessarily weakly continuous on H; this prevents us from proving the first step. Moreover, $t \mapsto \Phi(\mathbf{u}(t))$ is not sufficiently regular, as well, which prevents us from proving the second step.

Our strategy is thus to approximate Φ and $\mathbf{F}$ by functions with the appropriate regularity. For $m \in \mathbb{N}$, set $\Phi_m(\mathbf{u}) = \Phi(P_m\mathbf{u})$, and for $k \in \mathbb{N}$ define $\mathbf{F}_k \colon V \to V'$ by

$$\mathbf{F}_k(\mathbf{u}) = \mathbf{f} - \nu A\mathbf{u} - B(P_k\mathbf{u}) \quad \forall \mathbf{u} \in V.$$

We have the following estimate:

$$\begin{aligned}&|(\mathbf{F}(\mathbf{u}), \Phi_m'(\mathbf{u})) - (\mathbf{F}_k(\mathbf{u}), \Phi_m'(\mathbf{u}))| \\ &\quad= |b(\mathbf{u}, \mathbf{u}, \Phi_m'(\mathbf{u})) - b(P_k\mathbf{u}, P_k\mathbf{u}, \Phi_m'(\mathbf{u}))| \\ &\quad= |b(\mathbf{u} - P_k\mathbf{u}, \mathbf{u}, \Phi_m'(\mathbf{u})) + b(P_k\mathbf{u}, \mathbf{u} - P_k\mathbf{u}, \Phi_m'(\mathbf{u}))| \\ &\quad\le 2c_1\|\mathbf{u}\||\mathbf{u} - P_k\mathbf{u}||A^{3/4}\Phi_m'(\mathbf{u})| \\ &\quad\le 2c_1\|\mathbf{u}\|\lambda_{k+1}^{-1/2}\|(I - P_k)\mathbf{u}\|\|A^{1/4}P_m\Phi'(P_m\mathbf{u})\| \\ &\quad\le 2c_1\lambda_{k+1}^{-1/2}\|\mathbf{u}\|^2\lambda_m^{1/4}\|\Phi'(P_m\mathbf{u})\|\end{aligned}$$

for a suitable constant $c_1 \ge 0$. Since Φ' is bounded, we find

$$|(\mathbf{F}(\mathbf{u}), \Phi_m'(\mathbf{u})) - (\mathbf{F}_k(\mathbf{u}), \Phi_m'(\mathbf{u}))| \le c_2\lambda_{k+1}^{-1/2}\lambda_m^{1/4}\|\mathbf{u}\|^2 \quad \forall \mathbf{u} \in V \tag{B.17}$$

for another suitable constant $c_2 \ge 0$.

Assume for the moment that $t \mapsto \Phi_m(\mathbf{u}(t))$ is absolutely continuous for t on any compact inverval of $[0, \infty)$ with

$$\frac{d}{dt}\Phi_m(\mathbf{u}(t)) = \big(\mathbf{F}(\mathbf{u}(t)), \Phi_m'(\mathbf{u}(t))\big) \text{ a.e. } t \text{ on } [0, \infty). \tag{B.18}$$

Then,

$$\left|\frac{1}{T}\int_0^T \big(\mathbf{F}(\mathbf{u}(t)), \Phi_m'(\mathbf{u}(t))\big)\, dt\right| = \left|\frac{\Phi_m(\mathbf{u}(T)) - \Phi_m(\mathbf{u}(0))}{T}\right| \le \frac{c_3}{T} \quad \forall T > 0 \tag{B.19}$$

for some suitable constant $c_3 \geq 0$, since Φ (and hence Φ_m) is bounded on bounded subsets of H and since $\mathbf{u} = \mathbf{u}(t)$, $t \geq 0$, is bounded in H. Letting $T \to \infty$ in (B.19), we find that

$$\lim_{T\to\infty} \frac{1}{T} \int_0^T \big(\mathbf{F}(\mathbf{u}(t)), \Phi'_m(\mathbf{u}(t))\big)\, dt = 0. \tag{B.20}$$

From (B.17), we have

$$\begin{aligned}
&\left| \frac{1}{T} \int_0^T \big(\mathbf{F}_k(\mathbf{u}(t)), \Phi'_m(\mathbf{u}(t))\big)\, dt \right| \\
&\quad \leq \frac{1}{T} \left| \int_0^T \big(\mathbf{F}(\mathbf{u}(t)), \Phi'_m(\mathbf{u}(t))\big) - \big(\mathbf{F}_k(\mathbf{u}(t)), \Phi'_m(\mathbf{u}(t))\big)\, dt \right| \\
&\quad\quad + \left| \frac{1}{T} \int_0^T \big(\mathbf{F}(\mathbf{u}(t)), \Phi'_m(\mathbf{u}(t))\big)\, dt \right| \\
&\quad \leq c_3 \lambda_{k+1}^{-1/2} \lambda_m^{1/4} \frac{1}{T} \int_0^T \|\mathbf{u}(t)\|^2\, dt + \left| \frac{1}{T} \int_0^T \big(\mathbf{F}(\mathbf{u}(t)), \Phi'_m(\mathbf{u}(t))\big)\, dt \right|.
\end{aligned}$$

But from the estimate (II.A.40) we may deduce that

$$\frac{1}{T} \int_0^T \|\mathbf{u}(t)\|^2\, dt \leq \frac{1}{\nu T} |\mathbf{u}_0|^2 + \frac{1}{\nu^2 \lambda_1} |\mathbf{f}|^2.$$

Thus, using also (B.20), we find

$$\limsup_{T\to\infty} \left| \frac{1}{T} \int_0^T \big(\mathbf{F}_k(\mathbf{u}(t)), \Phi'_m(\mathbf{u}(t))\big)\, dt \right| \leq \lambda_{k+1}^{-1/2} \lambda_m^{1/4} \frac{c_3}{\nu^2 \lambda_1} |\mathbf{f}|^2.$$

Hence, properties (1.35) and (1.37) of the generalized limit imply that

$$\left| \operatorname*{LIM}_{T\to\infty} \frac{1}{T} \int_0^T \big(\mathbf{F}_k(\mathbf{u}(t)), \Phi'_m(\mathbf{u}(t))\big)\, dt \right| \leq \lambda_{k+1}^{-1/2} \lambda_m^{1/4} \frac{c_3}{\nu^2 \lambda_1} |\mathbf{f}|^2. \tag{B.21}$$

Now, since $\mathbf{u} \mapsto (\mathbf{F}_k(\mathbf{u}), \Phi'_m(\mathbf{u}))$ belongs to $\mathcal{C}(H_w)$, it follows from the definition of the time-average measures in dimension 3 that

$$\int (\mathbf{F}_k(\mathbf{u}), \Phi'_m(\mathbf{u}))\, d\mu(\mathbf{u}) = \operatorname*{LIM}_{T\to\infty} \frac{1}{T} \int_0^T \big(\mathbf{F}_k(\mathbf{u}(t)), \Phi'_m(\mathbf{u}(t))\big)\, dt.$$

Hence, from (B.21),

$$\left| \int (\mathbf{F}_k(\mathbf{u}), \Phi'_m(\mathbf{u}))\, d\mu(\mathbf{u}) \right| \leq \lambda_{k+1}^{-1/2} \lambda_m^{1/4} \frac{c_3}{\nu^2 \lambda_1} |\mathbf{f}|^2. \tag{B.22}$$

Using again (B.17) and also (1.29), we obtain

$$\begin{aligned}
\left| \int (\mathbf{F}(\mathbf{u}), \Phi'_m(\mathbf{u}))\, d\mu(\mathbf{u}) \right| &\leq \left| \int (\mathbf{F}_k(\mathbf{u}), \Phi'_m(\mathbf{u}))\, d\mu(\mathbf{u}) \right| \\
&\quad + \left| \int (\mathbf{F}(\mathbf{u}), \Phi'_m(\mathbf{u})) - (\mathbf{F}_k(\mathbf{u}), \Phi'_m(\mathbf{u}))\, d\mu(\mathbf{u}) \right| \\
&\leq \lambda_{k+1}^{-1/2} \lambda_m^{1/4} \frac{c_3}{\nu^2 \lambda_1} |f|^2 + c_3 \lambda_{k+1}^{-1/2} \lambda_m^{1/4} \int \|\mathbf{u}\|^2\, d\mu(\mathbf{u}) \\
&\leq c_4 \lambda_{k+1}^{-1/2} \lambda_m^{1/4}
\end{aligned}$$

for another constant $c_4 \geq 0$. Now let $k \to \infty$ to find that

$$\int (\mathbf{F}(\mathbf{u}), \Phi_m'(\mathbf{u}))\, d\mu(\mathbf{u}) = 0$$

or, plainly,

$$\int (\mathbf{F}(\mathbf{u}), P_m \Phi'(P_m \mathbf{u}))\, d\mu(\mathbf{u}) = 0 \quad \text{for any test functional } \Phi \in \mathcal{T}. \tag{B.23}$$

Since $P_m \mathbf{u} \to \mathbf{u}$ in V as $m \to \infty$, we have

$$(\mathbf{F}(\mathbf{u}), P_m \Phi'(P_m \mathbf{u})) \to (\mathbf{F}(\mathbf{u}), \Phi'(\mathbf{u}))$$

for μ-almost every u (since $\mu(H \setminus V) = 0$). Moreover, since Φ' is bounded in V,

$$(\mathbf{F}(\mathbf{u}), P_m \Phi'(P_m \mathbf{u})) \leq c_5(\|\mathbf{u}\| + \|\mathbf{u}\|^2 + |\mathbf{f}|)$$

for a suitable constant $c_5 \geq 0$. Since the RHS of this inequality is integrable, we can apply the Lebesgue dominated convergence theorem to (B.23) to find

$$\int (\mathbf{F}(\mathbf{u}), \Phi'(\mathbf{u}))\, d\mu(\mathbf{u}) = \lim_{m \to \infty} \int (\mathbf{F}(\mathbf{u}), P_m \Phi'(P_m \mathbf{u}))\, d\mu(\mathbf{u}) = 0,$$

which is (1.30).

It remains to prove (B.18). Let $I \subset [0, \infty)$ be a bounded interval and let $\{[s_j, t_j]\}_j$ be an arbitrary finite collection of nonoverlapping subintervals of I. Note that Φ_m is of class C^1 in H, so we can apply the mean value theorem to find that

$$\sum_j |\Phi_m(\mathbf{u}(t_j)) - \Phi_m(\mathbf{u}(s_j))| = \sum_j \left| \left(\Phi_m'(\theta_j \mathbf{u}(t_j) + (1 - \theta_j)\mathbf{u}(s_j)), \mathbf{u}(t_j) - \mathbf{u}(s_j) \right) \right|,$$

where $0 \leq \theta_j \leq 1$. Since Φ_m' is bounded in V, we find

$$\begin{aligned} \sum_j |\Phi_m(\mathbf{u}(t_j)) - \Phi_m(\mathbf{u}(s_j))| &\leq c_6 \sum_j \|\mathbf{u}(t_j) - \mathbf{u}(s_j)\|_{V'} \\ &\leq c_6 \sum_j \int_{s_j}^{t_j} \|\mathbf{u}'(\tau)\|_{V'}\, d\tau \end{aligned} \tag{B.24}$$

for some constant $c_6 \geq 0$. But $\mathbf{u}' \in L^1(I; V')$ (see (1.11)), so we may conclude from (B.24) that, for any given $\varepsilon > 0$, there exists a $\delta > 0$ such that, if

$$\sum_j |t_j - s_j| < \delta,$$

then

$$\sum_j |\Phi_m(\mathbf{u}(t_j)) - \Phi_m(\mathbf{u}(s_j))| < \varepsilon,$$

which means that Φ_m is absolutely continuous on I. Now, using again the mean value theorem, for $0 \leq t \leq t + h$ we have

$$\begin{aligned} \frac{\Phi_m(\mathbf{u}(t+h)) - \Phi_m(\mathbf{u}(t))}{h} &= \frac{1}{h}(\Phi_m'(\mathbf{u}(t + \theta h)), \mathbf{u}(t+h) - \mathbf{u}(t)) \\ &= \frac{1}{h} \int_t^{t+h} \left(\mathbf{F}(\mathbf{u}(\tau)), \Phi_m'(\mathbf{u}(t + \theta h)) \right) d\tau \end{aligned}$$

for some θ, $0 \le \theta \le 1$. Hence,

$$\begin{aligned}\frac{\Phi_m(\mathbf{u}(t+h)) - \Phi_m(\mathbf{u}(t))}{h} &= \frac{1}{h}\int_t^{t+h} \big(\mathbf{F}(\mathbf{u}(\tau)), \Phi_m'(\mathbf{u}(\tau))\big)\, d\tau \\ &\quad + \frac{1}{h}\int_t^{t+h} \big(\mathbf{F}(\mathbf{u}(\tau)), \Phi_m'(\mathbf{u}(t+\theta h)) - \Phi_m'(\mathbf{u}(\tau))\big)\, d\tau. \qquad \text{(B.25)}\end{aligned}$$

Because Φ_m' is bounded and $\mathbf{F}(\mathbf{u}(t))$ belongs to $L^1_{\text{loc}}(0, \infty; V')$, we have that

$$\tau \mapsto \big(\mathbf{F}(\mathbf{u}(\tau)), \Phi_m'(\mathbf{u}(t))\big)$$

belongs to $L^1_{\text{loc}}(0, \infty)$. Thus, for every Lebesgue point of this function (see Theorem A.5), that is, for a.e. $t \ge 0$,

$$\lim_{h\to 0} \frac{1}{h}\int_t^{t+h} \big(\mathbf{F}(\mathbf{u}(\tau)), \Phi_m'(\mathbf{u}(\tau))\big)\, d\tau = \big(\mathbf{F}(\mathbf{u}(t)), \Phi_m'(\mathbf{u}(t))\big). \qquad \text{(B.26)}$$

Moreover,

$$\begin{aligned}&\left|\frac{1}{h}\int_t^{t+h} \big(\mathbf{F}(\mathbf{u}(\tau)), \Phi_m'(\mathbf{u}(t+\theta h)) - \Phi_m'(\mathbf{u}(\tau))\big)\, d\tau\right| \\ &\quad \le \sup_{0\le\tau\le h} \|\Phi_m(\mathbf{u}(t+\theta h)) - \Phi_m(\mathbf{u}(\tau))\| \frac{1}{h}\int_t^{t+h} \|\mathbf{F}(\mathbf{u}(\tau))\|_{V'}\, d\tau. \qquad \text{(B.27)}\end{aligned}$$

Since $\tau \mapsto \mathbf{F}(\mathbf{u}(\tau))$ belongs to $L^1_{\text{loc}}(0, \infty; V')$, it follows that, for every Lebesgue point of this function – that is, for a.e. $t \ge 0$ – we have

$$\frac{1}{h}\int_t^{t+h} \|\mathbf{F}(\mathbf{u}(\tau))\|_{V'}\, d\tau \to \|\mathbf{F}(\mathbf{u}(t))\|_{V'} \text{ as } h \to 0.$$

On the other hand, by the continuity of Φ_m on V' and the continuity of $\mathbf{u}(\cdot)$ in V',

$$\sup_{0\le\tau\le h} \|\Phi_m(\mathbf{u}(t+\theta h)) - \Phi_m(\mathbf{u}(\tau))\| \to 0 \quad \forall t \ge 0$$

as h goes to zero. Thus, (B.27) gives

$$\lim_{h\to 0} \frac{1}{h}\int_t^{t+h} \big(\mathbf{F}(\mathbf{u}(\tau)), \Phi_m'(\mathbf{u}(t+\theta h)) - \Phi_m'(\mathbf{u}(\tau))\big)\, d\tau = 0 \quad \text{for a.e. } t \ge 0. \qquad \text{(B.28)}$$

Using now (B.26) and (B.28), we deduce from (B.25) that

$$\lim_{h\to 0} \frac{\Phi_m(\mathbf{u}(t+h)) - \Phi_m(\mathbf{u}(t))}{h} = \big(\mathbf{F}(\mathbf{u}(t)), \Phi_m'(\mathbf{u}(t))\big) \quad \text{for a.e. } t \ge 0,$$

which proves (B.18). This completes the proof of (1.30).

It remains to prove property (1.31). For that purpose we show that

$$\int \rho'(|\mathbf{u}|^2)[\nu\|\mathbf{u}\|^2 - (\mathbf{f}, \mathbf{u})]\, d\mu(\mathbf{u}) \le 0 \qquad \text{(B.29)}$$

for any $\rho \in \mathcal{C}^2([0,\infty),\mathbb{R})$ such that

$$\rho'(s) \geq 0 \quad \text{for all } s \geq 0 \tag{B.30}$$

and

$$\sup_{s\geq 0}\{|\rho(s)| + |\rho''(s)|\} < \infty. \tag{B.31}$$

Then, (1.31) follows from (B.29) if we approximate the step functions by functions of the form required for ρ in (B.30) and (B.31). The difficulty in proving (B.29) is that the functions $t \mapsto \mathbf{u}(t)$, $\mathbf{u} \mapsto |\mathbf{u}|$, and $\mathbf{u} \mapsto \|\mathbf{u}\|$ are not sufficiently regular, so we cannot show that

$$\begin{aligned}
\int \rho'(|\mathbf{u}|^2)[\nu\|\mathbf{u}\|^2 - (\mathbf{f},\mathbf{u})]\,d\mu(\mathbf{u}) \\
&= \operatorname{LIM}_{T\to\infty} \frac{1}{T}\int_0^T \rho'(|\mathbf{u}(t)|^2)[\nu\|\mathbf{u}(t)\|^2 - (\mathbf{f},\mathbf{u}(t))]\,dt \\
&\leq -\operatorname{LIM}_{T\to\infty} \frac{1}{T}\int_0^T \frac{d}{dt}\rho(|\mathbf{u}(t)|^2)\,dt \\
&= -\operatorname{LIM}_{T\to\infty} \frac{1}{T}[\rho(\mathbf{u}(T)) - \rho(\mathbf{u}(0))] \\
&= 0.
\end{aligned}$$

Hence, as before, we approximate $|\mathbf{u}|$ and $\|\mathbf{u}\|$ by (respectively) $|P_k\mathbf{u}|$ and $\|P_m\mathbf{u}\|$, $k, m \in \mathbb{N}$; the passages involving $\partial_t\rho(|\mathbf{u}(t)|^2)$ require more care.

Let then $\rho \in \mathcal{C}^2([0,\infty),\mathbb{R})$ satisfy (B.30) and (B.31). Let $I_0 \subset [0,\infty)$ be the "good" set for the energy inequality (II.7.5). Fix $T > 0$ for the moment, and for each $j \in \mathbb{N}$ let $l_j \in \mathbb{N}$ and $\{t_{j,l}\}_{l=0}^{l_j}$ be such that

$$0 = t_{j,0} < t_{j,1} < \cdots < t_{j,l_j} = T,$$

with $t_{j,l} \in I_0$ for $l = 1, \ldots, l_j - 1$, and

$$\lim_{j\to\infty} \max_{l=0,\ldots,l_j-1} (t_{j,l+1} - t_{j,l}) = 0. \tag{B.32}$$

Then, since ρ is of class $\mathcal{C}^2$,

$$\begin{aligned}
\rho(|\mathbf{u}(T)|^2) - \rho(|\mathbf{u}_0|^2) \\
&= \sum_{l=0}^{l_j-1}[\rho(|\mathbf{u}(t_{j,l+1})|^2) - \rho(|\mathbf{u}(t_{j,l})|^2)] \\
&= \sum_{l=0}^{l_j-1}\rho'(\theta_{j,l}|\mathbf{u}(t_{j,l+1})|^2 + (1-\theta_{j,l})|\mathbf{u}(t_{j,l})|^2)[|\mathbf{u}(t_{j,l+1})|^2 - |\mathbf{u}(t_{j,l})|^2],
\end{aligned}$$

with $\theta_{j,l}$ such that $0 < \theta_{j,l} < 1$ for $l = 0, \ldots, l_j - 1$, $j \in \mathbb{N}$.

Using (II.7.5) and (B.30) we deduce that

$$\rho(|\mathbf{u}(T)|^2) - \rho(|\mathbf{u}_0|^2)$$
$$\leq \sum_{l=0}^{l_j-1} \rho'(\theta_{j,l}|\mathbf{u}(t_{j,l+1})|^2 + (1-\theta_{j,l})|\mathbf{u}(t_{j,l})|^2) \int_{t_{j,l}}^{t_{j,l+1}} 2[(\mathbf{f}, \mathbf{u}(t)) - \nu\|\mathbf{u}(t)\|^2]\,dt$$
$$= \int_0^T 2[(\mathbf{f}, \mathbf{u}(t)) - \nu\|\mathbf{u}(t)\|^2]\rho'(|\mathbf{u}(t)|^2)\,dt$$
$$- \sum_{l=0}^{l_j-1} \int_{t_{j,l}}^{t_{j,l+1}} 2[(\mathbf{f}, \mathbf{u}(t)) - \nu\|\mathbf{u}(t)\|^2]$$
$$\{\rho'(|\mathbf{u}(t)|^2) - \rho'(\theta_{j,l}|\mathbf{u}(t_{j,l+1})|^2 + (1-\theta_{j,l})|\mathbf{u}(t_{j,l})|^2)\}\,dt.$$

Hence,

$$\rho(|\mathbf{u}(T)|^2) - \rho(|\mathbf{u}_0|^2) = 2\int_0^T \rho'(|\mathbf{u}(t)|^2)[(\mathbf{f}, \mathbf{u}(t)) - \nu\|\mathbf{u}(t)\|^2]\,dt$$
$$- 2\int_0^T [(\mathbf{f}, \mathbf{u}(t)) - \nu\|\mathbf{u}(t)\|^2]\beta_j(t)\,dt, \qquad \text{(B.33)}$$

where

$$\beta_j = \sum_{l=0}^{l_j-1} \chi_{[t_{j,l},t_{j,l+1})}(t)\{\rho'(|\mathbf{u}(t)|^2) - \rho'(\theta_{j,l}|\mathbf{u}(t_{j,l+1})|^2 + (1-\theta_{j,l})|\mathbf{u}(t_{j,l})|^2)\};$$

here $\chi_S = \chi_S(t)$ denotes the characteristic function of a set S.

From (II.7.23), the function $t \mapsto |\mathbf{u}(t)|$ is continuous from the right at any point in I_0. From the energy inequality (II.7.5) we also have, for any $t > 0$ and any sequence $\{t_j\}_j$ converging to t from the left with each $t_j \in I_0$, that $|u(t_j)|^2 \to |u(t)|^2$. Since ρ' is continuous, we therefore deduce, taking (B.32) into account, that

$$\beta_j(t) \to 0 \text{ as } j \to 0 \quad \text{for all } t \in [0, T). \qquad \text{(B.34)}$$

Moreover, since $\mathbf{u} = \mathbf{u}(t)$ is bounded in H, we see that $t \mapsto \rho'(|\mathbf{u}(t)|^2)$ belongs to $L^\infty(0, \infty)$ and that

$$\rho'(\theta_{j,l}|\mathbf{u}(t_{j,l+1})|^2 + (1-\theta_{j,l})|\mathbf{u}(t_{j,l})|^2)$$

is bounded uniformly in j and k. Hence, $\beta_j \in L^\infty(0, \infty)$ with

$$\|\beta_j\|_{L^\infty(0,\infty)} \text{ bounded uniformly in } j. \qquad \text{(B.35)}$$

Then, from (B.34), (B.35), and the fact that $t \mapsto (t, \mathbf{u}(t)) - \nu\|\mathbf{u}(t)\|^2$ belongs to $L^1(0, T)$, we can infer from Theorem A.3 that

$$\lim_{j\to\infty} \int_0^T 2[(\mathbf{f}, \mathbf{u}(t)) - \nu\|\mathbf{u}(t)\|^2]\beta_j(t)\,dt = 0;$$

hence, from (B.33) we find

$$\rho(|\mathbf{u}(T)|^2) - \rho(|\mathbf{u}_0|^2) \le 2\int_0^T [(\mathbf{f}, \mathbf{u}(t)) - \nu\|\mathbf{u}(t)\|^2]\rho'(|\mathbf{u}(t)|^2)\,dt \quad \forall T > 0.$$

Then, since ρ is bounded,

$$\liminf_{T\to\infty} \frac{1}{T}\int_0^T 2[(\mathbf{f}, \mathbf{u}(t)) - \nu\|\mathbf{u}(t)\|^2]\rho'(|\mathbf{u}(t)|^2)\,dt \ge \liminf_{T\to\infty} \frac{\rho(|\mathbf{u}(T)|^2) - \rho(|\mathbf{u}_0|^2)}{T} = 0. \quad \text{(B.36)}$$

Since ρ' is nonnegative and since $\|P_m\mathbf{u}\| \le \|\mathbf{u}\|$ for every $\mathbf{u} \in V$ and every $m \in \mathbb{N}$, we obtain from (B.36) that

$$\liminf_{T\to\infty} \frac{1}{T}\int_0^T 2[(\mathbf{f}, \mathbf{u}(t)) - \nu\|P_m\mathbf{u}(t)\|^2]\rho'(|\mathbf{u}(t)|^2)\,dt \ge 0. \quad \text{(B.37)}$$

Since ρ'' is assumed to be bounded, for $k \in \mathbb{N}$ we have

$$\begin{aligned}|\rho'(|\mathbf{u}(t)|^2) - \rho'(|P_k\mathbf{u}(t)|^2)| &\le c_1\big||\mathbf{u}(t)|^2 - |P_k\mathbf{u}(t)|^2\big| \\ &= c_1(\mathbf{u}(t) + P_k\mathbf{u}(t), \mathbf{u}(t) - P_k\mathbf{u}(t)) \\ &\le c_1|\mathbf{u}(t) + P_k\mathbf{u}(t)||\mathbf{u}(t) - P_k\mathbf{u}(t)| \\ &\le 2c_1|\mathbf{u}(t)||\mathbf{u}(t) - P_k\mathbf{u}(t)|\end{aligned}$$

for some constant $c_1 \ge 0$. Because $\mathbf{u} = \mathbf{u}(t)$ is bounded in H and belongs to V for almost every $t \ge 0$, using (1.14) we find that

$$\begin{aligned}|\rho'(|\mathbf{u}(t)|^2) - \rho'(|P_k\mathbf{u}(t)|^2)| &\le c_2|\mathbf{u}(t) - P_k\mathbf{u}(t)| \le c_2\lambda_k^{-1/2}\|\mathbf{u}(t) - P_k\mathbf{u}(t)\| \\ &\le c_2\lambda_k^{-1/2}\|\mathbf{u}(t)\| \quad \text{for a.e. } t \ge 0 \end{aligned} \quad \text{(B.38)}$$

for some other constant $c_2 \ge 0$. Using (B.38) and (1.15), we now obtain

$$\begin{aligned}&\left|\frac{2}{T}\int_0^T [\rho'(|\mathbf{u}(t)|^2) - \rho'(|P_k\mathbf{u}(t)|^2)]\{(\mathbf{f}, \mathbf{u}(t)) - \nu\|P_m\mathbf{u}(t)\|^2\}\,dt\right| \\ &\quad\le \frac{c_2}{\lambda_k^{1/2}T}\int_0^T \|\mathbf{u}(t)\|\{|\mathbf{f}||\mathbf{u}(t)| + \nu\lambda_m|\mathbf{u}(t)|^2\}\,dt \\ &\quad\le \frac{c_2(1+\lambda_m)}{\lambda_k^{1/2}T}\int_0^T \|\mathbf{u}(t)\|^2\,dt,\end{aligned}$$

but from the energy inequality (II.7.5) we have

$$\frac{1}{T}\int_0^T \|\mathbf{u}(t)\|^2\,dt \le \frac{1}{\nu T}|\mathbf{u}_0|^2 + \frac{1}{\nu^2\lambda_1}|\mathbf{f}|^2$$

(see (3.4)). Therefore, since $\mathbf{u} = \mathbf{u}(t)$ is bounded in H, we end up with

$$\left|\frac{2}{T}\int_0^T [\rho'(|\mathbf{u}(t)|^2) - \rho'(|P_k\mathbf{u}(t)|^2)]\{(\mathbf{f}, \mathbf{u}(t)) - \nu\|P_m\mathbf{u}(t)\|^2\}\,dt\right| \le c_3\frac{1+\lambda_m}{\lambda_k^{1/2}} \quad \text{(B.39)}$$

for a suitable constant $c_3 \geq 0$. From (B.37) and (B.39), we deduce that

$$\liminf_{T\to\infty} \frac{1}{T}\int_0^T 2[(\mathbf{f}, \mathbf{u}(t)) - \nu\|P_k\mathbf{u}(t)\|^2]\rho'(|P_m\mathbf{u}(t)|^2)\, dt \geq -c_3\frac{1+\lambda_m}{\lambda_k^{1/2}} \quad \forall T > 0. \tag{B.40}$$

Now, the function

$$\mathbf{u} \mapsto 2\rho'(|P_k\mathbf{u}|^2)[(\mathbf{f}, \mathbf{u}) - \nu\|P_m\mathbf{u}\|^2]$$

belongs to $\mathcal{C}(H_w)$ and μ is a time-average measure; thus, from (3.1) – and using (1.37) and (B.40) – we infer that

$$\begin{aligned}\int 2\rho'(|P_k\mathbf{u}|^2)&[(\mathbf{f}, \mathbf{u}) - \nu\|P_m\mathbf{u}\|^2]\, d\mu(\mathbf{u}) \\ &= \underset{T\to\infty}{\mathrm{LIM}} \frac{1}{T}\int_0^T 2\rho'(|P_k\mathbf{u}(t)|^2)[(\mathbf{f}, \mathbf{u}(t)) - \nu\|P_m\mathbf{u}(t)\|^2]\, dt \\ &\geq \liminf_{T\to\infty} \frac{1}{T}\int_0^T 2\rho'(|P_k\mathbf{u}(t)|^2)[(f, u(t)) - \nu\|P_m u(t)\|^2]\, dt \\ &\geq -c_3\frac{1+\lambda_m}{\lambda_k^{1/2}}.\end{aligned}$$

Letting $k \to \infty$ and using the Lebesgue dominated convergence theorem, we deduce that

$$2\int \rho'(|\mathbf{u}|^2)[(\mathbf{f}, \mathbf{u}) - \nu\|P_m\mathbf{u}\|^2]\, d\mu(\mathbf{u}) \geq 0.$$

By the Lebesgue dominated convergence theorem once again, we find (as $m \to \infty$) that the inequality (B.29) holds for every $\rho \in \mathcal{C}^2([0,\infty), \mathbb{R})$ satisfying (B.30) and (B.31). Then, as mentioned earlier, we may easily deduce (1.31) from (B.29). This completes the proof of Theorem 3.1.

B.3 Proof of Proposition 3.2

The proof of this proposition is simple. We approximate φ by the map $\mathbf{u} \mapsto \varphi(P_m\mathbf{u})$, where P_m is the Galerkin projector associated with the first m eigenvalues of the Stokes operator. This map is weakly continuous in H, so it follows from (3.1) that

$$\underset{T\to\infty}{\mathrm{LIM}} \frac{1}{T}\int_0^T \varphi(P_m\mathbf{u}(t))\, dt = \int_H \varphi(P_m\mathbf{u})\, d\mu(\mathbf{u}). \tag{B.41}$$

We now take the limit $m \to \infty$ in both sides of (B.41). For the term on the right-hand side, we note from the proof of Proposition 3.1 that μ is carried by K_0; that is, $\mu(H \setminus K_0) = 0$. Then, since $|P_m\mathbf{u}| \leq |\mathbf{u}|$ for any $\mathbf{u}$ in H, it follows that $\varphi(P_m\mathbf{u})$ is bounded μ-almost everywhere, uniformly in m. Moreover, since $P_m\mathbf{u} \to \mathbf{u}$ as m goes to infinity, it also follows that $\varphi(P_m\mathbf{u})$ converges to $\varphi(\mathbf{u})$ as m goes to infinity. Therefore, using again Theorem A.3 yields that

$$\int_H \varphi(P_m \mathbf{u})\, d\mu(\mathbf{u}) \to \int_H \varphi(\mathbf{u})\, d\mu(\mathbf{u})$$

as m goes to infinity.

For the LHS term of (B.41), we use the assumption that φ is Lipschitz continuous in K_0, the estimate (3.4) on the time average of $\|\mathbf{u}(t)\|^2$, and the properties of the Galerkin approximation. We find

$$\begin{aligned}\frac{1}{T}\int_0^T |\varphi(\mathbf{u}(t)) - \varphi(P_m\mathbf{u}(t))|\, dt &\le L\frac{1}{T}\int_0^T |\mathbf{u}(t) - \varphi(P_m\mathbf{u}(t))|\, dt \\ &\le \frac{L}{\lambda_m^{1/2}}\frac{1}{T}\int_0^T \|\mathbf{u}(t)\|\, dt \le \frac{c}{\lambda_m^{1/2}}\end{aligned}$$

for some constant c that is independent of T and m. Take the generalized limit to obtain

$$\left| \operatorname*{LIM}_{T\to\infty} \frac{1}{T}\int_0^T \varphi(\mathbf{u}(t))\, dt - \operatorname*{LIM}_{T\to\infty} \frac{1}{T}\int_0^T \varphi(P_m\mathbf{u}(t))\, dt \right| \le \frac{c}{\lambda_m^{1/2}}.$$

Let $m \to \infty$ to conclude the proof.

B.4 Average Transfer of Energy

We prove here the results that were stated without proof and used in Section 5 – namely, (5.22), (5.26), and (5.35). We give the details of the proof for (5.22). The proofs of (5.26) and (5.35) are similar (see the discussion at the end of this section).

From (5.14) we obtain the inequality

$$\frac{e(\mathbf{z}(T))}{T} + \nu\frac{1}{T}\int_0^T E(\mathbf{z}(t))\, dt \le \frac{e(\mathbf{z}(0))}{T} + \frac{1}{T}\int_0^T \Phi_{\mathbf{y}(t)}(\mathbf{z}(t))\, dt. \qquad \text{(B.42)}$$

Because $\mathbf{y}$ contains only a finite number of modes, we obtain the following bound (using (II.A.34) and (II.A.26)):

$$\begin{aligned}\Phi_{\mathbf{y}}(\mathbf{z}) &= -b(\mathbf{y},\mathbf{y},\mathbf{z}) + b(\mathbf{z},\mathbf{z},\mathbf{y}) \\ &= -b(\mathbf{y},\mathbf{y},\mathbf{z}) - b(\mathbf{z},\mathbf{y},\mathbf{z}) \le C_m|\mathbf{y}||\mathbf{z}|^2 \le C_m|\mathbf{u}|^3,\end{aligned} \qquad \text{(B.43)}$$

where C_m depends on $\lambda_m = \kappa_m^2$, which is the square of the frequency of the highest mode of $\mathbf{y}$. Since $|\mathbf{u}(t)|$ is uniformly bounded in time (see e.g. (II.7.24)), we are allowed to take the generalized limit in (B.42) and so we find

$$\nu \operatorname*{LIM}_{T\to\infty} \frac{1}{T}\int_0^T E(\mathbf{z}(t))\, dt \le \operatorname*{LIM}_{T\to\infty} \frac{1}{T}\int_0^T \Phi_{\mathbf{y}(t)}(\mathbf{z}(t))\, dt. \qquad \text{(B.44)}$$

As we did in (B.43), one can check that $\Phi_{\mathbf{y}}(\mathbf{z})$ satisfies the Lipschitz condition of Proposition 3.2. Thus, we obtain

$$\operatorname*{LIM}_{T\to\infty} \frac{1}{T}\int_0^T \Phi_{\mathbf{y}(t)}(\mathbf{z}(t))\, dt = \langle \Phi_{\mathbf{y}}(\mathbf{z}) \rangle.$$

It follows then from (B.44) that

$$\nu \operatorname*{LIM}_{T\to\infty} \frac{1}{T}\int_0^T E(\mathbf{z}(t))\,dt \le \langle \Phi_{\mathbf{y}}(\mathbf{z})\rangle. \tag{B.45}$$

Now, choose $n > m$ and consider the Galerkin projector P_n and its complement $Q_n = I - P_n$. We have $E(\mathbf{z}) \ge E(P_n\mathbf{z})$. Thus,

$$\operatorname*{LIM}_{T\to\infty} \frac{1}{T}\int_0^T E(\mathbf{z}(t))\,dt \ge \operatorname*{LIM}_{T\to\infty} \frac{1}{T}\int_0^T E(P_n\mathbf{z}(t))\,dt.$$

Moreover, since $E(P_n\mathbf{z})$ is continous in H_w, we find from (5.21) that

$$\operatorname*{LIM}_{T\to\infty} \frac{1}{T}\int_0^T E(P_n\mathbf{z}(t))\,dt = \langle E(P_n\mathbf{z})\rangle.$$

Therefore, (B.45) implies

$$\nu\langle E(P_n\mathbf{z})\rangle \le \langle \Phi_{\mathbf{y}}(\mathbf{z})\rangle.$$

Since $E(P_n\mathbf{z})$ converges monotonically to $E(\mathbf{z})$ as n goes to infinity, it follows from the monotone convergence theorem (see Theorem A.2) that

$$\lim_{n\to\infty}\langle E(P_n\mathbf{z})\rangle = \lim_{n\to\infty}\int_H E(P_n\mathbf{z})\mathbf{u}\,d\mu(\mathbf{u}) = \int_H E(\mathbf{z})\mathbf{u}\,d\mu(\mathbf{u}) = \langle E(\mathbf{z})\rangle.$$

Hence, taking the (usual) limit as n goes to infinity in the relation (B.46), we find

$$\nu\langle E(\mathbf{z})\rangle \le \langle \Phi_{\mathbf{y}}(\mathbf{z})\rangle,$$

which completes the proof of (5.22).

As mentioned in the beginning of this section, the proofs of (5.26) and (5.35) are similar; the only noticeable difference is that we obtain an equality instead of an inequality. This is due to two facts: first, that we start respectively from the equations (5.25) and (5.32) (instead of the inequality (5.14)); second, that $E(\mathbf{z})$ is replaced respectively by $E(\mathbf{y})$ and $E(\mathbf{w})$, which are continuous in H_w – in which case we can apply (5.21) directly.

Appendix C A Mathematical Complement: The Accretivity Property in Dimension 3

For any set $\omega \subset H$ let $\Sigma_t\omega$ $(t \ge 0)$ denote the set of all points $\mathbf{u}(t)$ such that $\mathbf{u} = \mathbf{u}(\cdot)$ is a weak solution on $[0, \infty)$ with initial condition $\mathbf{u}(0) \in \omega$. In the 2-dimensional case this set is exactly $\omega(t) = S(t)\omega$. In both the 2- and 3-dimensional cases, the following result holds.

Lemma C.1 *If ω is a Borel measurable set in H then, for any $t \ge 0$, the set $\Sigma_t\omega$ is measurable with respect to the Lebesgue extension of any Borel probability measure in H.*

The proof of this lemma is left to Section C.3. With this lemma in mind, we can make the following definition.

Definition C.1 *A probability measure μ on H is called* accretive *if*

$$\mu(\Sigma_t \omega) \geq \mu(\omega) \quad \forall t \geq 0$$

for every measurable subset $\omega \subset H$.

For any set $\omega \subset H$ we have

$$\Sigma_{t_2} \Sigma_{t_1} \omega \subset \Sigma_{t_1+t_2} \omega \quad \forall t_1, t_2 \geq 0,$$

so it follows that if μ is accretive then

$$\mu(\Sigma_t \omega) = \mu(\Sigma_{t-s+s} \omega) \geq \mu(\Sigma_{t-s} \Sigma_s \omega(s)) \geq \mu(\Sigma_s \omega)$$

for all $0 \leq s \leq t$ and all measurable subsets $\omega \subset H$, which means that the function $t \in [0, \infty) \mapsto \mu(\Sigma_t \omega)$ is nondecreasing. Our aim is now twofold. In the 2-dimensional case we want to prove a property stronger than accretivity – namely, that any invariant measure μ is such that $\mu(\Sigma_t \omega) = \mu(\omega)$ for all $t \geq 0$ and for every measurable set ω in H. In the 3-dimensional case, we want to show that any time-average measure is accretive.

C.1 The 2-Dimensional Case: A Stronger Result

In the 2-dimensional case, the following result, stronger than accretivity, follows from the injectivity of the associated semigroup.

Proposition C.1 *In the 2-dimensional case, any measure μ on H that is invariant for the NSE is such that*

$$\mu(S(t)\omega) = \mu(\omega) \quad \forall t \geq 0$$

for every measurable subset $\omega \subset H$.

Proof. Let μ be an invariant measure on H for $\{S(t)\}_{t \geq 0}$, and let $\omega \subset H$ be a measurable set. Since μ is invariant, we have

$$\mu(S(t)\omega) = \mu(S(t)^{-1} S(t)\omega).$$

Now, by the backward uniqueness theorem for the Navier–Stokes equations (Bardos and Tartar [1975], Ladyzhenskaya [1963]), the semigroup $\{S(t)\}_{t \geq 0}$ is one-to-one in H. Then, one can check that $S(t)^{-1} S(t)\omega = \omega$; thus,

$$\mu(S(t)\omega) = \mu(S(t)^{-1} S(t)\omega) = \mu(\omega),$$

which proves the proposition. ■

C.2 The 3-Dimensional Case: Time-Average Measures Are Accretive

For the 3-dimensional case we have the following result.

Proposition C.2 *In the 3-dimensional case, any time-average measure is accretive.*

For the proof of this proposition, we need the following lemma.

Lemma C.2 *Let $\mathbf{u} = \mathbf{u}(\cdot)$ be a weak solution of the Navier–Stokes equations defined for all $t \geq 0$, let K be a bounded subset of H that is closed in $D(A^{-1/2})$, and let $t_0 > 0$. For $\mathbf{v} \in H$ and $t \geq 0$, let $d_t(\mathbf{v})$ denote the distance in $D(A^{-1/2})$ from $\mathbf{v}$ to $\Sigma_t K$. Then, for each $t \geq 0$, the function $v \mapsto d_t(\mathbf{v})$ is continuous on the bounded subsets of H for the weak topology. Moreover, for every $\varepsilon > 0$, there exists a $\delta > 0$ such that*

$$d_0(\mathbf{u}(t)) \leq \delta \ \& \ t \geq 0 \implies d_{t_0}(\mathbf{u}(t + t_0)) \leq \varepsilon. \tag{C.1}$$

The proof of this lemma is also left to Section C.3.

Proof of Proposition C.2. Let $\mathbf{u} = \mathbf{u}(\cdot)$ be a weak solution of the NSE defined for all $t \geq 0$, and let μ be a time-average measure of μ. Let $t_0 > 0$. We need to show that $\mu(\Sigma_{t_0}\omega) \geq \mu(\omega)$ for any measurable set ω in H.

Since μ is a regular measure in the sense of (A.3), we have that

$$\mu(\omega) = \sup\{\mu(K);\ K \subset \omega,\ K \text{ weakly compact in } H\}. \tag{C.2}$$

Therefore, it suffices to show that $\mu(\Sigma_{t_0}\omega) \geq \mu(K)$ for any such K.

Let then $K \subset \omega$ be weakly compact in H. Hence, K is bounded in H and closed in H_w and in $D(A^{-1/2})$. For $\mathbf{v} \in H$ and $t \geq 0$, let $d_t(\mathbf{v})$ denote the distance in $D(A^{-1/2})$ from $\mathbf{v}$ to $\Sigma_t K$. From Lemma C.2, the function $v \mapsto d_t(\mathbf{v})$ is weakly continuous on H for each $t \geq 0$. Moreover, for each $\varepsilon > 0$ there exists a $\delta > 0$ for which (C.1) holds. Fix $\varepsilon > 0$ for the moment and let $\delta > 0$ be such that (C.1) holds. Let $\varphi_\varepsilon = \varphi_\varepsilon(\mathbf{u})$ and $\psi_\varepsilon = \psi_\varepsilon(\mathbf{u})$ be defined as follows:

$$\varphi_\varepsilon(\mathbf{u}) = \begin{cases} 1 & \text{if } d_0(\mathbf{u}) \geq \delta, \\ \dfrac{d_0(\mathbf{u})}{\delta} & \text{if } d_0(\mathbf{u}) < \delta; \end{cases}$$

$$\psi_\varepsilon(\mathbf{u}) = \begin{cases} 1 & \text{if } d_{t_0}(\mathbf{u}) \geq \varepsilon, \\ \dfrac{d_{t_0}(\mathbf{u})}{\varepsilon} & \text{if } d_{t_0}(\mathbf{u}) < \varepsilon. \end{cases}$$

Then, φ_ε and ψ_ε belong to $\mathcal{C}(H_w)$. Moreover, using (C.1), one can easily check that, for any $t \geq 0$,

$$\psi_\varepsilon(\mathbf{u}(t + t_0)) > \varepsilon \implies \varphi_\varepsilon(\mathbf{u}(t)) = 1. \tag{C.3}$$

Since $\varphi_\varepsilon, \psi_\varepsilon \in \mathcal{C}(H_w)$, it follows from the definition of the measure μ (see (3.1)) that

$$\int \varphi_\varepsilon(\mathbf{u})\, d\mu(\mathbf{u}) = \underset{T\to\infty}{\text{LIM}} \frac{1}{T} \int_0^T \varphi_\varepsilon(\mathbf{u}(t))\, dt \tag{C.4}$$

and

$$\int \psi_\varepsilon(\mathbf{u})\, d\mu(\mathbf{u}) = \underset{T\to\infty}{\text{LIM}} \frac{1}{T} \int_0^T \psi_\varepsilon(\mathbf{u}(t))\, dt.$$

Then, using property (1.40) of generalized limits, we obtain

$$\int \psi_\varepsilon(\mathbf{u})\,d\mu(\mathbf{u})$$
$$= \operatorname*{LIM}_{T\to\infty} \frac{1}{T+t_0}\int_0^{T+t_0} \psi_\varepsilon(\mathbf{u}(t))\,dt$$
$$= \operatorname*{LIM}_{T\to\infty} \left\{ \frac{T}{T+t_0}\frac{1}{T}\int_{t_0}^{T+t_0} \psi_\varepsilon(\mathbf{u}(t))\,dt + \frac{1}{T+t_0}\int_0^{t_0} \psi_\varepsilon(\mathbf{u}(t))\,dt \right\}$$
$$= \operatorname*{LIM}_{T\to\infty} \left\{ \frac{T}{T+t_0}\frac{1}{T}\int_0^{T} \psi_\varepsilon(\mathbf{u}(t+t_0))\,dt + \frac{1}{T+t_0}\int_0^{t_0} \psi_\varepsilon(\mathbf{u}(t))\,dt \right\}.$$

From this we deduce, since ψ_ε is bounded, that

$$\int \psi_\varepsilon(\mathbf{u})\,d\mu(\mathbf{u}) = \operatorname*{LIM}_{T\to\infty} \frac{1}{T}\int_0^{T} \psi_\varepsilon(\mathbf{u}(t+t_0))\,dt. \tag{C.5}$$

Using (C.3) and the facts that $\psi_\varepsilon \le 1$ and $\varphi_\varepsilon \ge 0$, we find for $T > 0$ that

$$\begin{aligned}\frac{1}{T}\int_0^T \psi_\varepsilon(\mathbf{u}(t+t_0))\,dt &\le \frac{1}{T}\int_0^T \max\{\psi_\varepsilon(\mathbf{u}(t+t_0)), \varepsilon\}\,dt \\ &\le \frac{1}{T}\operatorname{meas}\{t\in[0,T];\ \psi_\varepsilon(\mathbf{u}(t+t_0)) > \varepsilon\} + \varepsilon \\ &\le \frac{1}{T}\operatorname{meas}\{t\in[0,T];\ \varphi_\varepsilon(\mathbf{u}(t)) = 1\} + \varepsilon \\ &\le \varepsilon + \frac{1}{T}\int_0^T \varphi_\varepsilon(\mathbf{u}(t))\,dt,\end{aligned}$$

where meas$\{\cdot\}$ is used to denote the 1-dimensional Lebesgue measure. Then, using (C.4), (C.5), and (1.35), we find that

$$\int \psi_\varepsilon(\mathbf{u})\,d\mu(\mathbf{u}) \le \varepsilon + \int \varphi_\varepsilon(\mathbf{u})\,d\mu(\mathbf{u}).$$

From the definition of φ_ε we see that $\varphi_\varepsilon \le 1$ and that φ_ε vanishes on K. Hence, we deduce from the previous inequality that

$$\int \psi_\varepsilon(\mathbf{u})\,d\mu(\mathbf{u}) \le \varepsilon + \int_{H\setminus K} d\mu(\mathbf{u}) = \varepsilon + \mu(H\setminus K).$$

We let ε go to zero to find, using the Lebesgue dominated convergence theorem, that

$$\int \lim_{\varepsilon\to 0} \psi_\varepsilon(\mathbf{u})\,d\mu(\mathbf{u}) = \lim_{\varepsilon\to 0}\int \psi_\varepsilon(\mathbf{u})\,d\mu(\mathbf{u}) \le \mu(H\setminus K).$$

Since $\lim_{\varepsilon\to 0}\psi_\varepsilon(\mathbf{u})$ is a function equal to 1 on the complement of the closure K_{t_0} in $D(A^{-1/2})$ of $\Sigma_{t_0}K$ and 0 otherwise, it follows that

$$\mu(H\setminus K_{t_0}) \le \mu(H\setminus K).$$

From the regularity of the measure μ (see Section A.1), this implies that

$$\mu(H\setminus \Sigma_{t_0}K) = \mu(H\setminus K_{t_0}) \le \mu(H\setminus K),$$

which implies that

$$\mu(\Sigma_{t_0} K) \geq \mu(K).$$

Thus, since $K \subset \omega$, it is obvious that $\Sigma_{t_0} K \subset \Sigma_{t_0}\omega$, so that $\mu(\Sigma_{t_0}\omega) \geq \mu(\Sigma_{t_0} K) \geq \mu(K)$. Then, since K is an arbitrary weakly compact subset of ω, we deduce using (C.2) that $\mu(\Sigma_{t_0}\omega) \geq \mu(\omega)$. Since $t_0 > 0$ is arbitrary, the proof is complete. ■

We now address a property shared by all accretive probability measures. First, recall from Section 7 the map $S_V(t_0)$, which is defined, for each $t_0 \geq 0$, on the set

$$D_{S_V(t_0)} = \{\mathbf{u}_0 \in V;\ \text{there exists a weak solution } \mathbf{u}(t) \text{ regular on } [0, t_0] \text{ such that } \mathbf{u}(0) = \mathbf{u}_0\}$$

and which is given by $S_V(t_0)\mathbf{u}_0 = \mathbf{u}(t_0)$, with $\mathbf{u}_0$ and $\mathbf{u}$ as in the definition of $D_{S_V(t_0)}$. Define now a function $\theta = \theta(\mathbf{u})\colon H \to [0, \infty)$ by

$$\theta(\mathbf{u}) = \begin{cases} \sup\{t;\ \mathbf{u} \in D_{S_V(t)}\} & \text{if } \mathbf{u} \in V, \\ 0 & \text{if } \mathbf{u} \in H \setminus V, \end{cases} \tag{C.6}$$

which gives the maximal interval of regularity of a solution starting at $\mathbf{u}$. For any $t_0 > 0$, the set $\{\mathbf{u} \in H;\ \theta(\mathbf{u}) > t_0\}$ is equal to $D_{S_V(t_0)}$, which is open in V and hence is measurable. Thus, $\theta = \theta(\mathbf{u})$ is a measurable function on H (with respect to the Borel measure).

Proposition C.3 *Let μ be an accretive probability measure on H. Then*

$$\int \frac{1}{\theta(\mathbf{u})}\, d\mu(\mathbf{u}) = \infty \quad \textit{or} \quad \int \frac{1}{\theta(\mathbf{u})}\, d\mu(\mathbf{u}) = 0; \tag{C.7}$$

in the latter case, μ is invariant with respect to $\{S_V(t)\}_{t\geq 0}$.

Proof. Let us suppose that

$$\mathcal{I} = \int \frac{1}{\theta(\mathbf{u})}\, d\mu(\mathbf{u}) < \infty,$$

since otherwise there is nothing to prove. Let $t_0 > 0$ be fixed for the moment. Recall that

$$\mathcal{I} = \lim_{k\to\infty} \sum_{j=1}^{\infty} \frac{k}{jt_0} \mu(\{\mathbf{u} \in H;\ (j-1)t_0/k < \theta(\mathbf{u}) \leq jt_0/k\}).$$

For $k, j \in \mathbb{N}$, consider

$$\omega_{k,j} = \{\mathbf{u} \in H;\ (j+k-1)t_0/k < \theta(\mathbf{u}) \leq (j+k)t_0/k\}.$$

Note that if $\mathbf{u}_0 \in \omega_{k,j}$ then $\theta(\mathbf{u}_0) > t_0$. Thus, $\mathbf{u}_0 \in D_{S_V(t_0)}$ and

$$\begin{aligned} \theta(S_V(t_0)\mathbf{u}_0) = \theta(\mathbf{u}_0) - t_0 &\in \big((j+k-1)t_0/k - t_0, (j+k)t_0/k - t_0\big] \\ &= \big((j-1)t_0/k, jt_0/k\big]. \end{aligned}$$

Hence,

$$\Sigma_{t_0}\omega_{k,j} \subset \{\mathbf{u} \in H;\ (j-1)t_0/k < \theta(\mathbf{u}) \leq jt_0/k\}.$$

Since μ is accretive, we find then

$$\mu(\omega_{k,j}) \leq \mu(\Sigma_{t_0}\omega_{k,j}) \leq \mu(\{\mathbf{u} \in H;\ (j-1)t_0/k < \theta(\mathbf{u}) \leq jt_0/k\}).$$

Therefore,

$$\begin{aligned}
&\int_{\{\mathbf{u}\in H;\,\theta(\mathbf{u})>t_0\}} \frac{1}{\theta(\mathbf{u}) - t_0}\, d\mu(\mathbf{u}) \\
&\quad = \lim_{k\to\infty} \sum_{j=1}^{\infty} \frac{k}{jt_0} \mu(\{\mathbf{u} \in H;\ (j-1)t_0/k < \theta(\mathbf{u}) - t_0 \leq jt_0/k\}) \\
&\quad = \lim_{k\to\infty} \sum_{j=1}^{\infty} \frac{k}{jt_0} \mu(\{\mathbf{u} \in H;\ (j+k-1)t_0/k < \theta(\mathbf{u}) \leq (j+k)t_0/k\}) \\
&\quad = \lim_{k\to\infty} \sum_{j=1}^{\infty} \frac{k}{jt_0} \mu(\{\mathbf{u} \in H;\ (j-1)t_0/k < \theta(\mathbf{u}) \leq jt_0/k\}) \\
&\quad = \mathcal{I}.
\end{aligned}$$

As a result, we have

$$\int_{\{\mathbf{u}\in H;\,\theta(\mathbf{u})>t_0\}} \frac{1}{\theta(\mathbf{u}) - t_0}\, d\mu(\mathbf{u}) \leq \mathcal{I} < \infty \tag{C.8}$$

for any $t_0 > 0$. Let now $T > 0$ and integrate (C.8) from 0 to T with respect to t_0; this yields

$$\infty > T \cdot \mathcal{I} \geq \int_0^T \int_{\{\mathbf{u}\in H;\,\theta(\mathbf{u})>t_0\}} \frac{1}{\theta(\mathbf{u}) - t_0}\, d\mu(\mathbf{u})\, dt_0.$$

Denoting by χ_E the characteristic function of a set E (i.e., the function that is equal to 1 on E and to 0 on the complement of E), we find using Tonelli's theorem that

$$\begin{aligned}
\infty &> \int_0^T \int_H \frac{1}{\theta(\mathbf{u}) - t_0} \chi_{\{\mathbf{v}\in H;\,\theta(\mathbf{v})>t_0\}}(\mathbf{u})\, d\mu(\mathbf{u})\, dt_0 \\
&= \int_H \int_0^T \frac{1}{\theta(\mathbf{u}) - t_0} \chi_{\{\mathbf{v}\in H;\,\theta(\mathbf{v})>t_0\}}(\mathbf{u})\, d\mu(\mathbf{u})\, dt_0 \\
&= \int_H \int_0^{\min\{T,\theta(\mathbf{u})\}} \frac{1}{\theta(\mathbf{u}) - t_0}\, d\mu(\mathbf{u})\, dt_0 \\
&= \int_H \log \frac{1}{\theta(\mathbf{u}) - t_0}\bigg|_{t_0=0}^{\min\{T,\theta(\mathbf{u})\}} d\mu(\mathbf{u}) \\
&= \int_H \log \frac{\theta(\mathbf{u})}{\theta(\mathbf{u}) - \min\{T, \theta(\mathbf{u})\}}\, d\mu(\mathbf{u}).
\end{aligned}$$

Thus, the integral

$$\int_H \log \frac{\theta(\mathbf{u})}{\theta(\mathbf{u}) - \min\{T, \theta(\mathbf{u})\}}\, d\mu(\mathbf{u})$$

is finite. Since the integrand is nonnegative and is infinite on the set $\{\mathbf{u} \in H;\ \theta(\mathbf{u}) \leq T\}$, we conclude that $\mu(\{\mathbf{u} \in H\ \theta(\mathbf{u}) \leq T\}) = 0$; otherwise, the integral would not be finite. Because $T > 0$ is arbitrary, we infer that $\mu(\{\mathbf{u} \in H;\ \theta(\mathbf{u}) < \infty\}) = 0$. Thus, $\theta(\mathbf{u}) = \infty$ μ-almost everywhere. This implies that $\mathcal{I} = 0$, which proves (C.7).

Let us prove now the second statement. Assume that

$$\int \frac{1}{\theta(\mathbf{u})} d\mu(\mathbf{u}) = 0.$$

This means that $\theta(\mathbf{u}) = \infty$ for μ-almost every $\mathbf{u}$ in H. In particular, for any $t_0 > 0$, we have that $\theta(\mathbf{u}) > t_0$ μ-almost everywhere in H. Hence, $H \setminus D_{S_V(t_0)}$ is of null μ-measure:

$$\mu(H \setminus D_{S_V(t_0)}) = 0 \quad \forall t_0 > 0. \tag{C.9}$$

Let ω be a measurable subset of H and let $t_0 > 0$. Since $S_V(t_0)S_V(t_0)^{-1}\omega \subset \omega$ and since μ is accretive, we find that

$$\mu(\omega) \geq \mu(S_V(t_0)S_V(t_0)^{-1}\omega) \geq \mu(S_V(t_0)^{-1}\omega). \tag{C.10}$$

Similarly,

$$\mu(H \setminus \omega) \geq \mu(S_V(t_0)S_V(t_0)^{-1}(H \setminus \omega)) \geq \mu(S_V(t_0)^{-1}(H \setminus \omega)). \tag{C.11}$$

Since

$$H \setminus (S_V(t_0)^{-1}\omega \cup S_V(t_0)^{-1}(H \setminus \omega)) = H \setminus D_{S_V(t_0)},$$

it follows from (C.9) that

$$\mu\big(H \setminus (S_V(t_0)^{-1}\omega \cup S_V(t_0)^{-1}(H \setminus \omega))\big) = 0.$$

Therefore,

$$\begin{aligned}
1 &= \mu\big(H \setminus (S_V(t_0)^{-1}\omega)\big) + \mu\big(H \setminus (S_V(t_0)^{-1}(H \setminus \omega))\big) \\
&\quad - \mu\big(H \setminus (S_V(t_0)^{-1}\omega \cup S_V(t_0)^{-1}(H \setminus \omega))\big) \\
&= \mu\big(H \setminus (S_V(t_0)^{-1}\omega)\big) + \mu\big(H \setminus (S_V(t_0)^{-1}(H \setminus \omega))\big).
\end{aligned}$$

We then have

$$\begin{aligned}
\mu\big(H \setminus (S_V(t_0)^{-1}\omega)\big) &= 1 - \mu\big(H \setminus (S_V(t_0)^{-1}(H \setminus \omega))\big) \\
&= \mu\big(S_V(t_0)^{-1}(H \setminus \omega)\big) \\
&\leq \mu(H \setminus \omega) \quad \text{(using (C.11))}
\end{aligned}$$

and so

$$\mu(S_V(t_0)^{-1}\omega) = 1 - \mu(H \setminus (S_V(t_0)^{-1}\omega)) \geq 1 - \mu(H \setminus \omega) = \mu(\omega).$$

Together with (C.10), this shows that $\mu(\omega) = \mu(S_V(t_0)^{-1}\omega)$ for any measurable set ω in H and all $t_0 > 0$, which completes the proof. ∎

C.3 Proofs of Lemma C.1 and Lemma C.2

Proof of Lemma C.1. We use the result that a continuous map φ from a Polish space[13] X into another Polish space Y takes Borel sets in X into sets that are measurable with respect to the Lebesgue extension of any Borel measure on Y. (Note that

[13] Recall that a Polish space is a complete, separable, metric space.

the image of a Borel set by a continuous function is not necessarily a Borel set.) This result can be found in Bourbaki [1966, Sec. IX.6.9]; see also Bourbaki [1969].

Let ω be a Borel set in H. We can write

$$\omega = \bigcup_{n=1}^{\infty} \{u \in \omega;\ n-1 \le |u| < n\},$$

$$\Sigma_t \omega = \bigcup_{n=1}^{\infty} \Sigma_t \{u \in \omega;\ n-1 \le |u| < n\}.$$

Hence, if each set $\Sigma_t\{u \in \omega;\ n-1 \le |u| < n\}$ is measurable, then so is the countable union $\Sigma_t \omega$. Therefore, it suffices to assume that ω is bounded in H.

Suppose, then, that ω is contained in the closed ball in H of radius R_0 and centered at the origin. We denote this ball by $B_H(R_0)$, and we let $B_H(R_0)_w$ denote this set endowed with the weak topology of H. By the a priori estimate (A.41) in Chapter II, we know that every weak solution $\mathbf{u} = \mathbf{u}(t)$ satisfies

$$|\mathbf{u}(t)|^2 \le |\mathbf{u}(0)|^2 e^{-\nu\lambda_1 t} + \frac{1}{\nu^2\lambda_1^2}|\mathbf{f}|(1 - e^{-\nu\lambda_1 t}). \tag{C.12}$$

Hence, for $\mathbf{u}(0) \in B_H(R_0)$, the solution $\mathbf{u}$ is such that $|\mathbf{u}(t)| \le R$, where

$$R = R_0 + \frac{1}{\nu^2\lambda_1^2}|\mathbf{f}|.$$

The ball $B_H(R)$ is weakly compact in H, so $B_H(R)_w$ is compact and metrizable. For $T > 0$, the set

$$\mathcal{C}([0, T], B_H(R)_w) \tag{C.13}$$

is also complete, separable, and metrizable. We consider, then, the set

$$X(T, R) = \{\mathbf{u} \in \mathcal{C}([0, T], B_H(R)_w);\ \mathbf{u} \text{ is a weak solution with } \mathbf{u}(0) \in B_H(R_0)\}.$$

Owing to the a priori estimate (C.12), the set $X(T, R)$ contains all weak solutions with initial condition in $B_H(R_0)$. We endow $X(T, R)$ with the topology inherited from the space (C.13). The space $X(T; R)$ is compact in the space (C.13). Therefore, $B_H(R_0)_w$, $B_H(R)_w$, and $X(T, R)$ are all Polish spaces.

Now consider the functions

$$\delta_0 \colon X(T, R) \to B_H(R_0)_w, \quad \mathbf{u} \mapsto \delta_0 \mathbf{u} = \mathbf{u}(0),$$

and

$$\delta_t \colon X(T, R) \to B_H(R)_w, \quad \mathbf{u} \mapsto \delta_t \mathbf{u} = \mathbf{u}(t),$$

for all $0 < t \le T$. Because the topology in these spaces arises from the weak topology in H, the maps δ_0 and δ_t are continuous.

It is straighforward to check that $\Sigma_t = \delta_t \delta_0^{-1}$. Since ω is a Borel set and δ_0 is continuous, it follows that $\delta_0^{-1}\omega$ is a Borel set in $X(T, R)$. Since $X(T, R)$ and $B_H(R)_w$ are Polish spaces and δ_t is continuous, it follows by the result mentioned in the beginning of this proof that the image by δ_t of any Borel set in $X(T, R)$ is a set that is measurable with respect to the Lebesgue extension of any Borel probability measure in $B_H(R)_w$. In particular, $\Sigma_t\omega = \delta_t(\delta_0^{-1}\omega)$ is one such set. This completes the proof of Lemma C.1. ■

Proof of Lemma C.2. Since $d_t(\mathbf{v})$ denotes the distance from $\mathbf{v} \in H$ to $\Sigma_t K$ with respect to the norm of $D(A^{-1/2})$, it follows that, for each $t \geq 0$, the map $\mathbf{v} \mapsto d_t(\mathbf{v})$ is continuous on H with respect to the topology of $D(A^{-1/2})$. Thus, since on bounded subsets of H the weak topology coincides with the topology of $D(A^{-1/2})$, we deduce that $\mathbf{v} \mapsto d_t(\mathbf{v})$ is continous on bounded subsets of H with respect to the weak topology of H for each $t \geq 0$.

Now assume that (C.1) does not hold. Then there exist $\varepsilon > 0$ and a sequence $(t_j)_{j\in\mathbb{N}}$ in $[0, \infty)$ such that, as $j \to \infty$,

$$d_{t_0}(\mathbf{u}(t_j + t_0)) > \varepsilon \quad \text{and} \quad d_0(\mathbf{u}(t_j)) \to 0. \tag{C.14}$$

Because $t \mapsto \mathbf{u}(t)$ is weakly continuous and bounded in H, it follows from the first part of this proof that $t \mapsto d_0(\mathbf{u}(t))$ and $t \mapsto d_{t_0}(\mathbf{u}(t + t_0))$ are continuous. Thus, we can assume in view of (II.7.23) that $\mathbf{u}(t)$ is strongly continous from the right at each t_j, $j \in \mathbb{N}$. Then, $\mathbf{v}_j(t)$, defined as $\mathbf{u}(t + t_j)$, is a weak solution on $[0, \infty)$ for each $j \in \mathbb{N}$. Moreover, from (II.A.41) and (II.A.40) we have

$$|\mathbf{v}_j(t)|^2 \leq \rho = |\mathbf{u}_0|^2 + \frac{1}{\nu^2\lambda_1^2}|\mathbf{f}|^2 \quad \forall t \geq 0$$

and

$$\int_0^t \|\mathbf{v}_j(s)\|^2\, ds \leq \frac{\rho}{\nu} + \frac{1}{\nu^2\lambda_1}|\mathbf{f}|^2 t \quad \forall t > 0. \tag{C.15}$$

Thus, using also (1.12), we find for any $t_2 > t_1 \geq 0$ that

$$\begin{aligned}
&\|\mathbf{v}_j(t_2) - \mathbf{v}_j(t_1)\|_{D(A^{-1/2})} \\
&\leq \int_{t_1}^{t_2} \|\mathbf{v}_j'(t)\|_{D(A^{-1/2})}\, dt = \int_{t_1}^{t_2} \|\mathbf{F}(\mathbf{v}_j(t))\|_{D(A^{-1/2})}\, dt \\
&\leq \nu \int_{t_1}^{t_2} \|\mathbf{v}_j(t)\|\, dt + c\int_{t_1}^{t_2} |\mathbf{v}_j(t)|^{1/2}\|\mathbf{v}_j(t)\|^{3/2}\, dt + \frac{1}{\lambda_1^{1/2}}|\mathbf{f}|(t_2 - t_1) \\
&\leq \nu(t_2 - t_1)^{1/2}\left(\int_{t_1}^{t_2} \|\mathbf{v}_j(t)\|^2\, dt\right)^{1/2} + c\rho^{1/4}(t_2 - t_1)^{1/4}\left(\int_{t_1}^{t_2} \|\mathbf{v}_k(t)\|^2\, dt\right)^{3/4} \\
&\quad + \frac{1}{\lambda_1^{1/2}}|\mathbf{f}|(t_2 - t_1) \\
&\leq c_1(1 + t_2)(t_2 - t_1)^{1/4}
\end{aligned}$$

for some constant $c_4 > 0$. Therefore, the functions $\mathbf{v}_j = \mathbf{v}_j(t)$ are equicontinuous on any compact interval in $[0, \infty)$ as $D(A^{-1/2})$-valued functions. Since $\{\mathbf{u} \in H;\ |\mathbf{u}| \leq \rho^{1/2}\}$ is compact in $D(A^{-1/2})$, it follows from the Arzela–Ascoli theorem (see Theorem A.20) that there exists a subsequence $(\mathbf{v}_{j_k})_k$ converging in $D(A^{-1/2})$, uniformly on every compact interval in $[0, \infty)$. Let $\mathbf{v} = \mathbf{v}(t)$ be the limit of this subsequence. Since, on the bounded subsets of H, the topology of $D(A^{-1/2})$ and the weak topology of H coincide, it follows that $\mathbf{v} = \mathbf{v}(t)$ is an H-valued weakly continuous function on $[0, \infty)$. Moreover, taking (C.15) into account, we can assume that $(\mathbf{v}_{j_k}(t))_k$ is weakly convergent in $L^2(0, T; V)$ for every $T > 0$. Thus, as in the proofs of existence of solutions to the Navier–Stokes equations, we can deduce that $\mathbf{v} = \mathbf{v}(t)$ is a weak solution on $[0, \infty)$.

Since $\mathbf{v}_j(0) = \mathbf{u}(t_j)$ and $d_0(u(t_j)) \to 0$, we infer that $\mathbf{v}(0) \in K$. Therefore, $\mathbf{v}(t_0) \in \Sigma_{t_0} K$. Since $\mathbf{v}_{j_k}(t_0) \to \mathbf{v}(t_0)$ in $D(A^{-1/2})$, we see that

$$d_{t_0}(\mathbf{u}(t_{j_k} + t_0)) = d_{t_0}(\mathbf{v}_{j_k}(t_0)) \to 0,$$

which contradicts (C.14). Hence (C.1) holds, and this completes the proof of Lemma C.2. ■

V

Time-Dependent Statistical Solutions of the Navier–Stokes Equations and Fully Developed Turbulence

Introduction

This long and technical chapter aims at providing some basic connections between the mathematical theory of the Navier–Stokes equations (NSE) and the conventional theory of turbulence. As stated earlier, the conventional theory of turbulence (including the famous Kolmogorov spectrum law) is based principally on physical and scaling arguments, with little reference to the NSE. We believe that it is instructive to connect turbulence more precisely with the Navier–Stokes equations.

It is commonly accepted that turbulent flows are necessarily statistical in nature. Indeed, if a flow is turbulent, then all physical quantities are rapidly varying in space and time and we cannot determine the actual instantaneous values of these quantities. Instead, one usually measures the moments, or some averaged values of physical quantities; that is, only a statistical description of the flow is available. The first task in this chapter is to establish, in a more precise way, the time evolution of the probability distribution functions associated with the fluid flow – that is, the statistical solutions of the Navier–Stokes equations. Although the discussion is relevant to deterministic data (initial values of the velocities and volume forces), we extend our discussion to the case of random data; however, we will not examine the more involved case of very irregular forcing (such as white or colored volume forces), since deterministic or moderately irregular stochastic data suffice, in practice, to generate complex turbulent flows.

In Chapter IV we studied time-independent statistical solutions; in this chapter, we propose to extend the study to the time-dependent case. Thus we consider, for instance, the Navier–Stokes equations

$$\mathbf{u}_t - \nu\Delta\mathbf{u} + (\mathbf{u}\cdot\nabla)\mathbf{u} + \nabla p = \mathbf{f}(t), \qquad \nabla\cdot\mathbf{u} = 0, \tag{0.1}$$

and – as in previous chapters – rewrite them as a differential equation $\mathbf{u}' = \mathbf{F}(t, \mathbf{u})$ in the appropriate phase space H. We consider as the initial data a probability distribution (PD) μ_0; namely, for every measurable set $E \subset H$, we are given

$$\text{Probability}\{\mathbf{u}_0 \in E\} = \mu_0(E).$$

We want to study, for $t > 0$, the family of probability measures μ_t,

$$\mu_t(E) = \text{Probability}\{\mathbf{u}(t) \in E\}. \tag{0.2}$$

The first objective of this chapter is to provide the mathematical framework for the study of the family of measures $\{\mu_t\}_{t>0}$, given μ_0 and $\mathbf{f}$. Then we derive the Liouville-type equation (similar to the one in statistical mechanics) that this family satisfies.

In the framework of classical analysis and classical calculus, it is natural to study the evolution of certain moments of μ_t (in fact, this is done in laboratory experiments as well as in computing). The simplest moments are the linear ones corresponding to the average velocity:

$$\int \mathbf{u}\, d\mu(\mathbf{u})t \quad \text{or (componentwise)} \quad \int u_i\, d\mu(\mathbf{u})t, \quad i = 1, 2, 3.$$

One may also study scalar quantities, corresponding to the scalar product of $\mathbf{u} \in H$, with a given function $\mathbf{g} \in H$. For instance, for $\mathbf{x}_0 \in \Omega$ (the domain filled by fluid) and $\varepsilon > 0$ not too large (so that $B_\varepsilon(\mathbf{x}_0)$, the ball of $\mathbb{R}^3$ of radius ε centered at $\mathbf{x}_0$, is included in Ω), we can consider $\mathbf{g} = \mathbf{1}_{B_\varepsilon}(\mathbf{x}_0)\mathbf{e}_1/\text{Vol}(B_\varepsilon(\mathbf{x}_0))$, where $\mathbf{1}_{B_\varepsilon}(\mathbf{x}_0)$ is the characteristic function of $B_\varepsilon(\mathbf{x}_0)$ (equal to 1 on $B_\varepsilon(\mathbf{x}_0)$ and to 0 in $\Omega \setminus B_\varepsilon(\mathbf{x}_0)$) and $\mathbf{e}_1$ is the unit vector in the direction $0x_1$. Then $(\mathbf{u}, \mathbf{g})$ is the average of u_1 in $B_\varepsilon(\mathbf{x}_0)$:

$$\frac{1}{\text{Vol}(B_\varepsilon(\mathbf{x}_0))} \int_{B_\varepsilon(\mathbf{x}_0)} u_1(\mathbf{x})\, d\mathbf{x}.$$

More generally, we may consider several functions $\mathbf{g}_i \in H$ $(i = 1, \ldots, k)$, and we may study the evolution of their components

$$\int (\mathbf{u}, \mathbf{g}_i)_H\, d\mu(\mathbf{u})t, \quad i = 1, \ldots, k,$$

or of a quantity

$$\int \phi((\mathbf{u}, \mathbf{g}_1)_H, \ldots, (\mathbf{u}, \mathbf{g}_k)_H)\, d\mu(\mathbf{u})t. \tag{0.3}$$

Functions of the form (0.3) are called cylindrical functions of $\mathbf{u}$; they play a special role subsequently. One may also consider nonlinear moments such as

$$\int u_{i_1} \cdots u_{i_k}\, d\mu_t(\mathbf{u}).$$

In fact, we will consider "generalized" moments of the form

$$\int \Phi(\mathbf{u})\, d\mu_t(\mathbf{u}) \tag{0.4}$$

for suitable test functions Φ of $\mathbf{u}$ that map H (or a subspace of H) into $\mathbb{R}$.

In Section 1 we recall a few elements of the mathematical theory of the deterministic Navier–Stokes equations as developed in Chapters II and III, and we slightly modify or extend the results to make them suitable for this chapter. Then, our first task in Section 2 is to define precisely the statistical solutions μ_t of the NSE, informally defined in (0.1); we will do so by writing, for suitable test functions Φ, the evolution equations for generalized moments of type (0.4). This will lead to a linear evolution equation for the measures μ_t – namely, the Liouville equation of the problem, which will be discussed and will reappear several times in different forms.

Of course, in deriving this equation, many mathematical technicalities must be taken into account. In particular, we must consider separately flows in space dimensions 2 and 3, since the extent of completeness of the mathematical theory of the NSE is not the same in these two cases. We consider also separately (a) the flows in a bounded domain with no-slip boundary conditions and (b) space-periodic flows in the whole space $\mathbb{R}^2$ or $\mathbb{R}^3$ with zero or nonzero (space) average. We conclude Section 2 by considering the Hopf equation, corresponding to the case where

$$\Phi(\mathbf{u}) = e^{i(\mathbf{u},\mathbf{g})_H},$$

with $\mathbf{g} \in H$ and $i = \sqrt{-1}$. The Hopf equation is actually equivalent to the Liouville equation (its formal Fourier transform), and any existence and uniqueness result for the Liouville equation produces an existence and uniqueness result for the Hopf equation (and vice versa). For this reason we do not discuss existence and uniqueness for the Hopf equation.

Another problem is discussed in Section 3: the homogeneous statistical solutions of the Navier–Stokes equations for space-periodic flows, or in the whole space $\mathbb{R}^3$. This study is, of course, motivated by homogeneous turbulence, as we aim to give a mathematical meaning to the concept of homogeneity. Homogeneity here means invariance by translations in space. The deterministic NSE are invariant under the group of translations in space but, in the context of deterministic flows, homogeneous flows are space invariant and therefore essentially trivial. Nontrivial homogeneous flows appear in the statistical context; here we will use the notion of *homogeneous measures* on $\mathbb{R}^2$ or $\mathbb{R}^3$, which are measures that are invariant under translations (we shall give a more precise definition). The concept of homogeneous measure is not easy to internalize, and one may be tempted to believe that homogeneous measures are also trivial. This is not the case, and we will show how, by averaging, one can associate to any measure a nontrivial homogeneous one. Those familiar with stochastic processes may think also of stationary processes that are invariant under time translation and of which the homogeneous measures considered here are a spatial analog.

The theory of homogeneous flows as developed in Section 2 discusses the concept of homogeneous measures as well as the concepts of homogeneous statistical solutions in the periodic case and in the whole space. In fact, we remark that homogeneous statistical solutions in the whole-space case are directly obtained from homogeneous statistical solutions in the space-periodic case by letting the period go to infinity; hence, individual solutions in the whole space need not be discussed. Section 3 is devoted to the rigorous derivation of the Reynolds equation.

In this chapter, the statistical theory of the Navier–Stokes equations culminates in Section 4, where we investigate the question of self-similarity. After addressing the question of self-similar solutions in the deterministic and statistical cases, we show how to construct a 2-parameter family of homogeneous statistical solutions of the Navier–Stokes equations.

Finally, in Section 5 we establish a connection between the results of this chapter (especially those of Section 4) with the conventional theory of turbulence. To the best of our knowledge, the results discussed here – together with those of Chapter IV – serve to establish for the first time a rigorous connection between the conventional intuitive theory of turbulence and the mathematical theory of the Navier–Stokes equations. In this section the usefulness of the mathematical tools presented here becomes more evident. Specifically, the relationship between the so-called Kolmogorov spectrum ($\kappa^{-5/3}$) and the Navier–Stokes equations themselves is clarified. This is in contrast to the classical standard approach, which is based on purely dimensional arguments that do not refer to the underlying NSE. Interestingly enough, the mathematical results established here allow us to represent the statistics of decaying homogeneous turbulence in terms of a stationary measure. That, in turn, allows us to find some interesting and important properties of such turbulent flows without any extensive numerical effort.

Elements of the Mathematical Theory of the Navier–Stokes Equations

For the convenience of the reader, we now recall (from Chapter II) some elements of the mathematical theory of the NSE that will be needed in this chapter. We consider the Navier–Stokes equations

$$\begin{aligned} \mathbf{u}_t - \nu\Delta\mathbf{u} + (\mathbf{u}\cdot\nabla)\mathbf{u} + \nabla p = \mathbf{f}, \\ \nabla\cdot\mathbf{u} = 0, \end{aligned} \tag{0.5}$$

in $\Omega \subset \mathbb{R}^d$ ($d = 2$ or 3) with $\nu > 0$ and $\mathbf{f} = \mathbf{f}(x)$. We endow the system with either the no-slip boundary conditions or the space-periodic boundary conditions. In the first case, we assume more precisely that

$$\Omega \subset \mathbb{R}^d \text{ is open, bounded, and connected, with a } \mathcal{C}^2 \text{ boundary } \partial\Omega, \tag{0.6}$$

and require that

$$\mathbf{u} = 0 \text{ on } \partial\Omega. \tag{0.7}$$

In the space-periodic case, we assume that

$$\Omega = \prod_{i=1}^{d}\left(\frac{L_i}{2}, \frac{L_i}{2}\right), \quad L_i > 0,\ i = 1, \ldots, d, \tag{0.8}$$

and require that

$$\mathbf{u} = \mathbf{u}(\mathbf{x}, t),\ p = p(\mathbf{x}, t), \text{ and } \mathbf{f} = \mathbf{f}(\mathbf{x}) \text{ with } \mathbf{x} = (x_1, \ldots, x_d) \text{ are } L_i\text{-periodic in each variable } x_1, \ldots, x_d \tag{0.9}$$

and, for simplicity, that

$$\int_\Omega \mathbf{f}(\mathbf{x})\,d\mathbf{x} = \int_\Omega \mathbf{u}(\mathbf{x}, t)\,d\mathbf{x} = 0. \tag{0.10}$$

We now recall the mathematical setting of the problem. In the no-slip case, we denote by $\mathcal{V}$ the set of smooth (say, $\mathcal{C}^\infty$) divergence-free vector fields on Ω that are compactly supported in Ω. In the space-periodic case, $\mathcal{V}$ is the space of smooth ($\mathcal{C}^\infty$) divergence-free vector fields on $\mathbb{R}^d$ that are periodic with period Ω. In either case, we let H be the closure of $\mathcal{V}$ in $L^2(\Omega)^d$ and let V be the closure of $\mathcal{V}$ in $H^1(\Omega)^d$. In H and V, we consider (respectively) the scalar products

$$(\mathbf{u}, \mathbf{v}) = \int_\Omega \mathbf{u}(\mathbf{x}) \cdot \mathbf{v}(\mathbf{x})\, d\mathbf{x}, \qquad ((\mathbf{u}, \mathbf{v})) = \int_\Omega \sum_{i=1}^{d} \frac{\partial \mathbf{u}}{\partial x_i} \cdot \frac{\partial \mathbf{v}}{\partial x_i}\, d\mathbf{x}$$

and the associated norms denoted by

$$|\mathbf{u}| = (\mathbf{u}, \mathbf{u})^{1/2} \quad \text{for } \mathbf{u} \in H, \qquad \|\mathbf{v}\| = ((\mathbf{v}, \mathbf{v}))^{1/2} \quad \text{for } \mathbf{v} \in V.$$

Recall that the spaces H and V can be characterized as follows: in the no-slip case,

$$H = \{\mathbf{u} \in L^2(\Omega)^d;\ \nabla \cdot \mathbf{u} = 0,\ \mathbf{u} \cdot \mathbf{n}|_{\partial\Omega} = 0\}$$

and

$$V = \{\mathbf{u} \in H^1(\Omega)^d;\ \nabla \cdot \mathbf{u} = 0,\ \mathbf{u}|_{\partial\Omega} = 0\},$$

where $\mathbf{n}$ denotes the outward unit normal to $\partial\Omega$. We remark that $\mathbf{u} \cdot \mathbf{n}|_{\partial\Omega}$ makes sense when $\mathbf{u} \in L^2(\Omega)^d$ and $\nabla \cdot \mathbf{u} = 0$ in the distribution sense. In the space-periodic case,

$$H = \left\{\mathbf{u} \in L^2_{\text{per}}(\Omega)^d;\ \nabla \cdot \mathbf{u} = 0,\ \int_\Omega \mathbf{u}(\mathbf{x})\, d\mathbf{x} = 0\right\}$$

and

$$V = \left\{\mathbf{v} \in H^1_{\text{per}}(\Omega)^d;\ \nabla \cdot \mathbf{u} = 0,\ \int_\Omega \mathbf{u}(\mathbf{x})\, d\mathbf{x} = 0\right\},$$

where $H^1_{\text{per}}(\Omega)^d$ is the space of $\mathbb{R}^d$-valued functions $\mathbf{u}$ defined on $\mathbb{R}^d$ that are L_i-periodic in each variable x_i ($i = 1, \ldots, d$) and such that $\mathbf{u}|_{\mathcal{O}} \in H^1(\mathcal{O})^d$ for every bounded open set $\mathcal{O}$ in $\mathbb{R}^d$. The functions in $H^1_{\text{per}}(\Omega)^d$ are easily characterized by their Fourier series expansion

$$H^1_{\text{per}}(\Omega)^d = \left\{\mathbf{u} = \sum_{\mathbf{k} \in \mathbb{Z}^d} \hat{\mathbf{u}}_{\mathbf{k}} e^{2\pi i \frac{\mathbf{k}}{\mathbf{L}} \cdot \mathbf{x}};\ \hat{\mathbf{u}}_{-\mathbf{k}} = \bar{\hat{\mathbf{u}}}_{\mathbf{k}},\ \sum_{\mathbf{k} \in \mathbb{Z}^d} \left(\frac{1}{L^2} + 2\pi \left|\frac{\mathbf{k}}{\mathbf{L}}\right|^2\right) |\hat{\mathbf{u}}_{\mathbf{k}}|^2 < \infty\right\},$$

where $L = \min\{L_1, \ldots, L_d\}$ and

$$\frac{\mathbf{k}}{\mathbf{L}} = \left(\frac{k_1}{L_1}, \ldots, \frac{k_d}{L_d}\right).$$

Hence, we also have

$$V = \left\{\mathbf{u} = \sum_{\mathbf{k} \in \mathbb{Z}^d \setminus \{0\}} \hat{\mathbf{u}}_{\mathbf{k}} e^{2\pi i \frac{\mathbf{k}}{\mathbf{L}} \cdot \mathbf{x}};\ \hat{\mathbf{u}}_{-\mathbf{k}} = \bar{\hat{\mathbf{u}}}_{\mathbf{k}},\ \frac{\mathbf{k}}{\mathbf{L}} \cdot \hat{\mathbf{u}}_{\mathbf{k}} = 0,\ \sum_{\mathbf{k} \in \mathbb{Z}^d \setminus \{0\}} \left|\frac{\mathbf{k}}{\mathbf{L}}\right|^2 |\hat{\mathbf{u}}_{\mathbf{k}}|^2 < \infty\right\}$$

as well as

$$H = \left\{\mathbf{u} = \sum_{\mathbf{k} \in \mathbb{Z}^d \setminus \{0\}} \hat{\mathbf{u}}_{\mathbf{k}} e^{2\pi i \frac{\mathbf{k}}{\mathbf{L}} \cdot \mathbf{x}};\ \hat{\mathbf{u}}_{-\mathbf{k}} = \bar{\hat{\mathbf{u}}}_{\mathbf{k}},\ \frac{\mathbf{k}}{\mathbf{L}} \cdot \hat{\mathbf{u}}_{\mathbf{k}} = 0,\ \sum_{\mathbf{k} \in \mathbb{Z}^d \setminus \{0\}} |\hat{\mathbf{u}}_{\mathbf{k}}|^2 < \infty\right\}.$$

The problem posed in the weak formulation of the NSE with either no-slip or space-periodic boundary conditions is as follows: Given $T > 0$, $\mathbf{u}_0$ in H, and $\mathbf{f}$ in $L^2(0, T; H)$, find a function $\mathbf{u}$ in $L^\infty(0, T; H) \cap L^2(0, T; V)$ such that

$$\frac{d}{dt}(\mathbf{u}, \mathbf{v}) + \nu((\mathbf{u}, \mathbf{v})) + b(\mathbf{u}, \mathbf{u}, \mathbf{v}) = (\mathbf{f}, \mathbf{v}) \quad \text{for all } \mathbf{v} \in V, \tag{0.11}$$

in the distribution sense on $(0, T)$, and with $\mathbf{u}(0) = \mathbf{u}_0$ in some suitable sense. Here

$$b(\mathbf{u}, \mathbf{v}, \mathbf{w}) = \sum_{i,j=1}^{d} \int_\Omega u_i \frac{\partial v_j}{\partial x_i} w_j \, d\mathbf{x},$$

and if $\mathbf{f}$ is square integrable but not in H then we can replace it by its Leray projection on H, so that $\mathbf{f}$ is always assumed to be in H.

Note that the pressure term disappears in the weak formulation of the problem. This is because the gradient of a function is orthogonal to the space of solenoidal functions. More precisely, with respect to the inner product in $L^2(\Omega)^d$, we have

$$H^\perp = \{\mathbf{u} \in L^2(\Omega)^d;\ \mathbf{u} = \nabla p,\ p \in H^1(\Omega)^d\}$$

in the no-slip case and

$$H^\perp = \{\mathbf{u} \in L^2(\Omega)^d;\ \mathbf{u} = \nabla p,\ p \in H^1_{\text{per}}(\Omega)^d\}$$

in the space-periodic case. In any case, we have the so-called Leray projector

$$P_L \colon L^2(\Omega)^d \to H,$$

which is the orthogonal projector onto H in $L^2(\Omega)^d$.

The trilinear operator $b = b(\mathbf{u}, \mathbf{v}, \mathbf{w})$ defined previously can be extended to a continuous trilinear operator defined on V. Moreover,

$$b(\mathbf{u}, \mathbf{v}, \mathbf{v}) = 0 \quad \text{for } \mathbf{u}, \mathbf{v} \in V, \tag{0.12}$$

which implies that

$$b(\mathbf{u}, \mathbf{v}, \mathbf{w}) = -b(\mathbf{u}, \mathbf{w}, \mathbf{v}) \quad \text{for } \mathbf{u}, \mathbf{v}, \mathbf{w} \in V. \tag{0.13}$$

An equivalent formulation for the NSE is achieved with the definition of the Stokes operator

$$A\mathbf{u} = -P_L \Delta \mathbf{u} \quad \text{for all } \mathbf{u} \in D(A) = V \cap H^2(\Omega)^d. \tag{0.14}$$

The Stokes operator is a positive self-adjoint operator, so we can work with fractional powers of A. We will consider, as explained in Chapter II, the positive and negative powers of A, A^α with $\alpha \in \mathbb{R}$ and $\alpha \geq 0$ or $\alpha < 0$; we also explained in Chapter II how $D(A^\alpha)$ and $D(A^{-\alpha})$ are paired through the "duality" product (a bilinear form from $D(A^\alpha) \times D(A^{-\alpha})$ into $\mathbb{R}$). In particular, we have $D(A^{1/2}) = V$, and it follows that $A \colon D(A^{1/2}) \to D(A^{-1/2})$ with

$$(A\mathbf{u}, \mathbf{v}) = ((\mathbf{u}, \mathbf{v})) \quad \text{for all } \mathbf{u}, \mathbf{v} \in D(A^{1/2}),$$

where $(\cdot,\cdot)$ is the duality product between $D(A^{-1/2})$ and $D(A^{1/2})$. Moreover, the trilinear operator $b(\mathbf{u},\mathbf{v},\mathbf{w})$ leads to the bilinear operator $B\colon D(A^{1/2})\times D(A^{1/2})\mapsto D(A^{-1/2})$ defined by

$$(B(\mathbf{u},\mathbf{v}),\mathbf{w}) = b(\mathbf{u},\mathbf{v},\mathbf{w}) \quad \text{for all } \mathbf{u},\mathbf{v},\mathbf{w}\in D(A^{1/2}).$$

We then have the following functional formulation of the Navier–Stokes equations: Given $T>0$, $\mathbf{u}_0$ in H, and $\mathbf{f}$ in $L^2(0,T;H)$, find $\mathbf{u}$ in $L^\infty(0,T;H)\cap L^2(0,T;V)$ with $\mathbf{u}'\in L^1(0,T;D(A^{-1/2}))$ for all $T>0$ such that

$$\mathbf{u}' + \nu A\mathbf{u} + B(\mathbf{u}) = \mathbf{f} \tag{0.15}$$

and $\mathbf{u}(0)=\mathbf{u}_0$ in some suitable sense, where $B(\mathbf{u}) = B(\mathbf{u},\mathbf{u})$.

We will have the opportunity to use the following inequality for the bilinear operator:

$$\|B(\mathbf{u})\|_{D(A^{-1/2})} \le c|\mathbf{u}|^{1/2}\|\mathbf{u}\|^{3/2} \quad \forall \mathbf{u}\in V, \tag{0.16}$$

where c is a constant depending only on the shape of Ω.

Since the domain Ω is assumed to be bounded, it follows by Rellich's lemma that V is compactly embedded in H, so that A^{-1} is compact as a closed operator in H. Hence, there exists an orthonormal basis $\{\mathbf{w}_m\}_{m=1}^\infty$ in H such that

$$A\mathbf{w}_m = \lambda_m \mathbf{w}_m, \quad m=1,2,\ldots.$$

Moreover, we have that

$$0<\lambda_1\le\lambda_2\le\cdots\le\lambda_m\le\cdots, \quad \lambda_m\to+\infty \text{ as } m\to+\infty.$$

We also recall the Poincaré inequality

$$|\mathbf{u}|^2 \le \frac{1}{\lambda_1}\|\mathbf{u}\|^2 \quad \forall \mathbf{u}\in V. \tag{0.17}$$

In what follows we denote by P_m the orthogonal projector of H onto the space spanned by $\mathbf{w}_1,\mathbf{w}_2,\ldots,\mathbf{w}_m$. The following inequalities will be useful:

$$|\mathbf{u}-P_m\mathbf{u}|^2 \le \frac{1}{\lambda_m}\|\mathbf{u}-P_m\mathbf{u}\|^2 \quad \forall \mathbf{u}\in V,\ \forall m\in\mathbb{N}; \tag{0.18}$$

$$\|P_m\mathbf{u}\|^2 \le \lambda_m|\mathbf{u}|^2 \quad \forall \mathbf{u}\in H,\ \forall m\in\mathbb{N}. \tag{0.19}$$

In fact, as shown in Chapter II, we need to define solutions of (0.15) that possess further properties; in this way we introduced in Section II.7 the concepts of *weak* and *strong* solutions to the Navier–Stokes equations (0.15). We have seen also that the results of existence and uniqueness of solutions are different in space dimensions 2 and 3 (see Theorems 7.1–7.4 in Chapter II) and that, for technical reasons, even the notion of weak solution differs from one space dimension to the other. In space dimension 2, the weak solutions of (0.15) are also assumed to satisfy

$$\mathbf{u}'\in L^2(0,T;D(A^{-1/2})) \quad \text{for all } T>0, \tag{0.20}$$

and this suffices for all technical purposes. In space dimension 3, we can obtain only weak solutions of (0.15) that satisfy, instead of (0.20),

$$\mathbf{u}' \in L^{4/3}(0, T; D(A^{-1/2})) \quad \text{for all } T > 0. \tag{0.21}$$

Because this information is not sufficient, we also require that the 3-dimensional weak solutions satisfy further properties, which we proved in Chapter II, Theorem 7.1 (see also the discussion in the section entitled "Further properties of the solutions in dimension 3"). Particularly important for many purposes are the energy inequalities (II.7.5) and (II.7.7), together with their consequences (II.A.38) and (II.A.43).

On a more technical side, we will also need from Chapter II (specifically, Sections 7–9 and Appendix A) several estimates for the inertial term (either through $B(\mathbf{u}, \mathbf{v})$ or through $b(\mathbf{u}, \mathbf{v}, \mathbf{w})$).

1 Time-Dependent Statistical Solutions on Bounded Domains

Our aim in this section is to define rigorously the statistical solutions of the time-dependent NSE and to derive the (linear) evolution equation satisfied by the probability distribution.

We will study the bounded domain case with no-slip boundary conditions and the whole-space case with space-periodic boundary conditions. We intend to demonstrate the existence of time-dependent statistical solutions starting from an initial probability distribution μ_0. We consider first the no-slip boundary conditions, and then we consider the periodic case. We use the 2-dimensional case to motivate the definition of the statistical solutions before investigating the 3-dimensional case.

1.1 Statistical Solutions in the Case of No-Slip Boundary Conditions

Here we consider no-slip boundary conditions, so that, for instance, $H = H_{\text{nsp}}$ and $V = V_{\text{nsp}}$. In order to motivate the definition of time-dependent statistical solutions, we first consider the 2-dimensional Navier–Stokes equations. Then, we introduce the rigorous definition for the 2- and 3-dimensional cases and present the main existence and uniqueness results.

Motivation for the Definition of Statistical Solutions

Consider the 2-dimensional NSE and denote by $S(t; 0)$, $t \geq 0$, the solution operator in H that maps $\mathbf{u}_0 = \mathbf{u}(0)$ into $\mathbf{u}(t)$.[1]

Suppose we are given a probability distribution μ_0 as initial data. Then, since the solutions to the Navier–Stokes equations evolve according to

$$\mathbf{u}'(t) = \mathbf{F}(t, \mathbf{u}(t)), \quad \text{where } \mathbf{F}(t, \mathbf{u}) = \mathbf{f}(t) - \nu A\mathbf{u} - B(\mathbf{u}), \tag{1.1}$$

we expect the probability distribution to evolve as a time-dependent family of measures μ_t on H, depending on $t \geq 0$ and given by $\mu_t(E) = \mu_0(S(t; 0)^{-1}E)$, for every measurable set E in H. Indeed, we have

[1] Notice that $\{S(t; 0)\}_{t \geq 0}$ is not, in general, a semigroup, since the forcing term $\mathbf{f} = \mathbf{f}(t)$ may depend on time.

$$\begin{aligned}\mu_t(E) &= \text{Probability}\{S(t;0)\mathbf{u}_0 \in E\}\\ &= \text{Probability}\{\mathbf{u}_0 \in S(t;0)^{-1}E\}\\ &= \mu_0(S(t;0)^{-1}E).\end{aligned}$$

Note here that E is any measurable set; part of it may or may not be included in $S(t;0)H$, the image of H by $S(t;0)$. In other words, one may have $E \cap S(t;0)H \subsetneqq E$ or even $E \cap S(t;0)H = \emptyset$; such sets E are of no physical interest to us, but for the mathematical treatment we must handle them a priori.

At any time t, we can extract information from the system by using the (generalized) moments

$$\int \Phi(\mathbf{u})\, d\mu_t(\mathbf{u}),$$

where $\Phi \in L^1(\mu_t)$, the space of real-valued and μ_t-integrable functions on H. By working with simple functions, one can verify that

$$\int \Phi(\mathbf{u})\, d\mu_t(\mathbf{u}) = \int \Phi(S(t;0)\mathbf{u})\, d\mu_0(\mathbf{u}) \tag{1.2}$$

for every such Φ. Hence, the evolution in time of those moments is given by

$$\frac{d}{dt}\int \Phi(\mathbf{u})\, d\mu_t(\mathbf{u}) = \frac{d}{dt}\int \Phi(S(t;0)\mathbf{u})\, d\mu_0(\mathbf{u}). \tag{1.3}$$

Using the evolution equation $\mathbf{u}'(t) = \mathbf{F}(t, \mathbf{u}(t))$, we can formally differentiate (with respect to time) the term on the RHS of (1.3). For that purpose, we permute differentiation and integration and compute the time derivative of $\Phi(S(t;0)\mathbf{u})$ by the chain differentiation rule, which here takes the form

$$\begin{aligned}\frac{d}{dt}\Phi(S(t;0)\mathbf{u}) &= \left(\Phi'(S(t;0)\mathbf{u}),\, \frac{d}{dt}S(t;0)\mathbf{u}\right)\\ &= (\Phi'(S(t;0)\mathbf{u}),\, \mathbf{F}(t;\, S(t;0)\mathbf{u})).\end{aligned} \tag{1.4}$$

Hence, assuming that all quantities make sense, we find

$$\begin{aligned}\frac{d}{dt}\int \Phi(\mathbf{u})\, d\mu_t(\mathbf{u}) &= \int (\mathbf{F}(t, S(t;0)\mathbf{u}),\, \Phi'(S(t;0)\mathbf{u}))\, d\mu_0(\mathbf{u})\\ &= \int (\mathbf{F}(t,\mathbf{u}),\, \Phi'(\mathbf{u}))\, d\mu_t(\mathbf{u})\end{aligned}$$

for suitable test functions Φ.

The preceeding sequence of equalities leads to the Liouville-type equation

$$\frac{d}{dt}\int \Phi(\mathbf{u})\, d\mu_t(\mathbf{u}) = \int (\mathbf{F}(t,\mathbf{u}),\, \Phi'(\mathbf{u}))\, d\mu_t(\mathbf{u}). \tag{1.5}$$

Equation (1.5) makes sense even when the mappings $S(t;0)$ are not defined, as in the case of the 3-dimensional Navier–Stokes equations. Hence, equation (1.5) is taken as the basis for the equation for time-dependent statistical solutions for space dimensions 2 and 3. But in order to make the definition rigorous, we must first find conditions on $\{\mu_t\}_{t\in[0,T]}$ and define a class of functionals Φ for which (1.5) makes sense. This is our next task.

Definition of Statistical Solutions

The class of test functionals is defined as follows.

Definition 1.1 *We denote by $\mathcal{T}_{\rm nsp}$ the class of cylindrical* test functionals *consisting of the real-valued functionals $\Phi = \Phi(\mathbf{u})$ that depend only on a finite number k of components of $\mathbf{u}$, namely,*

$$\Phi(\mathbf{u}) = \phi((\mathbf{u}, \mathbf{g}_1), \ldots, (\mathbf{u}, \mathbf{g}_k)),$$

where ϕ is a $\mathcal{C}^1$ scalar function on $\mathbb{R}^k$ with compact support and where $\mathbf{g}_1, \ldots, \mathbf{g}_k$ belong to $V_{\rm nsp}$. For such a Φ we denote by Φ' its differential in $H_{\rm nsp}$, which has the form

$$\Phi'(\mathbf{u}) = \sum_{j=1}^{k} \partial_j \phi((\mathbf{u}, \mathbf{g}_1), \ldots, (\mathbf{u}, \mathbf{g}_k))\mathbf{g}_j,$$

where $\partial_j \phi$ denotes the derivative of ϕ with respect to the jth variable, $(\mathbf{u}, \mathbf{g}_j)$; since $\Phi'(\mathbf{u})$ is a linear combination of the $\mathbf{g}_j$, it obviously belongs to $V_{\rm nsp}$.

Let now $\{\mu_t\}_t$ denote a family of probability measures on H. We can extract statistical information from this family through the generalized moments

$$\int \varphi(\mathbf{u})\, d\mu_t(\mathbf{u})$$

with appropriate functionals φ. Hence, we assume that

$$t \mapsto \int \varphi(\mathbf{u})\, d\mu_t(\mathbf{u})$$

is measurable for all continuous and bounded real-valued functions φ on H. This assumption implies, in particular, that the functions

$$t \mapsto \int |\mathbf{u}|^2\, d\mu_t(\mathbf{u}) \quad \text{and} \quad t \mapsto \int \|\mathbf{u}\|^2\, d\mu_t(\mathbf{u}) \tag{1.6}$$

are measurable (see Appendix A.1).

Suppose now that

$$t \mapsto \int \|\mathbf{u}\|^2\, d\mu_t(\mathbf{u}) \tag{1.7}$$

is integrable. Notice that this implies $\mu_t(H \setminus V) = 0$ almost everywhere in t; that is, the measure μ_t is carried by V. It then follows (see Appendix A.1) that, for any test functional Φ in $\mathcal{T}_{\rm nsp}$,

$$\mathbf{u} \mapsto (\nu A\mathbf{u} + B(\mathbf{u}) - \mathbf{f}(t), \Phi'(\mathbf{u}))$$

is continuous in V (for almost every t since $\mathbf{f} \in L^2(0, T; H)$) and that

$$t \mapsto \int (\nu A\mathbf{u} + B(\mathbf{u}) - \mathbf{f}(t), \Phi'(\mathbf{u}))\, d\mu_t(\mathbf{u})$$

makes sense and is integrable. Then we can write the integral form of the Liouville-type equation (1.5):

$$\int \Phi(\mathbf{u})\, d\mu_t(\mathbf{u}) = \int \Phi(\mathbf{u})\, d\mu_0(\mathbf{u}) + \int_0^t (\mathbf{F}(s, \mathbf{u}), \Phi'(\mathbf{u}))\, d\mu_s(\mathbf{u})\, ds. \tag{1.8}$$

Finally, as in the case of individual solutions, we have the following energy-type inequality:

$$\int |\mathbf{u}|^2\, d\mu_t(\mathbf{u}) + 2\nu \int_0^t \|\mathbf{u}\|^2\, d\mu_s(\mathbf{u})\, ds \leq \int_0^t \int (\mathbf{f}(s), \mathbf{u})\, d\mu_s(\mathbf{u})\, ds + \int |\mathbf{u}|^2\, d\mu_0(\mathbf{u}) \quad \text{for all } t \in [0, T]. \tag{1.9}$$

In space dimension 2, both sides of (1.9) are equal, but it is necessary to weaken this relation to an inequality for space dimension 3 because the energy equation is not available for weak solutions in that dimension (see Section 7 in Chapter II).

In summary, then, in the 2- and 3-dimensional (no-slip) cases, given a probability measure μ_0 on H_{nsp} with finite kinetic energy

$$\int |\mathbf{u}|^2\, d\mu_0(\mathbf{u}) < \infty \tag{1.10}$$

and given $\mathbf{f}$ in $L^2(0, T; H_{\text{nsp}})$, where $0 < T < \infty$, we are led to define as follows a time-dependent statistical solution of the Navier–Stokes equations.

Definition 1.2 *A family* $\{\mu_t\}_{0 \leq t \leq T}$ *of probability measures on* H_{nsp} *is called a* (time-dependent) statistical solution *of the Navier–Stokes equations on* H_{nsp} *with initial data* μ_0 *and forcing term* $\mathbf{f}$ *if*: (a) *the* μ_t *satisfy* (1.8) *for all cylindrical test functionals* Φ *in* $\mathcal{T}_{\text{nsp}}$; *and* (b) *the energy inequality* (1.9) *holds. Furthermore*: *the function*

$$t \mapsto \int \varphi(\mathbf{u})\, d\mu_t(\mathbf{u}) \tag{1.11}$$

must be measurable on $[0, T]$ *for every bounded and continuous real-valued function* φ *on* H_{nsp}; *the function*

$$t \mapsto \int |\mathbf{u}|^2\, d\mu_t(\mathbf{u}) \tag{1.12}$$

is assumed to belong to $L^\infty(0, T)$; *and the function*

$$t \mapsto \int \|\mathbf{u}\|^2\, d\mu_t(\mathbf{u}) \tag{1.13}$$

is assumed to belong to $L^1(0, T)$.

Remark 1.1 One can show that property (1.13) actually follows from (1.11), (1.12), (1.9), and the hypothesis (1.10) of finite initial kinetic energy. Moreover, property (1.8) can be extended to more general test functionals Φ; this will be established as needed.

Existence and Uniqueness Results

The existence of statistical solutions was proved by Foias [1972], using a slight variation of the formulation just described, and by Vishik and Fursikov [1977b]. We state this result next and leave to Appendix A.1 a proof different from those in the given references.

Theorem 1.1 (Existence: Space Dimensions 2 and 3) *Let μ_0 be a probability measure on H_{nsp} with finite kinetic energy, that is,*

$$\int |\mathbf{u}|^2 \, d\mu_0(\mathbf{u}) < \infty.$$

Let $\mathbf{f}$ *be a forcing term in* $L^2(0, T; H_{\text{nsp}})$ *with* $T > 0$. *Then there exists a statistical solution* $\{\mu_t\}_{0\le t\le T}$ *of the Navier–Stokes equations on* H_{nsp} *in the sense of Definition 1.2 (i.e., satisfying* (1.9), (1.11), (1.12), *and* (1.13)).

Note that uniqueness of the statistical solution is not asserted in the 3-dimensional case. In the 2-dimensional case, similarly to the situation with individual solutions, the time-dependent statistical solutions are unique – provided they satisfy further regularity hypotheses. We state this result but leave its proof to Appendix A.1. However, we first remark that, in the 2-dimensional case, if the initial probability distribution μ_0 has its support bounded in H_{nsp} then a time-dependent statistical solution can be obtained with further regularity properties. For the sake of simplicity, we assume in the remainder of this section that the forcing term $\mathbf{f}$ is time-independent and belongs to H_{nsp}, which is the space H in the no-slip case. Then, one can obtain (see Appendix A.1) a time-dependent statistical solution with the following properties: for every φ in $\mathcal{C}(H_w)$, the map

$$t \mapsto \int \varphi(\mathbf{u}) \, d\mu_t(\mathbf{u}) \tag{1.14}$$

is continuous on $[0, \infty)$; the support $\operatorname{supp} \mu_t$ of μ_t is included in V and is bounded in H uniformly for $t \ge 0$ – that is, there exists $R > 0$, depending on (a bound in H for the support of) the initial measure μ_0, such that

$$\operatorname{supp} \mu_t \subset \{\mathbf{u} \in V;\ |\mathbf{u}| \le R\} \tag{1.15}$$

for all $t \ge 0$; and the Liouville-type equation

$$\int \Psi(t, \mathbf{u}) \, d\mu_t(\mathbf{u}) = \int \Psi(0, \mathbf{u}) \, d\mu_0(\mathbf{u}) + \int_0^t \int \{\Psi_s'(s, \mathbf{u}) + (\mathbf{F}(\mathbf{u}), \Psi_{\mathbf{u}}'(s, \mathbf{u}))\} \, d\mu_s(\mathbf{u}) \, ds \tag{1.16}$$

holds for all $t \ge 0$ and for all the following time-dependent test functions $\Psi = \Psi(t, \mathbf{u})$. Here, the allowed test functions Ψ are the continuous real-valued functions on $[0, \infty) \times V$ that are Fréchet differentiable in the sense that: (a) there exists $\Psi'(t, \mathbf{u}) = (\Psi_t'(t, \mathbf{u}), \Psi_{\mathbf{u}}'(t, \mathbf{u}))$ in $\mathbb{R} \times V$ such that

$$\frac{1}{|s|+|\mathbf{v}|}|\Psi(t+s,\mathbf{u}+\mathbf{v})-\Psi(t,\mathbf{u})-s\Psi_s'(t,\mathbf{u})-(\mathbf{v},\Psi_{\mathbf{u}}'(t,\mathbf{u}))|\to 0$$

as $|s|+|\mathbf{v}|$ goes to zero (where $|s|$ is the absolute value of the real number s and $|\mathbf{v}|$ is the H-norm of the vector field $\mathbf{v}$); and (b) $\Psi'(t,\mathbf{u})$ is continuous from $[0,\infty)\times V$ into $\mathbb{R}\times V$, where $\Psi_{\mathbf{u}}'(t,\mathbf{u})$ is uniformly bounded in V and $\Psi_t'(t,\mathbf{u})$ has at most a linear growth in $|\mathbf{u}|$.

Then, we have the following uniqueness result.

Theorem 1.2 (Uniqueness: Space Dimension 2) *In dimension $d=2$, let μ_0 be a probability measure on H_{nsp} with bounded support and let the forcing term $\mathbf{f}$ belong to H_{nsp}. Then, the time-dependent statistical solution $\{\mu_t\}_{t\geq 0}$ with initial distribution μ_0 and satisfying the conditions* (1.14)–(1.16) *is unique and is given by $\mu_t=S(t)\mu_0$ $(t\geq 0)$, where $\{S(t)\}_{t\geq 0}$ is the solution operator of the 2-dimensional Navier–Stokes equations.*

Under the assumptions of Theorem 1.2, since $\mu_t=S(t)\mu_0$ we obtain in a straightforward way the following energy equation for $\{\mu_t\}_{t\geq 0}$:

$$\frac{1}{2}\int|\mathbf{u}|^2\,d\mu_t(\mathbf{u})+\nu\int_0^t\int\|\mathbf{u}\|^2\,d\mu_s(\mathbf{u})\,ds$$
$$=\frac{1}{2}\int|\mathbf{u}|^2\,d\mu_0(\mathbf{u})+\int_0^t\int(\mathbf{f},\mathbf{u})\,d\mu_s(\mathbf{u})\,ds.$$

Actually, one can show that Theorem 1.2 holds under weaker hypotheses. Indeed, we need only (1.15) and (1.16) in the proof. On the other hand, in the 2-dimensional case one can show (a) that if the initial measure has bounded support in H_{nsp} then the statistical solutions constructed in the proof of Theorem 1.1 satisfy (1.14) and (1.15) and (b) that (1.16) follows from (1.14); hence, simply (1.14) and (1.15) may be assumed for the uniqueness. One can also show that the statistical solutions constructed in Theorem 1.1 satisfy a strengthened form of the energy equation, from which (1.15) can be deduced. We do not address these more technical issues here; the interested reader is referred to Foias [1972] for more details.

1.2 Statistical Solutions in the Space-Periodic Case

We now consider the periodic case with or without nonvanishing velocity average. The definitions of test functionals and stationary statistical solutions are essentially the same as in the no-slip case.

Definition 1.3 *We denote by $\mathcal{T}_{\text{per}}$ (resp., $\dot{\mathcal{T}}_{\text{per}}$) the class of cylindrical* test functionals *consisting of the real-valued functionals $\Phi=\Phi(\mathbf{u})$ that depend only on a finite number k of components of* $\mathbf{u}$:

$$\Phi(\mathbf{u})=\phi((\mathbf{u},\mathbf{g}_1)_{Q(L)},\ldots,(\mathbf{u},\mathbf{g}_k)_{Q(L)}),$$

where ϕ is a C^1 scalar function on $\mathbb{R}^k$ with compact support and where $\mathbf{g}_1, \dots, \mathbf{g}_k$ belong to $V(L)$ (resp., $\dot{V}(L)$). For such Φ we denote by Φ' its differential with respect to the inner product $(\cdot,\cdot) = L^d(\cdot,\cdot)_{Q(L)}$;[2] *this has the form*

$$\Phi'(\mathbf{u}) = \frac{1}{L^d} \sum_{j=1}^{k} \partial_j \phi((\mathbf{u}, \mathbf{g}_1)_{Q(L)}, \dots, (\mathbf{u}, \mathbf{g}_k)_{Q(L)}) \mathbf{g}_j,$$

where $\partial_j \phi$ denotes the derivative of ϕ with respect to its jth *argument. Since $\Phi'(\mathbf{u})$ is a linear combination of the g_j, it obviously belongs to $V(L)$ (resp., $\dot{V}(L)$).*

We emphasize that in this definition we consider the derivative of Φ with respect to $L^d(\cdot,\cdot)_{Q(L)}$. The need for introducing the factor L^d becomes apparent when we let $L \to \infty$ in view of obtaining the statistical solution on the whole space.

For the definition of time-dependent statistical solutions, we consider first the nonzero average case. In this case, we are given a probability measure μ_0 on $H(L)$ with finite kinetic energy

$$\int |\mathbf{u}|^2_{Q(L)} \, d\mu_0(\mathbf{u}) < \infty$$

as well as a forcing term $\mathbf{f}$ in $L^2(0, T, H(L))$, where $T > 0$. We are led to the following definition.

Definition 1.4 *A family $\{\mu_t\}_{0 \le t < T}$ of probability measures on $H(L)$ is called a* (time-dependent) statistical solution *of the Navier–Stokes equations on $H(L)$ with initial data μ_0 and forcing term $\mathbf{f}$ if the μ_t satisfy*: (a) *the Liouville-type equation*

$$\int \Phi(\mathbf{u}) \, d\mu_t(\mathbf{u}) = \int \Phi(\mathbf{u}) \, d\mu_0(\mathbf{u}) + L^d \int_0^t (\mathbf{F}(s, \mathbf{u}), \Phi'(\mathbf{u}))_{Q(L)} \, d\mu_s(\mathbf{u}) \, ds \tag{1.17}$$

for all $t \in [0, T]$ and for any test functional Φ in $\mathcal{T}_{\text{per}}$; and (b) *the energy inequality*

$$\begin{aligned} \int |\mathbf{u}|^2_{Q(L)} \, d\mu_t(\mathbf{u}) &+ 2\nu \int_0^t \|\mathbf{u}\|^2_Q \, d\mu_s(\mathbf{u}) \, ds \\ &\le \int_0^t \iint (\mathbf{f}(s), \mathbf{u})_{Q(L)} \, d\mu_s(\mathbf{u}) \, ds + \int |\mathbf{u}|^2_{Q(L)} \, d\mu_0(\mathbf{u}) \end{aligned} \tag{1.18}$$

for all $t \in [0, T]$. Furthermore, the function

$$t \mapsto \int \varphi(\mathbf{u}) \, d\mu_t(\mathbf{u}) \tag{1.19}$$

is required to be measurable on $[0, T]$ for every bounded and continuous real-valued function φ on $H(L)$, the function

$$t \mapsto \int |\mathbf{u}|^2_{Q(L)} \, d\mu_t(\mathbf{u}) \tag{1.20}$$

[2] That is, $|\Phi(\mathbf{u}+\mathbf{h}) - \Phi(\mathbf{u}) - (\mathbf{h}, \Phi'(\mathbf{u}))|/|\mathbf{h}| \to 0$ in the corresponding norm $|\mathbf{h}| = (\mathbf{h}, \mathbf{h})^{1/2} = L^{d/2}(\mathbf{h}, \mathbf{h})^{1/2}_{Q(L)}$.

is assumed to belong to $L^\infty(0,T)$*, and the function*

$$t \mapsto \int \|\mathbf{u}\|^2_{Q(L)}\, d\mu_t(\mathbf{u}) \tag{1.21}$$

is assumed to belong to $L^1(0,T)$.

Remark 1.2 As in the case of the no-slip boundary condition, one can show that property (1.21) actually follows from (1.19), (1.20), (1.18), and the assumption of finite initial kinetic energy. Moreover, property (1.17) can be extended to more general test functionals Φ, which will be introduced as needed.

Existence and Uniqueness Results

Theorem 1.3 (Existence: Space Dimensions 2 and 3) *Let* μ_0 *be a probability measure on* $H(L)$ *with finite kinetic energy, that is,*

$$\int |\mathbf{u}|^2_{Q(L)}\, d\mu_0(\mathbf{u}) < \infty. \tag{1.22}$$

Let the forcing term $\mathbf{f}$ *belong to* $L^2(0,T;H(L))$*, where* $T>0$*. Then there exists a family* $\{\mu_t\}_{0\le t<T}$ *of probability measures on* $H(L)$ *satisfying* (1.17)–(1.21).

In this case as well, if the dimension is $d=2$ then the time-dependent statistical solutions are unique.

Theorem 1.4 (Uniqueness: Space Dimension 2) *In dimension* $d=2$*, given a probability measure* μ_0 *on* $H(L)$ *with finite kinetic energy*

$$\int |\mathbf{u}|^2_{Q(L)}\, d\mu_0(\mathbf{u}) < \infty \tag{1.23}$$

and given a forcing term $\mathbf{f}$ *in* $L^2(0,T;H(L))$ *with* $T>0$*, the time-dependent statistical solution* $\{\mu_t\}_{0\le t<T}$ *provided by Theorem 1.3 is unique and is given by* $\mu_t = S(t;0)\mu_0$*, where* $\{S(t;0)\}_{0\le t<T}$ *is the solution operator of the Navier–Stokes equations on* $H(L)$.

Now, if the forcing term $\mathbf{f}$ has zero space average and if one starts with a measure concentrated on $\dot{H}(L)$, then one can obtain a statistical solution $\{\mu_t\}_t$ with each μ_t concentrated on $\dot{H}(L)$. This result can be stated as follows.

Theorem 1.5 (Existence: Space Dimensions 2 and 3) *Given a probability measure* μ_0 *on* $\dot{H}(L)$ *with finite kinetic energy*

$$\int |\mathbf{u}|^2_{Q(L)}\, d\mu_0(\mathbf{u}) < \infty, \tag{1.24}$$

and given a forcing term $\mathbf{f}$ *in* $L^2(0,T;\dot{H}(L))$ *with* $T>0$*, there exists a time-dependent statistical solution on* $\dot{H}(L)$ *with initial data* μ_0 *and forcing term* $\mathbf{f}$ *– that* *is, a family* $\{\mu_t\}_{0\le t<T}$ *of probability measures on* $\dot{H}(L)$ *satisfying* (1.17)–(1.21) *with* $H(L)$ *and* $V(L)$ *replaced by* $\dot{H}(L)$ *and* $\dot{V}(L)$*, respectively.*

The proof of Theorem 1.5 follows straightforwardly from Theorem 1.3 and is given in Appendix A.1.

1.3 Hopf Equation on Bounded Domains

Hopf studied the evolution of the probability distribution of a flow and arrived at an equation in infinitely many variables for the characteristic function of the probability distribution (see Hopf [1952]). This so-called Hopf equation can actually be derived from equations (1.8) (in the no-slip case) and (1.17) (in the periodic case) by choosing an appropriate test functional Φ. Moreover, by choosing this particular Φ in (1.8) and (1.17), a more useful functional formulation of the Hopf equation arises. Such a form of Φ is

$$\Phi(\mathbf{u}) = e^{i(\mathbf{u},\mathbf{g})},$$

defined for all $\mathbf{u}$ in H, where $\mathbf{g}$ belongs to V and is arbitrary. This Φ does not belong to the spaces $\mathcal{T}_{\text{nsp}}$ and $\mathcal{T}_{\text{per}}$ of test functionals, but we already observed in Remarks 1.1 and 1.2 that equations (1.8) and (1.17) hold for a larger class of test functionals, and that class includes this Φ.[3] Then, one obtains the equation

$$\int e^{i(\mathbf{u},\mathbf{g})}\,d\mu_t(\mathbf{u}) + i\int_0^t\int [\nu((\mathbf{u},\mathbf{g})) + ((\mathbf{u}\cdot\nabla)\mathbf{u},\mathbf{g})]e^{i(\mathbf{u},\mathbf{g})}\,d\mu_s(\mathbf{u})\,ds = \int e^{i(\mathbf{u},\mathbf{g})}\,d\mu_0(\mathbf{u}) \quad (1.25)$$

for all t in $[0,T)$. This is an equation for the characteristic function χ, which is the Fourier transform of the probability distribution μ_t,

$$\chi(t,\mathbf{g}) = \int e^{i(\mathbf{u},\mathbf{g})}\,d\mu_t(\mathbf{u}).$$

Indeed, denote by $\partial_{\mathbf{g}}$ the derivative with respect to $\mathbf{g}$ in H and notice that, formally,

$$\partial_{\mathbf{g}}\chi(t,\mathbf{g}) = i\int \mathbf{u}e^{i(\mathbf{u},\mathbf{g})}\,d\mu_t(\mathbf{u}).$$

Then, given the Stokes operator A, we can formally define an operator $A(\partial_{\mathbf{g}})$ on χ by setting

$$A(\partial_{\mathbf{g}})\chi(t,\mathbf{g}) = i\int A\mathbf{u}e^{i(\mathbf{u},\mathbf{g})}\,d\mu_t(\mathbf{u}),$$

so that

[3] We skip the technical details: this extension is obtained by (a) truncation and using the Lebesgue dominated convergence theorem (see Theorem A.3 in Chapter IV) and then (b) complexification of the space (as in Section II.8).

$$(A(\partial_{\mathbf{g}})\chi(t,\mathbf{g}),\mathbf{g}) = i\int(A\mathbf{u},\mathbf{g})e^{i(\mathbf{u},\mathbf{g})}\,d\mu_t(\mathbf{u}) = i\int((\mathbf{u},\mathbf{g}))e^{i(\mathbf{u},\mathbf{g})}\,d\mu_t(\mathbf{u}).$$

Similarly, considering the quadratic operator $B(\mathbf{u}) = (\mathbf{u}\cdot\nabla)\mathbf{u}$, we can define the operator $B(\partial_{\mathbf{g}})$ on $\chi(t,\mathbf{g})$ by

$$B(\partial_{\mathbf{g}})\chi(t,\mathbf{g}) = i\int B(\mathbf{u})e^{i(\mathbf{u},\mathbf{g})}\,d\mu_t(\mathbf{u}),$$

so that

$$(B(\partial_{\mathbf{g}})\chi(t,\mathbf{g}),\mathbf{g}) = i\int(B(\mathbf{u}),\mathbf{g})e^{i(\mathbf{u},\mathbf{g})}\,d\mu_t(\mathbf{u}).$$

Therefore, we can rewrite equation (1.25) in the form

$$\frac{d}{dt}\chi(t,\mathbf{g}) + \nu(A(\partial_{\mathbf{g}})\chi(t,\mathbf{g})\mathbf{g}) + (B(\partial_{\mathbf{g}})\chi(t,\mathbf{g}),\mathbf{g}) = 0 \qquad (1.26)$$

for $0 \le t < T$.

Equation (1.26) is the functional form of the Hopf equation. The operators $A(\partial_{\mathbf{g}})$ and $B(\partial_{\mathbf{g}})$ can be rigorously defined through the formulas

$$(A(\partial_{\mathbf{g}})\chi(t,\mathbf{g}),\mathbf{v}) = i\int(A\mathbf{u},\mathbf{v})e^{i(\mathbf{u},\mathbf{g})}\,d\mu_t(\mathbf{u}) \qquad (1.27)$$

and

$$(B(\partial_{\mathbf{g}})\chi(t,\mathbf{g}),\mathbf{v}) = i\int(B(\mathbf{u}),\mathbf{v})e^{i(\mathbf{u},\mathbf{g})}\,d\mu_t(\mathbf{u}) \qquad (1.28)$$

for all $\mathbf{v}$ in V. Moreover, by choosing for $\mathbf{g}$ the eigenfunctions $\{\mathbf{w}_m\}_{m\in\mathbb{N}}$ of the Stokes operator A, we obtain (as derived by Hopf [1952]) a system of equations in infinitely many variables for the components $u_m = (\mathbf{u},\mathbf{w}_m)$ of $\mathbf{u}$ in this basis, $\mathbf{u} = \sum_{m=1}^{\infty} u_m\mathbf{w}_m$.

2 Homogeneous Statistical Solutions

As discused earlier, turbulence far away from any wall is believed to be homogeneous (i.e., space-invariant); this belief leads to the physical concept of homogeneous turbulence. In this section and the next, we want to develop the mathematical tools adapted to this concept. That is achieved by means of the concept of homogeneous measures, which are measures invariant with respect to translations in physical space. The concept of homogeneous measures is not at all intuitively obvious, as it can be readily confused with homogeneous (space-invariant) functions. Homogeneous (space-invariant) functions are trivial because they are constant, whereas homogeneous measures are *not* trivial; indeed, we will construct some examples of nontrivial homogeneous measures.

In this section, we start by defining homogeneous probability measures and then present some of their properties and a few examples of nontrivial homogeneous measures. Next, we give the definition of homogeneous statistical solutions in the periodic case and state the main results concerning their existence. Finally, we consider the whole-space case. As we said before, the whole-space case is studied by considering the space-periodic case and letting the period L go to infinity, as in Foias and Temam [1980]. For the sake of clarity, in the periodic case we consider only domains of the form $Q(L) = (-L/2, L/2)^d$ with $L > 0$ and consider the spaces $H_{\text{per}} = H(L)$ and $V_{\text{per}} = V(L)$. We also consider the spaces H_{loc} and V_{loc} defined for the whole space. The reader is referred to Section 10 in Chapter II for the definition of these spaces.

Since we are interested in homogeneous turbulence, we assume throughout this section that the forcing term $\mathbf{f}$ is identically zero.

2.1 Homogeneous Measures

Consider a given probability measure μ on H_{loc}, and suppose this measure represents the statistics of a certain turbulent flow. The moments are obtained by means of averages, such as

$$\int \varphi(\mathbf{u})\, d\mu(\mathbf{u}),$$

for appropriate functionals φ on H_{loc}. If the statistical behavior of the turbulent flow is assumed to be independent of position, then the statistical averages do not change if we shift $\mathbf{u}$ in space. More precisely, if we consider the translation operator $\tau_{\mathbf{a}}$ defined by $\tau_{\mathbf{a}}\mathbf{u}(\mathbf{x}) = \mathbf{u}(\mathbf{x} - \mathbf{a})$ for all $\mathbf{x}$ in $\mathbb{R}^d$, where $\mathbf{a}$ belongs to $\mathbb{R}^d$ and is arbitrary, then we have

$$\int \varphi(\mathbf{u})\, d\mu(\mathbf{u}) = \int \varphi(\tau_{\mathbf{a}}\mathbf{u})\, d\mu(\mathbf{u}) \tag{2.1}$$

for all $\mathbf{a}$ in $\mathbb{R}^d$ and all appropriate functionals φ. Then we can define a measure $\tau_{\mathbf{a}}\mu$ by

$$\tau_{\mathbf{a}}\mu(E) = \mu(\tau_{-\mathbf{a}}E) \tag{2.2}$$

for all measurable sets E in H_{loc}, where $\tau_{-\mathbf{a}}$ is the inverse of $\tau_{\mathbf{a}}$; here, for $E \subset H_{\text{loc}}$, $\tau_{-\mathbf{a}}E$ is the set of functions $\tau_{-\mathbf{a}}\mathbf{u}$ for $\mathbf{u}$ in E. The RHS of (2.1) can then be written as the integral of φ with respect to the translated measure $\tau_{\mathbf{a}}\mu$, and the condition (2.1) can be rewritten as

$$\int \varphi(\mathbf{u})\, d\mu(\mathbf{u}) = \int \varphi(\mathbf{u})\, d\tau_{\mathbf{a}}\mu(\mathbf{u}) \tag{2.3}$$

for all $\mathbf{a}$ in $\mathbb{R}^d$ and all appropriate functionals φ. This is equivalent to saying that $\tau_{\mathbf{a}}\mu = \mu$ for every $\mathbf{a}$ in $\mathbb{R}^d$. A probability measure with this property is called a *homogeneous probability measure.*

Definition 2.1 *A probability measure μ on H_{loc} is said to be* homogeneous *if $\tau_{\mathbf{a}}\mu = \mu$ for every $\mathbf{a} \in \mathbb{R}^d$.*

Since $H(L)$ is contained in H_{loc}, we can consider homogeneous measures for the periodic case, that is, on $H(L)$. These are the homogeneous measures on H_{loc} that are carried by $H(L)$; by this we mean that $\mu(H_{\text{loc}} \setminus H(L)) = 0$. Hence, we can restrict μ to measurable subsets of $H(L)$ and have that (2.1) holds for all $\varphi \in \mathcal{C}_b(H(L))$.

Nontrivial Homogeneous Measures

As we mentioned earlier, homogeneous measures are not trivial, in general. Indeed, given any measure carried[4] by $H(L)$, we may easily associate to it a *homogeneous* measure carried by $H(L)$, yielding many nontrivial homogeneous measures. To obtain this, we average the measure over its period, as established in the following lemma.

Lemma 2.1 *If μ is a probability measure on H_{loc} that is carried[5] by $H(L)$, then*

$$\tilde{\mu} = \frac{1}{L^d} \int_{Q(L)} \tau_{\mathbf{a}}(\mu)\, d\mathbf{a}$$

is a homogeneous probability measure on H_{loc} that is carried by $H(L)$ (with $\tau_{\mathbf{a}}$ as defined in (2.2)).

Proof. It is clear that $\tilde{\mu}$ is a probability measure carried by $H(L)$. In order to see that $\tilde{\mu}$ is homogeneous, let $\varphi \in \mathcal{C}_b(H_{\text{loc}})$ and $\mathbf{b} \in \mathbb{R}^d$. Then

$$\int \varphi(\tau_{\mathbf{b}}\mathbf{u})\, d\tilde{\mu}(\mathbf{u}) = \int \frac{1}{L^d} \int_{Q(L)} \varphi(\tau_{\mathbf{a}+\mathbf{b}}\mathbf{u})\, d\mathbf{a}\, d\mu(\mathbf{u}).$$

For any $\mathbf{u} \in H(L)$, it now follows by periodicity of $\mathbf{u}$ that $\varphi(\tau_{\mathbf{a}+Le_i}\mathbf{u}) = \varphi(\tau_{\mathbf{a}}\mathbf{u})$ for $i = 1, \ldots, d$. Since μ is carried by $H(L)$, this relation holds for μ-almost every $\mathbf{u}$ in H_{loc}. Hence, the following relation also holds for μ-almost every $\mathbf{u}$ in H_{loc}:

$$\int_{Q(L)} \varphi(\tau_{\mathbf{a}+\mathbf{b}}\mathbf{u})\, d\mathbf{a} = \int_{Q(L)-\mathbf{b}} \varphi(\tau_{\mathbf{a}}\mathbf{u})\, d\mathbf{a} = \int_{Q(L)} \varphi(\tau_{\mathbf{a}}\mathbf{u})\, d\mathbf{a}.$$

Therefore,

$$\int \varphi(\tau_{\mathbf{b}}\mathbf{u})\, d\tilde{\mu}(\mathbf{u}) = \int \varphi(\mathbf{u})\, d\tilde{\mu}(\mathbf{u})$$

for all $\mathbf{b}$ in $\mathbb{R}^d$ and all φ in $\mathcal{C}_b(H_{\text{loc}})$, which shows that $\tilde{\mu}$ is homogeneous. ■

Properties of Homogeneous Measures

We now describe some properties of homogeneous probability measures that will be used in the sequel.

Consider, for instance, the ensemble average of the square of the norm of the velocity field of a given turbulent flow at some point $\mathbf{x}_0$. This is usually denoted in the literature by $\langle|\mathbf{u}(\mathbf{x}_0)|^2\rangle$. However, since the flow is homogeneous, the ensemble average does not depend on $\mathbf{x}_0$ and is equal to some value $\langle|\mathbf{u}|^2\rangle$. We can establish this here in a rigorous way. We may consider, more generally, a function $G = G(\mathbf{u})(\mathbf{x})$

[4] We say that a probability measure μ is *carried* by the set X if X has total μ-measure – that is, if $\mu(X) = 1$. We might say also "supported" by X, but we prefer to leave this term for the support in the sense of distribution theory.

[5] See footnote 4.

(in this case, $G(\mathbf{u})(\mathbf{x}) = |u(\mathbf{x})|^2$). If G commutes with translations and satisfies some other technical conditions (given in Lemma 2.2), then the function

$$\mathbf{u} \mapsto \int G(\mathbf{u})(\mathbf{x})\, d\mathbf{x}$$

is μ-integrable and, for every integrable function η, we find

$$\iint G(\mathbf{u})(\mathbf{x})\eta(\mathbf{x})\, d\mathbf{x}\, d\mu(\mathbf{u}) = \langle G \rangle \int \eta(\mathbf{x})\, d\mathbf{x}, \tag{2.4}$$

where $\langle G \rangle$ is independent of η. For obvious reasons we also denote the constant $\langle G \rangle$ by

$$\int G(\mathbf{u})(\mathbf{x})\, d\mu(\mathbf{u}) = \langle G \rangle,$$

as if we could interchange the integrals in (2.4). This is, of course, a formal expression because we cannot say in general that $G(\mathbf{u})(\mathbf{x})$ is μ-integrable for each $\mathbf{x}$.

The precise statement of this result is as follows.

Lemma 2.2 *Let μ be a homogeneous probability measure on H_{loc} (or on $H(L)$), and let G be a mapping defined μ-almost everywhere on H_{loc} (resp., $H(L)$) with values in $L^1_{\text{loc}}(\mathbb{R}^d)$. Assume that G commutes with the translations,*

$$G \circ \tau_{\mathbf{a}} = \tau_{\mathbf{a}} \circ G \tag{2.5}$$

for all $\mathbf{a}$ in $\mathbb{R}^d$, that

$$\mathbf{u} \mapsto \int_Q G(\mathbf{u})(\mathbf{x})\, d\mathbf{x} \tag{2.6}$$

is μ-measurable for all sets $Q = (\alpha, \beta)^d$ with $\alpha < \beta$, and finally that

$$\iint_{Q_0} |G(\mathbf{u})(\mathbf{x})|\, d\mathbf{x}\, d\mu(\mathbf{u}) < \infty \tag{2.7}$$

for some set $Q_0 = (\alpha_0, \beta_0)^d$.

Then, for every η in $L^1(\mathbb{R}^d)$, the integral

$$\iint G(\mathbf{u})(\mathbf{x})\eta(\mathbf{x})\, d\mathbf{x}\, d\mu(\mathbf{u})$$

makes sense and is equal to

$$\langle G \rangle \int \eta(\mathbf{x})\, d\mathbf{x},$$

where $\langle G \rangle$ is independent of η. We will also use the notation

$$\langle G \rangle = \int G(\mathbf{u})(\mathbf{x})\, d\mu(\mathbf{u})$$

for arbitrary $\mathbf{x}$ in $\mathbb{R}^d$.

Furthermore, if

$$G(\mathbf{u})(\mathbf{x}) = \mathcal{P}(\partial_{\mathbf{x}}) F(\mathbf{u})(\mathbf{x}), \tag{2.8}$$

where $\mathcal{P}(\partial_{\mathbf{x}})$ is a differential operator on the **x** *variables with no zero-order term*[6] *and if F maps H_{loc} (resp., $H(L)$) into $L^1_{\mathrm{loc}}(\mathbb{R}^d)$ and satisfies the same conditions* (2.5)–(2.7) *as G, then*

$$\langle G\rangle = \langle \mathcal{P}(\partial_{\mathbf{x}})F\rangle = 0.$$

The following result is a consequence of Lemma 2.2.

Lemma 2.3 *Under the assumptions of Lemma 2.2, if Q is a bounded measurable set in $\mathbb{R}^d$ with positive measure then the value of the integral*

$$\int \frac{1}{|Q|}\int_Q G(\mathbf{u})(\mathbf{x})\,d\mathbf{x}\,d\mu(\mathbf{u})$$

is independent of Q and is equal to $\langle G\rangle$.

If (2.8) *is satisfied, then this integral vanishes.*

Indeed, this follows immediately by taking η in (2.2) to be the characteristic function of a bounded, measurable set Q of positive measure. Then,

$$\iint_Q G(\mathbf{u})(\mathbf{x})\,d\mathbf{x}\,d\mu(\mathbf{u}) = \iint G(\mathbf{u})(\mathbf{x})\eta(\mathbf{x})\,d\mathbf{x}\,d\mu(\mathbf{u}) = \langle G\rangle \int \eta(\mathbf{x})\,d\mathbf{x} = \langle G\rangle |Q|,$$

so that

$$\int \frac{1}{|Q|}\int_Q G(\mathbf{u})(\mathbf{x})\,d\mathbf{x}\,d\mu(\mathbf{u}) = \langle G\rangle$$

is independent of Q.

This property of homogeneous measures will be used for the definition of the kinetic energy as

$$e(\mu) = \frac{1}{2}\langle |\mathbf{u}|^2\rangle = \frac{1}{2}\int \frac{1}{|Q|}\int_Q |\mathbf{u}(\mathbf{x})|^2\,d\mathbf{x}\,d\mu(\mathbf{u})$$

and of the enstrophy as

$$E(\mu) = \left\langle \sum_{i=1}^d |\partial_{x_i}\mathbf{u}|^2\right\rangle = \int \frac{1}{Q}\int_Q \left(\sum_{i=1}^d |\partial_{x_i}\mathbf{u}(\mathbf{x})|^2\right) d\mathbf{x}\,d\mu(\mathbf{u}),$$

for a statistical distribution of a given flow represented by a measure μ.

The proof of Lemma 2.2 appears in Appendix A.2.

2.2 Homogeneous Statistical Solutions on the Whole Space: The Periodic Case

In the periodic case, a homogeneous statistical solution is a statistical solution $\{\mu_t\}_{t\geq 0}$, in the sense of Definition 1.4, with the property that μ_t is homogeneous for each time $t \geq 0$. Here, we assume that the forcing term $\mathbf{f} = 0$.

For the sake of completeness, we restate the definition of homogeneous statistical solution taking into account the assumption that $\mathbf{f} = 0$. Note that the class of

[6] That is, a sum of monomials $\partial_{x_1}^{\alpha_1}\cdots\partial_{x_d}^{\alpha_d}$, $\alpha_1 + \cdots + \alpha_d > 0$.

test functionals remains the same – namely, the class $\mathcal{T}_{\text{per}}$ given in Definition 1.3. Note also that we still consider the differential Φ' of a test functional Φ to be defined through the inner product $L^d(\cdot, \cdot)_{Q(L)}$ (see footnote 2). Hence, we arrive at the following definition.

Definition 2.2 A homogeneous statistical solution *of the Navier–Stokes equations on* $H(L)$ *with initial data* μ_0 *is a time-dependent family* $\{\mu_t\}_{t\geq 0}$ *of homogeneous probability measures on* $H(L)$ *satisfying* (a) *the Liouville-type equation*

$$\int \Phi(\mathbf{u})\, d\mu_t(\mathbf{u}) + L^d \int_0^t \int [\nu((\mathbf{u}, \Phi'(\mathbf{u})))_{Q(L)} + (B(\mathbf{u}), \Phi'(\mathbf{u}))_{Q(L)}]\, d\mu_s(\mathbf{u})\, ds = \int \Phi(\mathbf{u})\, d\mu_0(\mathbf{u}) \quad (2.9)$$

for all $t \geq 0$ *and any test functional* Φ *in* $\mathcal{T}_{\text{per}}$ *and* (b) *the energy inequality*

$$\int |\mathbf{u}|^2_{Q(L)}\, d\mu_t(\mathbf{u}) + 2\nu \int_0^t \|\mathbf{u}\|^2_{Q(L)}\, d\mu_s(\mathbf{u})\, ds \leq \int |\mathbf{u}|^2_{Q(L)}\, d\mu_0(\mathbf{u}) \quad (2.10)$$

for all $t \geq 0$. *Furthermore*: *the function*

$$t \mapsto \int \varphi(\mathbf{u})\, d\mu_t(\mathbf{u}) \quad (2.11)$$

is required to be measurable on $[0, \infty)$ *for every bounded and countinuous real-valued function* φ *on* $H(L)$; *the function*

$$t \mapsto \int |\mathbf{u}|^2_{Q(L)}\, d\mu_t(\mathbf{u}) \quad (2.12)$$

is required to belong to $L^\infty(0, \infty)$; *and the function*

$$t \mapsto \int \|\mathbf{u}\|^2_{Q(L)}\, d\mu_t(\mathbf{u}) \quad (2.13)$$

is required to belong to $L^1(0, T)$ *for all* $T > 0$.

Remark 2.1 Properties (2.12) and (2.13) actually follow from (2.10), (2.11), and a finite initial kinetic energy assumption (see (2.14)). Moreover, property (2.9) can be extended to more general test functionals Φ, which will be introduced when needed.

The existence of homogeneous statistical solutions follows from the existence of statistical solutions given by Theorem 1.3. We state this existence result next; the proof is given in Appendix A.2.

Theorem 2.1 (Existence of Homogeneous Statistical Solutions: Space Dimensions 2 and 3) *Given a homogeneous probability measure* μ_0 *on* $H(L)$, $L > 0$, *with finite kinetic energy*

$$\int |\mathbf{u}|^2_{Q(L)}\, d\mu_0(\mathbf{u}) < \infty, \tag{2.14}$$

there exists a homogeneous statistical solution $\{\mu_t\}_{t\geq 0}$ *on* $H(L)$ *with initial data* μ_0.

In the 2-dimensional case, it follows immediately from the uniqueness of nonhomogeneous time-dependent statistical solutions (Theorem 1.4) that the homogeneous statistical solutions are also unique.

Theorem 2.2 (Uniqueness: Space Dimension 2) *In dimension* $d = 2$, *given a homogeneous probability measure* μ_0 *on* $H(L)$, $L > 0$, *with finite kinetic energy*

$$\int |\mathbf{u}|^2_{Q(L)}\, d\mu_0(\mathbf{u}) < \infty, \tag{2.15}$$

the homogeneous statistical solution $\{\mu_t\}_{t\geq 0}$ *provided by Theorem 2.1 is unique and is given by* $\mu_t = S(t)\mu_0$.[7]

2.3 Homogeneous Statistical Solutions on the Whole Space: The General Case

In the whole-space case, we work directly with homogeneous measures and rewrite the energy inequality in the definition of statistical solutions in a more appropriate form. More precisely, we define the kinetic energy $e(\mu)$ and enstrophy $E(\mu)$ of a given homogeneous probability measure μ by the numbers $\langle G\rangle$ given in Lemmas 2.2 and 2.3, with

$$G(\mathbf{u})(\mathbf{x}) = \frac{1}{2}|\mathbf{u}(\mathbf{x})|^2 \quad \text{and} \quad G(\mathbf{u})(\mathbf{x}) = \sum_{i=1}^{d} |\partial_{x_i}\mathbf{u}(\mathbf{x})|^2,$$

respectively. Hence,

$$e(\mu) = \frac{1}{2}\langle|\mathbf{u}|^2\rangle = \frac{1}{2}\int \frac{1}{|Q|}\int_Q |\mathbf{u}(\mathbf{x})|^2\, d\mathbf{x}\, d\mu(\mathbf{u}),$$

$$E(\mu) = \left\langle \sum_{i=1}^{d} |\partial_{x_i}\mathbf{u}|^2 \right\rangle = \int \frac{1}{|Q|}\int_Q \left(\sum_{i=1}^{d} |\partial_{x_i}\mathbf{u}(\mathbf{x})|^2\right) d\mathbf{x}\, d\mu(\mathbf{u}),$$

where $Q = (a, b)^d$ is arbitrary. The energy inequality for a homogeneous statistical solution in this case is then written as

$$e(\mu_t) + \nu \int_0^t E(\mu_s)\, ds \leq e(\mu_0) \quad \text{for all } t \geq 0. \tag{2.16}$$

Note that, in this case of homogeneous measures, we are assuming $\mathbf{f} = 0$. If $\{\mu_t\}_{t\geq 0}$ is a family of homogeneous probability measures concentrated on $H(L)$ and satisfying (2.9)–(2.13), then (2.16) is simply another way of writing (2.10).

[7] This is in contrast with the case of stationary statistical solutions, where uniqueness is not always a property of the solution.

For the rigorous formulation of the Liouville-type equation for $\{\mu_t\}_{t\geq 0}$, we need to introduce an appropriate class of test functionals as follows.

Definition 2.3 *We denote*[8] *by* $\mathcal{T}_{\mathrm{ws}}$ *the class of cylindrical* test functionals *consisting of the real-valued functional* $\Phi = \Phi(\mathbf{u})$ *of the form*

$$\Phi(\mathbf{u}) = \phi((\mathbf{u}, \mathbf{g}_1), \ldots, (\mathbf{u}, \mathbf{g}_k)),$$

where ϕ *is a scalar* C^1 *function on* $\mathbb{R}^k$ $(k \in \mathbb{N})$ *that is bounded together with its first-order derivatives and where* $\mathbf{g}_1, \ldots, \mathbf{g}_k$ *belong to* V_c *– that is, they belong to* V_{loc} *and have compact support on* $\mathbb{R}^d$. *For such* Φ *we denote by* Φ' *its differential in* H_{loc} *with respect to the inner product* $(\cdot,\cdot)$; *this has the form*

$$\Phi'(\mathbf{u}) = \sum_{j=1}^{k} \partial_j \phi((\mathbf{u}, \mathbf{g}_1), \ldots, (\mathbf{u}, \mathbf{g}_k))\mathbf{g}_j,$$

where $\partial_j \phi$ *denotes the derivative of* ϕ *with respect to its* jth *argument* $(1 \leq j \leq k)$. *Since* $\Phi'(\mathbf{u})$ *is a linear combination of the* $\mathbf{g}_j$, *it belongs to* V_c.

We then have the following definition for a homogeneous statistical solution in the whole-space case.

Definition 2.4 A homogeneous statistical solution *of the Navier–Stokes equations on* H_{loc} *with initial data* μ_0 *is a time-dependent family* $\{\mu_t\}_{t\geq 0}$ *of homogeneous probability measures on* H_{loc} *satisfying* (a) *the Liouville equation*

$$\int \Phi(\mathbf{u})\, d\mu_t(\mathbf{u}) + \int_0^t \int [\nu((\mathbf{u}, \Phi'(\mathbf{u}))) + ((\mathbf{u}\cdot\nabla)\mathbf{u}, \Phi'(\mathbf{u}))]\, d\mu_s(\mathbf{u})\, ds = \int \Phi(\mathbf{u})\, d\mu_0(\mathbf{u}) \quad (2.17)$$

for all $t \geq 0$ *and for any test functional* Φ *in* $\mathcal{T}_{\mathrm{ws}}$ *and* (b) *the energy inequality*

$$e(\mu_t) + \nu \int_0^t E(\mu_s)\, ds \leq e(\mu_0) \quad (2.18)$$

for all $t \geq 0$. *Furthermore: the function*

$$t \mapsto \int \varphi(\mathbf{u})\, d\mu_t(\mathbf{u}) \quad (2.19)$$

is required to be measurable on $[0, \infty)$ *for every bounded and continuous real-valued function* φ *on* $H(L)$; *the function*

$$t \mapsto e(\mu_t) = \frac{1}{2} \int |\mathbf{u}(\mathbf{x})|^2\, d\mu_t(\mathbf{u}) \quad (2.20)$$

is required to belong to $L^\infty(0, \infty)$; *and the function*

[8] The subscript "ws" stands for "whole space."

$$t \mapsto E(\mu_t) = \int \left(\sum_{i=1}^{d} |\partial_{x_i} \mathbf{u}(\mathbf{x})|^2 \right) d\mu_t(\mathbf{u}) \tag{2.21}$$

is required to belong to $L^1(0,T)$ *for all* $T > 0$.

Remark 2.2 Properties (2.20) and (2.21) actually follow from (2.18), (2.19), and the finite initial kinetic energy assumption, $e(\mu_0) < \infty$. Moreover, property (2.17) can be extended to more general test functionals Φ, which will be introduced when needed.

The existence of homogeneous statistical solutions on the whole space is obtained through a passage-to-the-limit ($L \to \infty$) process on the solutions of the periodic case on $Q(L) = (-L/2, L/2)^d$. We approximate the measure μ_0 by a measure concentrated on $H(L)$, obtain the corresponding homogeneous statistical solution on $H(L)$, and take the limit (in some sense) as L goes to infinity in order to find a solution on H_{loc}. We provide a sketch of the proof in Appendix A.2; a complete proof is given in Foias and Temam [1980]. The result can be stated as follows.

Theorem 2.3 *Given a homogeneous probability measure* μ_0 *on* H_{loc} *with finite kinetic energy* $e(\mu_0) < \infty$, *there exists a homogeneous statistical solution* $\{\mu_t\}_{t\geq 0}$ *on* H_{loc} *with initial data* μ_0.

2.4 Hopf Equation on the Whole Space

As in the bounded case, the Hopf equation for the whole-space case can be obtained from equation (2.17) by taking as the test functional

$$\Phi(\mathbf{u}) = e^{i(\mathbf{u},\mathbf{g})},$$

defined for all $\mathbf{u}$ in H_{loc}, where $\mathbf{g}$ belongs to V_c (i.e, $\mathbf{g}$ belongs to V_{loc} and has compact support on $\mathbb{R}^d$). In this case, however, the real and imaginary parts of Φ belong to the space $\mathcal{T}_{\text{ws}}$ of test functionals, so that we can substitute directly to find the complex-valued equation

$$\int e^{i(\mathbf{u},\mathbf{g})}\, d\mu_t(\mathbf{u}) + i \int_0^t \int [\nu((\mathbf{u},\mathbf{g})) + ((\mathbf{u}\cdot\nabla)\mathbf{u},\mathbf{g})] e^{i(\mathbf{g},\mathbf{u})}\, d\mu_s(\mathbf{u})\, ds = \int e^{i(\mathbf{u},\mathbf{g})}\, d\mu_0(\mathbf{u}),$$

which is valid for all t in $[0,T)$. Then, setting

$$\chi(t,\mathbf{g}) = \int e^{i(\mathbf{u},\mathbf{g})}\, d\mu_t(\mathbf{u})$$

and defining the operators $A(\partial_{\mathbf{g}})$ and $B(\partial_{\mathbf{g}})$ by

$$(A(\partial_{\mathbf{g}})\chi(t,\mathbf{g}),\mathbf{v}) = i \int ((\mathbf{u},\mathbf{v})) e^{i(\mathbf{u},\mathbf{g})}\, d\mu_t(\mathbf{u}) \tag{2.22}$$

and

$$(B(\partial_{\mathbf{g}})\chi(t,\mathbf{g}),\mathbf{v}) = i\int((\mathbf{u}\cdot\nabla)\mathbf{u},\mathbf{v})e^{i(\mathbf{u},\mathbf{g})}\,d\mu_t(\mathbf{u}) \tag{2.23}$$

for all $\mathbf{v}$ in the space V_c, we arrive – as in the bounded case – at the equation

$$\frac{d}{dt}\chi(t,\mathbf{g}) + \nu(A(\partial_{\mathbf{g}})\chi(t,\mathbf{g}),\mathbf{g}) + (B(\partial_{\mathbf{g}})\chi(t,\mathbf{g}),\mathbf{g}) = 0 \tag{2.24}$$

for all t in $[0,T)$, which is the functional form of Hopf equation. Observe that $\mathbf{g}\mapsto\chi(t,\mathbf{g})$ is a complex-valued function defined on V_c, and that $A(\partial_g)\chi(t,\mathbf{g})$ and $B(\partial_{\mathbf{g}})\chi(t,\mathbf{g})$ belong to the dual space V_c' because they make sense only when applied to v in V_c, through (2.22) and (2.23).

3 Reynolds Equation for the Average Flow

In practice, it is impossible to predict the behavior of a turbulent flow in detail. However, some gross characteristics of the flow, such as the mean velocity, behave in a more orderly manner, making its analysis more feasible. With this in mind, one may consider some ensemble average $\langle\cdot\rangle$ and then decompose the flow $\mathbf{u}$ into a mean part $\bar{\mathbf{u}} = \langle\mathbf{u}\rangle$ and a fluctuating part $\mathbf{u}' = \mathbf{u} - \langle\mathbf{u}\rangle$ (hence, $\mathbf{u} = \bar{\mathbf{u}} + \mathbf{u}'$ and $\langle\mathbf{u}'\rangle = 0$). Assuming that the averaging operator $\langle\cdot\rangle$ has some linearity properties that are compatible with an averaging process, one may apply this operator to the Navier–Stokes equations and arrive at the so-called Reynolds equations for the average flow $\bar{\mathbf{u}}$. It is our aim, here, to obtain a functional form for the Reynolds equations in a rigorous way, starting from the equations for the statistical solutions of the Navier–Stokes equations.

We consider the no-slip bounded case and the periodic case with vanishing space average. Hence, the space of finite kinetic energy H is either H_{nsp} or $\dot{H}_{\text{per}}$, the space of finite enstrophy V is either V_{nsp} or $\dot{V}_{\text{per}}$, and the space of test functionals $\mathcal{T}$ is either $\mathcal{T}_{\text{nsp}}$ or $\dot{\mathcal{T}}_{\text{per}}$. The periodic case with nonvanishing space average can be treated in a similar way. However, the homogeneous cases are not interesting because the average velocity field becomes constant in space.

If $\{\mu_t\}_{t\ge 0}$ is a statistical solution of the NSE, then

$$\int\Phi(\mathbf{u})\,d\mu_t(\mathbf{u}) = \int\Phi(\mathbf{u})\,d\mu_0(\mathbf{u}) + \int_0^t(\mathbf{F}(s,\mathbf{u}),\Phi'(\mathbf{u}))\,d\mu_s(\mathbf{u})\,ds \tag{3.1}$$

for all test functionals Φ in the class $\mathcal{T}$, where $\mathbf{F}(t,\mathbf{u}) = \mathbf{f}(t) - \nu A\mathbf{u} - B(\mathbf{u})$.

For a given $\mathbf{v}$ in V, the functional $\Phi(\mathbf{u}) = (\mathbf{u},\mathbf{v})$ is not in the class of test functionals $\mathcal{T}$; nonetheless, by a technical approximation argument (which we omit), we can show that (3.1) remains valid for such a Φ. In this case, the derivative of Φ is $\Phi'(\mathbf{u}) = \mathbf{v}$ and we arrive at the equation

$$\int(\mathbf{u},\mathbf{v})\,d\mu_t(\mathbf{u}) = \int(\mathbf{u},\mathbf{v})\,d\mu_0(\mathbf{u}) + \int_0^t(\mathbf{F}(s,\mathbf{u}),\mathbf{v})\,d\mu_s(\mathbf{u})\,ds,$$

which can be written as

$$\int (\mathbf{u}, \mathbf{v})\, d\mu_t(\mathbf{u}) + \int_0^t \int [\nu(A\mathbf{u}\mathbf{v}) + (B(\mathbf{u}), \mathbf{v})]\, d\mu_s(\mathbf{u})\, ds$$
$$= \int_0^t (\mathbf{f}(s), \mathbf{v})\, ds + \int (\mathbf{u}, \mathbf{v})\, d\mu_0(\mathbf{u}). \quad (3.2)$$

Since μ_t has finite kinetic energy for every $t \geq 0$, the function of $\mathbf{w}$ given by

$$\mathbf{w} \mapsto \int (\mathbf{u}, \mathbf{w})\, d\mu_t(\mathbf{u})$$

is a linear functional in the space of finite kinetic energy H. Hence, by the Riesz representation theorem (see Appendix A.1 in Chapter II), there exists an element in H, denoted $\overline{\mathbf{u}(t)}$, such that

$$(\overline{\mathbf{u}(t)}, \mathbf{w}) = \int (\mathbf{u}, \mathbf{w})\, d\mu_t(\mathbf{u})$$

for every $\mathbf{w}$ in H. This element $\overline{\mathbf{u}(t)}$ clearly plays the role of the mean velocity $\bar{\mathbf{u}}$ that we have already defined heuristically. Formally, we can express it as

$$\overline{\mathbf{u}(t)} = \int \mathbf{u}\, d\mu_t(\mathbf{u}).$$

Moreover, since for almost every t the distribution μ_t has finite enstrophy, we see that for all $\mathbf{w}$ in H,

$$|(\overline{\mathbf{u}(t)}, \mathbf{w})| = \left| \int (\mathbf{u}, \mathbf{w})\, d\mu_t(\mathbf{u}) \right| = \left| \int (\mathbf{w}, \mathbf{u})\, d\mu_t(\mathbf{u}) \right|$$
$$\leq \|\mathbf{w}\|_{D(A^{-1/2})} \int \|\mathbf{u}\|^2\, d\mu_t(\mathbf{u}),$$

so that, in fact, $\overline{\mathbf{u}(t)}$ has finite enstrophy and belongs to V for almost every $t > 0$. Therefore, we can apply the Stokes operator to $\overline{\mathbf{u}(t)}$ and obtain, for every $\mathbf{w}$ in V and almost every $t > 0$, that

$$(A\overline{\mathbf{u}(t)}, \mathbf{w}) = ((\overline{\mathbf{u}(t)}, \mathbf{w})) = ((\mathbf{w}, \overline{\mathbf{u}(t)})) = (A\mathbf{w}, \overline{\mathbf{u}(t)})$$
$$= \int (A\mathbf{w}, \mathbf{u})\, d\mu_t(\mathbf{u}) = \int ((\mathbf{w}, \mathbf{u}))\, d\mu_t(\mathbf{u}) = \int (A\mathbf{u}, \mathbf{w})\, d\mu_t(\mathbf{u}),$$

which is one of the terms in equation (3.2) for $\mathbf{w} = \mathbf{v}$. For the term in the equation (3.2) involving the bilinear operator $B(\mathbf{u}) = B(\mathbf{u}, \mathbf{u})$, we may simply write

$$\int (B(\mathbf{u}), \mathbf{v})\, d\mu_s(\mathbf{u})$$
$$= \int (B(\overline{\mathbf{u}(s)} + \mathbf{u} - \overline{\mathbf{u}(s)}, \overline{\mathbf{u}(s)} + \mathbf{u} - \overline{\mathbf{u}(s)}), \mathbf{v})\, d\mu_s(\mathbf{u})$$
$$= \int [(B(\overline{\mathbf{u}(s)}), \overline{\mathbf{u}(s)}), \mathbf{v}) + (B(\overline{\mathbf{u}(s)}, \mathbf{u} - \overline{\mathbf{u}(s)}), \mathbf{v})$$
$$+ (B(\mathbf{u} - \overline{\mathbf{u}(s)}, \overline{\mathbf{u}(s)}), \mathbf{v}) + (B(\mathbf{u} - \overline{\mathbf{u}(s)}, \mathbf{u} - \overline{\mathbf{u}(s)}), \mathbf{v})]\, d\mu_s(u).$$

Then, using the orthogonality property of the bilinear operator, which implies property (0.13), we find for almost every time $s > 0$ that

$$\begin{aligned}
\int (B(\overline{\mathbf{u}(s)}, \mathbf{u} - \overline{\mathbf{u}(s)}), \mathbf{v})\, d\mu_s(\mathbf{u})
&= -\int (B(\overline{\mathbf{u}(s)}, \mathbf{v}), \mathbf{u} - \overline{\mathbf{u}(s)})\, d\mu_s(\mathbf{u}) \\
&= -\int (B(\overline{\mathbf{u}(s)}, \mathbf{v}), \mathbf{u})\, d\mu_s(\mathbf{u}) + \int (B(\overline{\mathbf{u}(s)}, \mathbf{v}), \overline{\mathbf{u}(s)})\, d\mu_s(\mathbf{u}) \\
&= -(B(\overline{\mathbf{u}(s)}, \mathbf{v}), \overline{\mathbf{u}(s)}) + (B(\overline{\mathbf{u}(s)}, \mathbf{v}), \overline{\mathbf{u}(s)}) \\
&= 0.
\end{aligned}$$

Similarly,

$$\int (B(\mathbf{u} - \overline{\mathbf{u}(s)}, \overline{\mathbf{u}(s)}), \mathbf{v})\, d\mu_s(\mathbf{u}) = 0$$

for almost every time $s > 0$. Therefore, we can rewrite equation (3.2) in the form

$$\begin{aligned}
(\overline{\mathbf{u}(t)}, \mathbf{v}) &+ \int_0^t \left[\nu(A\overline{\mathbf{u}(t)}, \mathbf{v}) + (B(\overline{\mathbf{u}(t)}), \mathbf{v})\right] ds \\
&+ \int_0^t \int (B(\mathbf{u} - \overline{\mathbf{u}(s)}, \mathbf{u} - \overline{\mathbf{u}(s)}), \mathbf{v})\, d\mu_s(\mathbf{u})\, ds \\
&\qquad = \int_0^t (\mathbf{f}(s), \mathbf{v})\, ds + (\overline{\mathbf{u}(0)}, \mathbf{v}). \qquad (3.3)
\end{aligned}$$

Now one can check that, for almost every $t > 0$, the operator

$$\mathbf{w} \mapsto \int (B(\mathbf{u} - \overline{\mathbf{u}(t)}, \mathbf{u} - \overline{\mathbf{u}(t)}), \mathbf{w})\, d\mu_t(\mathbf{u})$$

is a bounded linear functional on the space of finite enstrophy V. Therefore, by the Riesz representation theorem, for almost every $t > 0$ there exists an element in the dual space $D(A^{-1/2})$, which we naturally denote by

$$\overline{B(\mathbf{u}(t)', \mathbf{u}(t)')},$$

with the property that

$$\int (B(\mathbf{u} - \overline{\mathbf{u}(t)}, \mathbf{u} - \overline{\mathbf{u}(t)}), \mathbf{w})\, d\mu_t(\mathbf{u}) = (\overline{B(\mathbf{u}(t)', \mathbf{u}(t)')}, \mathbf{w})$$

for all $\mathbf{w}$ in V. Therefore, we can rewrite equation (3.3) in the form

$$\begin{aligned}
(\overline{\mathbf{u}(t)}, \mathbf{v}) &+ \int_0^t \left[\nu(A\overline{\mathbf{u}(t)}, \mathbf{v}) + (B(\overline{\mathbf{u}(t)}), \mathbf{v}) + (\overline{B(\mathbf{u}(t)', \mathbf{u}(t)')}, \mathbf{v})\right] ds \\
&\qquad = \int_0^t (\mathbf{f}(s), \mathbf{v})\, ds + (\overline{\mathbf{u}(0)}, \mathbf{v}). \qquad (3.4)
\end{aligned}$$

Equation (3.4) holds for all $\mathbf{v}$ in V and for all $t > 0$. Therefore, we can write this equation as a functional equation in the space $D(A^{-1/2})$, which takes the form

$$\overline{\mathbf{u}(t)} + \int_0^t \left[\nu A\overline{\mathbf{u}(t)} + B(\overline{\mathbf{u}(t)}) + \overline{B(\mathbf{u}(t)', \mathbf{u}(t)')}\right] ds = \overline{\mathbf{u}(0)} + \int_0^t \mathbf{f}(s)\, ds. \qquad (3.5)$$

Equation (3.5) is the classical Reynolds averaged equation written in functional form.

4 Self-Similar Homogeneous Statistical Solutions

In this section we address the question of self-similarity in the context of the homogeneous statistical solutions to the Navier–Stokes equations. In Section 4.1, after recalling self-similarity for deterministic solutions, we study the similarity (rescaling) for homogeneous statistical solutions of the NSE in relation to the results of Sections 1 and 2. Then, in Section 4.2, we show how to obtain a 2-parameter family of self-similar homogeneous statistical solutions. Moreover, we transform the homogeneous statistical solutions and the self-similar homogeneous statistical solutions to stationary statistical solutions of a related equation. Throughout Section 4 we restrict ourselves to the 3-dimensional case ($d = 3$) and assume that $\mathbf{f} = 0$, since we are dealing with homogeneous flows.

4.1 Rescaling Properties of Homogeneous Statistical Solutions

We start by recalling self-similarity in the case of the individual solutions. The mathematical theory of deterministic self-similar solutions of the Navier–Stokes equations has been developed by Cannone [1995], Cannone and Meyer [1995], and Cannone, Meyer, and Planchon [1994] in the context of other function spaces (called Besov spaces; see also Barraza [1996] for a theory in weak L^p-spaces).

Rescaling in the Deterministic Case

Given a solution $\mathbf{u}(\mathbf{x}, t)$ to the Navier–Stokes equations with no forcing term,

$$\frac{\partial \mathbf{u}}{\partial t} - \nu\Delta\mathbf{u} + (\mathbf{u}\cdot\nabla)\mathbf{u} + \nabla p = 0, \qquad \nabla\cdot\mathbf{u} = 0, \tag{4.1}$$

we consider, for real positive numbers ξ, λ, α, the functions $\mathbf{u}_{\xi,\lambda,\alpha} = \sigma_{\xi,\lambda,\alpha}\mathbf{u}$ and $p_{\xi,\lambda,\alpha} = \bar{\sigma}_{\xi,\lambda,\alpha}p$ defined by

$$\begin{aligned} \mathbf{u}_{\xi,\lambda,\alpha}(\mathbf{x}, t) &= (\sigma_{\xi,\lambda,\alpha}\mathbf{u})(\mathbf{x}, t) = \xi\mathbf{u}(\lambda\mathbf{x}, \alpha t), \\ p_{\xi,\lambda,\alpha}(\mathbf{x}, t) &= (\bar{\sigma}_{\xi,\lambda,\alpha}p)(\mathbf{x}, t) = \xi^2 p(\lambda\mathbf{x}, \alpha^2 t). \end{aligned} \tag{4.2}$$

One can show that, if $\mathbf{u}$ is a solution of (4.1), then $\mathbf{u}_{\xi,\lambda,\alpha}$ satisfies

$$\frac{\xi\lambda}{\alpha}\frac{\partial \mathbf{u}_{\xi,\lambda,\alpha}}{\partial t} - \frac{\nu}{\lambda\xi}\Delta\mathbf{u}_{\xi,\lambda,\alpha} + (\mathbf{u}_{\xi,\lambda,\alpha}\cdot\nabla)\mathbf{u}_{\xi,\lambda,\alpha} + \nabla(\xi^2\lambda p) = 0, \quad \nabla\cdot\mathbf{u} = 0, \tag{4.3}$$

where the differential operators Δ and ∇ operate on the $\mathbf{x}$ variable. Depending on ξ, λ, and α, the function $\sigma_{\xi,\lambda,\alpha}\mathbf{u}$ satisfies an equation similar to (4.1), with rescaled pressure and (possibly) rescaled time and/or viscosity. For $\xi = \lambda$ and $\alpha = \lambda^2$,

$$\mathbf{u}_{\xi,\lambda,\alpha} = \lambda\mathbf{u}(\lambda\mathbf{x}, \lambda^2 t)$$

satisfies the Navier–Stokes equations (4.1) with the nondimensional viscosity ν/λ^2.

Classically, we consider then the case where λ depends on time, $\lambda = t^{-1/2}$. That is, we consider functions of the form

$$\tilde{\mathbf{u}}(\mathbf{x}, t) = \frac{1}{\sqrt{t}}\mathbf{u}\left(\frac{\mathbf{x}}{\sqrt{t}}, 1\right) = \frac{1}{\sqrt{t}}\mathbf{U}\left(\frac{\mathbf{x}}{\sqrt{t}}\right).$$

Independently of the question addressed in Cannone and Meyer [1995] and Barraza [1996] of defining such a function at $t = 0$, we substitute this expression in (4.1); renaming $\tilde{\mathbf{u}}$ for $\mathbf{u}$, we then seek under which conditions on $\mathbf{U}$ and P the functions

$$\mathbf{u}(\mathbf{x}, t) = \frac{1}{\sqrt{t}}\mathbf{U}\left(\frac{\mathbf{x}}{\sqrt{t}}\right) \quad \text{and} \quad p(\mathbf{x}, t) = \frac{1}{t}P\left(\frac{\mathbf{x}}{\sqrt{t}}\right)$$

satisfy (4.1). Equation (4.3) is not directly valid because λ is no longer a constant, but a simple similar calculation leads to the following equation, where $\mathbf{U} = \mathbf{U}(\mathbf{y})$, $P = P(\mathbf{y})$, and $\mathbf{y} = \mathbf{x}/\sqrt{t}$:

$$-\frac{1}{2}\mathbf{U}(\mathbf{y}) - \frac{1}{2}(\mathbf{y}\cdot\nabla_{\mathbf{y}})\mathbf{U} - \nu\Delta_{\mathbf{y}}\mathbf{U} + (\mathbf{U}\cdot\nabla_{\mathbf{y}})\mathbf{U} + \nabla_{\mathbf{y}}P = 0, \qquad \nabla_{\mathbf{y}}\cdot\mathbf{U} = 0. \tag{4.4}$$

For any suitable nonzero solution $\mathbf{U}$ of (4.4) – if such solution exists – the corresponding function $\mathbf{u}$ is called a *self-similar* solution of the NSE; for $t > 0$, the change of scale in (4.2) with $\xi = \lambda$ and $\alpha = \lambda^2$ changes the scale of $\mathbf{u}$ and p but not their shape.

Rescaling in the Statistical Case

In the statistical case we consider, for any $\xi, \lambda > 0$, the change of scale defined by

$$(\sigma_{\xi,\lambda}\mathbf{u})(\mathbf{x}) = \xi\mathbf{u}(\lambda\mathbf{x}) \tag{4.5}$$

for all $\mathbf{x}$ in $\mathbb{R}^3$, which takes vector fields in H_{loc} into rescaled vector fields in H_{loc}. We can extend this rescaling operation to measures by defining, for a given measure μ on H_{loc}, a rescaled measure $\sigma_{\xi,\lambda}\mu$ on H_{loc} given by

$$(\sigma_{\xi,\lambda}\mu)(E) = \mu(\sigma_{\xi,\lambda}^{-1}E) \tag{4.6}$$

for all measurable sets E in H_{loc}. Here, $\sigma_{\xi,\lambda}^{-1}$ is the inverse of $\sigma_{\xi,\lambda}$, which is $\sigma_{\xi,\lambda}^{-1} = \sigma_{\xi^{-1},\lambda^{-1}}$, and $\sigma_{\xi,\lambda}^{-1}E$ is the set of functions $\mathbf{u}$ such that $\sigma_{\xi,\lambda}\mathbf{u}$ belongs to E. This change of scale transforms statistical solutions into statistical solutions with viscosity and time rescaled. Indeed, let $\{\mu_t\}_{t\geq 0}$ be a homogeneous statistical solution of the Navier–Stokes equations with viscosity ν. Extending our calculations in the deterministic case, we want to deduce that $\{\sigma_{\xi,\lambda}\mu_{\alpha t}\}_{t\geq 0}$ is a homogeneous statistical solution of the NSE with a different viscosity and for a time rescaled by an appropriate factor $\alpha = \xi\lambda$; for that purpose, we refer to Definition 2.4. Observe first that – for an arbitrary, bounded, measurable set Q in $\mathbb{R}^3$ of positive measure and a given probability measure μ – we find using (4.5) and (4.6) that

$$\begin{aligned} e(\sigma_{\xi,\lambda}\mu) &= \frac{1}{2}\int |\mathbf{u}|_Q^2 \, d(\sigma_{\xi,\lambda}\mu)(\mathbf{u}) \qquad (4.7)\\ &= \frac{1}{2}\int |\sigma_{\xi,\lambda}\mathbf{u}|_Q^2 \, d\mu(\mathbf{u}) \\ &= \frac{1}{2}\xi^2\int |\mathbf{u}|_{\lambda Q}^2 \, d\mu(\mathbf{u}) = \xi^2 e(\mu), \qquad (4.8) \end{aligned}$$

where the last step comes from the fact that, for homogeneous flows, the kinetic energy (as well as the enstrophy) is defined independently of the averaging volume (Q or λQ in this case; see Section 2.3). Similarly, for the enstrophy we have

$$E(\sigma_{\xi,\lambda}\mu) = \int \|\mathbf{u}\|_Q^2 \, d(\sigma_{\xi,\lambda}\mu)(\mathbf{u}) = \int \|\sigma_{\xi,\lambda}\mathbf{u}\|_Q^2 \, d\mu(\mathbf{u}).$$

Then, observing the commutation property

$$\sigma_{\xi,\lambda}\frac{\partial}{\partial x_i} = \frac{1}{\lambda}\frac{\partial}{\partial x_i}\sigma_{\xi,\lambda}, \tag{4.9}$$

we have

$$\begin{aligned}
\|\sigma_{\xi,\lambda}u\|_Q^2 &= \frac{1}{|Q|}\int_Q\left(\sum_{i=1}^d\left|\frac{\partial}{\partial_{x_i}}(\xi\mathbf{u}(\lambda\mathbf{x}))\right|^2\right)d\mathbf{x} = \frac{1}{|Q|}\int_Q\left(\sum_{i=1}^d\left|\frac{\partial}{\partial_{x_i}}(\xi\mathbf{u}(\lambda\mathbf{x}))\right|^2\right)d\mathbf{x}\\
&= \xi^2\frac{1}{|Q|}\int_Q\left(\sum_{i=1}^d|\lambda\mathbf{u}_{\mathbf{x}_i}(\lambda\mathbf{x})|^2\right)d\mathbf{x} = \xi^2\lambda^2\frac{1}{|Q|}\int_{\lambda Q}\left(\sum_{i=1}^d|\mathbf{u}_{x_i}(\mathbf{x})|^2\right)\frac{d\mathbf{x}}{\lambda^d}\\
&= \xi^2\lambda^2\frac{1}{|\lambda Q|}\int_{\lambda Q}\left(\sum_{i=1}^d|\mathbf{u}_{x_i}(\mathbf{x})|^2\right)d\mathbf{x},
\end{aligned}$$

where $\mathbf{u}_{x_i} = \partial\mathbf{u}/\partial x_i$. Thus, we obtain

$$E(\sigma_{\xi,\lambda}\mu) = \xi^2\lambda^2\int\|\mathbf{u}\|_{\lambda Q}^2\,d\mu(\mathbf{u}) = \xi^2\lambda^2E(\mu). \tag{4.10}$$

For completeness we note that, with $\mu = \mu_{\alpha t}$,

$$e(\sigma_{\xi,\lambda}\mu_{\alpha t}) = \xi^2 e(\mu_{\alpha t}) \tag{4.11}$$

and

$$E(\sigma_{\xi,\lambda}\mu_{\alpha t}) = \xi^2\lambda^2E(\mu_{\alpha t}). \tag{4.12}$$

Hence, conditions (2.20) and (2.21) of Definition 2.4 hold also for $\{\sigma_{\xi,\lambda}\mu_{\alpha t}\}_{t\geq 0}$.

For (2.19), simply observe that

$$\int\varphi(\mathbf{u})\,d(\sigma_{\xi,\lambda}\mu_{\alpha t})(\mathbf{u}) = \int\varphi(\sigma_{\xi,\lambda}\mathbf{u})\,d\mu_{\alpha t}(\mathbf{u})$$

and that $\varphi\circ\sigma_{\xi,\lambda} = \varphi(\sigma_{\xi,\lambda}\cdot)$ belongs to $\mathcal{C}(H_{\text{loc}})$ for any φ in $\mathcal{C}(H_{\text{loc}})$.

Now, for any Φ in $\mathcal{T}_{\text{ws}}$, consider the composition $\tilde{\Phi} = \Phi\circ\sigma_{\xi,\lambda}$, that is, $\tilde{\Phi}(\mathbf{u}) = \Phi(\sigma_{\xi,\lambda}\mathbf{u}))$. Since $\sigma_{\xi,\lambda}$ is a bounded linear operator in H_{loc}, it follows that $\tilde{\Phi}$ belongs to $\mathcal{T}_{\text{ws}}$ as well, with

$$\tilde{\Phi}'(\mathbf{u}) = \Phi'(\sigma_{\xi,\lambda}\mathbf{u})\circ\sigma_{\xi,\lambda},$$

which yields

$$(\mathbf{v},\tilde{\Phi}'(\mathbf{u})) = (\sigma_{\xi,\lambda}\mathbf{v},\Phi'(\sigma_{\xi,\lambda}\mathbf{u}))$$

for all $\mathbf{u}$ and $\mathbf{v}$ in H_{loc}. For smooth $\mathbf{u}$ with compact support, and with elementary calculations similar to those leading to (4.10), we find that

$$
\begin{aligned}
((\mathbf{u}, \tilde{\Phi}'(\mathbf{u}))) &= -(\Delta \mathbf{u}, \tilde{\Phi}'(\mathbf{u})) = -(\sigma_{\xi,\lambda}\Delta \mathbf{u}, \Phi'(\sigma_{\xi,\lambda}\mathbf{u})) \\
&= -(\lambda^{-2}\Delta(\sigma_{\xi,\lambda}\mathbf{u}), \Phi'(\sigma_{\xi,\lambda}\mathbf{u})) = \frac{1}{\lambda^2}((\sigma_{\xi,\lambda}\mathbf{u}, \Phi'(\sigma_{\xi,\lambda}\mathbf{u})))
\end{aligned}
$$

and

$$
\begin{aligned}
((\mathbf{u}\cdot\nabla)\mathbf{u}, \tilde{\Phi}'(\mathbf{u})) &= ((\mathbf{u}\cdot\nabla)\mathbf{u}, \tilde{\Phi}'(\mathbf{u})) = (\sigma_{\xi,\lambda}(\mathbf{u}\cdot\nabla)\mathbf{u}, \Phi'(\sigma_{\xi,\lambda}\mathbf{u})) \\
&= \left(\xi^{-1}\lambda^{-1}((\sigma_{\xi,\lambda}\mathbf{u})\cdot\nabla)(\sigma_{\xi,\lambda}\mathbf{u}), \Phi'(\sigma_{\xi,\lambda}\mathbf{u})\right) \\
&= \frac{1}{\xi\lambda}\left(((\sigma_{\xi,\lambda}\mathbf{u})\cdot\nabla)(\sigma_{\xi,\lambda}\mathbf{u}), \Phi'(\sigma_{\xi,\lambda}\mathbf{u})\right).
\end{aligned}
$$

By an approximation argument, we can extend the previous relations (which are valid for smooth $\mathbf{u}$ with compact support) to all $\mathbf{u}$ in the space V_{loc}, and we conclude that

$$
((\mathbf{u}, \tilde{\Phi}'(\mathbf{u}))) = \frac{1}{\lambda^2}((\sigma_{\xi,\lambda}\mathbf{u}, \Phi'(\sigma_{\xi,\lambda}\mathbf{u}))) \tag{4.13}
$$

and

$$
((\mathbf{u}\cdot\nabla)\mathbf{u}, \tilde{\Phi}'(\mathbf{u})) = \frac{1}{\xi\lambda}\left(((\sigma_{\xi,\lambda}\mathbf{u})\cdot\nabla)(\sigma_{\xi,\lambda}\mathbf{u}), \Phi'(\sigma_{\xi,\lambda}\mathbf{u})\right) \tag{4.14}
$$

for all vector fields $\mathbf{u}$ in V_{loc}. Hence, since the measures μ_t are carried by V_{loc} for almost every $t \geq 0$ (see (2.21)), we deduce – using (2.17) with t and Φ replaced (respectively) by αt and $\tilde{\Phi}$ – that

$$
\begin{aligned}
&\int \Phi(\mathbf{u})\, d(\sigma_{\xi,\lambda}\mu_{\alpha t})(\mathbf{u}) - \int \Phi(\mathbf{u})\, d(\sigma_{\xi,\lambda}\mu_0)(\mathbf{u}) \\
&\quad = \int \Phi(\sigma_{\xi,\lambda}\mathbf{u})\, d\mu_{\alpha t}(\mathbf{u}) - \int \Phi(\sigma_{\xi,\lambda}\mathbf{u})\, d\mu_0(\mathbf{u}) \\
&\quad = \int \tilde{\Phi}(\mathbf{u})\, d\mu_{\alpha t}(\mathbf{u}) - \int \tilde{\Phi}(\mathbf{u})\, d\mu_0(\mathbf{u}) \\
&\quad = -\int_0^{\alpha t}\int [\nu((\mathbf{u}, \tilde{\Phi}'(\mathbf{u}))) + ((\mathbf{u}\cdot\nabla)\mathbf{u}, \tilde{\Phi}'(\mathbf{u}))]\, d\mu_s(\mathbf{u})\, ds \\
&\quad = -\int_0^{\alpha t}\int \Bigg[\frac{\nu}{\lambda^2}((\sigma_{\xi,\lambda}\mathbf{u}, \Phi'(\sigma_{\xi,\lambda}\mathbf{u}))) \\
&\qquad\qquad + \frac{1}{\xi\lambda}\left(((\sigma_{\xi,\lambda}\mathbf{u})\cdot\nabla)(\sigma_{\xi,\lambda}\mathbf{u}), \Phi'(\sigma_{\xi,\lambda}\mathbf{u})\right)\Bigg]\, d\mu_s(\mathbf{u})\, ds \\
&\quad = -\alpha\int_0^{t}\int \Bigg[\frac{\nu}{\lambda^2}((\sigma_{\xi,\lambda}\mathbf{u}, \Phi'(\sigma_{\xi,\lambda}\mathbf{u}))) \\
&\qquad\qquad + \frac{1}{\xi\lambda}\left(((\sigma_{\xi,\lambda}\mathbf{u})\cdot\nabla)(\sigma_{\xi,\lambda}\mathbf{u}), \Phi'(\sigma_{\xi,\lambda}\mathbf{u})\right)\Bigg]\, d\mu_{\alpha s}(\mathbf{u})\, ds \\
&\quad = -\int_0^{t}\int \left[\frac{\alpha\nu}{\lambda^2}((\mathbf{u}, \Phi'(\mathbf{u}))) + \frac{\alpha}{\xi\lambda}((\mathbf{u}\cdot\nabla)\mathbf{u}, \Phi'(\mathbf{u}))\right] d(\sigma_{\xi,\lambda}\mu_{\alpha s})(\mathbf{u})\, ds.
\end{aligned}
$$

Now, we take $\alpha = \xi\lambda$ and set

$$\tilde{\mu}_t = \sigma_{\xi,\lambda}\mu_{\xi\lambda t};$$

we then find that

$$\int \Phi(\mathbf{u})\, d\tilde{\mu}_t(\mathbf{u}) = -\int_0^t \int \left[\frac{\nu\xi}{\lambda}((\mathbf{u}, \Phi'(\mathbf{u}))) + ((\mathbf{u}\cdot\nabla)\mathbf{u}, \Phi'(\mathbf{u}))\right] d\tilde{\mu}_s(\mathbf{u})\, ds$$
$$+ \int \Phi(\mathbf{u})\, d\tilde{\mu}_0$$

for all $t \geq 0$. Therefore, the $\{\tilde{\mu}_t\}_{t\geq 0}$ satisfy the Liouville-type equation (2.17).

To prove that $\{\tilde{\mu}_t\}_{t\geq 0}$ is a homogeneous statistical solution according to Definition 2.4, it remains only to prove the energy inequality (2.18). For that purpose we use again the relations (4.11) and (4.12) as follows:

$$\begin{aligned} e(\tilde{\mu}_t) + \frac{\nu\xi}{\lambda}\int_0^t E(\tilde{\mu}_s)\, ds &= \xi^2 e(\mu_{\xi\lambda t}) + \frac{\nu\xi}{\lambda}\int_0^t \xi^2\lambda^2 E(\mu_{\xi\lambda s})\, ds \\ &= \xi^2 e(\mu_{\xi\lambda t}) + \nu\xi^3\lambda \int_0^{\xi\lambda t} E(\mu_s)\, ds \\ &= \xi^2\left(e(\mu_{\xi\lambda t}) + \nu\int_0^{\xi\lambda t} E(\mu_s)\, ds\right) \\ &\leq \xi^2 e(\mu_0) = e(\tilde{\mu}_0). \end{aligned}$$

This proves the energy inequality. Moreover, it is easy to see that if instead $\{\mu_t\}_{t\geq 0}$ satisfies an energy equation – namely,

$$e(\mu_t) + \nu\int_0^t E(\mu_s)\, ds = e(\mu_0) \quad \text{for all } t \geq 0 \tag{4.15}$$

– then so do the $\{\tilde{\mu}_t\}_{t\geq 0}$, with viscosity rescaled to $\nu\xi/\lambda$.

Thus, we have shown that $\{\tilde{\mu}_t\}_{t\geq 0}$ is a homogeneous statistical solution of the Navier–Stokes equations with viscosity $\nu\xi/\lambda$. We have therefore established the following result.

Proposition 4.1 *Let $\{\mu_t\}_{t\geq 0}$ be a homogeneous statistical solution of the Navier–Stokes equations with viscosity ν. For any $\xi, \lambda > 0$, define*

$$\tilde{\mu}_t = \sigma_{\xi,\lambda}\mu_{\xi\lambda t} \tag{4.16}$$

for $t \geq 0$. Then $\{\tilde{\mu}_t\}_{t\geq 0}$ is a homogeneous statistical solution of the NSE with viscosity

$$\tilde{\nu} = \frac{\nu\xi}{\lambda}. \tag{4.17}$$

Furthermore, the following relations hold for all $t \geq 0$:

$$e(\tilde{\mu}_t) = \xi^2 e(\mu_{\xi\lambda t}) \quad \text{and} \quad E(\tilde{\mu}_t) = \xi^2\lambda^2 E(\mu_{\xi\lambda t}). \tag{4.18}$$

Finally, if $\{\mu_t\}_{t\geq 0}$ satisfies the energy equation (4.15) then so does $\{\tilde{\mu}_t\}_{t\geq 0}$, with corresponding viscosity $\tilde{\nu}$.

4.2 Self-Similar Homogeneous Statistical Solutions

One of the major goals of turbulence theory is to find universal structures in fully developed turbulence. Hopf [1952] expected, in a sort of ergodic hypothesis, that under statistical equilibrium all the "relevant" statistical behavior of the flow should be determined by a 2-parameter family of stationary statistical solutions, depending on the viscosity ν and the energy dissipation rate ϵ. He also expected that this 2-parameter family should be exhausted by applying – to just one member of the family – an appropriate group transformation associated with scalings in the space variable $\mathbf{x}$ and the velocity $\mathbf{u}$ (i.e., a self-similarity property). However, since there is no force driving the fluid, he realized that average energy per unit mass must be infinite. This can be easily seen from the energy equation (4.15). Indeed, if $\mu_t \equiv \mu$ is a stationary statistical solution then, from the energy equation, we have that either the enstrophy $E(\mu)$ is zero or the energy $e(\mu)$ is infinite. In the first case, however, the energy dissipation is zero, so that the only case compatible with the loss of kinetic energy by viscosity is when the energy is infinite; but this, on the other hand, is in contradiction with empirical evidence. This difficulty is sometimes referred to as the Hopf paradox (for stationary statistical solutions).

Three decades later, Foias and Temam [1983] (see also Foias et al. [1987a]) noted that, by allowing the statistical solutions to be *time-dependent,* one can construct self-similar 2-parameter families of statistical solutions with finite energy and enstrophy and which seem to provide a good theoretical basis for the self-similar decaying turbulence. Our aim in what follows is to define those families in a precise way and to study some of their properties. In the following section, we discuss some applications to the study of turbulence.

Hence, we look for a suitable 2-parameter family $\{\mu^{\nu,\epsilon}\}_{\nu,\epsilon>0}$ of homogeneous probability distributions, where the parameters ν and ϵ are supposed to play the role of the viscosity and the energy dissipation rate, respectively. We want to relate the parameters ν and ϵ to their expected physical meanings. We therefore assume that this family satisfies the conditions

$$e(\mu^{\nu,\epsilon}) < \infty \quad \text{and} \quad \epsilon = \nu E(\mu^{\nu,\epsilon}) < \infty \quad \text{for all } \nu, \epsilon \geq 0, \tag{4.19}$$

and we require from this family the following two invariance properties. First, that for any μ^{ν,ϵ_0} in the family, there exists a homogeneous statistical solution $\{\mu_t\}_{t\geq 0}$ of the NSE with viscosity ν satisfying $\mu_0 = \mu^{\nu,\epsilon_0}$ and

$$\mu_t \in \{\mu^{\nu,\epsilon};\ 0 < \epsilon < \infty\} \quad \text{for all } t \geq 0, \tag{4.20}$$

that is,

$$\mu_t = \mu^{\nu,\epsilon(t)} \quad \text{for some } \epsilon = \epsilon(t) \in (0,\infty); \tag{4.21}$$

the usual energy equation is satisfied in the form

$$e(\mu_t) + \nu \int_0^t E(\mu_s)\,ds = e(\mu_0) \quad \text{for all } t \geq 0. \tag{4.22}$$

Second, in order to relate the different statistical solutions of the 2-parameter family and give to this family a relevant meaning, we require that it be invariant under the rescaling $\sigma_{\xi,\lambda}$. More precisely, we assume that, for all $\xi, \lambda > 0$, the operator $\sigma_{\xi,\lambda}$ takes the homogeneous statistical solution $\{\mu_t\}_{t\geq 0}$ associated to $\mu_0 = \mu^{\nu,\epsilon_0}$ ($\nu, \epsilon_0 > 0$, as before) into a homogeneous statistical solution $\{\tilde{\mu}_t\}_{t\geq 0}$ of the Navier–Stokes equations with viscosity $\tilde{\nu} = \nu\xi/\lambda$, given by $\tilde{\mu}_t = \sigma_{\xi,\lambda}(\mu_{\xi\lambda t})$ with $t \geq 0$, and satisfying both

$$\tilde{\mu}_t \in \{\mu^{\tilde{\nu},\epsilon};\ 0 < \epsilon < \infty\} \tag{4.23}$$

and

$$e(\tilde{\mu}_t) + \tilde{\nu} \int_0^t E(\tilde{\mu}_s)\, ds = e(\tilde{\mu}_0) \tag{4.24}$$

for all $t \geq 0$.

For convenience of terminology, we introduce the following definition.

Definition 4.1 *A family of probability measures $\mu^{\nu,\epsilon}$ satisfying conditions* (4.19)–(4.24) *is called a* self-similar homogeneous statistical solution *of the Navier–Stokes equations.*

The condition (4.22) is related to the energy equality (instead of inequality), which is usually valid in the deterministic case for smooth solutions. Hence, we view it here as a kind of regularity condition since we require equality and not only inequality (which is all that we can prove in the 3-dimensional case).

The self-similar homogeneous statistical solutions introduced in Definition 4.1 were first defined in Foias and Temam [1983], where they were called self-similar universal homogeneous statistical solutions. Besides the fact that the family depends on the two parameters ν and ϵ (as in the conventional theory of turbulence), the reason for the word "universal" comes from the conjecture made in Foias and Temam [1983] that "'most' homogeneous statistical solutions of the Navier–Stokes equations eventually 'approach' a superposition of solutions of [this] type." This conjecture was motivated by the comment in Hopf [1952] that the "relevant" probability distributions considered may be completely exhausted by applying a 2-parameter transformation to just one of those distributions.

Characterization of Self-Similar Statistical Solutions

The following result gives a more explicit characterization of self-similarity.

Proposition 4.2 *A homogeneous statistical solution of the Navier–Stokes equations $\{\mu^{\nu,\epsilon}\}_{\nu,\epsilon>0}$ is self-similar if and only if*

$$\sigma_{\xi,\lambda}(\mu^{\nu,\epsilon}) = \mu^{\nu\xi/\lambda,\,\xi^3\lambda\epsilon} \tag{4.25}$$

for all $\xi, \lambda, \nu, \epsilon > 0$.

Proof. Suppose $\{\mu^{\nu,\epsilon}\}_{\nu,\epsilon>0}$ is self-similar. Let $\{\mu_t\}_{t\geq 0}$ be a homogeneous statistical solution as in Definition 4.1, with $\mu_0 = \mu^{\nu,\epsilon_0}$. Then, by (4.20) and (4.21), we find that

$$\mu_t = \mu^{\nu,\epsilon(t)} \quad \text{with } \epsilon(t) = \nu E(\mu_t). \tag{4.26}$$

Since $\{\mu^{\nu,\epsilon}\}_{\nu,\epsilon>0}$ is self-similar, it follows that $\{\tilde{\mu}_t\}_{t\geq 0}$ given by (4.16) is a homogeneous statistical solution of the NSE with viscosity $\tilde{\nu} = \nu\xi/\lambda$ satisfying (see (4.23))

$$\tilde{\mu}_t = \mu^{\tilde{\nu},\tilde{\epsilon}(t)} \quad \text{with } \tilde{\epsilon}(t) = \tilde{\nu}E(\tilde{\mu}_t). \tag{4.27}$$

Using (4.17), (4.18), (4.26), and (4.27), we obtain

$$\tilde{\epsilon}(t) = \frac{\nu\xi}{\lambda}E(\tilde{\mu}_t) = \frac{\nu\xi}{\lambda}\xi^2\lambda^2 E(\mu_{\xi\lambda t}) = \xi^3\lambda\epsilon(\xi\lambda t). \tag{4.28}$$

At $t = 0$ we find $\tilde{\epsilon}(0) = \xi^3\lambda\epsilon_0$, so that

$$\sigma_{\xi,\lambda}(\mu^{\nu,\epsilon_0}) = \sigma_{\xi,\lambda}(\mu_0) = \tilde{\mu}_0 = \mu^{\tilde{\nu},\tilde{\epsilon}(0)} = \mu^{\nu\xi/\lambda,\xi^3\lambda\epsilon_0}.$$

For the converse implication, suppose $\{\mu^{\nu,\epsilon}\}_{\nu,\epsilon>0}$ satisfies (4.25). Take $\nu, \epsilon_0 > 0$ and $\{\mu_t\}_{t\geq 0}$ with $\mu_0 = \mu^{\nu,\epsilon_0}$ as in Definition 4.1. Let $\tilde{\mu}_t = \sigma_{\xi,\lambda}(\mu_{\xi\lambda t})$ for all $t \geq 0$. By Proposition 4.1, $\{\tilde{\mu}_t\}_{t\geq 0}$ is a homogeneous statistical solution of the Navier–Stokes equations with viscosity $\tilde{\nu}$ satisfying the energy equation (4.24). Hence, it remains only to show that (4.23) holds. For this purpose, we use (4.25) and the fact that

$$\mu_t = \mu^{\nu,\epsilon(t)} \quad \text{with } \epsilon(t) = \nu E(\mu_t),$$

which follows from (4.19) and (4.21), to obtain

$$\begin{aligned}\tilde{\mu}_t &= \sigma_{\xi,\lambda}(\mu_{\xi\lambda t}) = \sigma_{\xi,\lambda}(\mu^{\nu,\epsilon(\xi\lambda t)}) \\ &= \mu^{\nu\xi/\lambda,\xi^3\lambda\epsilon(\xi\lambda t)} = \mu^{\tilde{\nu},\xi^3\lambda\epsilon(\xi\lambda t)} \in \{\mu^{\tilde{\nu},\epsilon};\ 0 < \epsilon < \infty\}\end{aligned}$$

for all $t \geq 0$. This concludes the proof. ■

Evolution of the Kinetic Energy and the Enstrophy for Self-Similar Statistical Solutions

From the characterization given in Proposition 4.2, we may deduce a relation between the kinetic energy of the members of the family $\{\mu^{\nu,\epsilon}\}_{\nu,\epsilon>0}$ and also a more precise form for $\epsilon(t)$ in (4.26).

Proposition 4.3 *Let $\{\mu^{\nu,\epsilon}\}_{\nu,\epsilon>0}$ be a self-similar homogeneous statistical solution of the Navier–Stokes equations. Then,*[9]

$$e(\mu^{\nu,\epsilon}) = \gamma(\epsilon\nu)^{1/2}, \quad \text{where } \gamma = e(\mu^{1,1}). \tag{4.29}$$

Moreover, if $\{\mu_t\}_{t\geq 0}$ is a homogeneous statistical solution associated with $\mu_0 = \mu^{\nu,\epsilon_0}$ $(\nu, \epsilon_0 > 0)$ as in Definition 4.1, then for all $t \geq 0$ we have

$$\mu_t = \mu^{\nu,\epsilon(t)}, \quad \text{where } \epsilon(t) = \epsilon_0\left(1 + \frac{t\epsilon_0^{1/2}}{\gamma\nu^{1/2}}\right)^{-2}. \tag{4.30}$$

Hence

[9] We show in Section 5 that $2\gamma^2$ is the Reynolds number for the flow.

$$\frac{d}{dt}(\epsilon(t)^{1/2}) = -\frac{\epsilon}{\gamma\nu^{1/2}}, \tag{4.31}$$

with

$$e(\mu^{\nu,\epsilon(t)}) = e(\mu^{\nu,\epsilon_0})\left(1+\frac{t\epsilon_0^{1/2}}{\gamma\nu^{1/2}}\right)^{-1}. \tag{4.32}$$

Proof. By taking $\xi = (\nu\epsilon)^{-1/4}$ and $\lambda = \nu^{3/4}\epsilon^{-1/4}$ in (4.25), we see that

$$\sigma_{\xi,\lambda}\mu^{\nu,\epsilon} = \mu^{1,1}.$$

Hence, using (4.8) and (4.10), we obtain (4.29).

Let now $\{\mu_t\}_{t\geq 0}$ be a homogeneous statistical solution associated to $\mu_0 = \mu^{\nu,\epsilon_0}$ $(\nu, \epsilon_0 > 0)$, as in Definition 4.1. We infer from (4.19) and (4.21) that

$$\mu_t = \mu^{\nu,\epsilon(t)}, \quad \text{where } \epsilon(t) = \nu E(\mu^{\nu,\epsilon(t)}) = \nu E(\mu_t).$$

From (4.29), we find that

$$e(\mu_t) = e(\mu^{\nu,\epsilon(t)}) = \gamma(\epsilon(t)\nu)^{1/2}.$$

Hence, using the energy equation (4.22), we obtain

$$\gamma(\epsilon(t)\nu)^{1/2} + \int_0^t \epsilon(s)\,ds = \gamma(\epsilon_0\nu)^{1/2}.$$

Thus,

$$\frac{d(\epsilon(t)^{1/2})}{dt} + \frac{\epsilon(t)}{\gamma\nu^{1/2}} = 0, \qquad \epsilon(0) = \epsilon_0,$$

which proves (4.31). From this equation, it is straightforward to deduce (4.30). Finally, (4.32) follows from (4.30) and (4.29). ■

Remark 4.1 Formula (4.32) implies

$$e(\mu^{\nu,\epsilon(t)}) \sim \frac{\gamma^2\nu}{t} \quad \text{as } t \to \infty,$$

which represents a universal law of decay (i.e., a law that is independent of the initial data).

Connection with a Perturbed Form of the Navier–Stokes Equations

There is another important characterization of self-similar homogenous statistical solutions. It is a characterization that relates those solutions to rescalings of the stationary homogeneous statistical solutions of the following modified form of the NSE (note the presence of two additional terms):

$$\frac{\partial \mathbf{v}}{\partial s} - \frac{1}{2}\mathbf{v} - \frac{1}{2}(\mathbf{y}\cdot\nabla)\mathbf{v} - \Delta\mathbf{v} + (\mathbf{v}\cdot\nabla)\mathbf{v} + \nabla q = 0, \qquad \nabla\cdot\mathbf{v} = 0, \tag{4.33}$$

where $\mathbf{v} = \mathbf{v}(\mathbf{y}, s)$ and the differential operators ∇ and Δ are with respect to the spatial variable $\mathbf{y}$.

Specifically, that characterization relates the self-similar homogeneous statistical solution $\{\mu^{\nu,\epsilon}\}_{\nu,\epsilon>0}$ of the Navier–Stokes equations to a stationary homogeneous statistical solution μ of (4.33) satisfying $0 < e(\mu) = E(\mu) < \infty$ via the scaling

$$\mu^{\nu,\epsilon} = \sigma_{\xi,\lambda}(\mu), \quad \nu, \epsilon > 0,$$

where $\xi = \gamma^{-1/2}\epsilon^{1/4}\nu^{1/4}$ and $\lambda = \gamma^{-1/2}\epsilon^{1/4}\nu^{-3/4}$ with $\gamma = e(\mu)$. Note that ξ and $1/\lambda = \gamma^{1/2}\nu^{3/4}/\epsilon^{1/4}$ are the familiar velocity and length scales – with the generalization that here they are allowed to change with time (with $\epsilon = \epsilon(t)$ as in (4.30)), as they should in decaying turbulence.

We summarize this result in the following theorem; a proof can be found in Appendix A.3.

Theorem 4.1 *A 2-parameter family $\{\mu^{\nu,\epsilon}\}_{\nu,\epsilon>0}$ of homogeneous probability measures on H_{loc} is a self-similar homogeneous statistical solution of the Navier–Stokes equations if and only if it is of the form*

$$\mu^{\nu,\epsilon} = \sigma_{\gamma^{-1/2}\epsilon^{1/4}\nu^{1/4},\gamma^{-1/2}\epsilon^{1/4}\nu^{-3/4}}(\mu) \quad \textit{for all } \nu, \epsilon > 0 \tag{4.34}$$

for some stationary homogeneous statistical solution μ of equations (4.33) *satisfying*

$$e(\mu) = E(\mu) \ (= \gamma^2). \tag{4.35}$$

Moreover, the following relations hold:

$$\mu = \mu^{1,\gamma} \quad \textit{and} \quad \gamma = e(\mu^{1,1}). \tag{4.36}$$

Equations (4.33) can also be obtained from the NSE by making the following change of variables (inspired by the preceding dimensional considerations):

$$\begin{aligned} \mathbf{x} &= \mathbf{y}\left(\frac{\nu^3\gamma^2}{\epsilon(t)}\right)^{1/4}, \\ t &= \tau\left(\frac{\nu\gamma^2}{\epsilon(t)}\right)^{1/2}, \\ \mathbf{u} &= \mathbf{v}\left(\frac{\nu\epsilon(t)}{\gamma^2}\right)^{1/4}, \end{aligned} \tag{4.37}$$

where $\epsilon(t)$ is given by (4.30) and with γ and ϵ_0 arbitrary but fixed. This change of variables is valid for both the statistical and deterministic solutions of the Navier–Stokes equations.

With this change of variables, the NSE are transformed into

$$(1-\tau)\frac{\partial \mathbf{v}}{\partial \tau} - \frac{1}{2}\mathbf{v} - \frac{1}{2}(\mathbf{y}\cdot\nabla)\mathbf{v} - \Delta\mathbf{v} + (\mathbf{v}\cdot\nabla)\mathbf{v} + \nabla q = 0, \quad \nabla\cdot\mathbf{v} = 0. \tag{4.38}$$

Here again, the differential operators ∇ and Δ are to be understood with respect to the new spatial variable y. By taking $\tau = 1 - e^{-s}$, we recover equation (4.33).

As mentioned previously in the statistical case, the self-similar homogeneous statistical solutions of the Navier–Stokes equations are transformed into the stationary statistical solutions of (4.33), and vice versa. But, more generally, we can transform the non–self-similar solutions as well; they are transformed into the nonstationary homogeneous statistical solutions of (4.33), and vice versa. Indeed, take $\lambda = \lambda(t) = \gamma^{1/2}\nu^{3/4}/\epsilon(t)^{1/4}$, $\xi = \xi(t) = \gamma^{1/2}/(\nu\epsilon(t))^{1/4}$, and $s = -\log(1-\tau)$, and set

$$\tilde{\mu}_s = \sigma_{\xi(t),\lambda(t)}(\mu_t),$$

with τ as in (4.37) and $\epsilon(t)$ as in (4.30). For a given test functional Φ in $\mathcal{T}_{\text{ws}}$, define $\tilde{\Phi}_t = \Phi \circ \sigma_{\xi(t),\lambda(t)}$, which also belongs to $\mathcal{T}_{\text{ws}}$ for each t. Since $\{\mu_t\}_{t\geq 0}$ is a homogeneous statistical solution of the NSE and since $\tilde{\Phi}_t$ depends on t, we find that

$$\begin{aligned}\frac{d}{ds}&\int \Phi(\mathbf{v})\,d\tilde{\mu}_s(\mathbf{v})\\ &= \frac{dt}{ds}\frac{d}{dt}\int \tilde{\Phi}_t(\mathbf{v})\,d\mu_t(\mathbf{v})\\ &= \frac{dt}{ds}\int\left\{\frac{d}{dt}\tilde{\Phi}_t(\mathbf{v}) - [\nu((\mathbf{v}, D_{\mathbf{v}}\tilde{\Phi}(\mathbf{v}))) + ((\mathbf{v}\cdot\nabla)\mathbf{v}, D_{\mathbf{v}}\tilde{\Phi}(\mathbf{v}))]\right\} d\mu_t(\mathbf{v}),\end{aligned}$$

where $D_{\mathbf{v}}\tilde{\Phi}$ denotes the differential of $\tilde{\Phi}_t = \tilde{\Phi}_t(\mathbf{v})$ with respect to $\mathbf{v}$. Now,

$$\frac{d}{dt}\tilde{\Phi}_t(\mathbf{v}) = \left(\frac{d}{dt}\sigma_{\xi(t),\lambda(t)}\mathbf{v}, \Phi'(\sigma_{\xi(t),\lambda(t)})\right),$$

but

$$\frac{d}{dt}(\sigma_{\xi(t),\lambda(t)}\mathbf{v}) = \frac{\xi'(t)}{\xi(t)}\sigma_{\xi,\lambda}\mathbf{v} + \frac{\lambda'(t)}{\lambda(t)}\mathbf{y}\cdot\nabla\sigma_{\xi,\lambda}\mathbf{v},$$

with

$$\frac{\xi'(t)}{\xi(t)} = \frac{\lambda'(t)}{\lambda(t)} = \epsilon(t)^{1/4}\frac{d}{dt}\epsilon(t)^{-1/4} = -\frac{\epsilon'(t)}{4\epsilon(t)} = \frac{\epsilon(t)^{1/2}}{2\gamma\nu^{1/2}}. \tag{4.39}$$

Hence,

$$\frac{d}{dt}\tilde{\Phi}_t(\mathbf{v}) = \frac{\epsilon^{1/2}}{2\gamma\nu^{1/2}}(\sigma_{\xi,\lambda}\mathbf{v} + \mathbf{y}\cdot\nabla\sigma_{\xi,\lambda}\mathbf{v}, \Phi'(\sigma_{\xi,\lambda}\mathbf{v})).$$

Moreover, as in (4.13) and (4.14),

$$\begin{aligned}\int&[\nu((\mathbf{v}, D_{\mathbf{v}}\tilde{\Phi}(\mathbf{v}))) + ((\mathbf{v}\cdot\nabla)\mathbf{v}, D_{\mathbf{v}}\tilde{\Phi}(\mathbf{v}))]\,d\mu_t(\mathbf{v})\\ &= \int\left[\frac{\nu}{\lambda^2}((\sigma_{\xi,\lambda}\mathbf{v}, \Phi'(\sigma_{\xi,\lambda}\mathbf{v}))) + \frac{1}{\xi\lambda}\big(((\sigma_{\xi,\lambda}\mathbf{v})\cdot\nabla)(\sigma_{\xi,\lambda}\mathbf{v}), \Phi'(\sigma_{\xi,\lambda}\mathbf{v})\big)\right] d\mu_t(\mathbf{v}).\end{aligned}$$

Now, using (4.31), it is easy to see that

$$\frac{d\epsilon^{-1/2}}{d\tau} = \frac{1}{\gamma\nu^{1/2}}\frac{dt}{d\tau},$$

so that from (4.37) we obtain

$$\frac{dt}{d\tau} = \frac{\gamma\nu^{1/2}}{\epsilon^{1/2}} + \tau\frac{dt}{d\tau},$$

whence we deduce that

$$\frac{dt}{ds} = (1-\tau)\frac{dt}{d\tau} = \frac{\gamma\nu^{1/2}}{\epsilon^{1/2}}. \tag{4.40}$$

Therefore,

$$\frac{d}{ds}\int \Phi(\mathbf{v})\, d\tilde{\mu}_s(\mathbf{v})$$
$$= \int \left[\frac{1}{2}(\mathbf{v}+\mathbf{y}\cdot\nabla\mathbf{v}, \Phi'(\mathbf{v})) - ((\mathbf{v}, \Phi'(\mathbf{v}))) - ((\mathbf{v}\cdot\nabla)\mathbf{v}, \Phi'(\mathbf{v}))\right] d\tilde{\mu}_s(\mathbf{v}),$$

which shows that, indeed, $\{\tilde{\mu}_s\}_{s\geq 0}$ is a homogeneous statistical solution of (4.33).

Remarks on the Transformation of the Energy Equation

After the transformation (4.37), the energy equation takes a form that differs from (4.15). This can be anticipated from the fact that, for a stationary homogeneous statistical solution μ satisfying this energy equation, one would have $e(\mu) = \infty$ or $E(\mu) = 0$, which can be seen by taking $\mu_t = \mu$ in the energy equation. This is known as Hopf's paradox in the theory of universal equilibrium, which was first pointed out in Hopf [1952]. This paradox is resolved here by the premise – given in Foias and Temam [1983] and in Foias, Manley, and Temam [1987a] – that one should not seek any one specific stationary homogeneous statistical solution; rather, one should look for self-similar families of homogeneous statistical solutions.

In order to obtain the correct energy equation after the transformation, let us consider a homogeneous statistical solution $\{\mu_t\}_{t\geq 0}$ of the Navier–Stokes equations and the transformed probability distributions $\tilde{\mu}_s = \sigma_{\xi(t),\lambda(t)}\mu_t$, where $\lambda = \lambda(t) = \gamma^{1/2}\nu^{3/4}/\epsilon(t)^{1/4}$, $\xi = \xi(t) = \gamma^{1/2}/\nu\epsilon(t)^{1/4}$, and $s = -\log(1-\tau)$, with τ as in (4.37) and $\epsilon(t)$ as in (4.30).

We are interested in the evolution of the ensemble averaged kinetic energy after the transformation, $e(\tilde{\mu}_s)$, and its connection with the ensemble average enstrophy, $E(\tilde{\mu}_s)$. Using the relations (4.8) and (4.10) and the energy equation (4.15), one can see that

$$\frac{de(\tilde{\mu}_s)}{ds} = \frac{dt}{ds}\frac{de(\sigma_{\xi,\lambda}\mu_t)}{dt} = \frac{dt}{ds}\frac{d(\xi^2 e(\mu_t))}{dt}$$
$$= \frac{dt}{ds}\left(2\xi e(\mu_t)\frac{d\xi}{dt} - \xi^2\nu E(\mu_t)\right) = \frac{dt}{ds}\left(2e(\sigma_{\xi,\lambda}\mu_t)\frac{1}{\xi}\frac{d\xi}{dt} - \frac{\nu}{\lambda^2}E(\sigma_{\xi,\lambda}\mu_t)\right)$$
$$= \frac{dt}{ds}\left(2e(\tilde{\mu}_s)\frac{1}{\xi}\frac{d\xi}{dt} - \frac{\nu}{\lambda^2}E(\tilde{\mu}_s)\right).$$

Using (4.39), the fact that $\nu/\lambda^2 = \epsilon^{1/2}/(\gamma\nu^{1/2})$, and finally (4.40), we find that

$$\frac{de(\tilde{\mu}_s)}{ds} = \frac{dt}{ds}(e(\tilde{\mu}_s) - E(\tilde{\mu}_s))\frac{\epsilon(t)^{1/2}}{\gamma\nu^{1/2}} = e(\tilde{\mu}_s) - E(\tilde{\mu}_s).$$

Hence, the energy equation takes the form

$$e(\tilde{\mu}_s) + \int_0^s E(\tilde{\mu}_{s'})\, ds' = \int_0^s e(\tilde{\mu}_{s'})\, ds' + e(\tilde{\mu}_0). \tag{4.41}$$

The extra term on the right-hand side arises from the time-dependent change of scale; it may be interpreted, perhaps, as a stress accumulating in the strained transformed space.

5 Relation with and Application to the Conventional Theory of Turbulence

In this section we apply the preceding considerations to the case of decaying homogeneous isotropic turbulent flows in three dimensions. Such a flow can be generated and studied in detail in a suitably designed wind tunnel. The turbulence is induced by placing a grid at the upstream end of the tunnel. In a wind tunnel of sufficiently large diameter, the interior is well away from the boundary layers near the walls, and there the flow may be regarded as close to isotropic and homogeneous. As the flow proceeds downstream, the eddies generated by the upstream grid decay and are eventually damped out by viscosity, yielding a good approximation to the ideal of decaying isotropic homogeneous turbulence.

5.1 The Two-Point Correlation Function and the Energy Spectrum

An important quantity in the study of turbulence is the correlation matrix of a homogeneous flow: $R(\mathbf{y}) = (R_{jk}(\mathbf{y}))_{j,k=1}^{3}$. It is defined formally as the ensemble average

$$R_{jk}(\mathbf{y}) = \langle u_j(\mathbf{x}+\mathbf{y})u_k(\mathbf{x})\rangle, \quad \mathbf{y} \in \mathbb{R}^3, \; j,k = 1,\ldots,3.$$

Given a homogeneous probability measure μ, we can rigorously define the correlation matrix with the help of Lemmas 2.2 and 2.3. Indeed, if $e(\mu) < \infty$, then one can see that condition (2.7) is satisfied (with $G(\mathbf{u}) = G_{\mathbf{y},jk}(\mathbf{u}) = \tau_{-\mathbf{y}}u_j u_k$), so that we may define

$$R_{jk}(\mathbf{y}) = \int \frac{1}{|Q|}\int_Q u_j(\mathbf{x}+\mathbf{y})\mathbf{u}_k(\mathbf{x})\,d\mathbf{x}\,d\mu(\mathbf{u}), \quad \mathbf{y} \in \mathbb{R}^3, \; j,k = 1,\ldots,3, \tag{5.1}$$

where $Q = (a,b)^3$ and a, b are arbitrary, $0 < a < b$. Note that

$$e(\mu) = \tfrac{1}{2}\operatorname{Tr} R(0) = \tfrac{1}{2}(R_{11}(0) + R_{22}(0) + R_{33}(0)). \tag{5.2}$$

Moreover, if $E(\mu) < \infty$, then R_{jk} is of class C^2 for $j,k = 1,\ldots,3$, with

$$\partial^2_{y_i y_l} R_{jk}(\mathbf{y}) = -\int \frac{1}{|Q|}\int_Q \partial_{x_i} u_j(\mathbf{x}+\mathbf{y})\partial_{x_l} u_k(\mathbf{x})\,d\mathbf{x}\,d\mu(\mathbf{u}).$$

Hence,

$$E(\mu) = -\Delta \operatorname{Tr} R(\mathbf{y})|_{\mathbf{y}=0}. \tag{5.3}$$

From the homogeneity assumption, one finds that

$$R_{jk}(\mathbf{y}) = R_{kj}(-\mathbf{y}). \tag{5.4}$$

In particular,

$$\operatorname{Tr} R(\mathbf{y}) = \operatorname{Tr} R(-\mathbf{y}). \tag{5.5}$$

Furthermore, using the Cauchy–Schwarz inequality and homogeneity,

$$R_{jk}(\mathbf{y}) = \int \frac{1}{|Q|}\int_Q \mathbf{u}_j(\mathbf{x}+\mathbf{y})\mathbf{u}_k(\mathbf{x})\,d\mathbf{x}\,d\mu(\mathbf{u})$$
$$\leq \left(\int \frac{1}{|Q|}\int_Q |\mathbf{u}_j(\mathbf{x}+\mathbf{y})|^2\,d\mathbf{x}\,d\mu(\mathbf{u})\right)^{1/2}\left(\int \frac{1}{|Q|}\int_Q |\mathbf{u}_j(\mathbf{x})|^2\,d\mathbf{x}\,d\mu(\mathbf{u})\right)^{1/2}$$
$$= \left(\int \frac{1}{|Q|}\int_Q |\mathbf{u}_j(\mathbf{x})|^2\,d\mathbf{x}\,d\mu(\mathbf{u})\right)^{1/2}\left(\int \frac{1}{|Q|}\int_Q |\mathbf{u}_j(\mathbf{x})|^2\,d\mathbf{x}\,d\mu(\mathbf{u})\right)^{1/2}.$$

Therefore,

$$R_{jk}(\mathbf{y}) \leq R_{jj}(0)^{1/2}R_{kk}(0)^{1/2} \tag{5.6}$$

for all $j, k = 1, \ldots, 3$ and all $\mathbf{y}$ in $\mathbb{R}^3$. This implies, in turn, that

$$R_{jj}(\mathbf{y}) \leq R_{jj}(0) \tag{5.7}$$

for all $j = 1, \ldots, 3$ and for all $\mathbf{y}$ in $\mathbb{R}^3$. In particular,

$$\operatorname{Tr} R(\mathbf{y}) \leq \operatorname{Tr} R(0). \tag{5.8}$$

In words, the trace of the correlation matrix achieves its absolute maximum at the origin.

Suppose, now, that $\operatorname{Tr} R(\mathbf{y})$ is the inverse Fourier transform

$$\operatorname{Tr} R\,(\mathbf{y}) = \int_{\mathbb{R}^3} e^{i\boldsymbol{\kappa}\cdot\mathbf{y}}\mathcal{Q}(\boldsymbol{\kappa})\,d\boldsymbol{\kappa}, \quad \mathbf{y} \in \mathbb{R}^3, \tag{5.9}$$

of a continuous, integrable function $\mathcal{Q} = \mathcal{Q}(\boldsymbol{\kappa})$ on $\mathbb{R}^3$. The *energy spectrum* of μ is then defined for all $\kappa > 0$ by

$$\mathcal{S}(\kappa) = \int_{|\boldsymbol{\kappa}|=\kappa} \mathcal{Q}(\boldsymbol{\kappa})\,d\Sigma(\boldsymbol{\kappa}), \tag{5.10}$$

where $d\Sigma(\boldsymbol{\kappa})$ denotes the area element of the sphere (in $\mathbb{R}^3$) of radius κ. Note, then, that the kinetic energy $e(\mu)$ can be written as

$$e(\mu) = \frac{1}{2}\int_0^\infty \mathcal{S}(\kappa)\,d\kappa. \tag{5.11}$$

Moreover, if $E(\mu) < \infty$ then we also find, using (5.3), that

$$E(\mu) = -\Delta \operatorname{Tr} R(\mathbf{y})|_{\mathbf{y}=0} = \int_{\mathbb{R}^3} |\boldsymbol{\kappa}|^2 e^{i\boldsymbol{\kappa}\cdot\mathbf{y}}\mathcal{Q}(\boldsymbol{\kappa})\,d\boldsymbol{\kappa}\bigg|_{\mathbf{y}=0} = \int_0^\infty \kappa^2\mathcal{S}(\kappa)\,d\kappa.$$

The energy spectrum $\mathcal{S}(\kappa)$ at $\kappa > 0$ represents, up to the factor $1/2$, the amount of kinetic energy at this wavenumber κ, which is roughly the energy associated to the eddies of diameter of order κ^{-1}. The integral

$$\frac{1}{2}\int_{\kappa_1}^{\kappa_2} \mathcal{S}(\kappa)\,d\kappa$$

represents the amount of kinetic energy contained between the wavenumbers κ_1 and κ_2, $0 < \kappa_1 < \kappa_2$. We are interested in understanding the form of the spectrum

$\mathcal{S} = \mathcal{S}(\kappa)$. The Kolmogorov [1941a] theory, which is largely confirmed by many experiments, suggests that the behavior of the spectrum is universal in a certain range of length scales. More explicitly, the theory asserts that, over a certain range $\kappa_L > \kappa > \kappa_d$ of wavenumbers, viscosity does not play any role – the spectrum depends only on the wavenumber κ and the energy dissipation rate ϵ, so that the spectrum has the form $\mathcal{S}(\kappa) = \mathcal{S}(\kappa, \epsilon)$. The range (κ_L, κ_d) is called the *inertial subrange,* and within it the energy is simply dispersed among nearby length scales at rate determined by the inertial term. This assertion yields, on dimensional grounds that are divorced from the NSE, the well-known K41 spectrum. For $\kappa > \kappa_d$, however, dissipation dominates the effects of nonlinearity and one expects the spectrum to depend also on the viscosity, $\mathcal{S}(\kappa) = \mathcal{S}(\kappa, \nu, \epsilon)$. The range (κ_d, ∞), where the spectrum falls off rapidly, is called the *dissipation subrange.*

By considering a self-similar homogeneous statistical solution, we may rigorously obtain the picture just described for the inertial subrange. That is our aim in what follows. We start by deriving several properties for the two-point correlation function – first in the anisotropic and then in the isotropic case – followed by a consideration of the previously defined power spectrum (where "power" is energy per unit time).

The Two-Point Correlation Function in the Homogeneous Case

Consider a self-similar homogeneous satistical solution $\{\mu^{\nu,\epsilon}\}_{\nu,\epsilon>0}$ of the Navier–Stokes equations, and let $R(\mathbf{y}; \nu, \epsilon)$ be the correlation matrix associated with the measure $\mu^{\nu,\epsilon}$. From the characterization (4.25) and the fact that the ensemble average is independent of Q, we find that the rescaling $\sigma_{\xi,\lambda}$ transforms $R_{jk}(\mathbf{y}; \nu, \epsilon)$ according to

$$
\begin{aligned}
\sigma_{\xi,\lambda} R_{jk}(\mathbf{y}; \nu, \epsilon) &= \xi^2 R_{jk}(\lambda\mathbf{y}; \nu, \epsilon) \\
&= \xi^2 \int \frac{1}{|Q|} \int_Q u_j(\mathbf{x} + \lambda\mathbf{y}) \mathbf{u}_k(\mathbf{x})\, d\mathbf{x}\, d\mu^{\nu,\epsilon} \\
&= \xi^2 \int \frac{1}{|Q/\lambda|} \int_{Q/\lambda} u_j(\lambda\mathbf{x} + \lambda\mathbf{y}) \mathbf{u}_k(\lambda\mathbf{x})\, d\mathbf{x}\, d\mu^{\nu,\epsilon} \\
&= \xi^2 \int \frac{1}{|Q|} \int_Q \mathbf{u}_j(\lambda\mathbf{x} + \lambda\mathbf{y}) \mathbf{u}_k(\lambda\mathbf{x})\, d\mathbf{x}\, d\mu^{\nu,\epsilon} \\
&= \int \frac{1}{|Q|} \int_Q (\sigma_{\xi,\lambda}\mathbf{u})_j(\mathbf{x} + \mathbf{y}) (\sigma_{\xi,\lambda}\mathbf{u})_k(\mathbf{x})\, d\mathbf{x}\, d\mu^{\nu,\epsilon} \\
&= \int \frac{1}{|Q|} \int_Q u_j(\mathbf{x} + \mathbf{y}) \mathbf{u}_k(\mathbf{x})\, d\mathbf{x}\, d(\sigma_{\xi,\lambda}\mu^{\nu,\epsilon})(\mathbf{u}) \\
&= \int \frac{1}{|Q|} \int_Q \mathbf{u}_j(\mathbf{x} + \mathbf{y}) \mathbf{u}_k(\mathbf{x})\, d\mathbf{x}\, d\mu^{\nu\xi/\lambda, \xi^3\lambda\epsilon}(\mathbf{u}) \\
&= R_{jk}\left(\mathbf{y}; \frac{\nu\xi}{\lambda}, \xi^3\lambda\epsilon\right).
\end{aligned}
$$

Hence, by replacing $\mathbf{y}$ with $\mathbf{y}/\lambda$, we obtain the relation

$$R_{jk}(\mathbf{y}; \nu, \epsilon) = \xi^{-2} R_{jk}\left(\frac{\mathbf{y}}{\lambda}; \frac{\nu\xi}{\lambda}, \xi^3\lambda\epsilon\right). \tag{5.12}$$

Taking $\xi = (\nu\epsilon)^{-1/4}$ and $\lambda = \nu^{3/4}/\epsilon^{1/4}$ in (5.12), we find that

$$R(\mathbf{y}; \nu, \epsilon) = (\epsilon\nu)^{1/2} R\left(\frac{\epsilon^{1/4}}{\nu^{3/4}}\mathbf{y}; 1, 1\right). \tag{5.13}$$

If we introduce the Kolmogorov dissipation length

$$\ell_d = \left(\frac{\nu^3}{\epsilon}\right)^{1/4},$$

then we can rewrite (5.13) as

$$R(\mathbf{y}; \nu, \epsilon) = (\epsilon\nu)^{1/2} R\left(\frac{\mathbf{y}}{\ell_d}; 1, 1\right), \tag{5.14}$$

which gives us a scaling (invariance) law for the two-point double correlation function associated with each self-similar homogeneous statistical solution.

Finally, another scaling law can be obtained from (5.12) by taking $\xi^3\epsilon\lambda = 1$ and $\lambda = |\mathbf{y}|$. In this case, we find

$$R(\mathbf{y}; \nu, \epsilon) = (\epsilon|\mathbf{y}|)^{2/3} R\left(\frac{\mathbf{y}}{|\mathbf{y}|}; \left(\frac{\ell_d}{|\mathbf{y}|}\right)^{4/3}, 1\right). \tag{5.15}$$

Thus, we have proved the following result.

Theorem 5.1 *For a given self-similar homogeneous statistical solution $\{\mu^{\nu,\epsilon}\}_{\nu,\epsilon}$, the corresponding two-point double correlation functions $R(\mathbf{y}; \nu, \epsilon)$ satisfy the following scaling laws:*

$$R(\mathbf{y}; \nu, \epsilon) = (\epsilon\nu)^{1/2} R\left(\frac{\mathbf{y}}{\ell_d}; 1, 1\right) \tag{5.16}$$

and

$$R(\mathbf{y}; \nu, \epsilon) = (\epsilon|\mathbf{y}|)^{2/3} R\left(\frac{\mathbf{y}}{|\mathbf{y}|}; \left(\frac{\ell_d}{|\mathbf{y}|}\right)^{4/3}, 1\right), \tag{5.17}$$

where $\ell_d = \nu^{3/4}/\epsilon^{1/4}$.

The Reynolds Number

The nondimensional energy $\gamma = e(\mu^{1,1})$ defined in Proposition 4.3 is a fundamental quantity. In fact, we show in what follows that the square of γ is proportional to the Reynolds number.

Observe first that, from relation (5.16) and the fact that $e(\mu^{\nu,\epsilon}) = R(0; \nu, \epsilon)/2$ (see (5.2)), we find (see also Proposition 4.3)

$$\begin{aligned} e(\mu^{\nu,\epsilon}) &= \tfrac{1}{2}\operatorname{Tr} R(0; \nu, \epsilon) = \tfrac{1}{2}(\epsilon\nu)^{1/2}\operatorname{Tr} R(0; 1, 1) \\ &= (\epsilon\nu)^{1/2} e(\mu^{1,1}) = \gamma(\epsilon\nu)^{1/2}. \end{aligned} \tag{5.18}$$

Thus, we have recovered (5.2). A simple dimensional analysis now implies that the square of the nondimensional constant γ is proportional to the Reynolds number. Indeed, since $e(\mu^{\nu,\epsilon})$ represents the total average kinetic energy of the flow, it is natural to choose the characteristic velocity V as that given by

$$e(\mu^{\nu,\epsilon}) = \frac{V^2}{2}. \tag{5.19}$$

Then, from (5.18), we find that

$$V = \sqrt{2}\gamma^{1/2}(\epsilon\nu)^{1/4}. \tag{5.20}$$

Given that the energy dissipation rate[10] per unit mass $\epsilon \sim V^3/\ell_0$ for a characteristic length ℓ_0, we find (upon substituting for ϵ in the expression for V) that $V \sim \gamma^2\nu/\ell_0$ and hence that

$$\gamma^2 \sim \frac{V\ell_0}{\nu}, \tag{5.21}$$

which shows that γ^2 is, in fact, proportional to the Reynolds number.

Moreover, once we choose a characteristic length ℓ_0, we can find a more precise relation between γ^2 and Re. From (5.20) and (5.21) we see that the characteristic length ℓ_0 is proportional to the dissipation length ℓ_d, that is,

$$\ell_0 \sim \left(\frac{\nu^3}{\epsilon}\right)^{1/4}\gamma^{3/2}. \tag{5.22}$$

Thus, following Foias and Temam [1983], we choose

$$\ell_0 = \sqrt{2}\gamma^{3/2}\ell_d = \sqrt{2}\gamma^{3/2}\left(\frac{\nu^3}{\epsilon}\right)^{1/4}. \tag{5.23}$$

Hence, using again (5.20) and (5.23), we find that

$$\mathrm{Re} = \frac{V\ell_0}{\nu} = 2\gamma^2. \tag{5.24}$$

The Two-Point Correlation Function in the Homogeneous and Isotropic Case

In the isotropic case, the two-point correlation function depends only on the length $|\mathbf{y}|$ of the vector $\mathbf{y}$. Hence, (5.15) leads to

$$\operatorname{Tr} R(|\mathbf{y}|; \nu, \epsilon) = (\epsilon|\mathbf{y}|)^{2/3} \operatorname{Tr} R\left(1; \left(\frac{\ell_d}{|\mathbf{y}|}\right)^{4/3}, 1\right). \tag{5.25}$$

Therefore, for vanishing viscosity, we find the Kolmogorov $2/3$ power law in physical space:

$$\operatorname{Tr} R(|\mathbf{y}|; \nu, \epsilon) \sim C_K(\epsilon|\mathbf{y}|)^{2/3} \quad \text{as } \nu \to 0, \tag{5.26}$$

where the constant

$$C_K = \operatorname{Tr} R(1; 0^+, 1) = \lim_{\nu\to 0^+} \operatorname{Tr} R(1; \nu, 1), \tag{5.27}$$

[10] The term ϵ has dimension of velocity squared per unit time.

provided the limit exists. The constant C_K is is known as the Kolmogorov constant in physical space.

In the isotropic case it suffices to consider the longitudinal correlation function (see Taylor [1935], von Karman and Howarth [1938], and Dryden [1943]), which can be taken as

$$R_\ell(r;\nu,\epsilon) = R_{11}(r\mathbf{e}_1;\nu,\epsilon) = R_{22}(r\mathbf{e}_2;\nu,\epsilon) = R_{33}(r\mathbf{e}_3;\nu,\epsilon), \tag{5.28}$$

where $\mathbf{e}_1 = (1,0,0)$, $\mathbf{e}_2 = (0,1,0)$, and $\mathbf{e}_3 = (0,0,1)$ are the canonical basis vectors in $\mathbb{R}^3$. Indeed, from the incompressibility condition we may derive the following classical relation:

$$R_{22}(r\mathbf{e}_1;\nu,\epsilon) = R_{33}(r\mathbf{e}_1;\nu,\epsilon) = R_\ell(r;\nu,\epsilon) + \frac{r}{2}\frac{\partial R_\ell(r;\nu,\epsilon)}{\partial r}, \tag{5.29}$$

from which it follows that

$$\operatorname{Tr} R(r;\nu,\epsilon) = 3R_\ell(r;\nu,\epsilon) + r\frac{\partial}{\partial r}R_\ell(r;\nu,\epsilon). \tag{5.30}$$

Hence, we deduce that

$$R_\ell(0;\nu,\epsilon) = \tfrac{1}{3}\operatorname{Tr} R(0;\nu,\epsilon) = \tfrac{2}{3}e(\mu^{\nu,\epsilon}) = \tfrac{2}{3}\gamma(\epsilon\nu)^{1/2} \quad \text{(by (5.18))} \tag{5.31}$$

and, also from (5.30), that

$$\frac{\partial}{\partial r}R_\ell(0;\nu,\epsilon) = \frac{1}{4}\frac{\partial}{\partial r}\operatorname{Tr} R(0;\nu,\epsilon).$$

By the homogeneity of self-similar solutions, the trace of the correlation matrix achieves its maximum at the origin (see (5.8)), so that

$$\frac{\partial}{\partial r}R_\ell(0;\nu,\epsilon) = \frac{1}{4}\frac{\partial}{\partial r}\operatorname{Tr} R(0;\nu,\epsilon) = 0. \tag{5.32}$$

Finally, also from (5.30) we find that

$$\begin{aligned}\frac{\partial^2}{\partial r^2}R_\ell(0;\nu,\epsilon) &= \frac{1}{5}\frac{\partial^2}{\partial r^2}\operatorname{Tr} R(r;\nu,\epsilon)|_{r=0} \\ &= \frac{1}{15}\Delta \operatorname{Tr} R(\mathbf{y};\nu,\epsilon)|_{\mathbf{y}=0} \quad \text{(by (5.28))} \\ &= -\frac{\epsilon}{15\nu} \quad \text{(by (5.3))}.\end{aligned} \tag{5.33}$$

Expanding the longitudinal correlation in Taylor series, we then find, for r small enough,

$$R_\ell(r;\nu,\epsilon) \sim \frac{2}{3}\gamma(\nu\epsilon)^{1/2}\left(1 - \frac{1}{2}\frac{r^2}{10\ell_T^2}\right), \tag{5.34}$$

where

$$\ell_T = \gamma^{1/2}\ell_d = \gamma^{1/2}\left(\frac{\nu^3}{\epsilon}\right)^{1/4} = \frac{\ell_0}{\sqrt{2\gamma}}. \tag{5.35}$$

From the relation (5.34), we see that ℓ_T is related to the conventionally defined Taylor microscale ℓ_T' by a factor of $\sqrt{10}$; that is, $\ell_T' = \ell_T\sqrt{10}$. The conventional Taylor microscale as defined by Taylor [1935, 1937] represents the radius of curvature of the curve $1 - r^2/2\ell_T'$ at $r = 0$. Notice from (5.31) and (5.33) that we can also express the Taylor microscale as

$$\ell_T' = \ell_T\sqrt{10} = \left(\frac{R_\ell(0; \nu, \epsilon)}{-(\partial^2/\partial r^2)R_\ell(0; \nu, \epsilon)}\right)^{1/2}. \tag{5.36}$$

Because the correlation matrix $R_{ij}(\mathbf{y}, \nu, \epsilon)$ is twice continuously differentiable with respect to $\mathbf{y}$, it follows that the longitudinal correlation is twice continuously differentiable with respect to r. One can then check that the expansion (5.34) actually holds in the following sense:

$$R_\ell(r; \nu, \epsilon) = \frac{2}{3}\gamma(\nu\epsilon)^{1/2}\left(1 - \frac{1}{2}\frac{r^2}{10\ell_T^2}\right) + o(r^2) \text{ as } r \to 0, \tag{5.37}$$

where $o(r^2)$ means that the remainder (say, $g(r)$, goes to zero faster than r^2 in the sense that $g(r)/r^2$ goes to zero as r goes to zero.

Higher-Order Two-Point Correlation Function in the Homogeneous and Isotropic Case

Using the same techniques as before, one can easily deduce scaling laws for higher-order two-point correlation functions, for velocity gradient correlations, and so on. Indeed, take for example the triple correlation function

$$K_\ell(r; \nu, \epsilon) = \int \frac{1}{|Q|}\int_Q u_1^2(\mathbf{x})u_1(\mathbf{x} + r\mathbf{e}_1)\,d\mathbf{x}\,d\mu^{\nu,\epsilon}(\mathbf{u}). \tag{5.38}$$

Then, as for (5.12), we find that

$$K_\ell(r; \nu, \epsilon) = \xi^3 K_\ell\left(\frac{r}{\lambda}; \frac{\nu\xi}{\lambda}, \epsilon\lambda\xi^3\right).$$

Hence, letting $\lambda = r$ and $\xi^3\epsilon\lambda = 1$, we obtain

$$K_\ell(r; \nu, \epsilon) = \epsilon r K_\ell\left(1; \left(\frac{\ell_d}{r}\right)^{4/3}; 1\right).$$

For vanishing viscosity (i.e., as $\nu \to 0$), we recover the conventional result for the inertial subrange

$$K_\ell(r; \nu, \epsilon) = C_3(\epsilon r), \tag{5.39}$$

provided the limit $C_3 \equiv K_\ell(1; 0^+, 1)$ exists.

Similarly, any nth-order two-point correlation function transforms under $\sigma_{\xi,\lambda}$ as

$$M_n(r; \nu, \epsilon) = \xi^{-n}M_n\left(\frac{r}{\lambda}; \frac{\nu\xi}{\lambda}, \epsilon\lambda\xi^3\right),$$

whence we recover the scaling law

$$M_n(r; 0^+, \epsilon) = C_n(\epsilon r)^{n/3}$$

(obtained also by Kolmogorov), provided the limit $C_n = M_n(1; 0^+, 1)$ exists.

The Energy Spectrum in the Homogeneous Case

Using the expression for ℓ_d, we can rewrite (5.14) in the form

$$R(\mathbf{y};\nu,\epsilon) = (\epsilon\ell_d)^{2/3} R\left(\frac{\mathbf{y}}{\ell_d};1,1\right), \tag{5.40}$$

which, by writing the inverse of ℓ_d as

$$\kappa_d = \left(\frac{\epsilon}{\nu^3}\right)^{1/4},$$

can be rewritten as

$$R(\mathbf{y};\nu,\epsilon) = \epsilon^{2/3}\kappa_d^{-2/3} R(\kappa_d\mathbf{y};1,1). \tag{5.41}$$

Assume now that $R(\mathbf{y};1,1)$ is such that $\operatorname{Tr} R(\mathbf{y};1,1)$ is the inverse Fourier transform

$$\operatorname{Tr} R(\mathbf{y};1,1) = \int_{\mathbb{R}^3} e^{i\boldsymbol{\kappa}\cdot\mathbf{y}} \mathcal{Q}(\boldsymbol{\kappa})\,d\boldsymbol{\kappa}, \quad \mathbf{y}\in\mathbb{R}^3,$$

as described previously (see (5.9)). Then, using (5.41), we obtain

$$\begin{aligned}\operatorname{Tr} R(\mathbf{y};\nu,\epsilon) &= \epsilon^{2/3}\kappa_d^{-2/3} \operatorname{Tr} R(\kappa_d\mathbf{y};1,1)\\ &= \epsilon^{2/3}\kappa_d^{-2/3} \int_{\mathbb{R}^3} e^{i\kappa_d\boldsymbol{\kappa}\cdot\mathbf{y}} \mathcal{Q}(\boldsymbol{\kappa})\,d\boldsymbol{\kappa}\\ &= \epsilon^{2/3}\kappa_d^{-11/3} \int_{\mathbb{R}^3} e^{i\boldsymbol{\kappa}\cdot\mathbf{y}} \mathcal{Q}\left(\frac{\boldsymbol{\kappa}}{\kappa_d}\right) d\boldsymbol{\kappa}.\end{aligned}$$

Hence, $\mu^{\nu,\epsilon}$ has the energy spectrum $\mathcal{S}(\kappa;\nu,\epsilon)$ (see (5.10)) given by

$$\begin{aligned}\mathcal{S}(\kappa;\nu,\epsilon) &= \int_{|\boldsymbol{\kappa}|=\kappa} \epsilon^{2/3}\kappa_d^{-11/3} \mathcal{Q}\left(\frac{\boldsymbol{\kappa}}{\kappa_d}\right) d\Sigma(\boldsymbol{\kappa})\\ &= \epsilon^{2/3}\kappa_d^{-5/3} \int_{|\boldsymbol{\kappa}|=\kappa/\kappa_d} \mathcal{Q}(\boldsymbol{\kappa})\,d\Sigma(\boldsymbol{\kappa})\\ &= \epsilon^{2/3}\kappa_d^{-5/3} \mathcal{S}\left(\frac{\kappa}{\kappa_d}\right).\end{aligned}$$

The kinetic energy $e(\mu^{\nu,\epsilon})$ can be written as

$$e(\mu^{\nu,\epsilon}) = \frac{1}{2}\operatorname{Tr} R(0;\nu,\epsilon) = \frac{1}{2}\int_0^\infty \mathcal{S}(\kappa;\nu,\epsilon)\,d\kappa.$$

Finally, by setting

$$F(\kappa) = \kappa^{5/3}\mathcal{S}(\kappa), \quad \kappa > 0,$$

the energy spectrum $\mathcal{S}(\kappa;\nu,\epsilon)$ can be written in the form

$$\mathcal{S}(\kappa;\nu,\epsilon) = \epsilon^{2/3}\kappa^{-5/3} F\left(\frac{\kappa}{\kappa_d}\right), \quad \kappa > 0,$$

which is the usual form of the $-5/3$ Kolmogorov law for the energy spectrum (Kolmogorov [1941a,b], Landau and Lifshitz [1971]).

We may summarize the preceding results as follows.

Theorem 5.2 *Let $\{\mu^{\nu,\epsilon}\}_{\nu,\epsilon>0}$ be a self-similar homogeneous statistical solution of the Navier–Stokes equations. If $\mu^{1,1}$ has an energy spectrum $\mathcal{S} = \mathcal{S}(\kappa)$ then, for all $\nu, \epsilon > 0$, $\mu^{\nu,\epsilon}$ has an energy spectrum $\mathcal{S}(\kappa, \nu, \epsilon)$ given by*

$$\mathcal{S}(\kappa; \nu, \epsilon) = \epsilon^{2/3}\kappa^{-5/3}F\left(\frac{\kappa}{\kappa_d}\right), \quad \kappa > 0, \tag{5.42}$$

where

$$\kappa_d = \left(\frac{\epsilon}{\nu^3}\right)^{1/4} \tag{5.43}$$

and

$$F(\kappa) = \kappa^{5/3}\mathcal{S}(\kappa), \quad \kappa > 0. \tag{5.44}$$

5.2 The von Karman–Howarth–Dryden Equation

An equation for the self-similar evolution of the two-point double longitudinal correlation R_ℓ, derived by von Karman and Howarth [1938], reads

$$\frac{\partial R_\ell}{\partial t} - \frac{2\nu}{r^4}\frac{\partial}{\partial r}\left(r^4\frac{\partial R_\ell}{\partial r}\right) = \frac{1}{r^4}\frac{\partial}{\partial r}(r^4 K_\ell), \tag{5.45}$$

where K_ℓ is the longitudinal two-point triple correlation

$$K_\ell(r; \nu, \epsilon) = \langle u_1(\mathbf{x})^2 u_1(\mathbf{x} + r\mathbf{e}_1)\rangle$$

and where $\langle\cdot\rangle$ denotes an appropriate ensemble average. In this case, R_ℓ also denotes the longitudinal correlation with respect to some appropriate ensemble average, as in the conventional theory of turbulence (not necessarily as we defined it here). Subsequently, Dryden [1943] showed that if the double and triple correlations evolve in a self-similar fashion,

$$R_\ell(r; \nu, \epsilon) = e(t)b\left(\frac{r}{L(t)}\right) \quad \text{and} \quad K_\ell(r; \nu, \epsilon) = e(t)^{3/2}k\left(\frac{r}{L(t)}\right),$$

then we arrive at the equation

$$Ab + \frac{1}{2}A\psi\frac{db}{d\psi} + \frac{2}{\gamma}\frac{1}{\psi^4}\frac{d}{d\psi}\left(\psi^4\frac{db}{d\psi}\right) + \frac{1}{\psi^4}\frac{d}{d\psi}(\psi^4 k) = 0,$$

where $\psi = r/L(t)$. This equation is called the vKHD equation.

We aim now to obtain, from the von Karman–Howarth equation (5.45), a slight modification of the vKHD equation by (a) using the scaling laws obtained previously with the self-similar homogeneous statistical solutions and (b) assuming isotropy of the appropriate correlations. Then, from this modified vKHD equation, we obtain a slight correction to the conventionally accepted behavior $K_\ell \sim \epsilon r$ (see (5.39)) of the triple correlation in a certain region of the spectrum identified here as the inertial subrange.

Consider a self-similar homogeneous statistical solution $\{\mu^{\nu,\epsilon}\}_{\nu,\epsilon}$ of the Navier–Stokes equations. According to Proposition 4.3, it evolves in time as $\{\mu^{\nu,\epsilon(t)}\}_{t\geq 0}$ with $\epsilon(t)$ given by (4.30). We can then consider the spatial correlation associated with each measure $\mu^{\nu,\epsilon(t)}$. By assuming isotropy, the two-point correlations depend only on the distance $r = |\mathbf{y}|$ between the two points. In particular, we can consider the double and triple longitudinal correlations $R_\ell(r;\nu,\epsilon) = R_\ell(r;\nu,\epsilon(t))$ and $K_\ell(r;\nu,\epsilon) = K_\ell(r;\nu,\epsilon(t))$, as defined previously.

Consider then the structure function

$$D_\ell(r;\nu,\epsilon) = R_\ell(r;\nu,\epsilon) - R_\ell(0;\nu,\epsilon). \tag{5.46}$$

Assuming the von Karman–Howarth equation (5.45) holds for the evolution of R_ℓ, we find that

$$\frac{d}{dt}R_\ell(0;\nu,\epsilon) + \frac{\partial D_\ell}{\partial t} - \frac{1}{r^4}\frac{\partial}{\partial r}\left(r^4\left(2\nu\frac{\partial D_\ell}{\partial r} + K_\ell\right)\right) = 0.$$

But, for isotropic turbulence, $R(0;\nu,\epsilon) = 2e(t)/3$ with

$$\frac{dR(0;\nu,\epsilon)}{dt} = -\frac{2}{3}\epsilon(t).$$

Hence,

$$-\frac{2}{3}\epsilon + \frac{\partial D_\ell}{\partial t} - \frac{1}{r^4}\frac{\partial}{\partial r}\left(r^4\left(2\nu\frac{\partial D_\ell}{\partial r} + K_\ell\right)\right) = 0. \tag{5.47}$$

From (5.12) with $\nu\xi/\lambda = 1$ and $\xi^3\lambda\epsilon = \gamma^2$, we find that

$$D_\ell(r;\nu,\epsilon) = \left(\frac{\epsilon\nu}{\gamma^2}\right)^{1/2} D_\ell(\psi;1,\gamma^2) \tag{5.48}$$

and

$$K_\ell(r;\nu,\epsilon) = \left(\frac{\epsilon\nu}{\gamma^2}\right)^{3/4} K_\ell(\psi;1,\gamma^2), \tag{5.49}$$

where now

$$\psi = \frac{r}{\ell_T} = \frac{r\epsilon(t)^{1/4}}{\gamma^{1/2}\nu^{3/4}}.$$

Using (4.31), it follows that

$$\frac{d}{dt} = -\frac{\epsilon^{1/2}}{2\gamma\nu^{1/2}}\psi\frac{d}{d\psi}, \qquad \frac{d}{dr} = \frac{1}{\ell_T}\frac{d}{d\psi}.$$

Therefore, from (5.48) and (5.49) and using again (4.31), we find that

$$\begin{aligned}
\frac{\partial}{\partial t}D_\ell(r;\nu,\epsilon) &= \frac{\partial}{\partial t}\left(\left(\frac{\epsilon\nu}{\gamma^2}\right)^{1/2} D_\ell(\psi;1,\gamma^2)\right) \\
&= D_\ell(\psi;1,\gamma^2)\frac{\partial}{\partial t}\left(\frac{\epsilon\nu}{\gamma^2}\right)^{1/2} + \left(\frac{\epsilon\nu}{\gamma^2}\right)^{1/2}\frac{\partial}{\partial t}D_\ell(\psi;1,\gamma^2) \\
&= -\frac{\epsilon}{\gamma^2}D_\ell(\psi;1,\gamma^2) - \frac{\epsilon}{2\gamma^2}\psi\frac{d}{d\psi}D_\ell(\psi;1,\gamma^2)
\end{aligned}$$

and

$$\frac{1}{r^4}\frac{\partial}{\partial r}\left(r^4\left(2\nu\frac{\partial}{\partial r}D_\ell(r;\nu,\epsilon)+K_\ell(r;\nu,\epsilon)\right)\right)$$
$$=\frac{\epsilon}{\gamma^2}\frac{1}{\psi^4}\frac{d}{d\psi}\left(\psi^4\left(2\frac{d}{d\psi}D_\ell(\psi;1,\gamma^2)+K(\psi;1,\gamma^2)\right)\right).$$

Therefore, we can rewrite (5.47) in the form

$$\frac{2\gamma^2}{3}+\tilde{D}+\frac{\psi}{2}\frac{d\tilde{D}}{d\psi}+\frac{1}{\psi^4}\frac{d}{d\psi}\left(\psi^4\left(2\frac{d\tilde{D}}{d\psi}+\tilde{K}\right)\right)=0, \tag{5.50}$$

where (for notational simplicity) we have set $\tilde{D}(\psi)=D_\ell(\psi;1,\gamma^2)$ and $\tilde{K}(\psi)=K_\ell(\psi;1,\gamma^2)$.

By another similarity transformation, with $\lambda=\psi$ and $\xi=(\gamma^2\psi)^{-1/3}$ in (5.12), we find that

$$\tilde{D}=D_\ell(\psi;1,\gamma^2)=(\gamma^2\psi)^{2/3}D_\ell(1;(\gamma\psi^2)^{-2/3},1). \tag{5.51}$$

Since we are interested in the limiting case of large Reynolds number (i.e., for large γ), we consider the Kolmogorov-type constant

$$\bar{C}_K=-D_\ell(1;0^+,1)=-\lim_{\nu\to 0^+}D_\ell(1;\nu,1)$$

and assume the power law

$$\tilde{D}=-\bar{C}_K(\gamma^2\psi)^{2/3}. \tag{5.52}$$

Here, we neglect the effects of intermittency. Substituting for $\tilde{D}$ in (5.50), we obtain the equation

$$\frac{2\gamma^2}{3}-\frac{4}{3}\bar{C}_K\gamma^{4/3}\psi^{2/3}-\frac{44}{9}\bar{C}_K\gamma^{4/3}\frac{1}{\psi^{4/3}}+\frac{1}{\psi^4}\frac{d}{d\psi}(\psi^4\tilde{K})=0.$$

Solving for $\tilde{K}$ yields

$$\tilde{K}=\frac{c}{\psi^4}-\frac{2}{15}\gamma^2\psi+\frac{4}{17}\bar{C}_K\gamma^{4/3}\psi^{5/3}+\frac{4}{3}\bar{C}_K\gamma^{4/3}\frac{1}{\psi^{1/3}},$$

where c is an undetermined constant. Although we have means for fixing the constant $\bar{C}_K$ based on the observed value of the Kolmogorov constant, we have no obvious means for determining c. Therefore, in order to proceed further we set $c=0$, thus reducing the strength of the singularity of the triple correlation at $\psi=0$ $(r=0)$ and allowing us to determine the inertial range $[\kappa_0,\kappa_d]$ solely in terms of the data of the problem. With $c=0$ we obtain

$$\tilde{K}=-\frac{2}{15}\gamma^2\psi\left(1-\frac{30}{17}\bar{C}_K\left(\frac{\psi}{\gamma}\right)^{2/3}-10\bar{C}_K\left(\frac{1}{\gamma\psi^2}\right)^{2/3}\right).$$

By using (5.49) again, we can return to the physical variables and find

$$K_\ell(r;\nu,\epsilon)=-\frac{2}{15}\epsilon r\left(1-10\bar{C}_K\left(\frac{\ell_T}{\gamma^{1/2}r}\right)^{4/3}-\frac{30}{17}\bar{C}_K\left(\frac{r}{\gamma\ell_T}\right)^{2/3}\right).$$

Recall now that $\ell_T=\gamma^{1/2}\ell_d$ and that $\ell_0=\sqrt{2}\gamma\ell_T$. Hence, we obtain the following approximate relation for K_ℓ:

$$K_\ell(r;\nu,\epsilon) = -\frac{2}{15}\epsilon r\left(1 - 10\bar{C}_K\left(\frac{\ell_d}{r}\right)^{4/3} - \frac{30}{17}2^{1/3}\bar{C}_K\left(\frac{r}{\ell_0}\right)^{2/3}\right). \tag{5.53}$$

Notice that, since we are considering γ large, we have $\ell_0 \gg \ell_d$. Moreover, we have recovered in the leading term the well-known result that $K_\ell = -2\epsilon r/15$. In addition, we have also obtained a correction factor for this power law within the inertial subrange $\ell_0 \gg r \gg \ell_d$.

It is enlightening to analyze this correction factor, which we denote by

$$g(r) = 1 - 10\bar{C}_K\left(\frac{\ell_d}{r}\right)^{4/3} - \frac{30}{17}2^{1/3}\bar{C}_K\left(\frac{r}{\ell_0}\right)^{2/3}. \tag{5.54}$$

We can also write g in the form

$$g(r) = 1 - q\left(\frac{30}{17}\left(\frac{r}{\ell_T}\right)^{2/3} + 10\left(\frac{\ell_T}{r}\right)^{4/3}\right) \tag{5.55}$$

with $q = \bar{C}_K/\gamma^{2/3}$. The maximum of g occurs at $r_{\max} = \ell_T\sqrt{34/3}$, which is slightly above the classical Taylor microscale $\ell_T\sqrt{10}$. The maximum value is

$$g(r_{\max}) = 1 - 30\bar{C}_K\left(\frac{3}{34\gamma}\right)^{2/3} \approx 1 \quad \text{for } \gamma \gg 1.$$

It should be clear from the form of g that, for a sufficiently large Reynolds number, there exists a subinterval containing $r_{\max}$ within which $g \approx 1$. In fact, bearing in mind that $g(r)$ goes to negative infinity when r either decreases to zero or increases to infinity, it is natural to identify the region within which $0 < g < 1$ as the actual inertial subrange. Notice that, for a sufficiently large Reynolds number, the conventional endpoints ℓ_0 and ℓ_d of the equilibrium range are such that $g < 0$. Indeed, assuming the value $\bar{C}_K = 2.4$ for the Kolmogorov-type constant, we have

$$g(\ell_0) = 1 - \frac{30}{17}2^{1/3}\bar{C}_K - \frac{10}{\gamma^2 2^{1/3}}\bar{C}_K \approx 1 - \frac{30}{17}2^{1/3}\bar{C}_K \approx -4.3$$

and

$$g(\ell_d) = 1 - \frac{30\bar{C}_K}{17\gamma} - 10\bar{C}_K \approx 1 - 10\bar{C}_K = -23.$$

We can easily identify the endpoints where $g \approx 0$ by the fact that, at those endpoints, one of the terms in the expression for g is negligible for large enough γ. With that in mind, one can then easily check that $g \approx 0$ for $\tilde{\ell}_0 = \ell_0(17/30\bar{C}_K)^{3/2} = \ell_0/8.7$ and $\tilde{\ell}_d = \ell_d(10\bar{C}_K)^{3/4} = 10.8\ell_d$. We may now identify the interval $\tilde{\ell}_0 > r > \tilde{\ell}_d$ as the actual inertial subrange. It seems remarkable that parameters defining the equilibrium range – including the outer length ℓ_0 and exhibiting the separation of scales $\ell_0 \gg \ell_d$ – have arisen so naturally in the estimate of the triple correlation.

We may go further in the analysis of the correction term g and attempt to find conditions on the Reynolds number for the existence of the inertial subrange. For this, we recall the relation $\mathrm{Re} = \sqrt{2}\gamma^2$ between γ and the Reynolds number. Then,

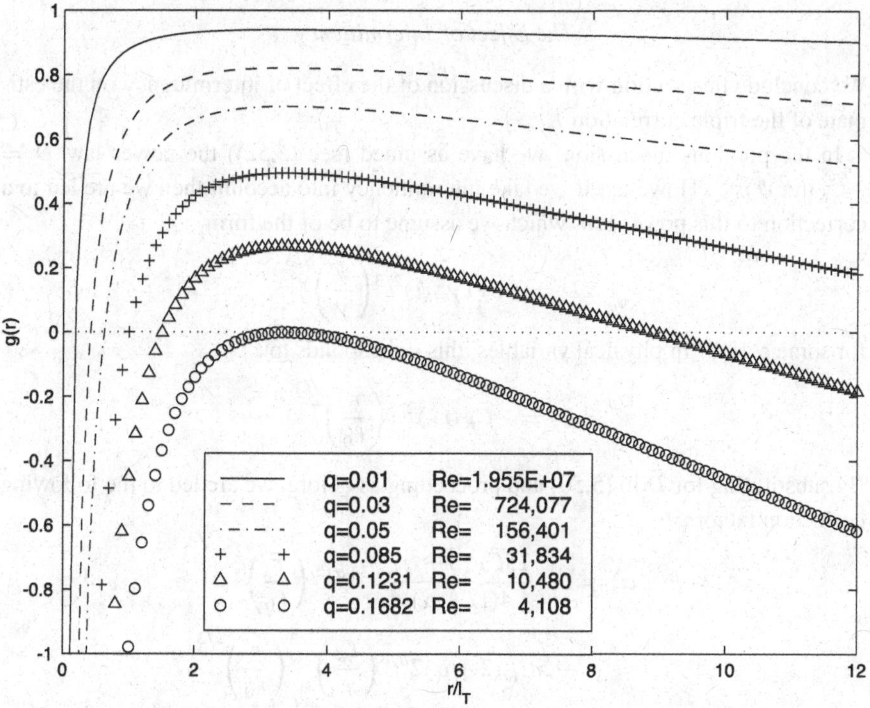

Figure 1. The correction factor g for various values of $q = \bar{C}_K/\gamma^{2/3} = \bar{C}_K 2^{1/3}/\mathrm{Re}^{1/3}$. Note that for $q \leq 0.05$ (i.e., for large values of the Reynolds number), $g(r)$ departs from constant value for $r/L_{\mathrm{Taylor}} < 1$, suggesting that the effects of viscosity cannot be neglected for $1/L_{\mathrm{Taylor}} < k < k_c$ (see Foias et al. [2001b]).

$q = \bar{C}_K/\gamma^{2/3} = \bar{C}_K 2^{1/3}/\mathrm{Re}^{1/3}$. In what follows, we take again the value $\bar{C}_K = 2.4$ and refer the reader to Figure 1 for plots of g for selected values of q.

There are four critical values of q, corresponding to four critical values of the Reynolds number. First, at $q_1 = 0.1682$, which corresponds to Re $\sim$4,108, we have $g(r_{\max}) = 0$ with $g < 0$ for $q > q_1$. Hence, according to our interpretation, there is no possibility for an inertial subrange below this Reynolds number. Next, at the slightly lower value $q_2 = 0.1679$, corresponding to Re $\sim$ 4,130, we have $g(\ell_T\sqrt{10}) = 0$ and so, for $q < q_2$, it follows that $g(\ell_T\sqrt{10}) > 0$ and the Taylor microlength lies inside the inertial subrange. Since the Taylor microlength must lie inside the inertial subrange, it is now indeed plausible that the inertial subrange exists. At the third critical value $q_3 = 0.1231$, corresponding to Re $\sim$ 10,480, we have $g = 0$ at the inflection point $r = 8.9\ell_T$. Therefore, as the Reynolds number increases, the maximum of g flattens out rapidly, lending credence to its presumed near-constant value in the inertial subrange. Finally, at the value $q_4 = 0.085$, corresponding to Re $= 31{,}834$, we have $g(\ell_T) = 0$; for $q < q_4$, there is an extensive range of values of r within which g is nearly constant and close to unity, indicating the presence of a well-defined inertial subrange.

The Effect of Intermittency

We conclude this section with a discussion of the effect of intermittency on the estimate of the triple correlation K_ℓ.

In the previous discussion, we have assumed (see (5.52)) the power law $\tilde{D} = -\bar{C}_K(\gamma^2\psi)^{2/3}$. However, if we take intermittency into account then we are led to a correction to this power law, which we assume to be of the form

$$\tilde{D} = -\bar{C}_K(\gamma^2\psi)^{2/3}\left(\frac{\psi}{\gamma}\right)^{\alpha/3}$$

for some $\alpha > 0$. In physical variables, this corresponds to

$$D_\ell = -\bar{C}_K(r\epsilon)^{2/3}\left(\frac{r}{\ell_0}\right)^{\alpha/3}.$$

By substituting for $\tilde{D}$ in (5.50) and proceeding as before, we are led to the following correction factor:

$$h(r,\alpha) = 1 - \frac{15\bar{C}_K(8+\alpha)}{4(17+\alpha)}2^{(2+\alpha)/6}\left(\frac{r}{\ell_0}\right)^{(2+\alpha)/3} - 5\bar{C}_K(2+\alpha)2^{\alpha/6}\left(\frac{\ell_d}{r}\right)^{2/3}\left(\frac{r}{\ell_0}\right)^{\alpha/3}.$$

For the value of h at $r = \ell_d$ there is no significant change, but at $r = \ell_d$ we have

$$h(\ell_d,\alpha) = 1 - \frac{15(8+\alpha)}{4(17+\alpha)}\bar{C}_K\frac{1}{\gamma^{(2+\alpha)/2}} - 5(2+\alpha)\bar{C}_K\frac{1}{\gamma^{\alpha/2}};$$

hence, for $\alpha < 1$ and large Reynolds numbers, intermittency causes the correction factor to be close to unity throughout most of the inertial subrange.

5.3 Intermittency

We discuss now some aspects concerning the nature of intermittency. We consider self-similar homogeous statistical solutions and show that, in homogeneous isotropic turbulence, intermittency is mostly caused by a kinematic constraint on the flow imposed by the Poincaré inequality.

Denote the local volume average over a cube Q by

$$\mathbf{u}_Q = \frac{1}{|Q|}\int_Q \mathbf{u}\,d\mathbf{x},$$

and let L be the linear dimension of Q. Poincaré's inequality asserts that

$$|\mathbf{u}-\mathbf{u}_Q|_Q^2 \leq \frac{L^2}{6}\|\mathbf{u}\|_Q^2.$$

On taking the ensemble average with respect to a $\mu^{\nu,\epsilon}$ and using (5.18), we obtain the following relation between the local average variance and the local average kinetic energy:

$$\int |\mathbf{u} - \mathbf{u}_Q|_Q^2 \, d\mu^{\nu,\epsilon}(\mathbf{u}) \leq \frac{L^2}{6} \int \|\mathbf{u}\|_Q^2 \, d\mu^{\nu,\epsilon}(\mathbf{u}) = \frac{L^2}{6} E(\mu^{\nu,\epsilon}) = \frac{L^2 \epsilon}{6\nu}$$

$$= 2\zeta^2 \gamma(\epsilon\nu)^2 = 2\zeta^2 \int |\mathbf{u}|_Q^2 \, d\mu^{\nu,\epsilon}(\mathbf{u}), \tag{5.56}$$

where

$$\zeta = \frac{L}{2\ell_T \sqrt{3}}.$$

Now, since

$$|\mathbf{u} - \mathbf{u}_Q|_Q^2 = |\mathbf{u}|_Q^2 - |\mathbf{u}_Q|^2, \tag{5.57}$$

we find for $\zeta < 1$ that

$$\int |\mathbf{u} - \mathbf{u}_Q|_Q^2 \, d\mu^{\nu,\epsilon}(\mathbf{u}) \leq \frac{\zeta^2}{1 - \zeta^2} \int |\mathbf{u}_Q|^2 \, d\mu^{\nu,\epsilon}(\mathbf{u}). \tag{5.58}$$

This relation means that, for very small averaging volumes (i.e., for $L \ll \ell_T$), the local average variance of the velocity field is only a small fraction of the square of the local mean velocity. On the other hand, when the linear size of the averaging volume approaches the Taylor microscale, the variance may become comparable with (or even exceed) the square of the local mean velocity. This property of the variance being dependent on the the local mean velocity, and hence on the averaging volume, is reminiscent of intermittency in turbulent flows. In order to explore this idea further, we estimate now the fraction of the flow moving with a velocity that is significantly different from the local mean velocity.

Consider a given cube Q. For each possible flow $\mathbf{u}$, denote by $Q_\delta(\mathbf{u})$ the portion of the flow within Q in which the difference of the velocity from the mean velocity exceeds some fraction δ of the mean velocity:

$$Q_\delta(\mathbf{u}) = \{\mathbf{x} \in Q;\ |\mathbf{u}(\mathbf{x}) - \mathbf{u}_Q| > \delta |\mathbf{u}_Q|\}.$$

Similarly, of all the flows $\mathbf{u}$ distributed according to the probability distribution $\mu^{\nu,\epsilon}$, we may consider those for which the portion $Q_\delta(\mathbf{u})$ exceeds in volume a fraction η of the averaging volume; that is, we may consider the family B of flows defined by

$$B = \{\mathbf{u} \in H;\ |Q_\delta(\mathbf{u})| > \eta |Q|\}.$$

For notational simplicity we may consider, for a given function $\Phi = \Phi(\mathbf{u})$, the ensemble average over B,

$$\langle \Phi(\mathbf{u}) \rangle_B = \int_B \Phi(\mathbf{u}) \, d\mu^{\nu,\epsilon}(\mathbf{u}),$$

as well as that over the complement G of B,

$$\langle \Phi(\mathbf{u}) \rangle_G = \int_G \Phi(\mathbf{u}) \, d\mu^{\nu,\epsilon}(\mathbf{u}).$$

We may also set $\langle \Phi(\mathbf{u}) \rangle = \langle \Phi(\mathbf{u}) \rangle_B + \langle \Phi(\mathbf{u}) \rangle_G$. This yields

$$\langle |\mathbf{u}-\mathbf{u}_Q|_Q^2\rangle_B > \left\langle \frac{1}{|Q|}\int_{Q_\delta(\mathbf{u})} |\mathbf{u}(\mathbf{x})-\mathbf{u}_Q|^2\,d\mathbf{x}\right\rangle_B > \delta^2 \frac{|Q_\delta(\mathbf{u})|}{|Q|}|\mathbf{u}_Q|^2$$
$$> \eta\delta^2\langle |\mathbf{u}_Q|^2\rangle_B. \tag{5.59}$$

Hence, using (5.58), we obtain

$$\eta\delta^2\langle |\mathbf{u}_Q|^2\rangle_B \le \langle |\mathbf{u}-\mathbf{u}_Q|_Q^2\rangle_B \le \langle |\mathbf{u}-\mathbf{u}_Q|_Q^2\rangle$$
$$\le \frac{\zeta^2}{1-\zeta^2}\langle |\mathbf{u}_Q|^2\rangle. \tag{5.60}$$

Now, using (5.57), (5.59), and (5.56), we find

$$\langle |\mathbf{u}|_Q^2\rangle_B = \langle |\mathbf{u}-\mathbf{u}_Q|_Q^2\rangle_B + \langle |\mathbf{u}_Q|^2\rangle_B \le \left(1+\frac{1}{\eta\delta^2}\right)\langle |\mathbf{u}-\mathbf{u}_Q|_Q^2\rangle_B$$
$$\le \left(1+\frac{1}{\eta\delta^2}\right)\langle |\mathbf{u}-\mathbf{u}_Q|_Q^2\rangle \le \left(1+\frac{1}{\eta\delta^2}\right)\zeta^2\langle |\mathbf{u}_Q|^2\rangle_B,$$

which implies that

$$\frac{\langle |\mathbf{u}|_Q^2\rangle_B}{\langle |\mathbf{u}|_Q^2\rangle} \le \zeta^2\left(1+\frac{1}{\eta\delta^2}\right). \tag{5.61}$$

This inequality means, loosely speaking, that when the averaging volume is small – that is, when $(L/\ell_T)^2 \ll 12\eta\delta^2/(1+\eta\delta^2)$ – most of the flow energy in that volume is carried by a set of velocities close to the local mean $\mathbf{u}_Q$.

There is another interesting physical interpretation for (5.61). Since departures from the local mean velocity occur only in a limited fraction of the averaging volume and since large velocity gradients occur at those points where large velocity changes occur, it follows that, for $L \ll \ell_T$, energy dissipation is distributed sparsely in the averaging volume. This phenomenon is approximated by the conventional phenomenological view of intermittency, which asserts that dissipation is intermittent throughout the inertial subrange. However, we have arrived at this result by examining rigorously the properties of some homogeneous statistical solutions of the Navier–Stokes equations.

As a final comment in this section, we draw attention to the pervasive presence of the Taylor microscale in the estimate of intermittency discussed here. It suggests the possibility that the intermittency correction to the $2/3$ power law should contain ℓ_T instead of the customary ℓ_0, that is, the correction should be of the form $(r/\ell_T)^{\alpha/3}$. Such intermittency scaling would introduce a slight Reynolds number dependence that may be barely perceptible in the dissipation correlation. For instance, with $\alpha \sim 0.4$ and the outer length fixed, such a dependence would vary approximately as $\mathrm{Re}^{1/5}$. On the other hand, in this event, two-point structure functions of all orders – when evaluated at $r = \ell_T$ – should be essentially independent of the intermittency exponent.

6 Some Concluding Remarks

An unstated underlying premise of this work is the conviction that a valid mathematical model of a physical phenomenon contains, almost by definition, all the properties and aspects of that phenomenon. To the extent that Navier–Stokes equations constitute a comprehensive model of incompressible viscous fluid flow, much of the current understanding of turbulent flow must be determinable from those equations without invoking additional, ad hoc assumptions. We believe that we have amply illustrated this point in the preceding pages of this work.

Moreover, we have shown that additional insights into the nature of turbulence can be gained from studying with adequate rigor the intrinsic properties of the underlying equations, some of which cannot be arrived at by purely intuitive considerations. For example, as we have shown rigorously, the attractors characterizing turbulent regimes have well-defined statistical properties – including the heretofore merely postulated equivalence of time and ensemble averages. Although the explicit forms of those statistics are still unavailable, enough is known about their properties to yield much practical knowledge about turbulent flows.

Of course, our work lays no claim to having solved all the puzzles presented by turbulent flows. Much remains to be done. Most notably, the existence of regular solutions of the Navier–Stokes equations in three dimensions that are consistent with the empirical evidence is still undemonstrated: at the time this book was completed, this mathematical problem has been recognized by the Clay Foundation as one among seven major mathematical problems for the twenty-first century.[11] Only after this flaw in NSE theory is resolved can the full resources of functional analysis be brought to bear on this problem. Many of our results are at best in the nature of sufficient conditions, and this is likewise due to limitations on the available tools for analysis – notably the Sobolev inequalities. Although we have succeeded in relating some of the more abstract mathematical results to the known (or expected) physical phenomena, some esoteric attributes of the NSE solutions established here continue to escape our full understanding. For example, what is the physical meaning and what are the practical consequences of our discovery that the stationary statistical measures of Navier–Stokes equations are accretive? Beyond this specific example, readers will certainly have found other instances that require further research to advance our understanding of turbulent flows. We fervently hope that the tools presented in this monograph will aid in these efforts.

We should also point out that – on the mathematical side and, to some extent, on the physical side – the methods presented in this work do pertain to some additional problems in fluid mechanics but are inappropriate (or insufficient) for others. For example, when the Mach number of a flow is not sufficiently small, incompressibility effects must be taken into account. Generally speaking, the tools and methods presented in this book do not apply to the compressible Navier–Stokes equations, which briefly appeared in (1.5) in Chapter I; for mathematical theory of the compressible NSE, see Hoff [1992a,b, 1995], Kanel [1968], Kazhikhov and Shelukhin [1977], Kazhikhov and Vaĭgant [1995], P. L. Lions [1998], Matsumura and Nishida [1983, 1989], Matsumura and Nishihara [1992], and the references therein. Also related to compressible flows

[11] See e.g. the forthcoming book to be published by the American Mathematical Society or ⟨www.clay-math.org⟩.

but not pertaining to the methods in this book are the equations for transonic, supersonic, and hypersonic flows.

On the other hand, a number of tools and methods presented in this book extend to such phenomena as thermohydraulics in the context of the Boussinesq approximation, magnetohydrodynamics (incompressible case), flows on a 2-dimensional manifold, and certain forms of the NSE with a viscosity coefficient depending on the symmetric part of the gradient of the velocity, as in the Smagorinsky modeling of turbulence (see e.g. J. L. Lions [1959], Smagorinsky [1963]). Some of our methods apply also to the primitive equations of the ocean, of the atmosphere, and of the coupled atmosphere–ocean (see e.g. Lions, Temam, and Wang [1992a,b, 1995]). For the latter, although the atmosphere is made up of a compressible fluid, it is noteworthy that (thanks to a remarkable, and classical, change of variables whereby the vertical variable is replaced by the pressure) the equation of conservation of mass becomes similar to the incompressibility equation $\operatorname{div} u = 0$ and so the methods of this book do apply to some extent – but of course not fully owing to a very different physical context.

Appendix A Proofs of Technical Results in Chapter V

We now present the more technical (mathematical) proofs omitted from the previous sections of this chapter.

A.1 Statistical Solutions on Bounded Domains

First of all, concerning Definition 1.2, let us show that if $\{\mu_t\}_{t\in[0,T]}$ is a family of probability measures on $H = H_{\text{nsp}}$ satisfying (1.11) – that is, if

$$t \mapsto \int \varphi(\mathbf{u})\, d\mu_t(\mathbf{u})$$

is measurable on $[0, T)$ for every φ in $\mathcal{C}_b(H)$ – then the functions

$$t \mapsto \int |\mathbf{u}|^2\, d\mu_t(\mathbf{u}) \quad \text{and} \quad t \mapsto \int \|\mathbf{u}\|^2\, d\mu_t(\mathbf{u})$$

are also measurable. Indeed, the functionals $\mathbf{u} \mapsto |\mathbf{u}|^2$ and $\mathbf{u} \mapsto \|\mathbf{u}\|^2$ do not belong to $\mathcal{C}_b(H)$; however, the functions just displayed are measurable because, by the monotone convergence theorem (Theorem A.2 in Chapter IV), they are the pointwise limit (for each t), as n goes to infinity, of the measurable functionals

$$t \mapsto \int \min\{n, |\mathbf{u}|^2\}\, d\mu_t(\mathbf{u}) \quad \text{and} \quad t \mapsto \int \min\{n, \|P_n\mathbf{u}\|^2\}\, d\mu_t(\mathbf{u}),$$

respectively. Here we have set, for convenience, $\|\mathbf{u}\| = \infty$ for all $\mathbf{u}$ in H but not in V, and P_n is the Galerkin projector on the space spanned by the eigenfunctions $\mathbf{w}_1, \dots, \mathbf{w}_n$ of the corresponding Stokes operator.

If, moreover, the energy inequality (1.9) is valid and the initial kinetic energy is finite (condition (1.10)), then one can deduce that

$$t \mapsto \int \|\mathbf{u}\|^2 \, d\mu_t(\mathbf{u})$$

is integrable. This implies that the set $H \setminus V$ of points in H but not in V has μ_t-measure zero, that is, $\mu_t(H \setminus V) = 0$ for almost every t. In other words, for almost every t, μ_t is carried by V and $\|\mathbf{u}\| = \infty$ only on a set of μ_t-measure zero. Then, for any test functional $\Phi \in \mathcal{T}_{\text{nsp}}$,

$$\mathbf{u} \mapsto (\nu A\mathbf{u} + B(\mathbf{u}) - \mathbf{f}(t), \Phi'(\mathbf{u}))$$

is continuous in V (for almost every t, since $\mathbf{f} \in L^2(0, T; H)$), and

$$\begin{aligned}
&|(\nu A\mathbf{u} + B(\mathbf{u}) - \mathbf{f}(t), \Phi'(\mathbf{u}))| \\
&\quad = |\nu((\mathbf{u}\Phi'(\mathbf{u}))) + b(\mathbf{u}, \mathbf{u}, \Phi'(\mathbf{u})) - (\mathbf{f}(t), \Phi'(\mathbf{u}))| \\
&\quad \le \nu\|\mathbf{u}\|\|\Phi'(\mathbf{u})\| + c_1|\mathbf{u}|^{1/2}\|\mathbf{u}\|^{3/2}\|\Phi'(u)\| + |\mathbf{f}(t)||\Phi'(\mathbf{u})| \\
&\quad \le (\nu\|\mathbf{u}\| + c_1\lambda_1^{1/4}\|\mathbf{u}\|^2 + \lambda_1^{1/2}|\mathbf{f}(t)|) \sup_{\mathbf{u}\in V} \|\Phi'(\mathbf{u})\|,
\end{aligned}$$

where we used the inequalities (0.16) and (0.17). Thus, since $\mu_t(H \setminus V) = 0$ almost everywhere in t, it follows from (1.7) and the foregoing estimate for $\mathbf{u} \in V$ that

$$t \mapsto \int (\nu A\mathbf{u} + B(\mathbf{u}) - \mathbf{f}(t), \Phi'(\mathbf{u})) \, d\mu_t(\mathbf{u})$$

makes sense and is integrable. Then we can write the integral form of the Liouville equation (1.5):

$$\int \Phi(\mathbf{u}) \, d\mu_t(\mathbf{u}) = \int \Phi(\mathbf{u}) \, d\mu_0(\mathbf{u}) + \int_0^t \int (\mathbf{F}(s, \mathbf{u}), \Phi'(\mathbf{u})) \, d\mu_s(\mathbf{u}) \, ds.$$

Proof of Theorem 1.1. First, we consider the case that the initial probability distribution μ_0 has a bounded support in H – that is, $\operatorname{supp} \mu_0 \subset B_H(R_0)$ for some $R_0 > 0$, where $B_H(R_0)$ denotes the closed ball in H of radius R_0 and centered at the origin. The general case will be obtained by writing $\mu_0 = \sum_{k=1}^{\infty} \mu_0^{(k)}$, where each $\mu_0^{(k)}$, defined by $\mu_0^{(k)}(E) = \mu_0\big(E \cap (B_H(k) \setminus B_H(k-1))\big)$ for any measurable set E in H, has a bounded support in H. The evolution equation for the statistical solutions is linear, so the solution may be obtained by summing up the solutions corresponding to each k.

The statistical solution having initial distribution μ_0 with bounded support is obtained via the Krein–Milman theorem (Theorem A.9 in Chapter IV), which implies that μ_0 is a (weak-star) limit of convex combinations of delta functions – in other words, the limit as n goes to infinity of measures of the form

$$\mu_0^{(n)} = \sum_{j=1}^{J(n)} \theta_j^{(n)} \delta_{\mathbf{u}_j^{(n)}}.$$

The measure $\delta_{\mathbf{u}_j^{(n)}}$ is given by

$$\delta_{\mathbf{u}_j^{(n)}}(E) = \begin{cases} 1 & \text{if } \mathbf{u}_j^{(n)} \text{ belongs to the measurable set } E, \\ 0 & \text{otherwise.} \end{cases}$$

Moreover, $\sum_{j=1}^{J(n)} \theta_j^{(n)} = 1$ for each n. The elements $\mathbf{u}_j^{(n)}$ belong to $B_H(R_0)$. The solution $\{\mu_t^{(n)}\}_{0\le t\le T}$ corresponding to the initial distribution $\mu_0^{(n)}$ is obtained by forming the convex combination of the solutions corresponding to each delta distribution. The statistical solution with initial distribution $\delta_{\mathbf{u}_j^{(n)}}$ is simply $\{\delta_{\mathbf{u}_j^{(n)}(t)}\}_{0\le t\le T}$, where $t \mapsto \mathbf{u}_j^{(n)}(t)$ is an arbitrary weak solution with initial condition $\mathbf{u}_j^{(n)}$. Finally, the solution with initial condition μ_0 is obtained by taking the (weak-star) limit of the sequence $\{\mu_t^{(n)}\}_{0\le t\le T}$. Having sketched the proof of Theorem 1.1, we now enter into the details.

We assume that the forcing term $\mathbf{f}$ belongs to $L^2(0, T; H)$, where $H = H_{\text{nsp}}$ (i.e., the space H in the no-slip case), and we consider an initial probability distribution μ_0 on H satisfying the finite kinetic energy assumption

$$\int_H |\mathbf{u}|^2 \, d\mu_0(\mathbf{u}) < \infty.$$

We consider first the case in which μ_0 has bounded support in H; that is, $\operatorname{supp} \mu_0 \subset B_H(R_0)$ for some $R_0 > 0$, where $B_H(R_0)$ denotes the closed ball in H of radius R_0 and centered at the origin. Consider the set

$$\mathcal{U}_T(R_0) = \{\mathbf{u} \in \mathcal{C}([0, T]; H_w);\ \mathbf{u} \text{ is a weak solution with } \mathbf{u}(0) \in B_H(R_0)\} \quad \text{(A.1)}$$

endowed with the topology inherited from $\mathcal{C}([0, T]; H_w)$, where H_w denotes the space H endowed with the weak topology. The reason for considering this set is that, as we will prove shortly, every measure carried by $\mathcal{U}_T(R_0)$ gives rise to a time-dependent statistical solution.

Thanks to the a priori estimate (A.39) in Chapter II, each trajectory $\mathbf{u} = \mathbf{u}(t)$ in $\mathcal{U}_T(R_0)$ is bounded in H – say it is included in $B_H(R)$, the closed ball of H of radius R and centered at the origin, where R is the same for all trajectories in $\mathcal{U}_T(R_0)$. Hence, $\mathbf{u} = \mathbf{u}(t)$ belongs to $\mathcal{C}([0, T]; B_H(R)_w)$, where $B_H(R)_w$ denotes the ball $B_H(R)$ endowed with the topology inherited from H_w (the weak topology in H). Thus, $\mathcal{U}_T(R_0)$ is included in $\mathcal{C}([0, T]; B_H(R)_w)$. Since H_w is metrizable, so is $\mathcal{C}([0, T]; H_w)$ and hence $\mathcal{U}_T(R_0)$. We infer from the a priori estimates (A.36), (A.38), and (A.39) in Chapter II that $\mathcal{U}_T(R_0)$ is a bounded set in the space

$$\mathcal{E}_T = \{\mathbf{u} \in L^\infty(0, T; H) \cap L^2(0, T; V);\ \mathbf{u}' \in L^{4/3}(0, T; V')\},$$

which is a Banach space when endowed with its natural norm. By the Aubin compactness theorem (see Theorem IV.A.11), bounded subsets of $\mathcal{E}_T$ are compactly embedded into $L^2(0, T; H)$; whereas, by the Arzelà–Ascoli theorem (see Theorem IV.A.10), bounded subsets of $\mathcal{E}_T$ are compactly embedded into $\mathcal{C}([0, T]; H_w)$. Moreover, as proved in the theory of existence of solutions of the Navier–Stokes equations, every limit point of weak solutions in $\mathcal{E}_T$ of the NSE is also a weak solution. This implies that the set $\mathcal{U}_T(R_0)$ is compact for the topology (metric) inherited from $\mathcal{C}([0, T], H_w)$.

Consider the space $\mathcal{M}_T(R_0)$ of Borel probability measures μ on $\mathcal{U}_T(R_0)$. We claim that each probability measure μ_0 in $\mathcal{M}_T(R_0)$ defines a statistical solution of the Navier–Stokes equations. First, each measure μ in $\mathcal{M}_T(R_0)$ defines a family $\{\mu_t\}_{t\in[0,T]}$ of Borel probability measures on H as follows. Each $\mathbf{u} = \mathbf{u}(t)$ in $\mathcal{U}_T(R_0)$ is weakly continuous in H and is bounded in H by some R independent of $\mathbf{u}$, as mentioned previously. We infer that, for each φ in $\mathcal{C}(B_H(R)_w)$, $\mathbf{u} \mapsto \varphi(\mathbf{u}(t))$ is a real-valued continuous map defined on $\mathcal{U}_T(R_0)$. Thus, the integral

$$\int_{\mathcal{U}_T(R_0)} \varphi(\mathbf{u}(t))\, d\mu(\mathbf{u})$$

makes sense and defines a positive continuous linear map on $\mathcal{C}(B_H(R)_w)$. Since $B_H(R)_w$ is compact, it follows by the Kakutani–Riesz representation theorem (see Theorem IV.A.1) that there exists a regular Borel measure on $B_H(R)_w$ – which we denote by μ_t, since it depends on t – such that

$$\int_H \varphi(\mathbf{u})\, d\mu_t(\mathbf{u}) = \int_{\mathcal{U}_T(R_0)} \varphi(\mathbf{u}(t))\, d\mu(\mathbf{u}).$$

The positivity of μ_t follows from that of μ, and by choosing φ identically equal to 1 we see that each μ_t is a probability measure. Moreover, we can extend the previous relation to all functions φ in $\mathcal{C}(H_w)$ simply by considering their restriction to $B_H(R)_w$. Thus we obtain

$$\int_H \varphi(\mathbf{u})\, d\mu_t(\mathbf{u}) = \int_{\mathcal{U}_T(R_0)} \varphi(\mathbf{u}(t))\, d\mu(\mathbf{u}) \tag{A.2}$$

for all functions φ in $\mathcal{C}(H_w)$ and all $t \in [0, T]$. We write $\mu = \{\mu_t\}_{t\in[0,T]}$. Notice that, in the left-hand side of (A.2), the dummy variable $\mathbf{u}$ belongs to H; on the right-hand side, it belongs to $\mathcal{U}_T(R_0)$.

We now show that the family $\mu = \{\mu_t\}_{t\in[0,T]}$ of Borel probability measures on H is a time-dependent statistical solution of the Navier–Stokes equations according to Definition 1.2. For that purpose we need to prove (1.8), (1.9), (1.11), and (1.12) (according to Remark 1.1, (1.13) follows from the other properties). We begin with (1.11). One approximates $\varphi(\mathbf{u})$ by $\varphi(P_n\mathbf{u})$, which belongs to $\mathcal{C}(H_w)$; here P_n is the Galerkin projector. It is straightforward to see, using (A.2) and the Lebesgue dominated convergence theorem (Theorem IV.A.3), that (1.11) holds for $\varphi(P_n\mathbf{u})$. Then the function in (1.11) is obtained as the pointwise (in t) limit (as n goes to infinity) of measurable functions, and thus it is measurable. This proves (1.11), and this limiting argument also proves that (A.2) holds with $\varphi(\mathbf{u}) = |\mathbf{u}|^2$ and $\varphi(\mathbf{u}) = \|\mathbf{u}\|^2$. The measurability of the function in (1.12) follows from (1.11). This function belongs to $L^\infty(0, T)$ because of (A.2) and the facts that μ is a probability measure and that every function in $\mathcal{U}_T(R_0)$ is uniformly bounded by R in H.

For (1.9), we start by observing that each $\mathbf{u}$ in $\mathcal{U}_T(R_0)$ is a weak solution of the NSE in the sense given in Section II.7; hence, it satisfies the energy inequality (II.A.21):

$$|\mathbf{u}(t)|^2 + 2\nu \int_0^t \|\mathbf{u}(s)\|^2\, ds \le |\mathbf{u}(0)|^2 + 2\int_0^t (\mathbf{f}(s), \mathbf{u}(s))\, ds$$

for all $t \in [0, T]$. Since μ is a measure on $\mathcal{U}_T(R_0)$, it follows that

$$\int_{\mathcal{U}_T(R_0)} \left(|\mathbf{u}(t)|^2 + 2\nu \int_0^t \|\mathbf{u}(s)\|^2\, ds \right) d\mu(\mathbf{u})$$
$$\le \int_{\mathcal{U}_T(R_0)} \left(|\mathbf{u}(0)|^2 + 2\int_0^t (\mathbf{f}(s), \mathbf{u}(s))\, ds \right) d\mu(\mathbf{u})$$

for all $t \in [0, T]$. Using Fubini's theorem (see Theorem IV.A.4), we see that the inequality just displayed can be written as

$$\int_{\mathcal{U}_T(R_0)} |\mathbf{u}(t)|^2\, d\mu(\mathbf{u}) + 2\nu \int_0^t \int_{\mathcal{U}_T(R_0)} \|\mathbf{u}(s)\|^2\, d\mu(\mathbf{u})\, ds$$

$$\leq \int_{\mathcal{U}_T(R_0)} |\mathbf{u}(0)|^2\, d\mu(\mathbf{u}) + 2 \int_0^t \int_{\mathcal{U}_T(R_0)} (\mathbf{f}(s), \mathbf{u}(s))\, d\mu(\mathbf{u})\, ds$$

for all $t \in [0, T]$. Using (A.2) (which holds also for $\varphi(\mathbf{u}) = |\mathbf{u}|^2 = \|\mathbf{u}\|^2$, as shown earlier in this proof), we obtain (1.9).

It remains to prove (1.8). Each $\mathbf{u}$ in $\mathcal{U}_T(R_0)$ is a weak solution and hence, according to (II.A.36),

$$\frac{d\mathbf{u}}{dt} = \mathbf{F}(t; \mathbf{u}(t)) \in L^{4/3}(0, T; V'),$$

where $F(t; \mathbf{u}(t))$ is given by (II.A.35). Thus, for any $\mathbf{g}$ in V,

$$\frac{d}{dt}(\mathbf{u}(t), \mathbf{g}) = (\mathbf{F}(t, \mathbf{u}(t)), \mathbf{g}) \in L^{4/3}(0, T).$$

If $\Phi(\mathbf{u}) = \phi((\mathbf{u}, \mathbf{g}_1), \ldots, (\mathbf{u}, \mathbf{g}_k))$ is a cylindrical test function in the sense of Definition 1.1, then

$$\begin{aligned}\frac{d}{dt}\Phi(\mathbf{u}(t)) &= \sum_{j=1}^k \partial_j \phi((\mathbf{u}, \mathbf{g}_1), \ldots, (\mathbf{u}, \mathbf{g}_k)) \frac{d}{dt}(\mathbf{u}(t), \mathbf{g}_j) \\ &= \sum_{j=1}^k \partial_j \phi((\mathbf{u}, \mathbf{g}_1), \ldots, (\mathbf{u}, \mathbf{g}_k))(\mathbf{F}(t, \mathbf{u}(t)), \mathbf{g}_j) \\ &= \big(\mathbf{F}(t, \mathbf{u}(t)), \Phi'(\mathbf{u}(t))\big) \in L^{4/3}(0, T).\end{aligned}$$

Therefore, $t \mapsto \Phi(\mathbf{u}(t)$ is almost everywhere equal to a continuous function, and

$$\Phi(\mathbf{u}(t)) = \Phi(\mathbf{u}(0)) + \int_0^t \big(\mathbf{F}(s, \mathbf{u}(s)), \Phi'(\mathbf{u}(s))\big)\, ds. \tag{A.3}$$

Since Φ belongs to $\mathcal{C}(H_w)$, the relation (A.2) holds with $\varphi = \Phi$. Using the Lebesgue dominated convergence theorem, one can show that

$$\lim_{n\to\infty} \int_H \big(\mathbf{F}(t, P_n\mathbf{u}(t)), \Phi'(\mathbf{u}(t))\big)\, d\mu_t(\mathbf{u}) = \int_H \big(\mathbf{F}(t, \mathbf{u}(t)), \Phi'(\mathbf{u}(t))\big)\, d\mu_t(\mathbf{u})$$

for almost every t in $[0, T]$. On the other hand, from (1.11) and the fact that $\varphi(\mathbf{u}) = (\mathbf{F}(t, P_n\mathbf{u}), \Phi'(\mathbf{u}))$ belongs to $\mathcal{C}(H_w)$, it follows that

$$t \mapsto \int_H \big(\mathbf{F}(t, P_n\mathbf{u}(t)), \Phi'(\mathbf{u}(t))\big)\, d\mu_t(\mathbf{u})$$

is continuous. Hence,

$$t \mapsto \int_H \big(\mathbf{F}(t, \mathbf{u}(t)), \Phi'(\mathbf{u}(t))\big)\, d\mu_t(\mathbf{u})$$

is measurable. One can then check that

$$t \mapsto \int_H \big(\mathbf{F}(t, \mathbf{u}(t)), \Phi'(\mathbf{u}(t))\big)\, d\mu_t(\mathbf{u}) \in L^{4/3}(0, T).$$

Now we use Fubini's theorem to write

$$\int_0^t \int_H \big(\mathbf{F}(s, \mathbf{u}(s)), \Phi'(\mathbf{u}(s))\big)\, d\mu_s(\mathbf{u})\, ds$$
$$= \int_{\mathcal{U}_T(R_0)} \int_0^t \big(\mathbf{F}(s, \mathbf{u}(s)), \Phi'(\mathbf{u}(s))\big)\, ds\, d\mu(\mathbf{u}).$$

Therefore, integrating equation (A.3) over $\mathcal{U}_T(R_0)$ and using the last relation, we obtain (1.8); this completes the proof that $\{\mu_t\}_{t\in[0,T]}$ is a statistical solution.

In particular, the previous results mean that, for any given weak solution $\mathbf{u} = \mathbf{u}(t)$, if $\delta_{\mathbf{u}(t)}$ denotes the Dirac delta distribution on H_w supported by $\mathbf{u}(t)$ then the family $\{\delta_{\mathbf{u}(t)}\}_{t\geq 0}$ is a time-dependent statistical solution and the corresponding initial probability distribution is $\delta_{\mathbf{u}(0)}$. The next step is to show that there exists a time-dependent statistical solution with the initial distribution μ_0 (with μ_0 carried by $B_H(R_0)$).

Let $\mathcal{M}_0(R_0)$ denote the set of Borel probability measures on H that are carried by $B_H(R_0)$. The set $\mathcal{M}_0(R_0)$ is a bounded subset of the dual space $\mathcal{C}(B_H(R_0)_w)'$. As such, $\mathcal{M}_0(R_0)$ is also convex. It can be proved that $\mathcal{M}_0(R_0)$ is weak-star closed. Since bounded sets of the dual space $\mathcal{C}(B_H(R_0)_w)'$ are weak-star precompact, it follows that $\mathcal{M}_0(R_0)$ is a convex (weak-star) compact subset of (the locally convex topological vector space) $\mathcal{C}(B_H(R_0)_w)'$. Therefore, by the Krein–Milman theorem (see Theorem IV.A.9), $\mathcal{M}_0(R_0)$ is the closed convex hull of the set of its extremal points. It can be shown that those extremal points are the delta distributions $\delta_{\mathbf{u}_0}$ with $\mathbf{u}_0 \in B_H(R_0)$. Since $B_H(R)_w$ is separable, metrizable, and compact, the space $\mathcal{C}(B_H(R_0)_w)$ is separable. Hence, the unit ball in the dual $\mathcal{C}(B_H(R_0)_w)'$ is metrizable. Thus $\mathcal{M}_0(R_0)$ is metrizable, which implies that compactness equals sequential compactness. Therefore, since μ_0 belongs to $\mathcal{M}_0(R_0)$, there exists a sequence of probability measures $\{\mu_{0n}\}_{n\in\mathbb{N}}$ on H of the form

$$\mu_{0n} = \sum_{j=1}^{J(n)} \theta_j^{(n)} \delta_{\mathbf{u}_{0j}^{(n)}},$$

with $J(n) \in \mathbb{N}$, $\mathbf{u}_{0j}^{(n)} \in B_H(R_0)$, $\theta_j^{(n)} \in (0, 1]$, and $\sum_{j=1}^{J(n)} \theta_j^{(n)} = 1$, such that

$$\mu_{0n} \overset{*}{\rightharpoonup} \mu_0 \text{ weak-star in } \mathcal{M}_0(R_0);$$

that is,

$$\mu_0(E) = \lim_{n\to\infty} \sum_{j=1}^{J(n)} \theta_j^{(n)} \delta_{\mathbf{u}_{0j}^{(n)}}(E)$$

for all Borel subsets E of $B_H(R_0)_w$.

Now, for each initial condition $\mathbf{u}_{0j}^{(n)}$ in $B_H(R_0)$, there exists a weak solution $\mathbf{u}_j^{(n)} = \mathbf{u}_j^{(n)}(t)$ in $\mathcal{U}_T(R_0)$ with $\mathbf{u}_j^{(n)}(0) = \mathbf{u}_{0j}^{(n)}$. Each corresponding probability distribution $\delta_{\mathbf{u}_j^{(n)}}$ belongs to $\mathcal{M}_T(R_0)$; since this set is convex, we also have the distributions

$$\mu_n = \sum_{j=1}^{J(n)} \theta_j^{(n)} \delta_{\mathbf{u}_j^{(n)}} \in \mathcal{M}_T(R_0).$$

Since $\mathcal{M}_T(R_0)$ is compact (endowed with the weak-star topology inherited from the dual space $\mathcal{C}(\mathcal{U}_T(R_0))'$), we find a weak-star convergent subsequence

$$\mu_{n'} \overset{*}{\rightharpoonup} \mu$$

for some μ in $\mathcal{M}_T(R_0)$. As explained earlier, this measure is associated with a time-dependent statistical solution $\{\mu_t\}_{t\in[0,T]}$. Moreover, it is straightforward to check that μ_t at $t = 0$ coincides with the initial probability distribution μ_0. This completes the proof of the existence of a time-dependent statistical solution for a given initial probability distribution with bounded support on H. It remains to consider the case when μ_0 is not carried by a bounded set in H.

For that case, we write

$$\mu_0 = \sum_{k=1}^{\infty} \mu_{0k},$$

where μ_{0k} is the restriction of μ_0 to the bounded sets $B_H(k) \setminus B_H(k-1)$ for $k = 2, 3, \ldots$ and to $B_H(1)$ for $k = 1$. Each μ_{0k} is a finite Borel probability measure on H with bounded support. We want to obtain corresponding time-dependent statistical solutions μ_k. However, we observe that the μ_{0k} are not probability distributions because $\mu_{0k}(H) < 1$. If $\mu_{0k} = 0$, we just set $\mu_k = 0$. If $0 < \mu_{0k}(H) < 1$ then we have two options: either we repeat the previous argument and obtain a family of Borel measures $\mu_k = \{\mu_{kt}\}$ satisfying (1.8), (1.9), (1.11), (1.12), and (1.13), with $\mu_{kt}(H) = \mu_{0k}(H)$ for all $t \in [0, T]$; or we simply normalize μ_{0k} to be a probability distribution, obtain the corresponding statistical solution, and divide the measures by the normalization factor. We then set

$$\mu_t = \sum_{k=1}^{\infty} \mu_{kt}$$

for each $t \in [0, T]$. One can check that $\{\mu_t\}_{t\in[0,T]}$ is a time-dependent statistical solution with initial distribution μ_0. We use here the remarkable fact that the Liouville-type equation (1.8) is linear in $\{\mu_t\}_{t\in[0,T]}$. This completes the proof of Theorem 1.1.

Remark A.1 See Foias and Temam [1980] for another proof of Theorem 1.1 using the Galerkin projections to obtain a sequence of time-dependent measures with a subsequence converging to the desired statistical solution. See further comments in Remark A.2.

As mentioned in Section 1, in the 2-dimensional case it is possible to obtain statistical solutions with further regularity properties. Then, under those additional regularity properties – and if the initial probability distribution has bounded support in H – a uniqueness result for the statistical solutions can be proved. Before we prove the uniqueness result, we show how to obtain more regular statistical solutions.

For simplicity, we consider the case in which the forcing term $\mathbf{f}$ is time-independent and belongs to $H = H_{\mathrm{nsp}}$, the space of finite kinetic energy in the no-slip boundary condition case. Let μ_0 be a given probability distribution on H. In the 2-dimensional case, the Navier–Stokes equations generate a semigroup $\{S(t)\}_{t\geq 0}$. We define the family of measures $\{\mu_t\}_{t\geq 0}$ by $\mu_t = S(t)\mu_0$. In the 2-dimensional case, the (individual) solutions $\mathbf{u}(t) = S(t)\mathbf{u}_0$, $t \geq 0$, are strong solutions, and the differential equation

$$\frac{d\mathbf{u}(t)}{dt} = \mathbf{F}(\mathbf{u}(t)),$$

where $F(\mathbf{u}) = \mathbf{f} - \nu A\mathbf{u} - B(\mathbf{u})$, is satisfied in the strong sense in the space $V' = D(A^{-1/2})$. It is then straightforward to show that $\{\mu_t\}_{t\geq 0}$ is a statistical solution with initial probability distribution μ_0. It is also straightforward to show that this statistical solution satisfies the regularity properties (1.14)–(1.16) if the initial probability distribution has bounded support in H.

Proof of Theorem 1.2. The aim is to show that μ_t is equal to $S(t)\mu_0$ for each $t > 0$. This is done by showing that

$$\int_H \Phi(\mathbf{u})\, d\mu_t(\mathbf{u}) = \int_H \Phi(S(t)\mathbf{u})\, d\mu_0(\mathbf{u}) \tag{A.4}$$

for all test functions Φ in the sense of Definition 1.1. Since those functions are dense in $C(H)$ and hence dense in $L^1(H)$, it follows from (A.4) that $\mu_t = S(t)\mu_0$. The proof of (A.4) is based on the stronger form of the Liouville-type equation given by (1.16). If we choose

$$\Psi(s, \mathbf{u}) = \Phi(S(t - s)\mathbf{u}) \quad \text{for } 0 \leq s \leq t,$$

then $\Psi(t, \mathbf{u}) = \Phi(\mathbf{u})$ and $\Psi(0, \mathbf{u}) = \Phi(S(t)\mathbf{u})$. If there were enough regularity for $\Psi = \Psi(s, \mathbf{u})$ to be a time-dependent test function in the sense allowed in (1.16), then by (1.16) we would have

$$\int \Phi(\mathbf{u})\, d\mu_t(\mathbf{u}) - \int \Phi(S(t)\mathbf{u})\, d\mu_0(\mathbf{u}) = \int_0^t \int \{\Psi_s'(s, \mathbf{u}) + (\mathbf{F}(\mathbf{u}), \Psi_{\mathbf{u}}'(s, \mathbf{u}))\}\, d\mu_s(\mathbf{u})\, ds. \tag{A.5}$$

Whence (A.4) would follow, provided the term in the RHS of (A.5) vanishes, which is formally what happens (see the derivation of (A.7)). For regularity reasons, however, we work with the Galerkin approximation of the Navier–Stokes equations. We denote by P_m the Galerkin projection onto the space spanned by the first m eigenmodes of the Stokes operator and by $\{S_m(t)\}_{t\geq 0}$ the solution operator associated with the Galerkin approximation. That is, $S_m P_m(t)\mathbf{u}_0 = \mathbf{u}_m(t)$ for $t \geq 0$, where $\mathbf{u}_m(t)$ is the solution of the following (finite-dimensional) equation on $P_m H$:

$$\frac{d\mathbf{u}_m}{dt} = P_m \mathbf{F}(\mathbf{u}_m), \quad \mathbf{u}_m(0) = P_m \mathbf{u}_0.$$

We set

$$\Psi_m(s, \mathbf{u}) = \Phi(S_m(t - s)P_m\mathbf{u}), \quad 0 \leq s \leq t, \quad \mathbf{u} \in H,$$

where Φ is an arbitrary test function in the sense of Definition 1.1. It is straightforward to check that $\Psi_m = \Psi_m(s, \mathbf{u})$ is a time-dependent test function in the sense allowed in (1.16). That $\Psi_m(s, \mathbf{u})$ is defined only for $s \in [0, t]$ is not a problem; in fact, the solutions of the Galerkin approximation are defined for all times t in $\mathbb{R}$, which can be seen by taking the scalar product of $\mathbf{u}_m$ with the equation for the Galerkin approximation and using that AP_m is a bounded operator. Hence $S_m(s)\mathbf{u}$ is defined for all real s, so that $\Psi_m(s, \mathbf{u})$ can be extended to all $s \in \mathbb{R}$ and in particular for all $s \geq 0$. This is not the reason for using the Galerkin approximation, since we could have considered test functions defined only on $[0, t]$. The reason for using the Galerkin approximation is the regularity of the solutions, which renders $\Psi_m(s, \mathbf{u})$ a time-dependent test function.

With this choice of time-dependent test function, we have

$$\Psi_m(t, \mathbf{u}) = \Phi(P_m\mathbf{u}), \qquad \Psi_m(0, \mathbf{u}) = \Phi(S_m(t)P_m\mathbf{u}).$$

From (1.16), we find

$$\int \Phi(P_m\mathbf{u})\, d\mu_t(\mathbf{u}) - \int \Phi(S_m(t)P_m\mathbf{u})\, d\mu_0(\mathbf{u})$$
$$= \int_0^t \int \{(\Psi_m)'_s(s, \mathbf{u}) + (\mathbf{F}(\mathbf{u}), (\Psi_m)'_{\mathbf{u}}(s, \mathbf{u}))\}\, d\mu_s(\mathbf{u})\, ds. \qquad (A.6)$$

On the other hand, from the Lebesgue dominated convergence theorem (see Theorem IV.A.3), it follows that

$$\lim_{m\to\infty} \int_H \Phi(P_m\mathbf{u})\, d\mu_t(\mathbf{u}) = \int_H \Phi(\mathbf{u})\, d\mu_t(\mathbf{u}),$$

and

$$\lim_{m\to\infty} \int_H \Phi(S_m(t)P_m\mathbf{u})\, d\mu_0(\mathbf{u}) = \int_H \Phi(S(t)\mathbf{u})\, d\mu_0.$$

Therefore, (A.4) follows if we can prove that the RHS of (A.6) goes to zero as m goes to infinity. Our first step in this direction is to rewrite the right-hand side of (A.6).

For each $\xi \geq -s$, we have

$$\Psi_m(s, S_m(s+\xi)P_m\mathbf{u}) = \Phi(S_m(t-s)S_m(s+\xi)P_m\mathbf{u}) = \Phi(S_m(t+\xi)P_m\mathbf{u}),$$

with the right-hand side independent of s. Hence, taking the derivative (with respect to s) of the term in the left-hand side, we find for all $\xi \geq -s$ that

$$(\Psi_m)'_s(s, S_m(s+\xi)P_m\mathbf{u}) + \big(P_m\mathbf{F}(S_m(s+\xi)P_m\mathbf{u}), (\Psi_m)'_{\mathbf{u}}(s, S_m(s+\xi)P_m\mathbf{u})\big) = 0.$$

Taking $\xi = -s$ yields

$$(\Psi_m)'_s(s, P_m\mathbf{u}) + (P_m\mathbf{F}(P_m\mathbf{u}), (\Psi_m)'_{\mathbf{u}}(s, P_m\mathbf{u})) = 0. \qquad (A.7)$$

Since $\Psi_m(s, P_m\mathbf{u}) = \Psi_m(s, \mathbf{u})$ for all $\mathbf{u}$ and s, it follows that

$$(\Psi_m)'_s(s, \mathbf{u}) + (P_m\mathbf{F}(P_m\mathbf{u}), (\Psi_m)'_{\mathbf{u}}(s, \mathbf{u})) = 0.$$

Thus, the integrand in the RHS of (A.6) can be written as

$$(\mathbf{F}(\mathbf{u}) - P_m\mathbf{F}(P_m\mathbf{u}), (\Psi_m)'_{\mathbf{u}}(s, \mathbf{u})).$$

Since $\mathbf{F}(\mathbf{u}) = \mathbf{f} - \nu A\mathbf{u} - B(\mathbf{u})$ and $(\Psi_m)'_{\mathbf{u}}(s, \mathbf{u})$ belongs to $P_m H$, the terms involving the forcing term $\mathbf{f}$ and the Stokes operator A cancel out. Hence, the expression just displayed simplifies to

$$(B(P_m\mathbf{u}) - B(\mathbf{u}), (\Psi_m)'_{\mathbf{u}}(s, \mathbf{u})),$$

and we can rewrite (A.6) as

$$\int \Phi(P_m\mathbf{u})\, d\mu_t(\mathbf{u}) - \int \Phi(S_m(t)P_m\mathbf{u})\, d\mu_0(\mathbf{u})$$

$$= \int_0^t \int (P_m B(P_m\mathbf{u}) - B(\mathbf{u}), (\Psi_m)'_{\mathbf{u}}(s, P_m\mathbf{u}))\, d\mu_s(\mathbf{u})\, ds. \quad \text{(A.8)}$$

We need now to show that the RHS of (A.8) goes to zero as m goes to infinity. This result we borrow from the proof of Theorem IV.2.2. From the assumption (1.15), the support of μ_s is bounded in H uniformly for $s \geq 0$. Hence, just as in the proof of Theorem IV.2.2, there exists $c(t) \geq 0$, depending on t, such that

$$|(B(\mathbf{u}) - B(P_m\mathbf{u}), (\Psi_m)'_{\mathbf{u}}(s, \mathbf{u}))| \leq \frac{c(t)}{\lambda_m^{1/4}}$$

for all $\mathbf{u}$ in supp μ_s and all s in $[0, t]$, where λ_m is the mth eigenvalue of the Stokes operator. Therefore, since λ_m goes to infinity with m, it follows that

$$\left| \int_0^t \int \{(P_m B(P_m\mathbf{u}) - B(\mathbf{u}), (\Psi_m)'_{\mathbf{u}}(s, P_m\mathbf{u}))\}\, d\mu_s(\mathbf{u})\, ds \right| \leq \frac{tc(t)}{\lambda_m^{1/4}} \to 0$$

as m goes to infinity. This shows that

$$\lim_{m\to\infty} \left[\int \Phi(P_m\mathbf{u})\, d\mu_t(\mathbf{u}) - \int \Phi(S_m(t)P_m\mathbf{u})\, d\mu_0(\mathbf{u}) \right] = 0,$$

which implies, as explained at the beginning of the proof, that $\mu_t = S(t)\mu_0$. This completes the proof of Theorem 2.1.

Proof of Theorem 1.3. Similar to the proof of Theorem 1.1.

Proof of Theorem 1.4. Similar to the proof of Theorem 1.2.

Proof of Theorem 1.5. Let μ_0 and $\mathbf{f}$ be given as in the statement of Theorem 1.5. Since $\dot{H}(L) \subset H(L)$, they satisfy the conditions of Theorem 1.3. Hence, we find a statistical solution $\{\mu_t\}_{0\leq t<T}$ on $H(L)$.

Consider the orthogonal projector $\dot{P}$ from $H(L)$ onto $\dot{H}(L)$, where $\dot{H}(L)$ is the subspace of $H(L)$ of vector fields with zero space average on the cube $Q(L)$. Define

$$\dot{\mu}_t = \dot{P}\mu_t \quad \forall t \geq 0.$$

Clearly, $\dot{\mu}_0 = \mu_0$. Moreover, each $\dot{\mu}_t$ is a probability measure on $\dot{H}(L)$. We need then to show that $\{\dot{\mu}_t\}_{0\leq t<T}$ satisfies conditions (1.17)–(1.20) with $H(L)$ replaced by $\dot{H}(L)$.

For each $\varphi \in \mathcal{C}_b(\dot{H}(L))$, it follows that $\varphi \circ \dot{P} \in \mathcal{C}_b(H(L))$. Moreover,

$$\int \varphi(\mathbf{u})\, d\dot{\mu}_t(\mathbf{u}) = \int \varphi(\dot{P}\mathbf{u})\, d\mu(\mathbf{u}).$$

Therefore, (1.19) for $\{\dot{\mu}_t\}_{0\le t<T}$ follows from (1.19) for $\{\mu_t\}_{0\le t<T}$.

Now, the function

$$t \to \int |\mathbf{u}|^2\, d\dot{\mu}_t(\mathbf{u}) = \int |\dot{P}\mathbf{u}|^2\, d\mu(\mathbf{u})$$

is measurable since, by the monotone convergence theorem (Theorem A.2 in Chapter IV), this function is the pointwise limit (as r goes to infinity) of the functions

$$t \to \int \min\{r, |\dot{P}\mathbf{u}|^2\}\, d\mu(\mathbf{u}),$$

which are measurable according to (1.19). Moreover, since $|\dot{P}\mathbf{u}| \le |\mathbf{u}|$ on $H(L)$, we deduce that (1.20) for $\{\mu_t\}_{0\le t<T}$ implies (1.20) for $\{\dot{\mu}_t\}_{0\le t<T}$, and similarly for (1.21).

In order to prove (1.18) for $\dot{\mu}_t$, we recall that $\dot{P}$ is an orthogonal projector in $H(L)$ and that $\mathbf{f}$ has its values in $\dot{H}(L)$; we then deduce from (1.18) for μ_t that

$$\begin{aligned}
\int |\mathbf{u}|^2\, d\dot{\mu}_t(\mathbf{u}) &+ 2\nu \int_0^t \int \|\mathbf{u}\|^2\, d\dot{\mu}_s(\mathbf{u})\, ds \\
&= \int |\dot{P}\mathbf{u}|^2\, d\mu_t(\mathbf{u}) + 2\nu \int_0^t \int \|\dot{P}\mathbf{u}|^2\, d\mu_s(\mathbf{u}) \\
&\le \int |\mathbf{u}|^2\, d\mu_t(\mathbf{u}) + 2\nu \int_0^t \int \|\mathbf{u}\|^2\, d\mu_s(\mathbf{u}) \\
&\le \int_0^t (\mathbf{f}(s), \mathbf{u})\, d\mu_s(\mathbf{u}) + \int |\mathbf{u}|^2\, d\mu_0(\mathbf{u}) \\
&= \int_0^t (\mathbf{f}(s), \dot{P}\mathbf{u})\, d\mu_s(\mathbf{u}) + \int |\mathbf{u}|^2\, d\mu_0(\mathbf{u}) \\
&= \int_0^t (\mathbf{f}(s), \mathbf{u})\, d\dot{\mu}_s(\mathbf{u}) + \int |\mathbf{u}|^2\, d\mu_0(\mathbf{u}).
\end{aligned}$$

Finally, for (1.17), take $\Phi \in \mathcal{T}_{\text{per}}$ and set $\tilde{\Phi} = \Phi \circ \dot{P}$, which clearly belongs to $\mathcal{T}_{\text{per}}$. Then

$$\int \Phi(\mathbf{u})\, d\dot{\mu}_t = \int \Phi(\dot{P}\mathbf{u})\, d\mu_t(\mathbf{u}) = \int \tilde{\Phi}(\mathbf{u})\, d\mu_t(\mathbf{u})$$

for every $t \ge 0$. Moreover, $\tilde{\Phi}'(\mathbf{u}) = \Phi(\dot{P}\mathbf{u}) \circ \dot{P}$ and so, for each $\mathbf{u} \in H(L)$,

$$(\mathbf{f}(t), \tilde{\Phi}'(\mathbf{u})) = (\dot{P}\mathbf{f}(t), \Phi'(\dot{P}\mathbf{u})) = (\mathbf{f}(t), \Phi'(\dot{P}\mathbf{u})),$$

where we have used that $\mathbf{f} = \mathbf{f}(t)$ is an $\dot{H}(L)$-valued function. Similarly,

$$\begin{aligned}
(A\mathbf{u}, \tilde{\Phi}'(\mathbf{u})) &= ((\mathbf{u}, \tilde{\Phi}'(\mathbf{u}))) \\
&= ((\dot{P}\mathbf{u}, \Phi'(\dot{P}\mathbf{u}))) \\
&= ((A\dot{P}\mathbf{u}, \Phi'(\dot{P}\mathbf{u})))
\end{aligned}$$

and

$$\begin{aligned}(B(\mathbf{u}), \tilde{\Phi}'(\mathbf{u})) &= (\dot{P}B(\mathbf{u}), \Phi'(\dot{P}\mathbf{u})) = (B(\mathbf{u}), \Phi'(\dot{P}\mathbf{u}))\\ &= b(\mathbf{u}, \mathbf{u}, \Phi'(\dot{P}\mathbf{u})) = b(\mathbf{u}, \dot{P}\mathbf{u}, \Phi'(\dot{P}\mathbf{u}))\\ &= b(\dot{P}\mathbf{u}, \dot{P}\mathbf{u}, \Phi'(\dot{P}\mathbf{u})) \quad \text{(by the spatial periodicity of } \mathbf{u}\text{)}\end{aligned}$$

for each $\mathbf{u} \in H(L)$. From these relations we deduce that (1.17) for $\{\mu_t\}_{0\le t<T}$ implies (1.17) for $\{\dot{\mu}_t\}_{0\le t<T}$, and this concludes the proof. ■

A.2 Homogeneous Statistical Solutions

Proof of Lemma 2.2. For the sake of completeness, we reproduce here the proof of this lemma borrowed from Foias and Temam [1980]. Because of the homogeneity assumption on the probability distribution μ, for every $\mathbf{a}$ in $\mathbb{R}^d$ and for Q_0 as in (2.7) we have

$$\begin{aligned}\iint_{\tau_{\mathbf{a}} Q_0} |G(\mathbf{u})(\mathbf{x})|\, d\mathbf{x}\, d\mu(\mathbf{u}) &= \iint_{Q_0} |G(\mathbf{u})(\mathbf{x}-\mathbf{a})|\, d\mathbf{x}\, d\mu(\mathbf{u})\\ &= \iint_{Q_0} |G(\tau_{\mathbf{a}}\mathbf{u})(\mathbf{x})|\, d\mathbf{x}\, d\mu(\mathbf{u})\\ &= \iint_{Q_0} |G(\mathbf{u})(\mathbf{x})|\, d\mathbf{x}\, d\mu(\mathbf{u}),\end{aligned}$$

which is finite thanks to the assumption (2.7). Since we can cover any bounded set Q by a finite number of cubes $\tau_{\mathbf{a}} Q_0$, we see that

$$\iint_{Q} |G(\mathbf{u})(\mathbf{x})|\, d\mathbf{x}\, d\mu(\mathbf{u}) < \infty.$$

Hence, for every ϕ in $\mathcal{D}(\mathbb{R}^d)$ (the set of $\mathcal{C}^\infty$ functions on $\mathbb{R}^d$ with compact support), the integral

$$I(\phi) = \iint G(\mathbf{u})(\mathbf{x})\phi(\mathbf{x})\, d\mathbf{x}\, d\mu(\mathbf{u})$$

exists, and the map $\phi \mapsto I(\phi)$ defines a distribution on $\mathbb{R}^d$. Denote this distribution by h. First we show that h is constant, which is achieved by proving that h is invariant under translation. Let ϕ belong to $\mathcal{D}(\mathbb{R}^d)$ and let $\mathbf{a} \in \mathbb{R}^d$. By definition, the translation $\tau_{\mathbf{a}} h$ of this distribution applied to ϕ is equal to h applied to $\tau_{-\mathbf{a}}\phi$. Using the commutativity of G with $\tau_{\mathbf{a}}$ and the homogeneity of μ, we find

$$\begin{aligned}(\tau_{\mathbf{a}} h, \phi) &= (h, \tau_{-\mathbf{a}}\phi)\\ &= \iint G(\mathbf{u})(\mathbf{x})\phi(\mathbf{x}+\mathbf{a})\, d\mathbf{x}\, d\mu(\mathbf{u})\\ &= \iint G(\mathbf{u})(\mathbf{x}-\mathbf{a})\phi(\mathbf{x})\, d\mathbf{x}\, d\mu(\mathbf{u}) =\end{aligned}$$

$$
\begin{aligned}
&= \iint (\tau_{\mathbf{a}} G(\mathbf{u}))(\mathbf{x})\phi(\mathbf{x})\,d\mathbf{x}\,d\mu(\mathbf{u}) \\
&= \iint G(\tau_{\mathbf{a}}\mathbf{u})(\mathbf{x})\phi(\mathbf{x})\,d\mathbf{x}\,d\mu(\mathbf{u}) \\
&= \iint G(\mathbf{u})(\mathbf{x})\phi(\mathbf{x})\,d\mathbf{x}\,d\tau_{\mathbf{a}}\mu(\mathbf{u}) \\
&= \iint G(\mathbf{u})(\mathbf{x})\phi(\mathbf{x})\,d\mathbf{x}\,d\mu(\mathbf{u}) \\
&= (h, \phi).
\end{aligned}
$$

This shows that $\tau_{\mathbf{a}} h = h$ for every $\mathbf{a}$ in $\mathbb{R}^d$. Therefore,

$$
\iint G(\mathbf{u})(\mathbf{x})\phi(\mathbf{x})\,d\mathbf{x}\,d\mu(\mathbf{u}) = \langle G\rangle \int \phi(\mathbf{x})\,d\mathbf{x}, \tag{A.9}
$$

where $\langle G\rangle$ is the (constant) value of h. Since $\mathcal{D}(\mathbb{R}^d)$ is dense in $L^1(\mathbb{R}^d)$, the relation (A.9) can be extended to all the ϕ in $L^1(\mathbb{R}^d)$.

It remains only to show that h vanishes under the assumption (2.8). If h_G and h_F are the distributions corresponding to G and F, then we have

$$
h_G = \mathcal{P}(\partial_{\mathbf{x}})h_F;
$$

since h_F is constant and $\mathcal{P}(\partial_{\mathbf{x}})$ has no zero-order term, it follows that $h_G = 0$. This completes the proof of Lemma 2.2. ■

Proof of Theorem 2.1. Consider $\mathbf{f} = 0$ and assume that μ_0 is a homogeneous probability measure on $H(L)$. By Theorem 1.1, there exists a time-dependent statistical solution $\{\mu_t\}_{t\geq 0}$ with initial data μ_0 and $\mathbf{f} = 0$. For every $t \geq 0$, define

$$
\tilde{\mu}_t = \frac{1}{L^d}\int_{Q(L)} \tau_{\mathbf{a}}(\mu_t)\,d\mathbf{a}.
$$

Thanks to Lemma 2.1, we know that each μ_t is a homogeneous probability measure. Since μ_0 is homogeneous (i.e., $\tau_{\mathbf{a}}\mu_0 = \mu_0$ for every $\mathbf{a} \in \mathbb{R}^d$), it follows that $\tilde{\mu}_0 = \mu_0$. It remains only to show that $\{\tilde{\mu}_t\}_{t\geq 0}$ is a statistical solution. For that purpose we need to check the conditions to (2.9)–(2.13).

For (2.11), note that

$$
\int \varphi(\mathbf{u})\,d\tilde{\mu}_t(\mathbf{u}) = \int \frac{1}{L^d}\int_{Q(L)} \varphi(\tau_{\mathbf{a}}\mathbf{u})\,d\mathbf{a}\,d\mu_t(\mathbf{u}) = \int \tilde{\varphi}(\mathbf{u})\,d\mu_t(\mathbf{u}),
$$

where

$$
\tilde{\varphi}(\mathbf{u}) = \frac{1}{L^d}\int_{Q(L)} \varphi(\tau_{\mathbf{a}}\mathbf{u})\,d\mathbf{a}.
$$

Since $\tau_{\mathbf{a}}$ is an isometry in $H(L)$ – that is,

$$
|\tau_{\mathbf{a}}\mathbf{u}|_{Q(L)} = |\mathbf{u}|_{Q(L)}
$$

for all $\mathbf{u}$ in $H(L)$ and all $\mathbf{a}$ in $\mathbb{R}^d$ – it follows that $\tilde{\varphi}$ belongs to $\mathcal{C}_b(H(L))$ whenever φ belongs to $\mathcal{C}_b(H(L))$. Hence, from (2.11) for $\{\mu_t\}_{t\geq 0}$ we deduce (2.11) for $\{\tilde{\mu}_t\}_{t\geq 0}$.

Similarly, since we also have

$$\|\tau_{\mathbf{a}}\mathbf{u}\|_{Q(L)} = \|\mathbf{u}\|_{Q(L)}$$

for all $\mathbf{u}$ in $V(L)$ and all $\mathbf{a}$ in $\mathbb{R}^d$, the relations (2.12), (2.13), and (2.10) for $\{\tilde{\mu}_t\}_{t\geq 0}$ follow promptly from the corresponding relations for $\{\mu_t\}_{t\geq 0}$.

Finally, for the Liouville equation (2.10) for $\{\tilde{\mu}_t\}_{t\geq 0}$, we consider (2.10) for $\{\mu_t\}_{t\geq 0}$ with Φ replaced by $\Phi_{\mathbf{a}} = \Phi \circ \tau_{\mathbf{a}}$. Clearly, $\Phi_{\mathbf{a}}$ belongs to $\mathcal{T}_{\text{per}}$ (with $\mathbf{g}_i$ replaced by $\tau_{-\mathbf{a}}\mathbf{g}_i$). Notice that, since $\tau_{\mathbf{a}}$ is linear, $\Phi_{\mathbf{a}}$ belongs to $\mathcal{T}_{\text{per}}$ with

$$\Phi'_{\mathbf{a}}(\mathbf{u}) = \Phi'(\tau_{\mathbf{a}}\mathbf{u}) \circ \tau_{\mathbf{a}},$$

which means that

$$(\Phi'_a(\mathbf{u}), \mathbf{v})_{Q(L)} = (\Phi'(\tau_{\mathbf{a}}\mathbf{u}), \tau_{\mathbf{a}}\mathbf{v})_{Q(L)}$$

for all $\mathbf{u}$ and $\mathbf{v}$ in $H(L)$. Since $\Phi'(\mathbf{u})$ belongs to $V(L)$, this relation can (by an approximation argument) be extended to $\mathbf{v}$ in the dual $V(L)'$ of $V(L)$. Hence,

$$(\mathbf{v}, \Phi'_{\mathbf{a}}(\mathbf{u}))_{Q(L)} = (\tau_{\mathbf{a}}\mathbf{v}, \Phi'(\tau_{\mathbf{a}}\mathbf{u}))_{Q(L)}$$

for all $\mathbf{u} \in V(L)$ and all $\mathbf{v} \in V(L)'$. Then, we also have the relation

$$\begin{aligned}((\mathbf{u}, \Phi'_{\mathbf{a}}(\mathbf{u})))_{Q(L)} &= (A(L)\mathbf{u}, \Phi'_{\mathbf{a}}(\mathbf{u}))_{Q(L)} \\ &= (\tau_{\mathbf{a}}A(L)\mathbf{u}, \Phi'(\tau_{\mathbf{a}}\mathbf{u}))_{Q(L)} \\ &= (A(L)\mathbf{u}, \tau_{-\mathbf{a}}\Phi'(\tau_{\mathbf{a}}\mathbf{u}))_{Q(L)} \\ &= ((\mathbf{u}, \tau_{-\mathbf{a}}\Phi'(\tau_{\mathbf{a}}\mathbf{u})))_{Q(L)} \\ &= ((\tau_{\mathbf{a}}\mathbf{u}, \Phi'(\tau_{\mathbf{a}}\mathbf{u})))_{Q(L)},\end{aligned}$$

where we have used that the transpose of $\tau_{\mathbf{a}}$ is $\tau_{-\mathbf{a}}$. Similarly, by working with regular enough functions we find that, for smooth Φ,

$$\begin{aligned}(B(\mathbf{u}), \Phi'(\mathbf{u}))_{Q(L)} &= (\tau_{\mathbf{a}}B(\mathbf{u}), \Phi'(\tau_{\mathbf{a}}\mathbf{u}))_{Q(L)} \\ &= (B(\mathbf{u}), \tau_{-\mathbf{a}}\Phi'(\tau_{\mathbf{a}}\mathbf{u}))_{Q(L)} \\ &= \frac{1}{L^d}\int_{Q(L)} [(\mathbf{u}\cdot\nabla)\mathbf{u}]\tau_{-\mathbf{a}}\Phi'(\tau_{\mathbf{a}}\mathbf{u})\,d\mathbf{x} \\ &= \frac{1}{L^d}\int_{Q(L)} [(\tau_{\mathbf{a}}\mathbf{u}\cdot\nabla)\tau_{\mathbf{a}}\mathbf{u}]\Phi'(\tau_{\mathbf{a}}\mathbf{u})\,d\mathbf{x} \\ &= (B(\tau_{\mathbf{a}}\mathbf{u}), \Phi'(\tau_{\mathbf{a}}\mathbf{u}))_{Q(L)}.\end{aligned}$$

The result is then extended by density and continuity to all the desired functions Φ. Thus, we arrive at the equation

$$\begin{aligned}&\int \Phi(\tau_{\mathbf{a}}\mathbf{u})\,d\mu_t(\mathbf{u}) \\ &\quad + L^d \int_0^t \int [\nu((\tau_{\mathbf{a}}\mathbf{u}, \Phi'(\tau_{\mathbf{a}}\mathbf{u})))_{Q(L)} + (B(\tau_{\mathbf{a}}\mathbf{u}), \Phi'(\tau_{\mathbf{a}}\mathbf{u}))_{Q(L)}]\,d\mu_s(\mathbf{u})\,ds \\ &\qquad = \int \Phi(\tau_{\mathbf{a}}\mathbf{u})\,d\mu_0(\mathbf{u}).\end{aligned}$$

Integrating this relation with respect to $\mathbf{a}$ over the cube $Q(L)$, we obtain (2.9) for $\{\tilde{\mu}_t\}_{t\geq 0}$. This completes the proof. ■

A.3 Self-Similar Homogeneous Statistical Solutions

First of all, let us make precise what we mean by a stationary homogeneous statistical solution of (4.33).

Definition A.1 *A stationary homogeneous statistical solution of the equations* (4.33) *is a homogeneous probability measure μ on H_{loc} such that*

$$E(\mu) < \infty \tag{A.10}$$

and

$$\int \left[\frac{1}{2}(-\mathbf{u} - (\mathbf{x}\cdot\nabla)\mathbf{u}, \Phi'(\mathbf{u})) + ((\mathbf{u}, \Phi'(\mathbf{u}))) + ((\mathbf{u}\cdot\nabla)\mathbf{u}, \Phi'(\mathbf{u}))\right] d\mu(\mathbf{u}) = 0 \tag{A.11}$$

for all test functions Φ in $\mathcal{T}_{\text{ws}}$.

Proof of Theorem 4.1. Suppose first that $\{\mu^{\nu,\epsilon}\}_{\nu,\epsilon>0}$ is a self-similar universal homogeneous statistical solution of the Navier–Stokes equations. Take

$$\mu = \mu^{1,\gamma^2}$$

with $\gamma > 0$ and apply Proposition 4.2 to find that

$$\sigma_{\gamma^{-1/2}\epsilon^{1/4}\nu^{1/4},\gamma^{-1/2}\epsilon^{1/4}\nu^{-3/4}}(\mu) = \mu^{\nu,\epsilon},$$

which establishes (4.34). For (4.35), we set $\gamma = e(\mu^{1,1})$ as in Proposition 4.3 and apply (4.29); we find

$$e(\mu) = e(\mu^{1,\gamma^2}) = \gamma^2.$$

Then, apply (4.19) to obtain

$$E(\mu) = E(\mu^{1,\gamma^2}) = \gamma^2, \tag{A.12}$$

and we arrive at (4.35). It remains only to show that μ is a stationary homogeneous statistical solution of equations (4.33). For that purpose, note first that (A.10) follows from (A.12) and (4.19). Set now

$$\tau = \frac{\gamma\nu^{1/2}}{\epsilon^{1/2}}. \tag{A.13}$$

Hence, $(\nu/\tau)^{1/2} = \nu^{1/4}\epsilon^{1/4}/\gamma^{1/2}$ and $(\nu\tau)^{1/2} = \epsilon^{1/4}/\gamma^{1/2}\nu^{3/4}$, so that by (4.34) we have

$$\mu^{\nu,\epsilon} = \sigma_{(\nu/\tau)^{1/2},(\nu\tau)^{-1/2}}(\mu). \tag{A.14}$$

Thus, if $\{\mu^{\nu,\epsilon(t)}\}_{t\geq 0}$ is the homogeneous statistical solution considered in Proposition 4.3, then

$$\{\sigma_{(\nu/\tau(t))^{1/2},(\nu\tau(t))^{-1/2}}(\mu)\}_{t\geq 0}$$

is a homogeneous statistical solution of the NSE with viscosity ν, and

$$\tau(t) = \frac{\gamma\nu^{1/2}}{\epsilon(t)^{1/2}} = \frac{\gamma\nu^{1/2}}{\epsilon_0^{1/2}} + t \quad \text{(by (4.30))}$$
$$= \tau_0 + t,$$

where $\epsilon_0 = \epsilon(0)$ and $\tau_0 = \tau(0)$. Observe that

$$\frac{d}{dt}(\sigma_{(\nu/\tau(t))^{1/2},(\nu\tau(t))^{-1/2}}\mathbf{u})$$
$$= -\frac{1}{2\tau(t)}[(\sigma_{(\nu/\tau(t))^{1/2},(\nu\tau(t))^{-1/2}}\mathbf{u}) + (\mathbf{x}\cdot\nabla)(\sigma_{(\nu/\tau(t))^{1/2},(\nu\tau(t))^{-1/2}}\mathbf{u})], \quad \text{(A.15)}$$

for all $t > 0$ and all $\mathbf{u} \in V_{\text{loc}}$, where we have used in particular that $\tau'(t) = 1$. Then, for any test functional $\Phi \in \mathcal{T}_{\text{ws}}$, since Φ and its derivative Φ' are bounded in H_{loc}, it is not difficult to see that

$$t \mapsto \int \Phi(\mathbf{u})\, d(\sigma_{(\nu/\tau(t))^{1/2},(\nu\tau(t))^{-1/2}}\mu)(\mathbf{u})$$

is differentiable and that

$$\frac{d}{dt}\int \Phi(\mathbf{u})\, d(\sigma_{(\nu/\tau(t))^{1/2},(\nu\tau(t))^{-1/2}}(\mu))(\mathbf{u})$$
$$= \frac{d}{dt}\int \Phi(\sigma_{(\nu/\tau(t))^{1/2},(\nu\tau(t))^{-1/2}}\mathbf{u})\, d\mu$$
$$= -\int \frac{1}{2\tau(t)}\big((\sigma_{\xi(t),\lambda(t)}\mathbf{u}) + (\mathbf{x}\cdot\nabla)(\sigma_{\xi(t),\lambda(t)}\mathbf{u}),\, \Phi'(\sigma_{\xi(t),\lambda(t)}\mathbf{u})\big)\, d\mu$$
$$= -\frac{1}{2\tau(t)}\int (\mathbf{u} + (\mathbf{x}\cdot\nabla)\mathbf{u},\, \Phi'(\mathbf{u})\, d(\sigma_{(\nu/\tau(t))^{1/2},(\nu\tau(t))^{-1/2}}(\mu))(\mathbf{u}). \quad \text{(A.16)}$$

On the other hand, since $\{\sigma_{(\nu/\tau(t))^{1/2},(\nu\tau(t))^{-1/2}}(\mu)\}_{t\geq 0}$ is a homogeneous statistical solution of the NSE with viscosity ν, we have

$$\frac{d}{dt}\int \Phi(\mathbf{u})\, d(\sigma_{(\nu/\tau(t))^{1/2},(\nu\tau(t))^{-1/2}}(\mu))(\mathbf{u})$$
$$= -\int [\nu(\!(\mathbf{u}, \Phi'(\mathbf{u}))\!) + ((\mathbf{u}\cdot\nabla)\mathbf{u}, \Phi'(\mathbf{u}))]\, d(\sigma_{(\nu/\tau(t))^{1/2},(\nu\tau(t))^{-1/2}}(\mu))(\mathbf{u}) \quad \text{(A.17)}$$

for all $t > 0$.

Now take $\nu = 1$ and $\epsilon_0 = \gamma^2$. Hence, $\tau(t) = 1 + t$ for all $t \geq 0$. Set also $t = 0$ in (A.16) to find, using (A.17) and the fact that $\sigma_{1,1}(\mu) = \mu$,

$$\frac{1}{2}\int (\mathbf{u} + (\mathbf{x}\cdot\nabla)\mathbf{u}, \Phi'(\mathbf{u}))\, d\mu(\mathbf{u}) = \int \Big[(\!(\mathbf{u}, \Phi'(\mathbf{u}))\!) + ((\mathbf{u}\cdot\nabla)\mathbf{u}, \Phi'(\mathbf{u}))\Big]\, d\mu(\mathbf{u});$$

this proves that μ is a stationary homogeneous statistical solution of equation (4.33).

For the converse, assume that μ is a stationary homogeneous statistical solution of equations (4.33) such that $e(\mu) = E(\mu)$. Take $\gamma^2 = e(\mu)$ and define, according to (4.34), a 2-parameter family $\{\mu^{\nu,\epsilon}\}_{\nu,\epsilon>0}$ of homogeneous probability measures on H_{loc}. We need to show that this family is a self-similar universal statistical solution of the Navier–Stokes equations. From the rescaling of the enstrophy (see (4.10)), we find that

$$\begin{aligned}\nu E(\mu^{\nu,\epsilon}) &= \nu E(\sigma_{\gamma^{-1/2}\epsilon^{1/4}\nu^{1/4},\gamma^{-1/2}\epsilon^{1/4}\nu^{-3/4}}(\mu))\\ &= \nu(\gamma^{-1/2}\epsilon^{1/4}\nu^{1/4})^2(\gamma^{-1/2}\epsilon^{1/4}\nu^{-3/4})^2 E(\mu)\\ &= \epsilon\gamma^{-2}E(\mu) = \epsilon,\end{aligned}$$

where we have used that $\gamma^2 = e(\mu) = E(\mu)$. This proves (4.19). Now, for any $\nu, \epsilon_0 > 0$, consider $\{\mu^{\nu,\epsilon(t)}\}_{t\geq 0}$ with $\epsilon(t)$ given by (4.30). Set

$$\begin{aligned}\tau(t) &= \gamma\nu^{1/2}\epsilon(t)^{-1/2}\\ &= \tau_0 + t \quad \text{(by (4.30))},\end{aligned}$$

where $\tau_0 = \gamma\nu^{1/2}\epsilon_0^{-1/2}$. Then

$$(\nu/\tau(t))^{1/2} = \nu^{1/4}\epsilon(t)^{1/4}/\gamma^{1/2} \quad \text{and} \quad (\nu\tau(t))^{1/2} = \epsilon(t)^{1/4}/\gamma^{1/2}\nu^{3/4},$$

so that by (4.34) we see that

$$\mu^{\nu,\epsilon(t)} = \sigma_{(\nu/\tau(t))^{1/2},(\nu\tau(t))^{-1/2}}(\mu) \quad \forall t \geq 0. \tag{A.18}$$

For a given $\Phi \in \mathcal{T}_{\text{ws}}$, consider $\tilde{\Phi}_t = \Phi \circ \sigma_{(\nu/\tau(t))^{1/2},(\nu\tau(t))^{-1/2}}$, which clearly belongs to $\mathcal{T}_{\text{ws}}$ for each t, since it is a composition of Φ with a bounded linear operator in H_{loc}. Moreover, its differential with respect to $\mathbf{u}$ is

$$D_{\mathbf{u}}\tilde{\Phi}_t(\mathbf{u}) = \Phi'(\sigma_{(\nu/\tau(t))^{1/2},(\nu\tau(t))^{-1/2}}\mathbf{u}) \circ \sigma_{(\nu/\tau(t))^{1/2},(\nu\tau(t))^{-1/2}} \quad \forall t \geq 0.$$

Since μ is a stationary statistical solution of equations (4.33), we have

$$\begin{aligned}\int\Bigg[&\frac{1}{2}(-\mathbf{u} - (\mathbf{x}\cdot\nabla)\mathbf{u}, D_{\mathbf{u}}\tilde{\Phi}_t(\mathbf{u}))\\ &+ ((\mathbf{u}, D_{\mathbf{u}}\tilde{\Phi}_t(\mathbf{u}))) + ((\mathbf{u}\cdot\nabla)\mathbf{u}, D_{\mathbf{u}}\tilde{\Phi}_t(\mathbf{u}))\Bigg]\, d\mu(\mathbf{u}) = 0\end{aligned} \tag{A.19}$$

for each $t \geq 0$. On the other hand, using (A.15) (as in (A.16)) we find

$$\begin{aligned}&\frac{1}{2}\int(-\mathbf{u} - (\mathbf{x}\cdot\nabla)\mathbf{u}, D_{\mathbf{u}}\tilde{\Phi}_t(\mathbf{u}))\, d\mu(\mathbf{u})\\ &\quad = -\frac{1}{2}\int[(\sigma_{(\nu/\tau(t))^{1/2},(\nu\tau(t))^{-1/2}}\mathbf{u}, \Phi'(\sigma_{(\nu/\tau(t))^{1/2},(\nu\tau(t))^{-1/2}}\mathbf{u}))\\ &\qquad\qquad + ((\mathbf{x}\cdot\nabla)(\sigma_{(\nu/\tau(t))^{1/2},(\nu\tau(t))^{-1/2}}\mathbf{u}), \Phi'(\sigma_{(\nu/\tau(t))^{1/2},(\nu\tau(t))^{-1/2}}\mathbf{u}))]\, d\mu(\mathbf{u})\\ &\quad = \tau(t)\frac{d}{dt}\int \Phi(\sigma_{(\nu/\tau(t))^{1/2},(\nu\tau(t))^{-1/2}}\mathbf{u})\, d\mu(\mathbf{u})\\ &\quad = \tau(t)\frac{d}{dt}\int \Phi(\mathbf{u})\, d(\sigma_{(\nu/\tau(t))^{1/2},(\nu\tau(t))^{-1/2}}\mu)(\mathbf{u})\\ &\quad = \tau(t)\frac{d}{dt}\int \Phi(\mathbf{u})\, d\mu^{\nu,\epsilon(t)}(\mathbf{u}).\end{aligned} \tag{A.20}$$

Moreover, owing to (4.13) and (4.14),

$$\int [((\mathbf{u}, D_{\mathbf{u}}\tilde{\Phi}_t(\mathbf{u}))) + ((\mathbf{u}\cdot\nabla)\mathbf{u}, D_{\mathbf{u}}\tilde{\Phi}_t(\mathbf{u}))]\, d\mu(\mathbf{u})$$

$$= \int [\nu\tau(t)((\mathbf{u}, \Phi'(\mathbf{u}))) + \tau(t)((\mathbf{u}\cdot\nabla)\mathbf{u}, \Phi'(\mathbf{u}))]\, d(\sigma_{(\nu/\tau(t))^{1/2},(\nu\tau(t))^{-1/2}}\mu)(\mathbf{u})$$

$$= \tau(t)\int [\nu((\mathbf{u}, \Phi'(\mathbf{u}))) + ((\mathbf{u}\cdot\nabla)\mathbf{u}, \Phi'(\mathbf{u}))]\, d\mu^{\nu,\epsilon(t)}(\mathbf{u}). \qquad \text{(A.21)}$$

From (A.19)–(A.21) we deduce that

$$\frac{d}{dt}\int \Phi(\mathbf{u})\, d\mu^{\nu,\epsilon(t)}(\mathbf{u}) + \int [\nu((\mathbf{u}, \Phi'(\mathbf{u}))) + ((\mathbf{u}\cdot\nabla)\mathbf{u}, \Phi'(\mathbf{u}))]\, d\mu^{\nu,\epsilon(t)}(\mathbf{u}) = 0.$$

For the energy equation, we infer from (4.8) and our assumption of $\gamma^2 = e(\mu)$ that

$$e(\mu^{\nu,\epsilon(t)}) = \frac{\nu}{\tau(t)} e(\mu) = \frac{\gamma^2\nu}{\tau(t)};$$

from (4.10) and $\gamma^2 = E(\mu)$, we see that

$$E(\mu^{\nu,\epsilon(t)}) = \frac{1}{\tau(t)^2} E(\mu) = \frac{\gamma^2}{\tau(t)^2}.$$

We thus derive, using $\tau'(t) = 1$, the energy equation (4.15) for $\{\mu^{\nu,\epsilon(t)}\}_{t\geq 0}$.

This shows that $\{\mu^{\nu,\epsilon(t)}\}_{t\geq 0}$ is a homogeneous statistical solution of the NSE with viscosity ν and satisfying the energy equation (4.15). It also readily implies (4.21). Hence, we have proven that $\{\mu^{\nu,\epsilon}\}_{\mu,\epsilon>0}$ is a universal homogeneous statistical solution of the Navier–Stokes equations. The self-similarity follows now from the characterization given in Proposition 4.2, since (4.25) follows directly from (4.34). ■

Remark A.2 During the past years, the stochastic Navier–Stokes equations have been extensively studied (forcing by a white noise, in contrast to the present statistical case in which the forcing is made by random volume forces and initial data). Alternate proofs of Theorem 1.1 can be derived using these approaches, since the statistical case is somewhat simpler than the stochastic one. We refer here to the work of Vishik and Fursikov [1978] on the one hand and, on the other hand, to the work of M. Viot (following Bensoussan and Temam [1973]) and of G. Da Prato, A. Debussche, F. Flandoli, D. Gatarek, E. Pardoux, B. Schmalfuss, and J. Zabczyk. See, for example, Da Prato and Zabczyk [1992], Flandoli and Gatarek [1995], and the references therein.

References

ABERGEL, F. [1989]. Attractor for a Navier–Stokes flow in an unbounded domain. Attractors, Inertial Manifolds and Their Approximation (Marseille-Luminy, 1987). *RAIRO Modél. Math. Anal. Numér.* **23**, 359–70.

ABERGEL, F. [1990]. Existence and finite dimensionality of the global attractor for evolution equations on unbounded domains. *J. Differential Equations* **83**, 85–108.

ADAMS, R. S. [1975]. *Sobolev Spaces.* Academic Press, New York.

AGMON, S., A. DOUGLIS, and L. NIRENBERG [1959]. Estimates near the boundary for solutions of elliptic partial differential equations satisfying general boundary conditions, I. *Comm. Pure Appl. Math.* **12**, 623–727.

AGMON, S., A. DOUGLIS, and L. NIRENBERG [1964]. Estimates near the boundary for solutions of elliptic partial differential equations satisfying general boundary conditions, II. *Comm. Pure Appl. Math.* **17**, 35–92.

AUBIN, J. P. [1963]. Un théorème de compacité. *C. R. Acad. Sci. Paris Sér. I Math.* **256**, 5042–4.

BABIN, A. V. [1992]. The attractor of a Navier–Stokes system in an unbounded channel-like domain. *J. Dynam. Differential Equations* **4**, 555–84.

BABIN, A. V., and M. I. VISHIK [1983]. Attractors of evolution partial differential equations and estimates of their dimension (Russian). *Uspekhi Mat. Nauk* **38**, 133–87. English translation: *Russian Math Surveys* **38**, 151–213.

BABIN, A. V., and M. I. VISHIK [1986]. Maximal attractors of semigroups corresponding to evolution differential equations (Russian). *Mat. Sb.* (N.S.) 126. English translation: *Math. USSR-Sb.* **54**, 387–408.

BALL, J. [1997]. Continuity properties and global attractors of generalized semiflows and the Navier–Stokes equations. *J. Nonlinear Sci.* **7**, 475–502. Erratum: *J. Nonlinear Sci.* **8**, 233 (1998).

BARDOS, C., and L. TARTAR [1975]. Sur l'unicité rétrograde des équations paraboliques et quelques équations voisines. *Arch. Rational Mech. Anal.* **50**, 10–25.

BARENBLATT, G. I., and A. J. CHORIN [1998a]. Scaling laws and vanishing viscosity limits in turbulence theory. Recent Advances in Partial Differential Equations (Venice 1996). *Proc. Sympos. Appl. Math.* **54**, 1–25. Amer. Math. Soc., Providence, RI.

BARENBLATT, G. I., and A. J. CHORIN [1998b]. New perspectives in turbulence: Scaling laws, asymptotics, and intermittency. *SIAM Rev.* **40**, 265–91.

BARENBLATT, G. I., A. J. CHORIN, and V. M. PROSTOKISHIN [2000a]. Characteristic length scale of the intermediate structure in zero-pressure-gradient boundary layer flow. *Proc. Natl. Acad. Sci. USA* **97**, 3799–3802.

BARENBLATT, G. I., A. J. CHORIN, and V. M. PROSTOKISHIN [2000b]. Self-similar intermediate structures in turbulent boundary layers at large Reynolds numbers. *J. Fluid Mech.* **410**, 263–83.

BARENBLATT, G. I., A. J. CHORIN, and V. M. PROSTOKISHIN [2000c]. A note on the intermediate region in turbulent boundary layers. *Phys. Fluids* **12**, 2159–61.

BARRAZA, O. A. [1996]. Self-similar solutions in weak L^p-spaces of the Navier–Stokes equations. *Rev. Mat. Iberoamericana* **12**, 411–39.

BATCHELOR, G. K. [1959]. *The Theory of Homogeneous Turbulence* (Cambridge Monographs Mech. Appl. Math.). Cambridge University Press.

BATCHELOR, G. K. [1988]. *An Introduction to Fluid Dynamics.* Cambridge University Press.

BEN-ARTZI, M. [1994]. Global solutions of two-dimensional Navier–Stokes and Euler equations. *Arch. Rational Mech. Anal.* **128**, 329–58.

BENSOUSSAN, A., and R. TEMAM [1973]. Équations stochastiques du type Navier–Stokes. *J. Funct. Anal.* **13**, 195–222.

BERCOVICI, H., P. CONSTANTIN, C. FOIAS, and O. P. MANLEY [1995]. Exponential decay of the power spectrum of turbulence. *J. Statist. Phys.* **80**, 579–602.

BEWLEY, T., P. MOIN, and R. TEMAM [1999a]. Control of turbulent flows. Systems Modelling and Optimization (Detroit, 1997). *Chapman & Hall / CRC Res. Notes Math.* **396**, 3–11. CRC, Boca Raton, FL.

BEWLEY, T., P. MOIN, and R. TEMAM [1999b]. DNS-based predictive control of turbulence: An optimal benchmark for feedback algorithms. *J. Fluid Mech.* (submitted).

BEWLEY, T. R., R. TEMAM, and M. ZIANE [2000]. A general framework for robust control in fluid mechanics, *Phys. D* **138**, 360–92.

BILICKI, Z., C. DAFERMOS, J. KESTIN, G. MAJDA, and D.-L. ZEING [1987]. Trajectories and singular points in steady state models of two-phase flows. *Int. J. Multiphase Flow* **13**, 511.

BLEHER, P. M., and M. I. VISHIK [1971]. A certain class of pseudodifferential operators with an infinite number of variables, and their applications. *Mat. Sb.* (N.S.) **86**, 446–94.

BOURBAKI, N. [1966]. *Elements of Mathematics, General Topology,* part 2. Addison-Wesley, Reading, MA.

BOURBAKI, N. [1969]. *Éléments de mathématique. Fasc. XXXV. Livre VI: Intégration. Chapitre IX: Intégration sur les espaces topologiques séparés,* Actualités Scientifiques et Industrielles, no. 1343. Hermann, Paris.

BOWEN, R., and D. RUELLE [1975]. The ergodic theory of Axiom A flows. *Invent. Math.* **29**, 181–202.

BRÉZIS, H. and T. GALLOUET [1980]. Nonlinear Schrödinger evolution equation. *Nonlinear Analysis TMA* **4**, 677.

BUCHAHAVE, P., W. K. GEORGE, JR., and J. L. LUMLEY [1979]. The measurement of turbulence with the laser-Doppler anemometer. *Annual Rev. Fluid Mech.* **11**, 443–503.

CAFFARELLI, L., R. KOHN, and L. NIRENBERG [1982]. Partial regularity of suitable weak solutions of the Navier–Stokes equations. *Comm. Pure Appl. Math.* **35**, 771–831.

CALGARO, C., J. LAMINIE, and R. TEMAM [1997]. Dynamic multi-level schemes for the solution of evolution equations by hierarchical finite element discretization. *Appl. Numer. Math.* **21**, 1–40.

CANNONE, M. [1995]. *Ondelettes, paraproduits et Navier–Stokes.* Diderot Editeur, Paris.

CANNONE, M., and Y. MEYER [1995]. Littlewood–Paley decomposition and Navier–Stokes equations. *Methods Appl. Anal.* **2**, 307–19.

CANNONE, M., Y. MEYER, and F. PLANCHON [1994]. Solutions auto-similaires des équations de Navier–Stokes. *Séminaire sur les Équations aux Dérivées Partielles,* 1993–1994, exp. no. VIII. École Polytech., Palaiseau.

CAO, C., M. A. RAMMAHA, and E. S. TITI [1999]. The Navier–Stokes equations on the rotating 2-D sphere: Gevrey regularity and asymptotic degrees of freedom. *Z. Angew. Math. Phys.* **50**, 341–60.

CATTABRIGA, L. [1961]. Su un problema al contorno relativo al sistema di equazioni di Stokes (Italian). *Rend. Sem. Mat. Univ. Padova* **31**, 308–40.

CHAE, D., and C. FOIAS [1994]. On the homogeneous statistical solutions of the 2-D Navier–Stokes equations. *Indiana Univ. Math. J.* **43**, 177–85.

CHEN, S., G. DOOLEN, J. R. HERRING, R. H. KRAICHNAN, S. A. ORSZAG, and Z. S. SHE [1993]. Far dissipation range of turbulence. *Phys. Rev. Lett.* **70**, 3051–4.

CHEN, S., C. FOIAS, D. D. HOLM, E. OLSON, E. S. TITI, and S. WYNNE [1999a]. A connection between the Camassa–Holm equations and turbulent flows in channels and pipes. The International Conference on Turbulence (Los Alamos, 1998). *Phys. Fluids* **11**, 2343–53.

CHEN, S., C. FOIAS, D. D. HOLM, E. OLSON, E. S. TITI, and S. WYNNE [1999b]. The Camassa–Holm equations and turbulence. *Phys. D* **133**, 49–65.

CHING, E., P. CONSTANTIN, L. P. KADANOFF, A. LIBCHABER, I. PROCACCIA, and X.-Z. WU [1991]. Transitions in convective turbulence: The role of thermal plumes. *Phys. Rev. A* **44**, 8091–8102.

CHORIN, A. J. [1988]. Scaling laws in the vortex lattice model of turbulence. *Comm. Math. Phys.* **114**, 167–76.

CHORIN, A. J. [1991]. Equilibrium statistics of a vortex filament with applications. *Comm. Math. Phys.* **141**, 619–31.

CHORIN, A. J. [1994]. *Vorticity and Turbulence* (Appl. Math. Sci. **103**). Springer-Verlag, New York.

CHORIN, A. J. [1998]. New perspectives in turbulence. Current and Future Challenges in the Applications of Mathematics (Providence, 1997). *Quart. Appl. Math.* **56**, 767–85.

CONSTANTIN, P. [1994]. Geometric statistics in turbulence. *SIAM Rev.* **36**, 73–98.

CONSTANTIN, P., and C. R. DOERING [1992]. Energy dissipation in shear driven turbulence. *Phys. Rev. Lett.* **69**, 1648–51.

CONSTANTIN, P., and C. R. DOERING [1994]. Variational bounds on energy dissipation in incompressible flows: Shear flow. *Phys. Rev. E* (3) **49**, 4087–99.

CONSTANTIN, P., and C. R. DOERING [1995]. Variational bounds on energy dissipation in incompressible flows: Channel flow. *Physical Rev. E* (3) **51**, 3192–8.

CONSTANTIN, P., C. R. DOERING, and E. S. TITI [1996]. Rigorous estimates of small scales in turbulent flows. *J. Math. Phys.* **37**, 6152–6.

CONSTANTIN, P., and C. FOIAS [1985]. Global Lyapunov exponents, Kaplan–Yorke formulas and the dimension of the attractor for the 2D Navier–Stokes equations. *Comm. Pure Appl. Math.* **38**, 1–27.

CONSTANTIN, P., and C. FOIAS [1988]. *Navier–Stokes Equations.* University of Chicago Press.

CONSTANTIN, P., C. FOIAS, and O. MANLEY [1994]. Effects of the forcing function spectrum on the energy spectrum in 2-D turbulence. *Phys. Fluids* **6**, 427–9.

CONSTANTIN, P., C. FOIAS, O. MANLEY, and R. TEMAM [1985]. Determining modes and fractal dimension of turbulent flows. *J. Fluid Mech.* **150**, 427–40.

CONSTANTIN, P., C. FOIAS, and R. TEMAM [1985]. Attractors representing turbulent flows. *Mem. Amer. Math. Soc.* 53.

CONSTANTIN, P., C. FOIAS, and R. TEMAM [1988]. On the dimension of the attractors in two-dimensional turbulence. *Phys. D* **30**, 284–96.

COSTA, B., L. DETTORI, D. GOTTLIEB, and R. TEMAM [2001]. Time marching multi-lever techniques for evolutionary dissipative equations. *SIAM J. Statist. Sci. Comput.* (submitted).

COURANT, R., and D. HILBERT [1953]. *Methods of Mathematical Physics.* Interscience, New York.

DA PRATO, G., and J. ZABCZYK [1992]. *Stochastic Equations in Infinite Dimension.* Cambridge University Press.

DEBUSSCHE, A., and T. DUBOIS [1994]. Approximation of exponential order of the attractor of a turbulent flow. *Phys. D* **72**, 372–89.

DEBUSSCHE, A., and R. TEMAM [1994]. Convergent families of approximate inertial manifolds. *J. Math. Pures Appl.* (9) **73**, 489–522.

DINCULEANU, N. [1967]. *Vector Measures* (Internat. Ser. Monogr. Pure Appl. Math. **95**). Pergamon, Oxford.

DOERING, C. R., and J. D. GIBBON [1995]. *Applied Analysis of the Navier–Stokes Equations.* Cambridge University Press.

DOERING, C. R., and E. S. TITI [1995]. Exponential decay rate of the power spectrum for solutions of the Navier–Stokes equations. *Phys. Fluids* **7**, 1384–90.

DOERING, C. R., and X. WANG [1998]. Attractor dimension estimates for two-dimensional shear flows. *Phys. D* **123**, 206–22.

DRYDEN, H. L. [1943]. A review of the statistical theory of turbulence. *Quart. Appl. Math.* **1**, 7–42.

DUBOIS, T., F. JAUBERTEAU, and R. TEMAM [1998]. Incremental unknowns, multilevel methods and the numerical simulation of turbulence. *Comput. Methods Appl. Mech. Engrg.* **159**, 123–89.

DUBOIS, T., F. JAUBERTEAU, and R. TEMAM [1999]. *Dynamic Multilevel Methods and the Numerical Simulation of Turbulence.* Cambridge University Press.

DUNFORD, N., and J. T. SCHWARTZ [1958]. *Linear Operators, I. General Theory* (Pure Appl. Math. **7**). Interscience, New York.

EDEN, A., C. FOIAS, B. NICOLAENKO, and R. TEMAM [1994]. *Exponential Attractors for Dissipative Evolution Equations.* Masson, Paris.

EDEN, A., C. FOIAS, and R. TEMAM [1991]. Local and global Lyapunov exponents. *J. Dynam. Differential Equations* **3**, 133–77.

FALCONER, K. J. [1985]. *The Geometry of Fractal Sets.* Cambridge University Press.

FARMER, J., E. OTT, and J. A. YORKE [1983]. The dimension of chaotic attractors. *Phys. D* **7**, 153–80.

FAURE, S., J. LAMINIE, and R. TEMAM [2001]. Implementation of dynamical multilevel methods in the context of finite differences and in the presence of stochastic noise (in preparation).

FEDERER, H. [1969]. *Geometric Measure Theory.* Springer-Verlag, New York.

FEIGENBAUM, M. J. [1980]. The transition to aperiodic behavior in turbulent systems. *Comm. Math. Phys.* **77**, 65–86.

FERZIGER, J. H., U. B. MEHTA, and W. C. REYNOLDS [1977]. Large eddy simulation of homogeneous isotropic turbulence. Symposium on Turbulent Shear Flows (University Park, PA).

FLANDOLI, F., and D. GATAREK [1995]. Martingale and stationary solutions for stochastic Navier–Stokes equations. *Probab. Theory Related Fields* **102**, 367–91.

FOIAS, C. [1972]. Statistical study of Navier–Stokes equations, I. *Rend. Sem. Mat. Univ. Padova* **48**, 219–348.

FOIAS, C. [1973]. Statistical study of Navier–Stokes equations, II. *Rend. Sem. Mat. Univ. Padova* **49**, 9–123.

FOIAS, C. [1974]. A functional approach to turbulence. *Russian Math. Surveys* **29**, 293–326.

FOIAS, C. [1975]. On the statistical study of the Navier–Stokes equations. Partial Differential Equations and Related Topics (New Orleans, 1974). *Lecture Notes in Math.* **446**, 184–97. Springer-Verlag, Berlin.

FOIAS, C. [1997]. What do the Navier–Stokes equations tell us about turbulence? Harmonic Analysis and Nonlinear Differential Equations (Riverside, 1995). *Contemp. Math.* **208**, 151–80. Amer. Math. Soc., Providence, RI.

FOIAS, C., M. S. JOLLY, I. G. KEVREKIDIS, and E. S. TITI [1988a]. On the computation of inertial manifolds. *Phys. Lett. A* **131**, 433–6.

FOIAS, C., O. P. MANLEY, M. S. JOLLY, and R. ROSA [2001a]. Statistical estimates for the Navier–Stokes equations and the Kraichnan theory of 2-D fully developed turbulence (in preparation).

FOIAS, C., O. P. MANLEY, R. ROSA, and R. TEMAM [2001b]. Cascade of energy in turbulent flows. *C. R. Acad. Sci. Paris Sér. I* **332**, 1–6.

FOIAS, C., O. P. MANLEY, R. ROSA, and R. TEMAM [2001c]. Estimates for the energy cascade in three-dimensional turbulent flows. *C. R. Acad. Sci. Paris Sér. I* (to appear).

FOIAS, C., O. P. MANLEY, and L. SIROVICH [1989a]. Empirical and Stokes eigenfunctions and the far-dissipative turbulent spectrum. *Phys. Fluids A* **2**, 464–7.

FOIAS, C., O. P. MANLEY, and R. TEMAM [1983a]. New representation of Navier–Stokes equations governing self-similar homogeneous turbulence. *Phys. Rev. Lett.* **51**, 617–20.

FOIAS, C., O. P. MANLEY, and R. TEMAM [1986]. Physical estimates of the number of degrees of freedom in free convection. *Phys. Fluids* **29**, 3101–3.

FOIAS, C., O. P. MANLEY, and R. TEMAM [1987a]. Self-similar invariant families of turbulent flows. *Phys. Fluids* **30**, 2007–20.

FOIAS, C., O. P. MANLEY, and R. TEMAM [1987b]. An estimate of the Hausdorff dimension of the attractor for homogeneous decaying turbulence. *Phys. Lett. A* **122**, 140–4.

FOIAS, C., O. P. MANLEY, and R. TEMAM [1987c]. Sur l'interaction des petits et grands tourbillons dans les écoulements turbulents, *C. R. Acad. Sci. Paris Sér. I Math.* **305**, 497–500.

FOIAS, C., O. P. MANLEY, and R. TEMAM [1988b]. Modelling of the interaction of small and large eddies in two-dimensional turbulent flows. *RAIRO Modél. Math. Anal. Numér.* **22**, 93–118.

FOIAS, C., O. P. MANLEY, and R. TEMAM [1991]. Approximate inertial manifolds and effective viscosity in turbulent flows. *Phys. Fluids A* **3**, 898–911.

FOIAS, C., O. P. MANLEY, and R. TEMAM [1993]. Bounds for the mean dissipation of 2-D enstrophy and 3-D energy in turbulent flows. *Phys. Lett. A* **174**, 210–15.

FOIAS, C., O. P. MANLEY, R. TEMAM, and Y. M. TRÈVE [1983b]. Asymptotic analysis of the Navier–Stokes equations. *Phys. D* **9**, 157–88.

FOIAS, C., and E. OLSON [1996]. Finite fractal dimension and Hölder–Lipschitz parametrization. *Indiana Univ. Math. J.* **45**, 603–16.

FOIAS, C., and G. PRODI [1967]. Sur le comportement global des solutions non-stationnaires des équations de Navier–Stokes en dimension 2. *Rend. Sem. Mat. Univ. Padova* **39**, 1–34.

FOIAS, C., and G. PRODI [1976]. Sur les solutions statistiques des équations de Navier–Stokes. *Ann. Mat. Pura Appl.* (4) **111**, 307–30.

FOIAS, C., and J.-C. SAUT [1987]. Linearization and normal form of the Navier–Stokes equations with potential forces. *Ann. Inst. H. Poincaré Anal. Non Linéaire* **4**, 1–47.

FOIAS, C., G. R. SELL, and R. TEMAM [1985]. Variétés inertielles des équations différentielles dissipatives. *C. R. Acad. Sci. Paris Sér. I Math.* **301**, 139–41.

FOIAS, C., G. R. SELL, and R. TEMAM [1988c]. Inertial manifolds for nonlinear evolutionary equations. *J. Differential Equations* **73**, 309–53.

FOIAS, C., G. R. SELL, and E. TITI [1989b]. Exponential tracking and approximation of inertial manifolds for dissipative nonlinear equations. *J. Dynam. Differential Equations* **1**, 199–243.

FOIAS, C., and R. TEMAM [1978]. Remarques sur les équations de Navier–Stokes stationnaires et les phénomènes successifs de bifurcation. *Ann. Scuola Norm. Sup. Pisa Cl. Sci.* (4) **5**, 29–63.

FOIAS, C., and R. TEMAM [1979]. Some analytic and geometric properties of the solutions of the evolution Navier–Stokes equations. *J. Math. Pures Appl.* (9) **58**, 339–68.

FOIAS, C., and R. TEMAM [1980]. Homogeneous statistical solutions of Navier–Stokes equations. *Indiana Univ. Math. J.* **29**, 913–57.

FOIAS, C., and R. TEMAM [1982]. Homogeneous statistical solutions of Navier–Stokes equations. *Selected Studies: Physics-Astrophysics, Mathematics, History of Science,* pp. 175–96. North-Holland, Amsterdam.

FOIAS, C., and R. TEMAM [1983]. Self-similar universal homogeneous statistical solutions of the Navier–Stokes equations. *Comm. Math. Phys.* **90**, 187–206.

FOIAS, C., and R. TEMAM [1984]. Determination of the solutions of the Navier–Stokes equations by a set of nodal values. *Math. Comp.* **43**, 117–33.

FOIAS, C., and R. TEMAM [1987]. The connection between the Navier–Stokes equations, dynamical systems, and turbulence theory. Directions in Partial Differential Equations (Madison, 1985). *Publ. Math. Res. Center Univ. Wisconsin* **54**, 55–73. Academic Press, Boston.

FOIAS, C., and R. TEMAM [1989]. Gevrey class regularity for the solutions of the Navier–Stokes equations. *J. Funct. Anal.* **87**, 359–69.

FOIAS, C., and R. TEMAM [1994]. Approximation of attractors by algebraic or analytic sets. *SIAM J. Math. Anal.* **25**, 1269–1302.

FRISCH, U. [1995]. *Turbulence: The Legacy of A. N. Kolmogorov.* Cambridge University Press.

FRISCH, U., and R. MORF [1981]. Intermittency in nonlinear dynamics and singularities at complex times. *Phys. Rev. A* **23**, 2673–2705.

FRISCH, U., M. NELKIN, and P.-L. SULEM [1978]. A simple dynamical model of intermittent fully developed turbulence. *J. Fluid Mech.* **87**, 719–36.

FRIZ, K., and J. C. ROBINSON [2000]. Parametrizing the attractor of the two-dimensional Navier–Stokes equations with a finite number of nodal values. *Phys. D* (to appear).

GARCÍA-ARCHILLA, B., J. NOVO, and E. S. TITI [1999]. An approximate inertial manifolds approach to postprocessing the Galerkin method for the Navier–Stokes equations. *Math. Comp.* **6**, 893–911.

GARCÍA-ARCHILLA, B., and E. S. TITI [2000]. Postprocessing the Galerkin method: The finite-element case. *SIAM J. Numer. Anal.* **37**, 470–99.

GHIDAGLIA, J.-M. [1984]. Régularité des solutions de certains problèmes aux limites linéaires liés aux équations d'Euler. *Comm. Partial Differential Equations* **9**, 1265–98.

GHIDAGLIA, J.-M., and R. TEMAM [1991]. Lower bound on the dimension of the attractor for the Navier–Stokes equations in space dimension 3. *Mechanics, Analysis and Geometry: 200 Years after Lagrange* (M. Francaviglia, D. Holmes, eds.). Elsevier, Amsterdam.

GIHMAN, I. I., and A. V. SKOROHOD [1974]. *The Theory of Stochastic Processes.* Springer-Verlag, New York.

GREENBERG, O. W., and Y. M. TRÈVE [1960]. Shock wave and solitary wave structure in a plasma. *Phys. Fluids* **3**, 786.

GRISVARD, P. [1985]. *Elliptic Problems in Nonsmooth Domains* (Monogr. Stud. Math. **24**). Pitman, Boston.

GRISVARD, P. [1992]. *Singularities in Boundary Value Problems* (Res. Appl. Math. **22**). Springer-Verlag, Berlin.

GRUJIC, Z., and I. KUKAVICA [1998]. Space analyticity for the Navier–Stokes and related equations with initial data in L^p. *J. Funct. Anal.* **152**, 447–66.

GRUJIC, Z., and I. KUKAVICA [1999]. Space analyticity for the nonlinear heat equation in a bounded domain. *J. Differential Equations* **154**, 42–54.

HALE, J. K. [1988]. *Asymptotic Behavior of Dissipative Systems* (Math. Surveys Monogr. **25**). Amer. Math. Soc., Providence, RI.

HEWITT, E., and K. STROMBERG [1975]. *Real and Abstract Analysis: A Modern Treatment of the Theory of Functions of a Real Variable,* 3rd ed. (Grad. Texts in Math. **25**). Springer-Verlag, New York.

HINZE, J. O. [1975]. *Turbulence.* McGraw-Hill, New York.

HOFF, D. [1992a]. Spherically symmetric solutions of the Navier–Stokes equations for compressible, isothermal flow with large, discontinuous initial data. *Indiana Univ. Math. J.* **41**, 1225–1302.

HOFF, D. [1992b]. Global well-posedness of the Cauchy problem for the Navier–Stokes equations of nonisentropic flow with discontinuous initial data. *J. Differential Equations* **95**, 33–74.

HOFF, D. [1995]. Global solutions of the Navier–Stokes equations for multidimensional compressible flow with discontinuous initial data. *J. Differential Equations* **120**, 215–54.

HOPF, E. [1951]. Über die Aufanswertaufgabe für die hydrodynamischen Grundgleichungen. *Math. Nachr.* **4**, 213–31.

HOPF, E. [1952]. Statistical hydromechanics and functional calculus. *J. Rational Mech. Anal.* **1**, 87–123.

ILYIN, A. A. [1996]. Attractors for Navier–Stokes equations in domains with finite measure. *Nonlinear Anal.* **27**, 605–16.

IOOSS, G. [1969]. Application de la théorie des semi-groupes à l'étude de la stabilité des écoulements laminaires. *J. Mécanique* **8**, 477–507.

IOOSS, G. [1988]. Local techniques in bifurcation theory and nonlinear dynamics. *Chaotic Motions in Nonlinear Dynamical Systems* (CISM Courses and Lectures **298**), pp. 137–93. Springer, Vienna.

IOOSS, G., and M. ADELMEYER [1998]. *Topics in Bifurcation Theory and Applications,* 2nd ed. (Adv. Ser. Nonlinear Dynam. **3**). World Scientific, River Edge, NJ.

JAUBERTEAU, F., F. ROSIER, and R. TEMAM [1990]. A nonlinear Galerkin method for the Navier–Stokes equations *Comput. Methods Appl. Mech. Engrg.* **80**, 245–60.

JONES, D. A., and E. S. TITI [1992]. Determination of the solutions of the Navier–Stokes equations by finite volume elements. *Phys. D* **60**, 165–74.

JONES, D. A., and E. S. TITI [1993]. Upper bounds on the number of determining modes, nodes, and volume elements for the Navier–Stokes equations. *Indiana Univ. Math. J.* **42**, 875–87.

KALTENBACH, H.-J., M. FATICA, R. MITTAL, T. S. LUND and P. MOIN [1999]. Study of flow in a planar asymmetric diffuser using large-eddy simulation. *J. Fluid Mech.* **390**, 151–85.

KANEL, YA. I. [1968]. A model system of equations for the one-dimensional motion of a gas. *Differencial'nye Uravnenija* **4**, 721–34.

KAZHIKHOV, A. V., and V. V. SHELUKHIN [1977]. Unique global solution with respect to time of initial-boundary value problems for one-dimensional equations of a viscous gas (Russian). *Prikl. Mat. Mekh.* **41**, 282–91. English translation: J. Appl. Math. Mech. **41**, 273–82.

KAZHIKHOV, A. V., and V. A. VAĬGANT [1995]. On the existence of global solutions of two-dimensional Navier–Stokes equations of a compressible viscous fluid (Russian). *Sibirsk. Mat. Zh.* **36**, 1283–1316. English translation: Siberian Math. J. **36**, 1108–41.

KELLOGG, R. B., and J. E. OSBORN [1976]. A regularity result for the Stokes problem in a convex polygon. *J. Funct. Anal.* **21**, 397–431.

KHINCHIN, A. I. [1949]. *Mathematical Foundations of Statistical Mechanics.* Dover, New York.

KIM, J., P. MOIN, and R. MOSER [1987]. Turbulence statistics in fully developed channel flow at low Reynolds number. *J. Fluid Mech.* **177**, 133–66.

KOLMOGOROV, A. N. [1941a]. The local structure of turbulence in incompressible viscous fluid for very large Reynolds numbers. *C. R. (Doklady) Acad. Sci. URSS* (N.S.) **30**, 301–5.

KOLMOGOROV, A. N. [1941b]. On degeneration of isotropic turbulence in an incompressible viscous liquid. *C. R. (Doklady) Acad. Sci. URSS* (N.S.) **31**, 538–40.

KOLMOGOROV, A. N. [1962]. A refinement of previous hypotheses concerning the local structure of turbulence in a viscous incompressible fluid at high Reynolds number. *J. Fluid Mech.* **13**, 82–5.

KRAICHNAN, R. H. [1967]. Inertial ranges in two-dimensional turbulence. *Phys. Fluids* **10**, 1417–23.

KRAICHNAN, R. H. [1972]. Some modern developments in the statistical theory of turbulence. *Statistical Mechanics: New Concepts, New Problems, New Applications* (S. A. Rice, K. F. Freed, J. C. Light, eds.), pp. 201–27. University of Chicago Press.

KREISS, H.-O., and J. LORENZ [1989]. *Initial Boundary Value Problems and the Navier–Stokes Equations.* Academic Press, Boston.

LADYZHENSKAYA, O. [1963]. *The Mathematical Theory of Viscous Incompressible Flow,* revised English edition (translated from the Russian by Richard A. Silverman). Gordon & Breach, New York.

LADYZHENSKAYA, O. A. [1967]. New equations for the description of the motions of viscous incompressible fluids and global solvability for their boundary value problems. *Trudy Mat. Inst. Steklov.* **102**, 85–104.

LADYZHENSKAYA, O. [1972]. On the dynamical system generated by the Navier–Stokes equations (Russian). *Zap. Nauchn. Sem. LOMI* **27**, 91–114. English translation: *J. Soviet Math.* **3** (1975).

LADYZHENSKAYA, O. [1973]. Dynamical system generated by the Navier–Stokes equations. *Soviet Phys. Dokl.* **17**, 647–9.

LADYZHENSKAYA, O. [1975]. A dynamical system generated by the Navier–Stokes equations. *J. Soviet Math.* **3**, 458–79.

LADYZHENSKAYA, O. [1991]. *Attractors for Semigroups and Evolution Equations* (Lezioni Lincei). Cambridge University Press.

LADYZHENSKAYA, O. [1992]. First boundary value problem for the Navier–Stokes equations in domains with nonsmooth boundaries. *C. R. Acad. Sci. Paris Sér. I Math.* **314**, 253–8.

LADYZHENSKAYA, O. A., and A. M. VERSHIK [1977]. Sur l'évolution des mesures déterminées par les équations de Navier–Stokes et la résolution du problème de Cauchy pour l'équation statistique de E. Hopf. *Ann. Scuola Norm. Sup. Pisa Cl. Sci.* (4) **4**, 209–30.

LAMB, H. [1957]. *Hydrodynamics,* 6th ed. Cambridge University Press.

LANDAU, L., and E. LIFSHITZ [1971]. *Mécanique des Fluides, Physique Théorique,* tome 6. Éditions Mir, Moscow.

LERAY, J. [1933]. Etude de diverses équations intégrales non linéaires et de quelques problèmes que pose l'hydrodynamique. *J. Math. Pures Appl.* **12**, 1–82.

LERAY, J. [1934a]. Essai sur les mouvements plans d'un liquide visqueaux que limitent des parois. *J. Math. Pures Appl.* **13**, 331–418.

LERAY, J. [1934b]. Essai sur les mouvements d'un liquide visqueux emplissant l'espace. *Acta Math.* **63**, 193–248.

LESIEUR, M. [1997]. *Turbulence in Fluids,* 3rd ed. (Fluid Mech. Appl. **40**). Kluwer, Dordrecht.

LESLIE, D. C. [1973]. *Developments in the Theory of Turbulence.* Clarendon Press, Oxford.

LIEB, E., and W. THIRRING [1976]. Inequalities for the moments of the eigenvalues of the Schrödinger equations and their relation to Sobolev inequalities. *Studies in Mathematical Physics: Essays in Honor of Valentine Bargmann* (E. Lieb, B. Simon, A. S. Wightman, eds.), pp. 269–303. Princeton University Press, Princeton, NJ.

LIONS, J. L. [1959]. Quelques résultats d'existence dans des équations aux dérivés partielles non linéaires. *Bull. Soc. Math. France* 87, 245–73.

LIONS, J. L. [1969]. *Quelques méthodes de résolution des problémes aux limites non linéaires.* Dunod, Paris.

LIONS, J. L., and E. MAGENES [1972]. *Nonhomogeneous Boundary Value Problems and Applications.* Springer-Verlag, New York.

LIONS, J. L., and G. PRODI [1959]. Un théorème d'existence et d'unicité dans les équations de Navier–Stokes en dimension 2. *C. R. Acad. Sci. Paris Sér. I Math.* **248**, 3519–21.

LIONS, J. L., R. TEMAM, and S. WANG [1992a]. New formulations of the primitive equations of atmosphere and applications. *Nonlinearity* **5**, 237–88.

LIONS, J. L., R. TEMAM, and S. WANG [1992b]. On the equations of the large-scale ocean. *Nonlinearity* **5**, 1007–53.

LIONS, J. L., R. TEMAM, and S. WANG [1995]. Mathematical theory for the coupled atmosphere–ocean models (CAO III). *J. Math. Pures Appl.* (9) **74**, 105–63.

LIONS, P. L. [1993]. Existence globale de solutions pour les équations de Navier–Stokes compressibles isentropiques. *C. R. Acad. Sci. Paris Sér. I Math.* **316**, 1335–40.

LIONS, P. L. [1998]. *Mathematical Topics in Fluid Mechanics,* vol. 2: *Compressible Models* (Oxford Lecture Ser. Math. Appl. **10**). Clarendon Press, Oxford.

LIU, V. X. [1993]. A sharp lower bound for the Hausdorff dimension of the global attractors of the 2D Navier–Stokes equations. *Comm. Math. Phys.* **158**, 327–39.

LIU, V. X. [1994]. Remarks on the Navier–Stokes equations on the two and three dimensional torus. *Comm. Partial Differential Equations* **19**, 873–900.

LORENZ, E. N. [1963]. Deterministic nonperiodic flow. *J. Atmospheric Sci.* **20**, 130–41.

MANDELBROT, B. B. [1982]. *The Fractal Geometry of Nature* (Schriftenreihe für den Referenten [Series for the Referee]). Freeman, San Francisco.

MAÑE, R. [1981]. On the dimension of the compact invariant sets of certain nonlinear maps. *Lecture Notes in Math.* **898**, 230–42. Springer-Verlag, New York.

MANLEY, O. P. [1992]. The dissipation range spectrum. *Phys. Fluids A* **4**, 1320–1.

MARCHIORO, C. [1986]. An example of absence of turbulence for any Reynolds number. *Comm. Math. Phys.* **105**, 99–106.

MARION, M., and R. TEMAM [1989]. Nonlinear Galerkin methods. *SIAM J. Numer. Anal.* **26**, 1139–57.

MARION, M., and R. TEMAM [1990]. Nonlinear Galerkin methods: The finite element case. *Numer. Math.* **57**, 205–26.

MARION, M., and R. TEMAM [1998]. *Navier–Stokes Equations: Theory and Approximation* (Handbook Numer. Anal. **6**). North-Holland, Amsterdam.

MASUDA, K. [1967]. On the analyticity and the unique continuation theorem for solutions of the Navier–Stokes equations. *Proc. Japan Acad. Ser. A Math. Sci.* **43**, 827–32.

MATSUMURA, A., and T. NISHIDA [1983]. Initial-boundary value problems for the equations of motion of compressible viscous and heat-conductive fluids. *Comm. Math. Phys.* **89**, 445–64.

MATSUMURA, A., and T. NISHIDA [1989]. Periodic solutions of a viscous gas equation. Recent Topics in Nonlinear PDE, IV (Kyoto, 1988). *North-Holland Math. Stud.* **160**, 49–82. North-Holland, Amsterdam.

MATSUMURA, A., and K. NISHIHARA [1992]. Global stability of the rarefaction wave of a one-dimensional model system for compressible viscous gas. *Comm. Math. Phys.* **144**, 325–35.

MAZJA, V. G. [1985]. *Sobolev Spaces.* Springer-Verlag, New York.

MÉTIVIER, G. [1978]. Valeurs propres d'opérateurs définis par la restriction de systèmes variationelles a des sous-espaces. *J. Math. Pures Appl.* (9) **57**, 133–56.

MIRANVILLE, A. [1993]. Shear layer flow in a channel: Estimate on the dimension of the attractor. *Phys. D* **65**, 135–53.

MIRANVILLE, A., and X. WANG [1996]. Upper bound on the dimension of the attractor for nonhomogeneous Navier–Stokes equations. *Discrete Contin. Dynam. Systems* **2**, 95–110.

MOISE, I., R. TEMAM, and M. ZIANE [1997]. Asymptotic analysis of the Navier–Stokes equations in thin domains. *Topol. Methods Nonlinear Anal.* **10**, 249–82.

MONIN, A. S., and A. M. YAGLOM [1975]. *Statistical Fluid Mechanics: Mechanics of Turbulence.* MIT Press, Cambridge, MA.

NAYLOR, A. W., and G. R. SELL [1982]. *Linear Operator Theory in Engineering and Science,* 2nd ed. (Appl. Math. Sci., **40**). Springer-Verlag, Berlin.

NOVIKOV, E. A., and R. V. STEWART [1964]. The intermittency of turbulence and the spectrum of energy dissipation. *Izv. Akad. Nauk SSSR Ser. Geoffiz.* **3**, 408–13.

OHKITANI, K. [1989]. Log-corrected energy spectrum and dimension of attractor in two-dimensional turbulence. *Phys. Fluids A* **1**, 451–2.

ONSAGER, L. [1945]. The distribution of energy in turbulence. *Phys. Rev.* **68**, 285.

ORSZAG, S. A. [1970]. Analytical theories of turbulence. *J. Fluid Mech.* **41**, 363–86.

ORSZAG, S. A., and G. S. PATTESON [1972]. Numerical simulation of turbulence: Statistical models and turbulence. *Lecture Notes in Phys.* **12**, 127–47. Springer-Verlag, Berlin.

H. POINCARÉ [1889]. *Sur le probéme des trois corps et les équations de la dynamique.* Memoiré présenté à l'Academie Royale de Suède.

PRANDTL, L. [1904]. Über Fluessigkeitbewegung bei sehr kleine Reibung. Third International Math. Congress (Heidelberg), pp. 484–91.

PRODI, G. [1961]. On probability measures related to the Navier–Stokes equations in the 3-dimensional case. *Air Force Res. Div. Contract* AF61(052)-414, Tech. Note 2, Trieste.

PROKHOROV, Y. V. [1956]. Convergence of random processes and limit theorems in probability theory. *Theory Probab. Appl.* **1**, 157–214.

REYNOLDS, O. [1883]. On the experimental investigation of the circumstances which determine whether the motion of water shall be direct or sinuous. *Philos. Trans. Roy. Soc. London Ser. A* **74**, 935–82.

REYNOLDS, O. [1895]. On the dynamical theory of incompressible viscous fluids and the determination of the criterion. *Philos. Trans. Roy. Soc. London Ser. A* **186**, 123–64.

ROBINSON, J. [1998]. All possible chaotic dynamics can be approximated in three dimensions. *Nonlinearity* **11**, 529–45.

ROBINSON, J. [1999]. Global attractors: Topology and finite-dimensional dynamics. *J. Dynam. Differential Equations* **11**, 557–81.

ROSA, R. [1995]. Approximate inertial manifolds of exponential order. *Discrete Contin. Dynam. Systems* **1**, 421–48.

ROSA, R. [1998]. The global attractor for the 2D Navier–Stokes flow on some unbounded domains. *Nonlinear Anal.* **32**, 71–85.

ROSE, H. A., and P. L. SULEM [1978]. Fully developed turbulence and statistical mechanics. *J. Physique* **39**, 441–84.

RUDIN, W. [1987]. *Real and Complex Analysis,* 3rd ed. McGraw-Hill, New York.

RUELLE, D. [1976]. A measure associated with Axiom A attractors. *Trans. Amer. Math. Soc.* **98**, 619–54.

RUELLE, D. [1998]. Smooth dynamics and new theoretical ideas in nonequilibrium statistical mechanics. Lecture notes, Rutgers University.

RUELLE, D., and F. TAKENS [1971]. On the nature of turbulence. *Comm. Math. Phys.* **20**, 167–92.

SCHEFFER, V. [1977]. Hausdorff measures and the Navier–Stokes equations. *Comm. Math. Phys.* **55**, 97–112.

SCHLICHTING, H. [1979]. *Boundary Layer Theory,* 7th ed. (translated by J. Kestin). McGraw-Hill, New York.

SCHWARTZ, L. [1950/51]. *Théorie des distributions I, II.* Hermann, Paris.

SCHWARTZ, L. [1957/58]. Distributions à valeurs vectorielles I, II. *Ann. Inst. Fourier* **7**, 1–141; **8**, 1–209.

SCHWARTZ, L. [1973]. *Radon Measures on Arbitrary Topological Spaces and Cylindrical Measures* (Tata Institute of Fundamental Research Studies in Mathematics **6**). Oxford University Press, London.

SELL, G. R. [1996]. Global attractors for the three-dimensional Navier–Stokes equations. *J. Dynam. Differential Equations* **8**, 1–33.

SERRE, D. [1983]. Équations de Navier–Stokes stationnaires avec données peu régulières. *Ann. Scuola Norm. Sup. Pisa Cl. Sci.* (4) **10**, 543–59.

SIGETI, D. E. [1995]. Exponential decay of power spectra at high frequency and positive Lyapunov exponents. *Physica D* **82**, 136–53.

SINAI, YA. G. [1972]. Gibbsian measures in ergodic theory. *Russian Math. Surveys* **27**, 21–69.

SMAGORINSKY, J. [1963]. General circulation experiments with the primitive equations, I: The basic experiment. *Monthly Weather Review* **91**, 99–164.

SMALE, S. [1967]. Differentiable dynamical systems. *Bull. Amer. Math. Soc.* **73**, 747–817.

SMITH, L. M., and W. C. REYNOLDS [1991]. The dissipation-range spectrum and the velocity-derivative skewness in turbulent flows. *Phys. Fluids A* **3**, 992.

SOLONNIKOV, V. A. [1964]. Estimates of solutions of nonstationary linearized systems of Navier–Stokes equations (Russian). *Trudy Mat. Inst. Steklov.* **70**, 213–317. English translation: *Amer. Math. Soc. Transl. Ser. 2* **75**, 1–116 (1968).

TAKENS, F. [1981]. Detecting strange attractors in turbulence. *Lecture Notes in Math.* **898**, 366–81. Springer-Verlag, New York.

TAKENS, F. [1985]. On the numerical determination of the dimension of an attractor. Dynamical Systems and Bifurcations (Groningen, 1984). *Lecture Notes in Math.* **1125**, 99–106. Springer-Verlag, Berlin.

TAYLOR, G. I. [1935]. Statistical theory of turbulence. *Proc. Roy. Soc. London Ser. A* **151**, 421–78.

TAYLOR, G. I. [1937]. The spectrum of turbulence. *Proc. Roy. Soc. London Ser. A* **164**, 476–90.

TEMAM, R. [1982]. Behaviour at time $t = 0$ of the solutions of semilinear evolution equations. *J. Differential Equations* **43**, 73–92.

TEMAM, R. [1983]. *Navier–Stokes Equations and Nonlinear Functional Analysis* (CBMS-NSF Regional Conf. Ser. in Appl. Math. **66**). SIAM, Philadelphia; 2nd ed. published in 1995.

TEMAM, R. [1985]. Attractors for Navier–Stokes equations. *Nonlinear Partial Differential Equations and Their Applications* (H. Brézis, J. L. Lions, eds.) (Séminaire du Collège de France **7**), pp. 272–92. Pitman, Boston.

TEMAM, R. [1986]. Infinite dimensional dynamical systems in fluid mechanics. *Nonlinear Functional Analysis and Its Applications* (F. Browder, ed.) (Proc. Sympos. Pure Math. **45**), pp. 413–45. Amer. Math. Soc., Providence, RI.

TEMAM, R. [1989]. Induced trajectories and approximate inertial manifolds. *Math. Model. Numer. Anal.* **23**, 541–61.

TEMAM, R. [1991]. Approximation of attractors, large eddy simulations and multiscale methods. *Proc. Roy. Soc. London Ser. A* **434**, 23–39.

TEMAM, R. [1995]. *Navier–Stokes Equations and Nonlinear Functional Analysis,* 2nd ed. SIAM, Philadelphia.

TEMAM, R. [1997]. *Infinite Dimensional Dynamical Systems in Mechanics and Physics,* 2nd ed. (Appl. Math. Sci. **68**). Springer-Verlag, New York; 1st ed. published in 1988.

TEMAM, R. [2000]. Some developments on Navier–Stokes equations in the second half of the 20th Century. *Développments des mathématiques au cours de la seconde moitié du XXeme siècle* (J. P. Pier, ed.). Birkhäuser, Basel; reprinted in Temam [2001, Apx. III].

TEMAM, R. [2001]. *Navier–Stokes Equations: Theory and Numerical Analysis.* Amer. Math. Soc., Providence, RI; 1st ed. published in 1979.

TEMAM, R., and X. WANG [1995]. Asymptotic analysis of the linearized Navier–Stokes equations in a 2-D channel. *J. Differential Integral Equations* **8**, 1591–1618.

TEMAM, R., and X. WANG [1997]. The convergence of the solutions of the Navier–Stokes equations to that of the Euler equations. *Appl. Math. Lett.* **10**, 29–33.

TEMAM, R., and M. ZIANE [1996]. Navier–Stokes equations in three-dimensional thin domains with various boundary conditions. *Adv. Differential Equations* **1**, 499–546.

TENNEKES, H., and J. L. LUMLEY [1972]. *A First Course in Turbulence.* MIT Press, Cambridge, MA.

TITI, E. [1988]. Une varieté approximante de l'attracteur universel des équations de Navier–Stokes non linéaires de dimension finie. *C. R. Acad. Sci. Paris Sér. I Math.* **307**, 383–5.

Y. M. TRÈVE and O. P. MANLEY [1981]. Minimum number of modes in approximate solutions of equations of hydrodynamics. *Phys. Lett. A* **82**, 88.

Y. M. TRÈVE and O. P. MANLEY [1982]. Energy conserving Galerkin approximations for 2-D hydrodynamic and MHD Bénard convection. *Phys. D* **4**, 319–42.

VIANA, M. [1997]. Multidimensional nonhyperbolic attractors. *Inst. Hautes Études Sci. Publ. Math.* **85**, 63–96.

VISHIK, M. I., and A. V. FURSIKOV [1977a]. L'équation de Hopf, les solutions statistiques, les moments correspondants aux systémes des équations paraboliques quasilinéaires. *J. Math. Pures Appl.* (9) **56**, 85–122.

VISHIK, M. I., and A. V. FURSIKOV [1977b]. Solutions statistiques homogènes des systèmes differentiels paraboliques et du système de Navier–Stokes. *Ann. Scuola Norm. Sup. Pisa Cl. Sci.* (4) **4**, 531–76.

VISHIK, M. I., and A. V. FURSIKOV [1978]. Translationally homogeneous statistical solutions and individual solutions with infinite energy of a system of Navier–Stokes equations. *Sibirsk. Mat. Zh.* **19**, 1005–31.

VISHIK, M. I., and A. V. FURSIKOV [1988]. *Mathematical Problems of Statistical Hydrodynamics.* Kluwer, Dordrecht.

VON KARMAN, T. [1911]. Über den Mechanismus des Widerstandes den ein bewegter Körper in einer Fluesigkeiterzeugt. *Nachr. Akad. Wiss. Göttingen Math.-Phys. Kl. II,* 509–17.

VON KARMAN, T. [1912]. Über den Mechanismus des Widerstandes den ein bewegter Körper in einer Fluesigkeiterzeugt. *Nachr. Akad. Wiss. Göttingen Math.-Phys. Kl. II,* 547–56.

VON KARMAN, T. and L. HOWARTH [1938]. On the statistical theory of isotropic turbulence. *Proc. Roy. Soc. London Ser. A* **164**, 192–215.

WOUK, A. [1979]. *A Course of Applied Functional Analysis.* Wiley, New York.

WU, X.-Z., L. P. KADANOFF, A. LIBCHABER, and M. SANO [1990]. Frequency power spectrum of temperature fluctuations in free convection. *Phys. Rev. Lett.* **64**, 2140–3.

YOUNG, L.-S. [1997]. Ergodic theory of chaotic dynamical systems. Lecture, International Congress of Mathematical Physics.

ZIANE, M. [1997]. Optimal bounds on the dimension of the attractor of the Navier–Stokes equations. *Phys. D* **105**, 1–19.

Index